Strong Motion Instrumentation for Civil Engineering Structures

edited by

M. Erdik

Department of Earthquake Engineering,
Kandilli Observatory and Earthquake Research Institute,
Department of Earthquake Engineering,
Boğaziçi University,
Çengelköy, Istanbul, Turkey

M. Celebi

U.S. Geological Survey,
Menlo Park, CA, U.S.A.

V. Mihailov

Institute of Earthquake Engineering and Engineering Seismology,
University of 'St. Cyril and Methodius',
Republic of Macedonia

and

N. Apaydın

Motorway Bridge Department,
General Directorate of Turkish Highways,
Zincirlikuyu, Istanbul, Turkey

Kluwer Academic Publishers

Dordrecht / Boston / London

Published in cooperation with NATO Scientific Affairs Division

Proceedings of the NATO Advanced Research Workshop on
Strong Motion Instrumentation for Civil Engineering Structures
Istanbul, Turkey
June 2–5, 1999

A C.I.P. Catalogue record for this book is available from the Library of Congress.

ISBN 0-7923-6916-5 (HB)
ISBN 0-7923-6917-3 (PB)

Published by Kluwer Academic Publishers,
P.O. Box 17, 3300 AA Dordrecht, The Netherlands.

Sold and distributed in North, Central and South America
by Kluwer Academic Publishers,
101 Philip Drive, Norwell, MA 02061, U.S.A.

In all other countries, sold and distributed
by Kluwer Academic Publishers,
P.O. Box 322, 3300 AH Dordrecht, The Netherlands.

Printed on acid-free paper

Printed in the Netherlands.

TABLE OF CONTENTS

Preface ix

Timothy K. Ahern
State-of The-Art Technology For Data Storage and Dissemination 1

Moh J. Huang, Anthony F. Shakal
Structure Instrumentation In The California Strong Motion Instrumentation Program 17

Farzad Naeim
Learning From Seismic Response of Instrumented Buildings During The 1994 Northridge Earthquake 33

Robert Nigbor, John Fort
New Developments In Health Monitoring For Civil Structures 53

Dario Rinaldis
Experimental Dynamic Analysis of Historic Monuments and Buildings 61

William U. Savage
Utilization of Rapid Post-Earthquake Data By Utilities 81

Erdal Şafak
Analysis of Earthquake Records From Structures: An Overview 91

R.T. Severn
Structural Monitoring of Bridges - An Overview 109

M.D. Trifunac, M.I. Todorovska
Recording and Interpreting Earthquake Response of Full-Scale Structures 131

Robert F. Ballard Jr.
The U.S. Army Corps of Engineers Seismic Strong-Motion Instrumentation Program 157

R.D. Borcherdt, H.P. Liu, R.E. Westerlund, C. Dietel, J.F. Gibbs, R.E. Warrick
Integrated Surface and Borehole Strong-Motion, Soil-Response Arrays in San Francisco, California 167

Mehmet Çelebi
Current and New Trends In Utilization of Data From Instrumented Structures 179

G.R. Darbre
Instrumentation of Dam Structures In Switzerland 195

E. Durukal, M. Erdik, S. Cimilli
Strong Motion Networks: A Tool for the Assessment of Earthquake Response of Historical Monuments 209

John R. Evans
Wireless Monitoring and Low-Cost Accalerometers For Structures And Urban Sites 229

Asterios A. Liolios
The Need for Data From Instrumented Structures for an Optimal Control Approach to the Seismic Interaction between Adjacent Buildings 243

C. Richard Liu, Lanling Zhou, Xuemin Chen, S.T. Mau
Wireless Sensors for Structural Monitoring 253

José A. Martínez-Cruzado, Esteban L. Llop-Ramirez
Puerto Rico Strong Motion Network and Instrumented Buildings 267

Vladimir Mihailov, Dragi Dojcinovski
Strong Motion Instrumentations of Dams In Macedonia Some Experience and Results 275

Yutaka Nakamura
An Effective Earthquake Monitoring Process For Emergency Response 293

Yutaka Nakamura, Dilek E. Gurler
Estimation of Dynamic Characteristics of Ground and Structures With Microtremor Measurements - A Supportive Tool for Strong Ground Motion Instrumentation 303

Carlos E. Ventura, Yuming Ding
Strong Motion Instrumentation Of Buildings 313

Chris Wood, Andy Viksne, Jon Ake, David Copeland
Current Status of Strong-Motion Monitoring and Notification At The United States Bureau of Reclamation 331

Nurdan Apaydin, Mustafa Erdik
Structural Vibration Monitoring System for the Bosporus Suspension Bridges 343

M. Baur, O. Novak, J. Eibl, D. Lungu
Soil-Structure-Interaction and Seismic Isolation, an Inter-Disciplinary Investigation at the Multidisciplinary Seismic Test Site Incerc, Bucharest, Romania 369

Mehmet Çelebi
GPS In Dynamic Monitoring of Long-Period Structures 383

R.N. Celik
Real Time Large Structure Monitoring Using The Inclination Sensor 397

R.N. Celik, T. Ayan, H. Denli, T. Özludemir, S. Erol, B. Ozoner, N. Apaydin, M. Erincer, S. Leinen, E. Groten
Monitoring Deformation On Karasu Viaduct Using GPS & Precise Leveling Techniques 407

D.M. Dojcinovski, D.J. Mamucevski and V.P. Mihailov
Seismic Monitoring of Nuclear Power Plants; An Approach To Optimal and More Accurate Seismic Data Processing and Interpretation Procedure 417

W.D.L. Finn, E. Zhai, T. Thavaraj, X.-S. Hao and C.E. Ventura
Analysis of Data From Strong Motion Network In Fraser Delta, British Columbia, Canada 433

S.S. Ivanovic, M.D. Trifunac and M.D. Todorovska
On Identification of Damage In Structures Via Wave Travel Times 447

Toshihide Kashima, Izuru Okawa and Shin Koyama
Earthquake Motion Observation In and Around 8-Story SRC Building 469

V.A. Lekidis, C.Z. Karakostas, D.G. Talaslidis
Instrumentation, Measurements and Numerical Analysis Bridges: An Example of The Cable-Stayed Bridge on Evripos Channel, Greece 481

J. Leonov
Damage Detection In Semi-Rigid Joint, RC Frames Subjected To Strong Motion Excitation 495

D. Lungu, C. Arion, A. Aldea, S. Demetriu
Assessment of Seismic Hazard In Romania Based on 25 Years of Strong Ground Motion Instrumentation 505

Chikahiro Minowa, Michio Iguchi and Masanori Iiba
Measurement of Lateral Earth Pressures on an Embedded Foundation During Earthquakes 519

Shahram Pezeshk, Mehmet Çelebi, Greg Steiner, Charles V. Camp, Howard Hwang
Seismic Instrumentation of The I-40 Mississippi River Bridge in Memphis, Tennessee 533

T. Rashidov
Engineering and Seismometric Service in the Buildings of Tashkent and Tashkent Regions 545

K.M. Rasmussen, S.R.K. Nielsen, P.H. Kirkegaard
Stress Wave Propagation Due to a Moving Force: Comparison of FEM and BEM Solutions 551

V. Zaalishvili
Strong Motion in Absorbing Nonlinear Medium and Problems of Their Registration 561

Y. Zaslavsky, J. Leonov, A. Shapira
Seismic Response Study of Two-Storey Building in Eilat Using Weak and Strong Motion Data 573

V. Zaalishvili, I. Timchenko, V. Kacharava, Z. Zaalishvili
Strong Motion Instrumentation For Structures of Civil Engineering and Economical Aspects of Planning of Territory of Big Cities 593

Subject Index 603

PREFACE

The provision of earthquake resistant structures and facilities is one of the main concerns of Civil and Earthquake Engineering Profession. As such, the main objective of seismic instrumentation program for civil engineering structural systems is to improve our understanding of the behavior and potential for damage of structures under the dynamic loads of earthquakes. As a result of this understanding, design and construction practices can be modified so that future earthquake damage is minimized. Therefore, there are significant implications in (a) hazard reduction, (b) improvement of codes, (c) identification of seismic response characteristics of structural system that may be used in determination of strategies for improvement of their performances.

There are two main approaches to evaluate seismic behavior and performance of structural systems. One requires a laboratory in which subsystems, components, or (if the facility is large enough) prototypes or large, scaled models of complete systems are tested under static, quasi-static, or dynamic loading. This approach does not necessarily demand a time-dependent testing scheme, such as a shaking table or hydraulically powered and electronically controlled loading systems; however, testing of structural systems under controlled simulated environments is desirable. Since the early 1950's such laboratory research has increased both in quantity and quality, with engineering colleges in the United States and private and governmental laboratories in Japan playing a key role. Laboratory testing has also contributed substantially to our understanding of dynamic soil properties and the interaction phenomenon between the soil and structure.

The second approach to evaluate behavior and performance of structural systems is to use the natural laboratory of the Earth, by observing and studying damage to structures from earthquakes. By determining why specific designs lack earthquake resistance and then by using extensive laboratory testing of modified designs, significant progress in improved designs can be achieved. The validity of the models of analysis used in the assessment of the earthquake behavior of structures can only be checked by way of comparison with relevant field observations. For such design studies, a natural laboratory would be a seismically prone area that offers a variety of structural systems. Integral to the "natural laboratory" approach is the advance instrumentation of selected structures so that their responses can be recorded during future earthquakes. Thus, it is essential that integrated arrays of instrumentation be planned and installed to assess thoroughly the relation of ground motion that starts at a source and is transmitted through various soils to a substructure and finally to a superstructure. The direction for seismologists and engineers working together is clear; to develop integrated networks which measure the seismic source, the transmittal of ground motion, and the structural response processes.

Strong motion instrumentation of structures has been utilized since 1940's. Throughout the world, strong motion instrumentation networks have been installed on buildings, monumental and historic structures, bridges, dams, tunnels, pipelines and power plants. Recent strong earthquakes, Mexico City (1985), Loma Prieta (1989), Landers (1992), Northridge (1994), and Kobe (1995) have yielded a wealth of structural response data from instrumented structures. These data have contributed to the evolution and enhancement of seismic analysis and design methodologies, seismic building codes and practices. In the last decade, there have been significant advances in the development of digital seismic monitoring systems, data retrieval, processing, storage and dissemination capabilities. The new strong motion recording instruments with advanced technologies allow for: High dynamic range (24-bit with micro-g acceleration resolution); Large memory capacity (few hours); On-board real-time processing of data and; Extensive communication options. These developments have reduced the initial and the maintenance cost of the strong motion instrumentation and paved the path for their increased utilization.

Recent changes in instrumentation technology and the need by public officials as well as the public for early damage assessment and disaster response have added a new dimension to these efforts. The ideal venue for obtaining strong-motion response data is now being considered in terms of real-time acquisition and processing. Monitoring the dynamic behavior of structures for quick damage assessment and early warning are new and active areas of research. Although still in development stage, several developments in the structural health monitoring technology is slowly taking place within several strong-motion programs, for research and on "as needed and feasible basis". Developments for real-time (or near-real-time) serviceability assessment for important lifelines (bridges, dams and pipelines) are also being implemented.

It should be noted that, most of the existing strong motion instrumentation on civil engineering structures are installed and operated as federal, state, university, industry or private applications. In many cases, there is very little or no coordination between them. Some of them are managed as closed systems. Limited co-operation programs and data exchange arrangements hinder difficulties and delays in acquisition of strong motion and structural data, even within the same country, by engineers and researchers external to the system. As a result of this lack of coordination and information, the strong-motion data obtained from the instrumented structures are not fully utilized by the engineering profession.

In the on-going globalization movement there exist a strong desire and need for co-operation in strong motion instrumentation programs and enhanced data exchange programs. Similarly, there exist a strong need to inform practicing engineers on the use of existing strong-motion data and to improve the worldwide accessibility of data. The NATO ARW held and this publication is an attempt to answer this need through dissemination of the developments on strong motion instrumentation of civil engineering structures.

The editors wish to express their deep appreciation to NATO, other sponsoring organizations and several individuals whose personal efforts made the workshop and, therefore, this book possible. NATO has provided the opportunity, encouragement and the bulk of the funding towards the realization of the workshop and this publication. Dr.F.Pedrazzini, Program Director, of the NATO-Scientific and Environmental Affairs Division patiently guided us throughout the entire process of support application to the publication of this book. Bogazici University and General Directorate of Turkish Highways acted as the host organizations. U.S.National Science Foundation (through COSMOS), Akashi Co., Kinemetrics Inc. and Leica Inc. have generously provided support to host several participants and social functions. Ms.Gülüm Birgoren, Dr.Oğuz Özel, Dr.Kemal Beyen, Mr.Muzaffer Gül, Mr.Cem Ozbey, Mr.Hakan Akman, Mr Aydın Mert were of great assistance during the planning and conduct of the workshop. Finally special thanks are due to Ms.Nurdan Apaydin who was instrumental in all efforts of this undertaking. She has personally supervised all of the workshop activities and almost singe handedly prepared all the papers for this publication.

MUSTAFA ERDİK,
ISTANBUL, December, 2000

STATE-OF-THE-ART TECHNOLOGY FOR DATA STORAGE AND DISSEMINATION

TIMOTHY K. AHERN
IRIS Data Management Center
1408 NE 45th Street #201
Seattle, WA 98105 USA
e-mail:tim@iris.washington.edu

Abstract

Seismology presents unique challenges in data storage requirements and the need to rapidly access data and disseminate information. This paper will briefly summarize some current technological trends in each of these areas.

In general, seismological data sets can vary in size from minute to massive. Seismology is somewhat unique in that, not only can the data volumes be large, there still exists a requirement to recover and manipulate data at the level of individual samples. These competing requirements make data storage solutions more complicated. This paper will review some of the current technologies that exist and how they can be applied to seismological data archives.

There exists a need to access data from remote seismological data recorders in real time. Often the telecommunications infrastructure at these remote sites does not exist and this presents unique challenges to operators of seismic networks. This paper will discuss several communication techniques including use of telephone circuits, Internet Service Providers (ISPs), and satellite systems that can be exploited to benefit seismology on local, regional and global scales.

New opportunities have emerged that allow easy dissemination of data and information to outside users. This paper will also discuss some of the efforts taking place within the Comprehensive Test Ban Treaty Organization and the Federation of Digital Broadband Seismographic Networks in this area.

KEYWORDS: seismology, data storage, broadband, strong-motion, data dissemination, data communication, satellite data transmission

M. Erdik et al. (eds.), Strong Motion Instrumentation for Civil Engineering Structures, 1–15.

1. Introduction

The oil exploration industry, the national security agencies, the aerospace industry and remote sensing initiatives are examples of disciplines that generate massive amounts of data. However, seismology is somewhat unique in that, not only are the data volumes large, but there still exists a requirement to recover and manipulate data at the level of individual samples.

Not only is the volume of data generated by seismologists large, the need to have real time access to the data is somewhat unique. Earthquake hazard applications require that the seismic event be characterized within tens of seconds to a few minutes in order to minimize societal impact. The stopping of trains, orderly shut down of utilities, and similar proactive behavior requires near real time data transfer from distributed seismic recorders as well as the dissemination of this information to a large number of organizations quickly. For this reason the issue of data telemetry to remote and distributed locations is also a seismological problem. Nevertheless solutions are beginning to immerge that will provide computer connectivity to locations throughout the world.

Seismological events, earthquakes, are of general interest to the worldwide earth science community as well as the general public. It is becoming increasingly more important for data generating organizations to make these data available to a wide audience. The widespread proliferation of the Internet and the World Wide Web (WWW) now offer opportunities for information from all branches of seismology to be made available to scientific users and the general public as well.

2. Mass Storage Systems for Scientific Data Sets

Although seismic data sets can be massive, many uses call for rapidly accessing individual samples inside a large archive. Driven by such activities as the EOSDIS and the requirements of the national security organizations, the computer industry and specifically the mass storage industry have been able to produce systems that can now store petabytes (1×10^{15} bytes) of information in a cost effective manner. There now exist systems that can store several petabytes of data in a single system. Systems vary from the complex to the relatively simple but in general all large systems utilize magnetic tape media and all petabyte storage systems fall in this class. Mid-range systems can sometimes exist as both disk based and optical disk based systems and capacities can reach the few terabyte ($1x10^{12}$ bytes) in a relatively cost effective manner. Smaller systems can be either disk based or tape based systems that store tens to hundreds of gigabytes for a few tens of thousands of dollars.

Experience has shown that the requirement to store large amounts of information in on-line or near-line (robotically accessible) systems is basically solved. Even small to mid-sized organizations can afford to purchase multi-terabyte mass storage systems with the storage capacities and input/output bandwidth meeting their needs. Table 1 shows estimates of data generating rates for a variety of data types, including several seismic data sources. Figure 1 shows how the capacity of disk drives has increased and the costs per gigabyte of disk space have decreased with time.

The Kyoshin net in Japan is one of the largest deployments of strong motion accelerometers in existence. Since it generates only 125 megabytes of data per year, clearly all of the data can be stored in on-line disk systems. In general strong motion networks do not have a data storage problem that can not be solved with limited resources. Seismological networks such as the system being deployed by the United Nations to monitor the Comprehensive Test Ban Treaty (CTBT) do require higher capacity systems. When the data volume increases beyond a few terabytes, mass storage systems that use magnetic tape are generally used. These systems are normally found at NASA data centers, National Security Agency installations and at medium sized data centers such as the IRIS DMC. Table 2 provides some rough estimates of the cost per terabyte for a variety of different storage technologies. The IRIS DMC selected StorageTek Redwood/Wolfcreek mass storage system to meet its needs for data archiving and storage. Figure 2 presents a picture of this system. At the present time IRIS manages approximately 9 terabytes of seismic waveform data in a total of more than 2,500,000 files. We anticipate that we will be managing about 50 terabytes of information by the year 2002.

IRIS has found that the more difficult problem is that of correctly specifying the characteristics of the various recording systems that digitize the ground motion data. These metadata (data that describes the primary observational data) are now routinely being stored in commercial relational database management systems. In general the data management problem experienced by seismic data centers is not really the ability to handle the numbers of bytes of data. It is instead managing the metadata correctly. Primary observational data are of little use if the necessary metadata that scientists and engineers need to interpret the time series data are incorrect. We have found Oracle to provide a very effective platform for the relational database system that IRIS needs. IRIS decouples the time series data from the metadata by simply inserting the name of the mass storage system file in one field of one of the Oracle tables.

To a certain extent, the IRIS Data Management Center located at the University of Washington pioneered the concept of combining the metadata in a commercial database management system with a pointer to waveform files that reside in a mass storage system. The concept of data localization has also been exploited to its fullest. Data localization, as its name implies, means that data are organized in the tape based mass storage system in a manner such that when one file is recovered the probability of needing another file on the same piece of media is much higher. The need for efficiency in servicing requests for waveform data in seismology can lead one to the conclusion that data should be stored in both a time based sort order as well as a station based sort order.

Ingenious software allows the small to large mass storage systems to attach to simple UNIX based computers that are affordable to most scientists. Additionally NT has made inroads into the enterprise class of storage systems primarily in the area of large disk RAID systems. Several ports of file managers are also being made to computers running NT but it still appears that the richest suite of software for managing mass storage systems exists in the MVS and UNIX environments.

3. Technological Trends in Computer Communications to Remote Sensors

Inherently, seismic data must come from geographically distributed sensors. This presents unique communication requirements. Many different technologies to transfer information from remote sensors to data processing centers now exist. One of the more common methods of transmitting data from remote recording systems to a data center is standard telephone lines. Phone circuits are supplemented by spread spectrum radio modems, frame relay data transmission circuits, public Internet, as well as complex systems involving satellite circuits. All of these methods have been applied to the problem of transmitting seismic data in near real time to seismological data centers. For instance, Caltech uses commercially available frame relay circuits to recover data from seismic stations throughout southern California. IRIS now transfers all seismological data generated by US stations that are part of the International Monitoring System (IMS) for the Comprehensive Test Ban Treaty (CTBT) from Patrick Air Force Base in Florida to the IRIS DMC using a commercial frame relay circuit. This will allow seismic data from a global network to arrive at the IRIS DMC in real time. Spread spectrum radio modems are used to recover data from regional networks in the United States and throughout the world. Some of the more ambitious communication systems presently being developed involve Low Earth Orbiting Satellites (LEO) that will provide fiber like delays for globally distributed computer information systems.

For many years, IRIS has been developing a system called SPYDER® (System to Provide You with Data from Earthquakes Rapidly). Figure 3 shows the locations of the 132 globally distributed seismic stations that can be accessed by SPYDER®.

Communication systems generally do not need to be developed by operators of seismological networks. In general operators of seismic networks just need to contact a local telecommunications services provider, identify the specific data communication problem, and these companies will propose solutions that meet or exceed your requirements. On a global basis the situation is a bit more complex and at times operators of seismic networks must develop the infrastructure in the remote location to transfer the data to a central collection point with connectivity to the Internet. For instance the Kyrgystan Telemetered Network uses a system of spread spectrum modems to transmit data from individual stations, to relay stations and finally into the capital city of Bishkek.

In Bishkek, a leased circuit transmits the data to Obninsk, the national center for earthquake information in Russia. Leased circuits then transfer the data to Moscow where it is placed on the commercial Internet. Data from Kyrgystan arrive in the United States in basically real time. Transfer of information in this network fully utilizes TCP/IP protocols from station to final data center and where possible leverages the commercial Internet.

The amount of effort required in establishing networks such as the one in Kyrgystan is significant. The near future holds a variety of exciting possibilities in the area of global data communication systems. One of the more exciting possibilities is the development of the Internet in the Sky as proposed by Teledesic. The Teledesic Network will consist of 288 operational satellites, divided into 12 planes, each with 24 satellites. As the satellite planes orbit north-to-south and south-to-north, the Earth rotates underneath. This provides global coverage. Figure 4 shows a graphic of the proposed Teledesic constellation. The Teledesic satellites will orbit at less than 1,400 kilometers and as such will offer very low latency. In fact the time it will take for a byte of information to reach half way around the globe will be less than the time it would take for a byte to transfer across an optical fiber. We feel that when the availability of the Internet-in-the-sky system being developed by Teledesic becomes available in the year 2003, it will have obvious relevance to operators of seismic networks. Not only does such a system have global coverage they might be less prone to damage and failure than terrestrial systems.

4. Dissemination of Data

In addition to the telemetry problem of sending data from the data recording systems in the field to data centers, the dissemination of data to government agencies and the public is becoming more important. Most notable among the new methods of data dissemination is the Internet. Government funded projects have developed powerful and novel methods of using the Internet to transfer information from data centers to end users. There exists an increasingly well known World Wide Web link at

URL= http://www.geophys.washington.edu/seismosurfing.html

that lists some of the better known WWW sites that provide access to seismological information. WWW sites are very effective in disseminating visually appealing information about a variety of seismological topics. Unfortunately, the WWW is not well suited for the distribution of voluminous information such as seismic waveforms.

In the Comprehensive Test Ban Treaty Organization (CTBTO) and The Federation of Digital Broadband Seismographic Networks (FDSN) communities a great deal of effort has been expended in developing standardized data request mechanisms and standard data exchange formats. Both of these organizations work in the global seismological arena and some of their choices may have been influenced by this fact.

4.1 DATA REQUEST MECHANISMS

4.1.1 *AutoDRM Requests*

Although a variety of FDSN data centers have implemented several different methods of requesting data, there are two that are gaining very wide spread acceptance. The first is the autoDRM request mechanism developed, by the Swiss, for the United Nations, for use in the area of non-proliferation of nuclear weapons . This email based system standardizes the request format as shown in Figure 5. The request is interpreted by the receiving computer system, data extracted from the data center's archive, reformatted into a compressed ASCII format, and the results are normally emailed back to the email address specified in the request. A more complete description of the autoDRM format and data centers supporting this format can be found at

URL= http://orfeus.knmi.nl/autodrm/autodrm_help.html

4.1.2 *Networked Data Centers*

The FDSN has recently begun implementing a truly distributed data center system. This system was developed primarily at the IRIS DMC but has been implemented at the IRIS DMC, the University of California, Berkeley, the GEOSCOPE data Center in Paris, and the ORFEUS Data Center in de Bilt, The Netherlands. This system is called the Networked Data Center System or NetDC.

Some of the more difficult activities in the area of dissemination of seismic data between data centers involve concepts that eliminate the traditional centralized data center in favor of distributed centers. In the distributed model, data are held at geographically dispersed data centers. The distributed model has a variety of distinct advantages,

- data are held at the center that best understands the data set,
- the resource available to the entire system equals the sum of the parts and helps balance the load across the distributed network,
- multiple sources of funding are available to support the networked data centers,
- distributed data centers are politically more viable as funding agencies can see the need to fund a local center more easily than an international center, and
- data are as accessible to the users in the same manner as from a centralized data center.

Figure 6 shows a model now being developed within the FDSN for the development of a network data center model. From a user's perspective the system appears to be one homogenous data center. In reality the Internet is used to service a user's request at many different participating data centers. This model might prove to be quite useful to the strong motion community as it allows each data generating facility to retain control of the data they record while still providing easy access to the general community of seismologists.

Although the NetDC system is being used within the FDSN, for primarily broadband, weak motion seismology, it can easily be extended to other seismological data types, such as strong motion data. In fact, most FDSN stations installed in areas with strong ground motion potential are already recording strong motion data, and those time series are available through the NetDC system.

Figure 7 provides a brief example of a NetDC request.

4.2 DATA FORMATS

The need for standardization in request methods, data formats, and format of disseminated information is essential. Previous sections have briefly discussed some concepts in standardization of request mechanisms. Just as crucial is the need to standardize the format of data being supplied by seismological data centers.

The FDSN has formed a working group on data formats that is sanctioned by IASPEI. It is possible to place strong motion time series data in this format and this is routinely done by members of the FDSN. The format is the Standard for Exchange of Earthquake Data (SEED). More information about the SEED format can be found from the IRIS DMC or through the WWW at

URL= http://www.iris.washington.edu/NEW/HTM/manuals.htm

If the strong motion community adopts such a standard format, the members of the FDSN would be prepared to work with the strong motion community to make any necessary changes to the SEED format as required.

5. Conclusions

This article attempted to provide an overview of many of the initiatives presently taking place in the seismological community. A brief overview of storage systems including:

- small to large disk RAID technology,
- medium sized tape based solutions, and
- large multi-terabyte tape based mass storage systems

was presented.

In general there appear to be no real problems with data storage for the strong motion community. For under $4000 , fully redundant RAID disks with a capacity of 10 gigabytes could be acquired. This seems to exceed the storage capacity needed. For larger data centers such as the IRIS DMC, large multiterabyte tape based storage solutions are available that can store several tens of terabytes of data and still return data in reasonable times. These larger tape based solutions are provided by companies such as Storage Tek and IBM.

Data communication options including:

2. telephone based telemetry systems,
- Internet based systems,
- frame relay based systems, and
- satellite based systems,

both geo-synchronous and low earth orbit were discussed.

The standard telephone is in most cases a good solution for data telemetry. In the case where Internet exists close to a station then the flexibility gained by using Internet communication protocols warrant its use. For very remote installations, where access to the data in a timely fashion is important, companies such as Teledesic and Motorola are developing low earth orbiting satellite solutions that can provide very high bandwidths to any location on the earth, and in general at affordable costs. Data Centers such as the Southern California Data Center and the IRIS DMC, have begun to exploit Frame Relay technology to move very large amounts of data around the United States. This technology is available directly from long distance data communication companies such as Sprint, MCI and ATT with relatively easy integration within a system.

Finally dissemination of information to end users and government agencies was discussed in areas such as:

- WWW,
- Networked Data Centers

It is obvious that the WWW is one of the best ways to present data and information to outside parties. The strong motion community might provide a useful service by trying to coordinate the appearance of how strong motion data are provided.

The Strong Motion Community could develop a Networked Data Center model for providing access to data by the outside community. These developments are now taking place within the global broadband seismological community through the FDSN as well as the Council of the National Seismic System that operate the various short period networks around the United States.

Table 1. Volume of Data Generated by a Variety of Geophysical and Scientific Applications. The amount of data being generated for a variety of different scientific data acquisition activities. Seismic sources produce tens of terabytes of data per year but the mass storage industry are driven more by EOSDIS and particle accelerator experiments where the data volumes are measured in terms of hundreds of terabytes up to a few petabytes.

DATA SOURCE	**per**	**terabytes**
Seismic Exploration Industry		
Marine Seismic Vessel	year	31
Vibroseis® Crew	year	42
IRIS Global Seismic Network	year	0.8
International Monitoring System of CTBTO		
1999	year	3
2002	year	15
Strong Motion Seismic Networks		
Kyoshin Net	year	0.000125
Particle Accelerator		
1994	experiment	1
2000	experiment	1000
NASA EOSDIS	year	350

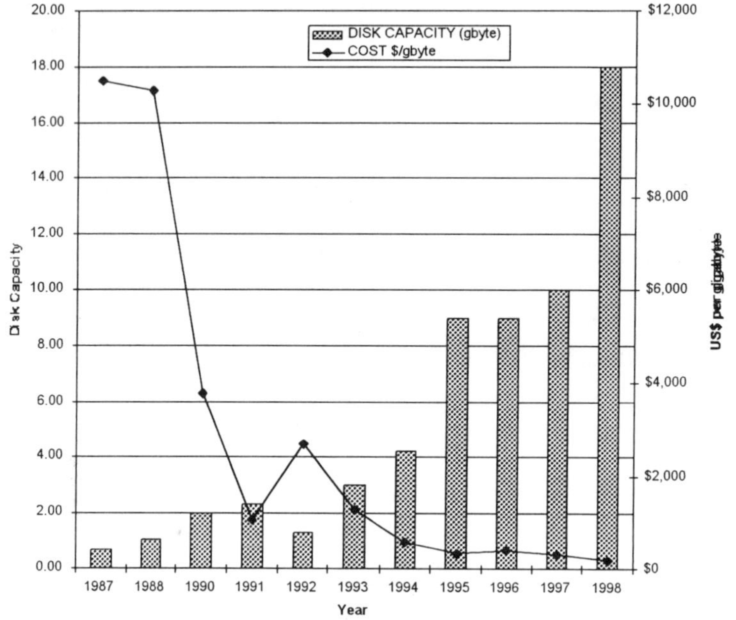

Figure 1. A Graph Showing the Size in Gigabytes of Large Disk Drives and their Cost per Gigabyte by Year. The ability to store large quantities of data in computer systems has become very affordable over the past decade. This figure shows that as disk capacities have gone from 1 gigabyte per disk in 1987 to 18 gigabytes per disk in 1998, the costs per byte have dropped by more than a factor of 55. In fact in 1987 1 gigabyte of disk cost about $10,500, whereas in 1998 the cost for one gigabyte of disk dropped to $190. These values were calculated from actual purchases made by the IRIS Data Management Center over the past 10 years.

Table 2. Representative Prices for Multi-terabyte Mass Storage Systems. This table provides recent prices that have been quoted for a variety of mass storage solutions. The IRIS Data Management Center selected the Storage Tek DS3 solution and has been using this system for one and one-half years.

TECHNOLOGY	$ per tbyte	Capacity
RAID Disk	$150,000	any
DLT Tape 7000	$20,000	3.5 tbyte
IBM Magstar/Grau Robot	$17,578	38.4 tbyte
Storage Tek DS3 Redwood/Wolfcreek	$14,000	50 tbyte

Figure 2. The StorageTek Wolfcreek Mass Storage System in Operation at the IRIS DMC. This system is roughly 72" tall and covers a 10 feet square floor area. It has a capacity of 50 terabytes of storage and an aggregate input/output rate of about 45 megabytes/second. At the right of the picture are three racks that contain 1.5 terabytes of RAID that is used as a temporary buffer to the larger tape based mass storage system.

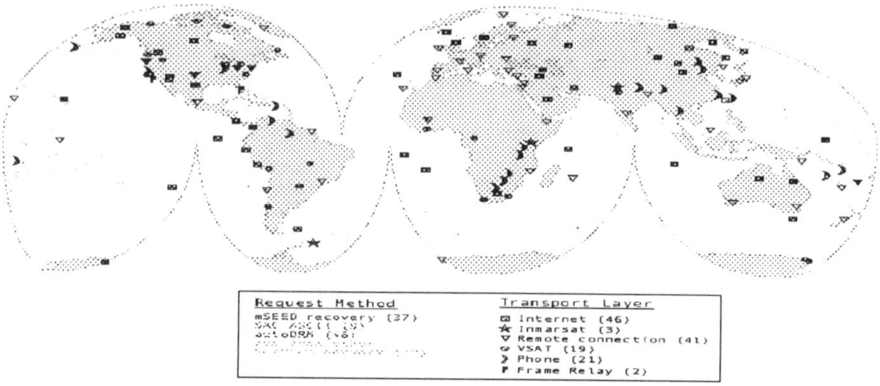

Figure 3. Stations Accessed by the IRIS SPYDER® System.
This diagram shows the locations of stations that are accessed by the IRIS SPYDER® system. As you can see, in today's climate it takes resourcefulness and flexibility to connect to stations on a global basis. 46 stations are accessible over the Internet, 3 using Inmarsat satellites, 41 by telephone connections from cooperating data centers, 19 via VSAT Satellite circuits, 21 from phone calls made by the IRIS DMC, and 2 are available over Frame Relay circuits.

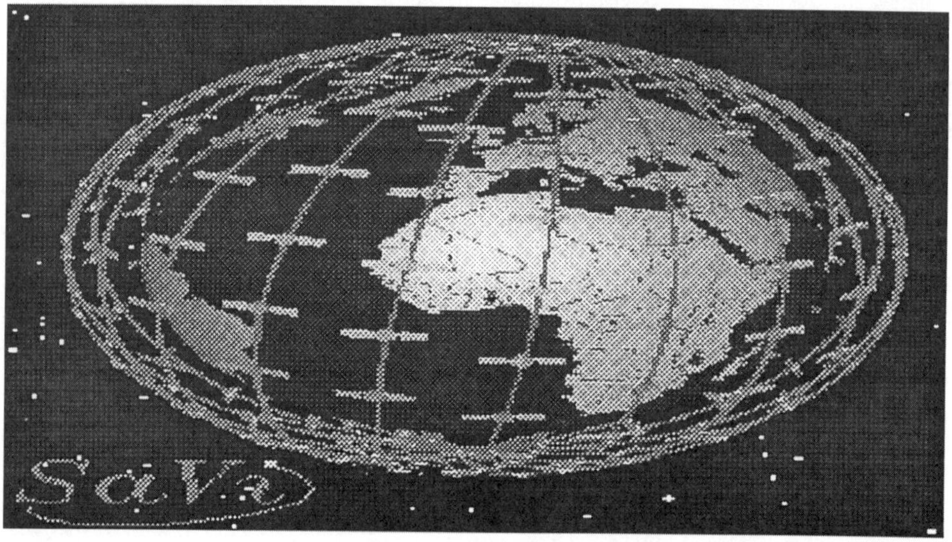

Figure 4. The Teledesic Satellite Constellation. This diagram shows the Teledesic constellation of satellites. Each satellite will be capable of transmitting data at 64 megabits per second from over 95% of the world's landmass. Taken from the Teledesic WWW site found at www.teledesic.com,

```
GSE2.0                                              Begin of message
MSG_TYPE      request                               This is a request
MSG_ID        your_specific_ID  DEU_NDC                     Your ID
TIME                    1994/12/18 22:54:4.0 TO 1994/12/18 22:56:12.0
        Define Times
STA_LIST      APL                                   Define station(s)
CHAN_LIST     SHZ, SLZ                                      Define channel(s)
WAVEFORM      GSE2.0                                You want waveforms
STATION       GSE2.0                                Station-info needed
CHANNEL       GSE2.0                                Channel-info needed
RESPONSE      GSE2.0                                Send calibration-info
OUTAGE        GSE2.0                                Send outage info
E-MAIL        fish@sdac.bgr.hannover.de                     Your e-mail address
STOP                                                        End of request
```

Figure 5. AutoDRM Request Format. This is an example of the autoDRM formatted request message. It is used to request small waveform segments from Data Centers that support this format. This format was developed by Urs Kradolfer at ETH in Zurich, Switzerland.

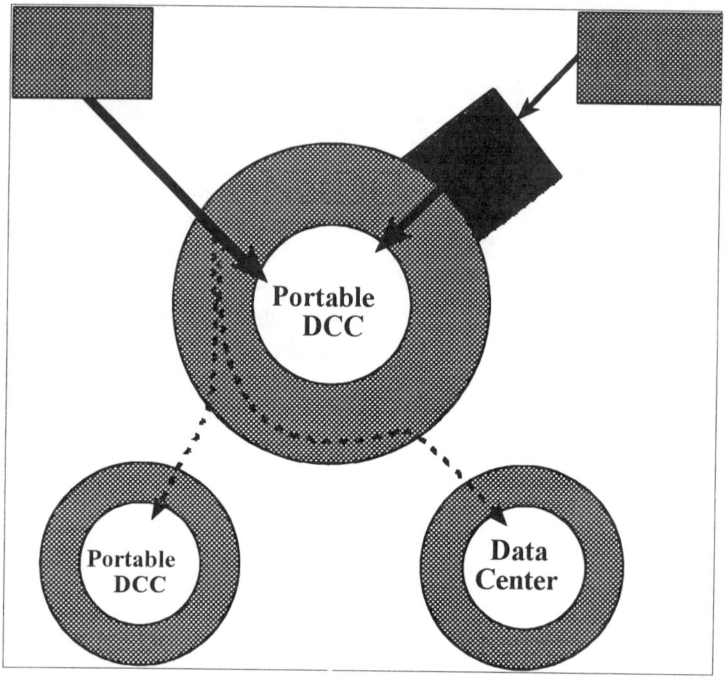

Figure 6. The Networked Data Center Model Developed by the FDSN. This system allows for a truly distributed data center model to be implemented. The NetDC software is a layer of software that exists around the outside of the existing operational software at a data center. It is software that receives information requests, routes portions of requests to other appropriate data centers, and reformats information responses into a standard response format.

```
.NETDC_REQUEST
.HUB_ID                  <machine assigned request label>
.NAME                    <name of user requesting data>
.INST                    <name of institution>
.MAIL                    <return mailing address>
.EMAIL                   <return email address>
.PHONE                   <phone number>
.FAX                     <fax number>
.LABEL                   <user-assigned label for request>
.MEDIA                   <primary media selection>
.ALTERNATE MEDIA         <alternate media selection>
.FORMAT_WAVEFORM         <what format to receive waveform traces in>
.FORMAT_RESPONSE         <what format to receive response information >
.MERGE_DATA              <YES or NO and waiting time>
.DISPOSITION             <instructions for data transfer through network>
.END
```

```
.DATA * AA ORCA * "BHE LH? E*" "1995 06 22 04 00 23.4522" "1995 06 22 05 30 00"
.RESP * AA ORCA * "BHE LH? E*" "1995 06 22 04 00 23.4522" "1995 06 22 05 30 00"
.INV * IU * * * "1994 09 01 02 00 00" "1994 09 30 23 59 59.9999"
```

Figure 7. An Example NetDC Request. The above example shows both the header structure and information request structure that has been implemented within the NetDC system. Fundamentally requests can be made in a standard manner, for data or information existing in any cooperating data center. At the present time three types of requests have been defined, 1) a request for waveforms, 2) a request for response of the recording channel to ground motion, and 3) inventory information about what networks, stations, seismic channels, and time periods exist.

STRUCTURE INSTRUMENTATION IN THE CALIFORNIA STRONG MOTION INSTRUMENTATION PROGRAM

MOH J. HUANG, ANTHONY F. SHAKAL
California Strong Motion Instrumentation Program
Division of Mines and Geology
California Department of Conservation
801 K Street, MS 13-35 Sacramento, California 95814, U.S.A
e-mail:mhuang@consrv.ca.gov

Abstract

In the last 27 years the California Strong Motion Instrumentation Program (CSMIP) in the California Department of Conservation, Division of Mines and Geology has instrumented over 240 structures, including 160 buildings, 58 bridges, and 20 dams. In addition to these structures, CSMIP has installed over 500 ground response stations. Many of these stations have been installed since the 1994 Northridge earthquake. These modern stations have modern digital recording and communication systems that can transmit the recorded data to a central facility in Sacramento within a few minutes after the shaking occurs. The recorded data are automatically processed in the central facility and then disseminated rapidly to the users. These near-real-time data include acceleration, velocity, and displacement time series and response spectra.

The near-real-time strong-motion data provide information on ground shaking and response of instrumented structures, and are useful not only for improving seismic design practices but also for post-earthquake damage evaluation of structures. This paper describes the current status and future plan of CSMIP structure instrumentation projects, and discusses quick application of strong-motion data to post-earthquake response and damage evaluation of structures.

Introduction

The California Strong Motion Instrumentation Program (CSMIP) was established following the destructive 1971 San Fernando earthquake to increase data on strong earthquake shaking. The program installs and maintains strong-motion instruments in representative structures and geological environments throughout California. Strong-motion data recovered from the instruments are processed and made available to engineers and seismologists involved in

M. Erdik et al. (eds.), Strong Motion Instrumentation for Civil Engineering Structures, 17–31.

designing for earthquake shaking and assessing seismic hazards. The goal of the program is to provide the data necessary to improve seismic design codes and increase seismic safety.

Since the inception of the program over 700 stations of various types have been completed. Ground-response sites and structures are selected for instrumentation on the basis of the recommendation of an advisory committee which is part of the California Seismic Safety Commission. The committee is made up of leading engineers and seismologists from universities, government, and private industry. Various organizations and professional groups also provide input to the advisory committee. Current program funding is provided by an assessment on construction costs for building permits issued by cities and counties in California, with additional funding from the California Department of Transportation (Caltrans), the California Office of Statewide Health Planning and Development (OSHPD), and the California Department of Water Resources.

New stations include modern digital recording and utilize a communication system which can transmit the records to a central facility in Sacramento within a few minutes after the shaking occurs. The recorded data are automatically processed in the central facility and then disseminated rapidly to the users.

In January 1997, a joint project,TriNet, between CDMG/CSMIP, Caltech and the U.S. Geological Survey, was funded by the U.S. Federal Emergency Management Agency (FEMA) through the California Office of Emergency Services (OES). The TriNet project will install or upgrade over 600 stations in southern California by the year 2002. In the event of moderate to large earthquakes, TriNet will produce a map, called "ShakeMap", of ground motion distribution within minutes. This map shows contoured and colored presentations of ground shaking parameters for the affected areas so that areas that experienced potentially damaging ground shaking can be quickly determined.

In general, structures located in the areas that experienced damaged ground shaking will not have been instrumented. In addition to the "ShakeMap", one can also interpret a particular ground motion record obtained in the vicinity of the structure to determine whether inspection of these structures in the area is needed. Quick interpretation of near-real-time ground motion records will provide a very useful tool for post-earthquake evaluation of critical structures, such as major buildings, hospitals, bridges and dams. For a particular structure that is instrumented, the recorded response data will help engineers determine whether the structure was damaged, although it may not tell which parts of the structure were damaged.

Building Instrumentation

More than 160 buildings have been instrumented by CSMIP. Typically, 12 to 20 sensors are installed in a building. The sensors are positioned in the building so that important modes of vibration of each specific building will be recorded and specific measurement objectives will be achieved. An example is given in Figure 1 for a 52-story office building in downtown Los Angeles.

In 1989, the California Office of Statewide Health Planning and Development (OSHPD) initiated a program providing funding for instrumentation of selected hospitals by CSMIP. In addition, base-isolated hospital buildings are required by OSHPD to be instrumented by the owner and maintained by CSMIP. A total of 24 hospital buildings are now part of the CSMIP network.

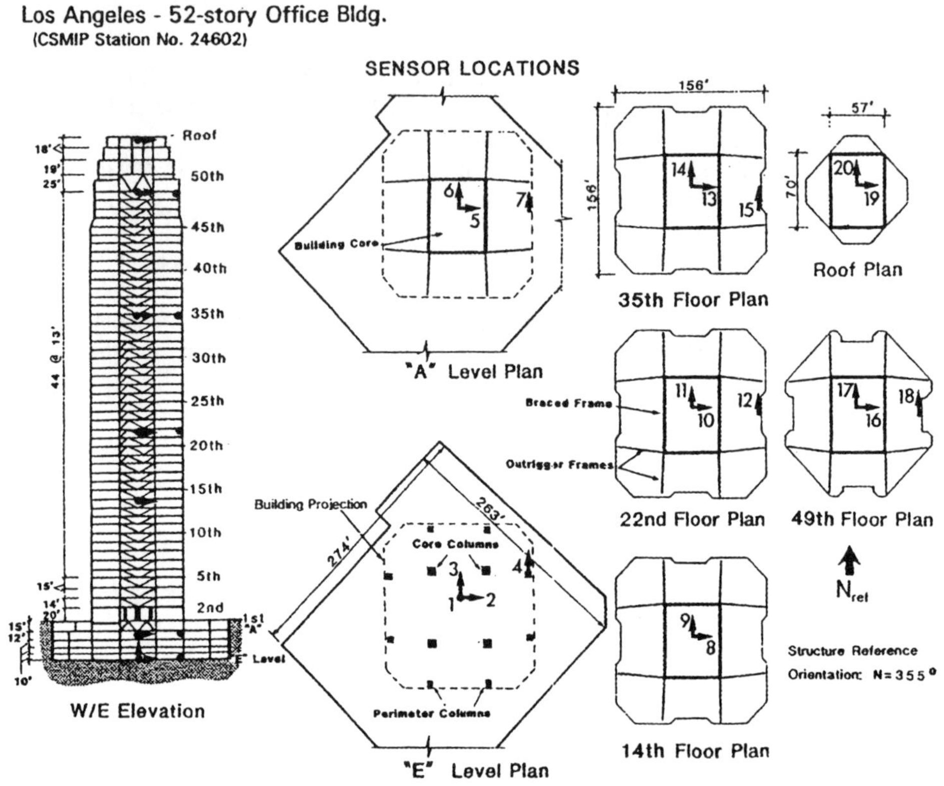

Figure 1. Locations of sensors installed in a 52-story office building in Los Angeles.

The primary goals of CSMIP building instrumentation are to obtain data to improve engineering design practices and to assess and mitigate the hazards posed by existing buildings. Buildings are selected for instrumentation on the basis of recommendations of an advisory committee. It is planned that a total of 400 buildings will eventually be instrumented by CSMIP. All representative building types (frames, shear walls and base isolation), construction techniques and materials (steel, concrete, masonry and wood) are instrumented. These buildings are selected at locations where significant ground shaking is likely to occur.

Instrumentation of a building involves the installation of small sensors (accelerometers in most cases) distributed throughout the structure, all connected to a central recorder by low-voltage electrical cabling. Optimal locations for sensors installed in a building are developed by CSMIP engineering staff by studying lateral force resisting systems from the drawings. Sensors are installed at several key locations so the important modes of building will be recorded and specific instrumentation objectives can be achieved. CSMIP instrumentation of buildings always includes installation of sensors at the base and the roof of the building. The instrumentation includes as many intermediate levels as appropriate for a particular building and a reference free-field site if possible.

Records have been obtained in several buildings damaged by earthquakes (Huang et al., 1996) and they have provided important information on the response of buildings to damaging levels of motion. For example, a hotel in Van Nuys, a 7-story concrete frame structure, suffered structural damage during the 1994 Northridge earthquake. The building was instrumented and suffered minor structural damage in the 1971 San Fernando earthquake and was later instrumented more extensively with 16 sensors by CSMIP. Figure 2 shows the accelerations recorded at the roof, the 6th, the 3rd and the ground floors in the longitudinal direction. Analyses of the records indicate that the fundamental period of the building was lengthened from about 1.5 seconds in the first 10 seconds of the record to about 2 seconds later.

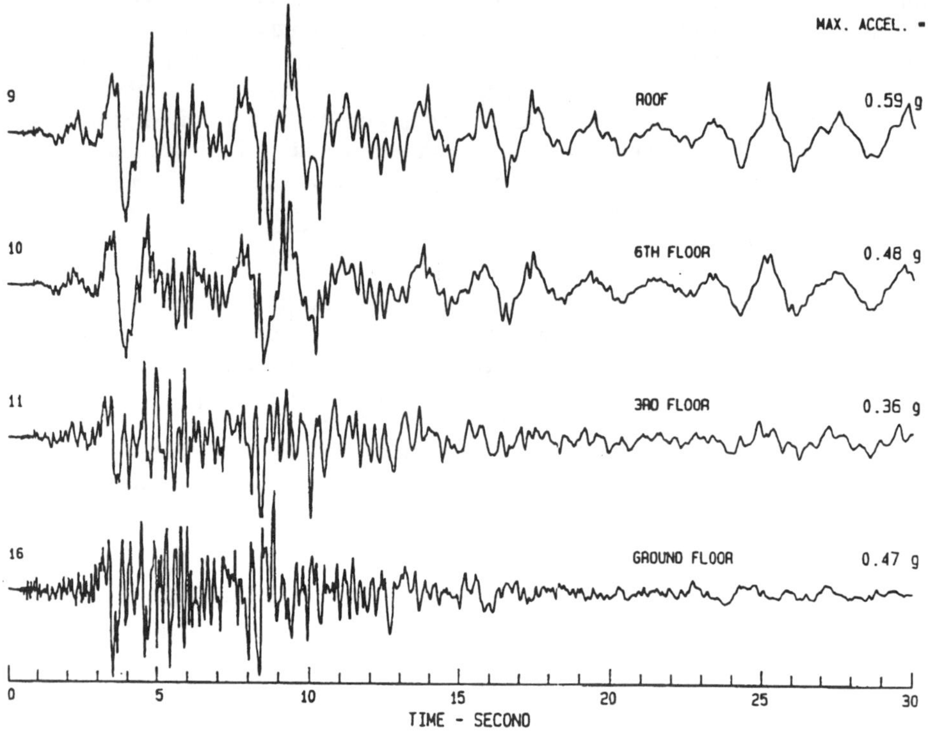

Figure 2. Acceleration in the longitudinal direction recorded at a 7-story concrete building during the 1994 Northridge Earthquake. The building had column failure between the 4th and 5th floors.

Lifeline Structure Instrumentation

Lifeline structures instrumented by CSMIP include regular highway bridges, large toll bridges, dams, power plants, wharves, and airport control towers. As of January 1999, CSMIP has instrumented 58 bridges, 20 dams, 2 power plans, 1 wharf and 1 airport control tower. Most lifeline structures in California were designed according to specific criteria that are quite different from those for buildings. In general, the purposes of instrumentation are to record response data of lifeline structures for better understanding of their seismic response and for improving seismic design and analysis procedures.

After the 1989 Loma Prieta earthquake, a comprehensive program was initiated by the Department of Transportation (Caltrans) and CSMIP to instrument more Caltrans bridges throughout the State. This bridge strong motion instrumentation program was in response to recommendations by the Governor's Board of Inquiry that Caltrans implement a comprehensive program of seismic instrumentation to provide measurements of the excitation and response of transportation structures during earthquakes (Housner, 1990). As of January 1999, a total of 58 bridges have been instrumented. The number of sensors installed at these bridges ranges from 6 (lightly instrumented) to 42 sensors (extensively instrumented) per bridge (Hipley et al., 1998). The locations of instrumented bridges are shown in Figure 3.

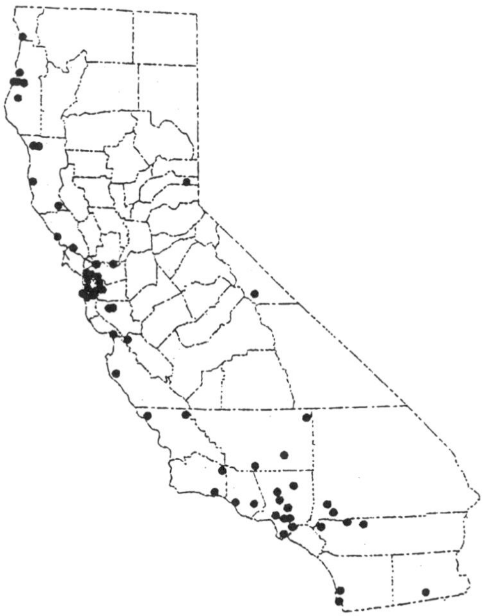

Figure 3. Locations of Caltrans bridges instrumented by CSMIP in California, U.S.A.

Currently, the bridge instrumentation project includes light and extensive instrumentation of representative regular highway bridges. (There are about 12,000 Caltrans bridges in California and it is not practical to instrument every bridge) Under this project, various types of bridges located near major faults have been instrumented. These bridges range from the straight 2-span bridge illustrated in Figure 4 to multi-span curved bridges. Some are newly constructed and some have been strengthened.

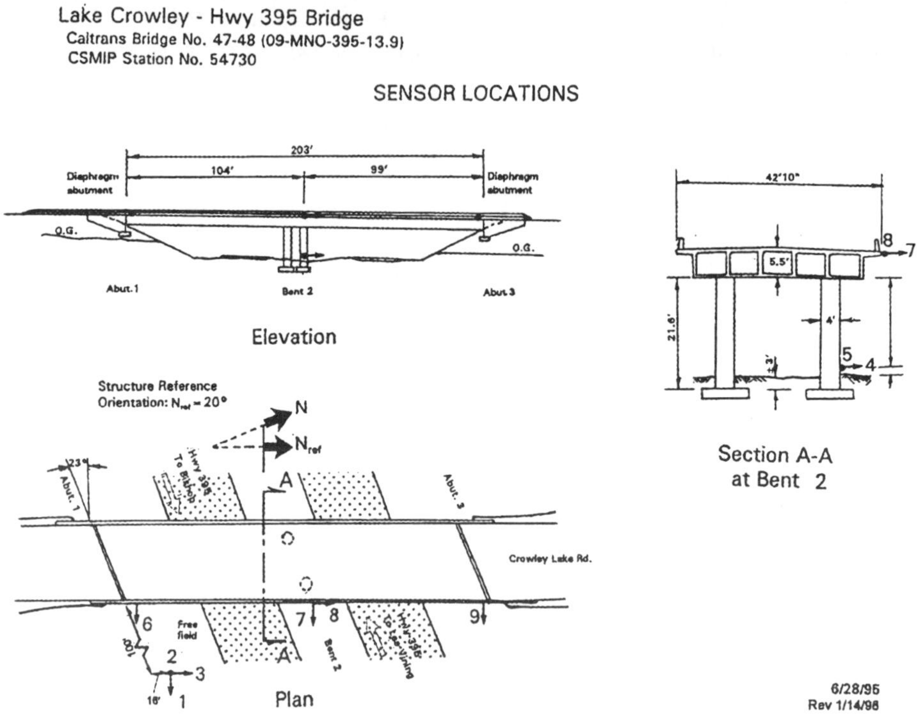

Figure 4. Locations of sensors installed on a Highway 395 bridge near Lake Crowley in eastern California (the arrows and dots denote the measuring direction of the sensors).

In addition to regular highway bridges, the bridge instrumentation project will extensively instrument 7 toll bridges and 2 underground tubes. Caltrans has been designing the retrofit and strengthening of the toll bridges. CSMIP staff have been working with Caltrans engineers and their consulting engineers in developing instrumentation plans so that the instrumentation can be incorporated as part of the retrofit work. As many as 120 sensors will be installed on some bridges to measure the bedrock motions, free-field ground motions, substructure and superstructure responses in future earthquakes.

Funded by the Golden Gate Bridge Transportation District, the Golden Gate Bridge, including the suspension structure, the north and south approach structures, was instrumented in 1995 with a total of 75 sensors. Locations of sensors installed on the suspension structure are shown in Figure 5. Additional sensors will be installed after the bridge is strengthened.

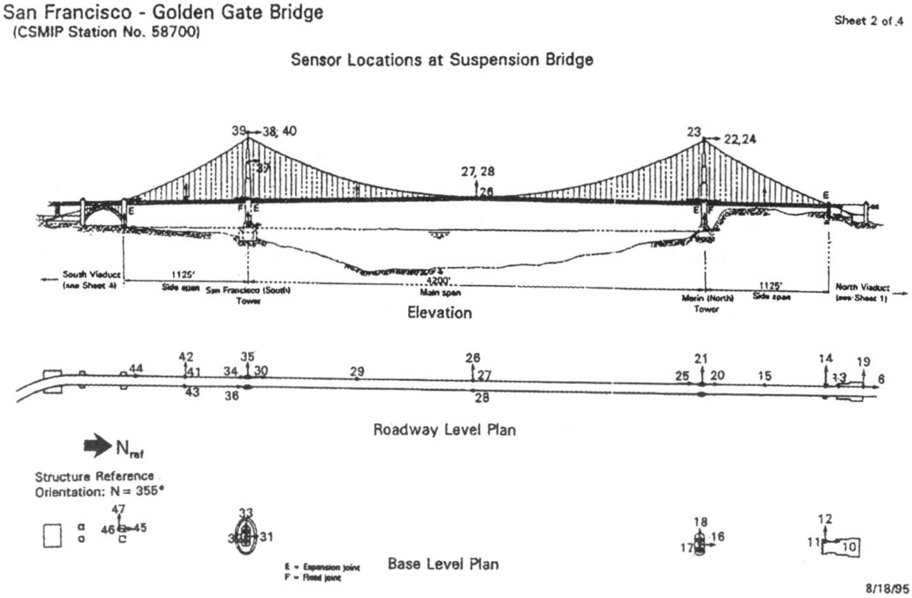

Figure 5. Locations of sensors installed on the suspension structure of the Golden Gate Bridge in San Francisco, California.

Near-Real-Time Data

Developments in accelerographic instruments and communication technology have made possible significant advances in the monitoring and reporting of earthquake strong motion. Since 1995 CSMIP has developed and implemented a system for near-real-time data recovery from strong-motion stations (Shakal et al., 1995 and Shakal et al., 1997). The data recovered are automatically processed to produce the ground motion parameters that are most useful for engineering assessment of the earthquake impact. As an example of the near-real-time data, Figures 6 and 7 show the record recovered from the Mammoth Lakes station, approximately 8 km west of the epicenter of a magnitude 3.6 earthquake that occurred on January 3, 1998. Three components of band-passed acceleration, velocity and displacement, and the acceleration response spectra were automatically calculated and plotted after the event.

The TriNet project will rapidly produce maps of potentially damaging ground shaking within minutes of a damaging earthquake, called "ShakeMap." By year 2002, there will be 670 stations in southern California that record the ground motions. The "ShakeMap" will use these stations to produce contoured maps of ground shaking parameters for the affected areas so that the heavily impacted areas can be determined from these maps. For example, TriNet will generate a map such as the one shown in Figure 8 for the 1994 Northridge earthquake. The ground motion parameters include peak ground acceleration, velocity, and spectral acceleration at 0.3, 1 and 3 seconds. The maps are still under development to meet a variety of needs. Current maps are now available on the Worldwide Web at *http://www.trinet.org* after an event. ShakeMaps for previous events are also available there. Details on the early development of these maps are presented in a paper by Wald et al. (1998).

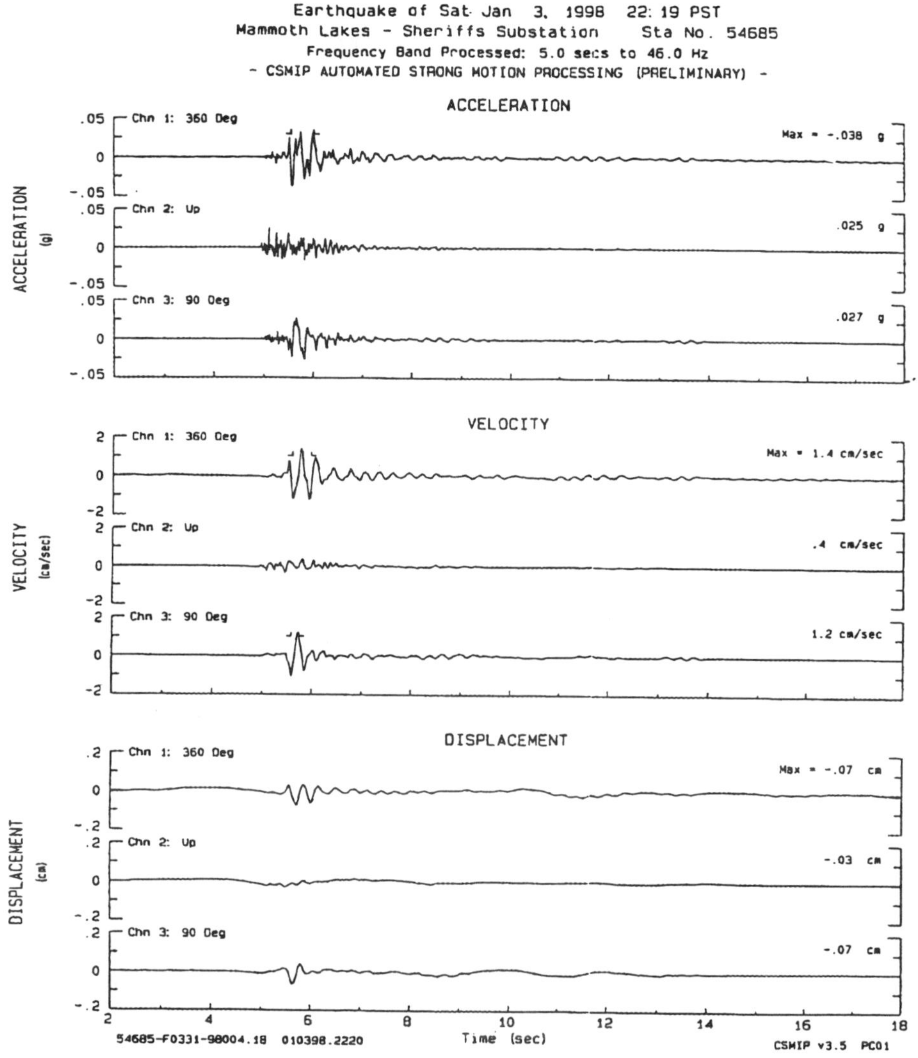

Figure 6. Near-real-time data - 3 components of band-passed acceleration, velocity and displacement at Mammoth Lakes from the magnitude 3.6 earthquake of January 3, 1998.

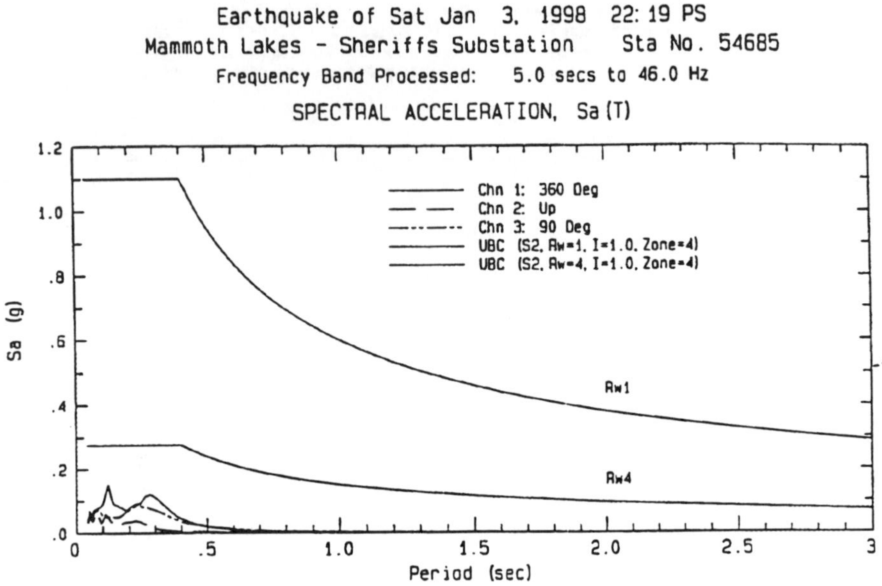

Figure 7. Near-real-time data - three components of spectral acceleration (5% damping) at mammoth Lakes from the Magnitude 3.6 earthquake of January 3, 1998. The design spectra from the 1994 Uniform Building Code (UBC) are plotted for convenient comparison.

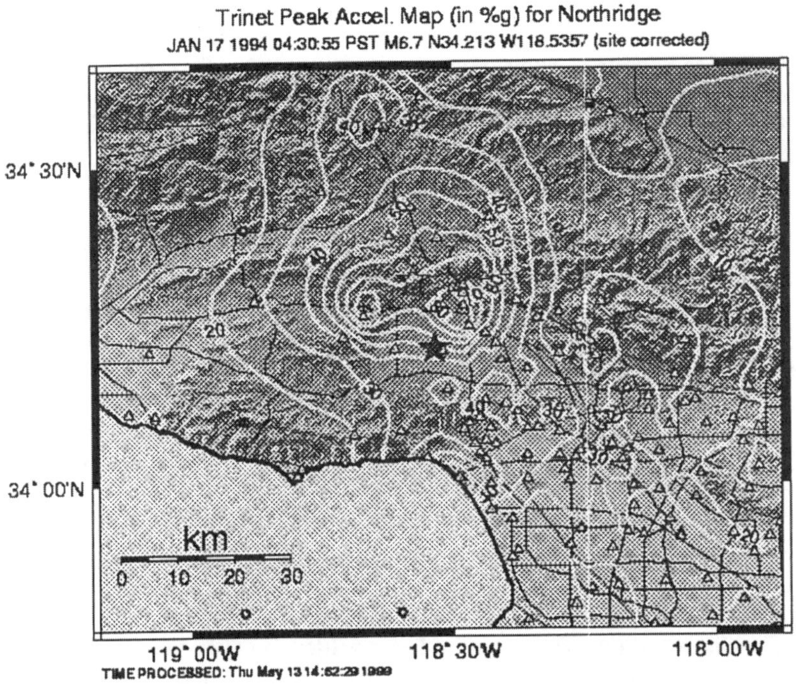

Figure 8. TriNet "ShakeMap" (peak ground acceleration) for the 1994 Northridge Earthquake. The ShakeMaps are available on the Web site *http://www.trinet.org*

The TriNet ShakeMap will be useful for quick determination of which areas experienced damaging ground shaking. In addition, the ground motion records in those areas affected can be quickly studied and compared with design spectra and other design parameters to determine whether inspection of the bridges in the area is needed. For those structures that are instrumented, the recorded response data can be quickly interpreted without complex analyses to facilitate the determination.

Post-Earthquake Application

In California and Nevada, locations and magnitudes of recent earthquakes are available on the Worldwide Web at *http://quake.wr.usgs.gov* right after the event. Response personnel can also receive the information via e-mail and from paging systems. Generally, earthquakes smaller than 5 do not cause damage to engineering structures. Therefore, the response personnel will not need to proceed further for earthquakes smaller than 5.

Many of the instrumented structures have a recording system from which the recorded data can be recovered via phone lines and processed immediately after an event as discussed above, and the data can be used by engineers to quickly assess the integrity of that individual structure. However, for the structures that are not instrumented, engineers can utilize the near-real-time data recorded at free-field sites in the area. From the ground motion information and the characteristics of that structure, the earthquake force experienced by the structure can be estimated. Based on the design information and the experience from past earthquakes, the engineer can then determine whether inspection of that structure should be performed.

For highway bridges, Caltrans currently has a post-earthquake investigation team procedure. In this procedure, the team coordinator will determine the area of damage and develop a list of bridges to be investigated based on the magnitude and location of the earthquake. A GIS-based software package and the database of bridges are used to create the map and the list. This procedure can be expanded to include the ground shaking information produced by TriNet. Depending on the year of design, design forces and the characteristics of each bridge, Caltrans engineers can determine whether the earthquake forces are larger than the design forces for each bridge.

For the near-real-time data to be fully useful for post-earthquake response and damage evaluation of structures, the data need to be interpreted quickly. Performance of structures depends on the characteristics of the ground shaking as well as the characteristics of the structures. In general, important parameters of the ground shaking are peak amplitudes, frequency content and duration of strong shaking. Material and quality of construction, design criteria, structural system, and non-structural components also affect the building performance during earthquakes. Case studies of 25 buildings by the California Seismic Safety Commission (1996) attempted to correlate the damage predicted by the analyses with the actual observed performance of the buildings in the 1994 Northridge earthquake. However, more studies are needed to correlate the building performance with the ground motion.

Summary

The California Strong Motion Instrumentation Program continues to instrument more structures to measure the ground motion and the response of the structure to these motions. The strong-motion data recorded at these structures provide valuable information for understanding the seismic response of the structures and for improving seismic design and analysis procedures. In addition to these long-term objectives, the near-real-time recovery and processing of the recorded data is useful for post-earthquake evaluation of structures. Continued efforts towards utilizing the near-real-time data for post-earthquake response and structural integrity assessments will be ongoing and expanding.

References

1. Housner, G., chairman (1990). Competing Against Time, Report to Governor George Deukmejian by the Governor's Board of Inquiry on the 1989 Loma Prieta Earthquake, State of California, Office of Planning and Research, May.
2. Huang, M., Malhotra, P. and Shakal, A. (1996). Strong-Motion Records from Buildings Damaged in Earthquakes, in SMIP96 Seminar Proceedings, California Strong Motion Instrumentation Program, May.
3. Hipley P., Huang, M. and Shakal, A. (1998). Bridge Instrumentation and Post-Earthquake Evaluation of Bridges, in SMIP98 Seminar Proceedings, California Strong Motion Instrumentation Program, September.
4. Seismic Safety Commission (1996). 1994 Northridge Earthquake Building Case Studies Project, Edited by Rutherford & Chekene.
5. Shakal, A., Petersen, C., Cramlet, A. and Darragh, R. (1995). CSMIP Near-Real-Time Strong Motion Monitoring System: Rapid Recovery and Processing for Event Response, in SMIP95 Seminar Proceedings, California Strong Motion Instrumentation Program, May.
6. Shakal, A., Graizer, V., Petersen, C., and Darragh, R. (1997). Near-Real-Time Strong Motion, TriNet Data and Data Dissemination through the Internet, in SMIP97 Seminar Proceedings, California Strong Motion Instrumentation Program, May.
7. Wald, D., Quitoriano, V.,.Heaton, T., Kanamori, H., and Scrivner, C. (1998) TriNet ShakeMaps: Rapid Generations of Peak Ground Motion and Intensity Maps for Earthquakes in Southern California, in SMIP98 Seminar Proceedings, California Strong Motion Instrumentation

LEARNING FROM SEISMIC RESPONSE OF INSTRUMENTED BUILDINGS DURING THE 1994 NORTHRIDGE EARTHQUAKE

FARZAD NAEIM
John A. Martin and Associates, Inc.
e-mail:farzad@johnmartin.com

Abstract

In the aftermath of the January 17, 1994 Northridge earthquake hundreds of strong ground motion and building response accelerograms were retrieved from stations throughout the greater Los Angeles basin. Particularly important among the building response records were the data obtained from instrumented buildings which experienced relatively large ground motions. This paper provides a summary of the results obtained from an elaborate two-year project which included inspection of the buildings, damage assessment, performance evaluations. The forces, displacements, and dynamic characteristics interpreted from recorded data are contrasted with those suggested by building codes. Key response parameters and characteristics of each building are studied and where necessary observations are provided which may be used to improve future editions of the building codes.

1. Introduction

According to Nehru every theory must be tempered with reality. The January 17, 1994 Northridge earthquake (M_L = 6.4, M_W = 6.7, M_S = 6.8) provided ample opportunity for earthquake engineers to test their theories and practices of structural design and seismic performance against the realities of strong ground shakings. Hundreds of strong ground motion and building response accelerograms were recorded by and retrieved from instruments installed by California Division of Mines and Geology, Strong Motion Instrumentation Program (Shakal and others, 1994; CSMIP 1994-95), United States Geological Survey (USGS, 1994) and other agencies throughout the greater Los Angeles basin.

Particularly important among the building response records were the data obtained from 17 instrumented buildings distributed throughout the Los Angeles area which experienced peak

M. Erdik et al. (eds.), Strong Motion Instrumentation for Civil Engineering Structures, 33–52.

base accelerations greater than 0.25 g, two instrumented downtown skyscrapers which experienced ground level accelerations of about 0.18g, and a two-story base isolated Fire Command Control building which experienced a peak base acceleration of about 0.22g.

As a part of this investigation, the above buildings were inspected to the extent possible and their performance were evaluated relative to various aspects of recorded ground motion and building configuration. Building superintends and structural engineers who had examined the buildings were consulted and their observations were summarized. Detailed information on building structural systems, nonstructural systems, contents, construction history, extent and location of damage, and loss estimates were gathered.

For each building the code specified values for natural periods design base shears and drift indices were calculated. Two sets of code values were developed: one corresponding to the edition of the building code used in the actual design of each building , and the other based on the 1994 edition of the Uniform Building Code (ICBO, 1994). These values were compared with natural periods and maximum base shears interpreted from the earthquake records.

A unique feature of this project is development of a CD-ROM based interactive information system which contains all text, photos, sketches, earthquake records and most importantly all of the analytical tools which were developed and utilized for this study (Naeim, 1996). The companion SMIP Information System is a Microsoft Windows based system and is built around an open-architecture relational database system which can be modified and expanded by the users.

2. Description Of The Buildings

The following buildings were studied as a part of this investigation. The acronyms used for identification of these buildings in the rest of the paper are given in parenthesis:

Burbank 10-story residential building with 16 sensors (BURBANK 10) This building was designed and constructed in 1974. Its vertical load carrying system consists of precast and poured-in-place concrete floor slabs supported by precast concrete bearing walls. The lateral load resisting system consists of precast concrete shear walls in both direction. The foundation system includes concrete caissons which are 25 to 35 feet deep. The largest peak horizontal accelerations recorded at the base and at the roof are 0.34g and 0.77g, respectively. The peak velocity at the roof is about 63 cm/sec.

Burbank, 6-story commercial building with 13 sensors (BURBANK 6) This steel moment frame building was designed in 1976 and constructed in 1977. The vertical load carrying system consists of 3" concrete slab over metal deck supported by steel frames. The lateral load resisting moment frames are located at the perimeter of the building. The foundation system includes concrete caissons approximately 32 feet deep. The largest peak horizontal acceleration recorded at the base and at the roof are 0.36g and 0.47g, respectively. The peak velocity at the roof is about 48 cm/sec.

Los Angeles, 17-story residential building with 14 sensors (LARES 17) This building was designed in 1980 and constructed in 1982. Its vertical load carrying system consists of 4" or 8" precast, pretensioned concrete slabs supported by precast concrete walls. The lateral load resisting system consists of distributed precast concrete shear walls in both direction. The foundation system includes 44" diameter and 54 feet long drilled piles. The largest peak horizontal acceleration recorded at the base and at the roof are 0.26g and 0.58g, respectively. The peak velocity at the roof is about 48 cm/sec.

Los Angeles, 19-story office building with 15 sensors (LAOFFI 19) This office building has 19 stories above the ground and 4 stories of parking structure below the ground. It was designed in 1966-67 and constructed in 1967. The vertical load carrying system consists of 4.5" reinforced concrete slabs supported on steel frames. The lateral load resisting system consists of moment resisting steel frames in the longitudinal and X-braced steel frames in the transverse direction. The foundation system consists of 72 feet long, driven, steel I-beam piles. The largest peak horizontal acceleration recorded at the base, ground floor and roof are 0.32g, 0.53g and 0.65g, respectively. The peak velocity at the roof is about 65 cm/sec.

Los Angeles, 2-story Fire Command Control building with 16 sensors (LACC 2) This is a 2 story seismic isolated building. The isolation system is composed of elastomeric bearings. The vertical load carrying system is steel vented roof decking and steel decking with 3 to 4 inches of concrete fill at the first and second floors. The floor system is supported by steel frames and rubber bearings. The lateral load resisting system is perimeter chevron braced frames above the isolation interface. The foundation system is composed of spread footings. The building was designed in 1988 and constructed in 1989-90. In the E-W direction, the largest peak horizontal accelerations recorded below the isolation plane, at the floor directly above the isolation plane, and the roof are 0.22g, 0.35g and 0.77g, respectively. In the N-S direction, the largest peak horizontal accelerations vary from 0.18g at the base to 0.07g directly above the isolation system and 0.09g at the roof.

Los Angeles, 3-story commercial building with 15 sensors (LACOMM 3) This department store building has three stories above and two parking levels below the ground. The building was designed in 1974 and constructed in 1975-76. The vertical load carrying system consist of

3.25 inches of light-weight concrete slab over metal deck in upper three floors and 18 inches thick waffle slabs in the basement floors. The lateral load resisting system is steel braced frames in the upper three stories and concrete shear walls in parking floors. The foundation system consists of spread footings and drilled bell caissons. The largest peak horizontal accelerations recorded at the base is 0.33g. At the roof peak horizontal acceleration of 0.97g and peak velocity of 57 cm/sec were recorded.

Los Angeles, 5-story Warehouse with13 sensors (LAWH 5) This is a 5-story reinforced concrete building was constructed in 1970 with perimeter ductile concrete frames acting as its lateral system. The largest peak horizontal accelerations recorded at the base and at the roof are 0.25g and 0.28g, respectively. The peak velocity at the roof is about 34 cm/sec.

Los Angeles, 52-story office building with 20 sensors (LAOFFI 52) This office building has 52 stories above and 5 levels below the ground. It was designed in 1988 and constructed in 1988-90. The vertical load carrying system consists of 3 to 7 inches of concrete slabs on steel deck supported by steel frames. The lateral force resisting system consists of concentrically braced steel frames at the core with moment resisting connections and outrigger moment frames in both directions. The foundation is composed of spread footings of 9 to 11 feet thickness. The largest peak horizontal accelerations recorded at the basement and at the roof are 0.15g and 0.41g, respectively. The peak velocity at the roof is about 40 cm/sec.

Los Angeles, 54-story office building with 20 sensors (LAOFFI 54) This office building has 54 stories above and 4 levels below the ground. It was designed in 1988 and constructed in 1988-90. The vertical load carrying system consists of 2.5 inches of concrete slabs on a 2inche metal deck supported by steel frames. The lateral force resisting system consists of perimeter tubular moment resisting frames which step in twice in elevation.. The foundation system consists of a 9 feet deep mat foundation. The largest peak horizontal accelerations recorded at the basement and at the roof are 0.14g and 0.19g, respectively. The peak velocity at the roof is about 34 cm/sec.

Los Angeles, 6-story office building with 14 sensors (LAOFFI 6) This office building has five stories above and one level below the ground. It was designed in 1988 and constructed in 1989. The vertical load carrying system consists of composite construction of concrete slabs over metal decks supported by steel frames. The lateral load resisting system is a combination of Chevron braced and moment resisting steel frames. Mat foundations are utilized beneath the four towers and spread footings are used elsewhere. The largest peak horizontal accelerations recorded at the base and at the roof are 0.24g and 0.48g, respectively. The peak velocity at the roof is about 70 cm/sec.

Los Angeles, 6-story parking structure with 14 sensors (LAPARK 6) The first three stories of this concrete parking structure were constructed in 1977. The upper three floors were added in 1979 based on designs dated 1975 and 1978. The vertical load carrying system consists of 5.75 in. concrete slabs and 5 in. post-tensioned concrete slabs supported by precast concrete beams and columns (see the Information System photos). The lateral force resisting system consists of six cast-in-place reinforced concrete shear walls in the North-South direction and two in the E-W direction. The foundation system is made of drilled concrete caissons. The largest peak horizontal acceleration recorded at the base, near the north-east shear is 0.29g. Channel 1 at the base of the North shear wall recorded a peak vertical acceleration of 0.22g. At the roof, the sensor placed on the mid-span of a girder recorded a peak vertical acceleration of 0.52g. Elsewhere on the roof, a sensor attached to a parapet on the North side recorded peak horizontal and vertical accelerations of 1.21g and 0.52g, respectively.

Los Angeles, UCLA Math-Science building with 18 sensors (UCLA7) This Math-Science addition to the engineering school building at UCLA is a 7 story building with no basement. It is separated by seismic joints from adjacent wings of the building. The vertical load carrying system for the upper floors (third and higher) consists of 2.5 inches of concrete slab over metal deck supported by steel frames. At the third floor a thick concrete slab supported by concrete walls make up the gravity system. The lateral load resisting system consists of concrete shear walls between the base and the third floor and moment resisting steel frames from the third floor to the roof. The largest peak horizontal accelerations recorded at the base and roof are 0.29g and 0.76g, respectively. The maximum velocity recorded at the roof is about 73 cm/sec.

Los Angeles, 7-story hospital building in with 24 sensors (LAHOSP 7) This structure is the first base isolated hospital building in the United States. It was designed in 1988 and constructed between 1989 to 1991. The vertical load carrying system consists of concrete slabs on metal decks supported by steel frames and rubber isolators. The lateral force resisting system consists of diagonally braced perimeter steel frames isolated by lead-rubber and elastomeric isolator units. Foundation system consists of continuous and isolated spread footings.. The largest free-field peak horizontal acceleration recorded adjacent to the building is 0.49g in the N-S direction. The largest horizontal peak acceleration recorded at the foundation, immediately above the isolation plane, and at the roof of the building are 0.37g (N-S), 0.14g (E-W), and 0.21g (N-S).

Los Angeles, 9-story office building with 18 sensors (LAOFFI 9) This office building was designed and constructed in 1923. It consists of concrete frames with unreinforced masonry infill walls. It consists of one level of basement and 9 floors above the ground. The largest peak horizontal accelerations recorded at the basement and roof are 0.18g and 0.34g , respectively. The maximum velocity recorded at the roof is about 45 cm/sec.

Los Angeles, Hollywood Storage Building with 12 sensors (HWSTOR) This building has 14 stories above and one level below the ground. It was designed in 1925. The vertical load carrying system consists of 8 in. thick concrete slabs supported by concrete frames. The lateral load resisting system, consists of reinforced concrete frames in both directions. The deep foundation system consists of concrete piles. The "free-field" station adjacent to the building recorded peak accelerations of 0.41g in the N-S direction, 0.19g in the E-W direction, and 0.19g in the vertical direction. The maximum peak horizontal accelerations recorded at the base and at the roof are 0.28g and 0.49g, respectively. The uncorrected trace of Channel 11 at the roof shows a peak acceleration of 1.61g. However, at the time of publishing this paper the corrected version of this record was not available. It is possible that this sensor was not properly calibrated at the time of the earthquake since it has high frequency content which is not corroborated by other instruments. The peak velocity at the roof is about 38 cm/sec.

North Hollywood, 20-story hotel with 16 sensors (NHHOTEL 20) This hotel has 20 stories above and one level below the ground. It was designed in 1967 and constructed in 1968. The vertical load carrying system consists of 4.5 to 6 inches thick concrete slabs supported by concrete beams and columns. The lateral load resisting system consists of ductile moment resisting concrete frames in the upper stories and concrete shear walls in the basement. The exterior frames in the transverse direction are infilled between the second and the 19th floors. The building rests on spread footings. The largest peak horizontal accelerations recorded at the basement and at the roof are 0.33g and 0.66g, respectively. The largest velocity recorded at the roof is about 77 cm/sec.

Sherman Oaks, 13-story commercial building with 15 sensors (SHERMAN 13) This office building has 13 stories above and two floors below the ground. It was designed in 1964. The vertical load carrying system consists of 4.5 inches thick one-way concrete slabs supported by concrete beams, girders and columns. The lateral load resisting system consists of moment resisting concrete frames in the upper stories and concrete shear walls in the basements. The foundation system consists of concrete piles. The first floor spandrel girders were modified by post-tensioning after the 1971 San Fernando earthquake. The largest peak horizontal accelerations recorded at the basement and at the roof are 0.46g and 0.65g, respectively. The middle floors experienced large acceleration in the neighborhood of 0.6g. The largest velocity recorded at the roof is about 68 cm/sec.

Sylmar, 6-story hospital building with 13 sensors (SYLMAR) The Sylmar County Hospital Building is a unique building built on the site of the old Olive View hospital building which suffered major and irreparable damage during the 1979 San Fernando earthquake. This six story cruciform shaped building has no basements. It was designed in 1976 and was constructed during the period of 1977 to 1986. Its vertical load carrying system consists of concrete slabs over metal deck supported by steel frames. The lateral load resisting system consists of concrete shear walls in lower two floors and steel shear walls encased in concrete at the perimeter of the upper four floors. The building rests on spread footings. The "free-field" station located at the parking lot adjacent to the building recorded 0.91g, 0.61g, and 0.60g in the N-S, E-W, and vertical directions, respectively. The largest peak horizontal accelerations recorded at the ground floor and at the roof of the building are unprecedented at 0.80g and 1.71g, respectively. The largest velocity recorded at the roof was as large as 140 cm/sec.

Van Nuys, 7-story hotel with 16 sensors (VAN NUYS 7) This reinforced concrete structure with no basements was designed in 1965 and constructed in 1966. Its vertical load carrying system consists of 8 in. and 10 in. concrete slabs supported by concrete columns, and spandrel beams at the perimeter. The lateral load resisting system consists of interior column-slab frames and exterior column-spandrel beam frames. The foundations consist of 38 inch deep pile caps, supported by groups of two to four poured-in-place 24 inch diameter reinforced concrete friction piles. The largest peak horizontal accelerations recorded at the basement and at the roof are 0.45g and 0.58g, respectively. The largest velocity recorded at the roof is about 77 cm/sec.

3. Results Obtained

The 20 instrumented buildings exhibited structural and nonstructural damages ranging from *None* to *High* based on the ATC-38 post-earthquake evaluation procedure. Hundreds of photos exhibiting various types of damage to these buildings and a wide variety of analytical tools developed as a part of this project are available on the CD-ROM based information system which was developed as a part of this investigation (Naeim, 1996). A few sample representatives are reproduced in Figures 1 to 11.

Interpreted maximum direct (N-S or E-W) and differential (torsional) base shears and drift indices are presented in Table 1 where interpreted base shears are compared with recommended code strength design values. Overall levels of structural and nonstructural damage are also indicated on this table.

Interpreted vibration periods of these buildings are compared to code recommended period estimates in Table 2 where predominant periods --if significantly different from fundamental periods-- are identified. This table also shows the shifts in building periods during and after the earthquake. In light of the results of this investigation the following observations are made:Building code estimates of building periods are consistently less than both the initial and final fundamental periods obtained from interpretation of recorded data. UBC-94 estimates, however, are much better than the estimates provided by the older editions of the code. It may be necessary to further calibrate code period estimation formulas to reduce this inconsistency.

1. Except for the two base isolated buildings and the two downtown skyscrapers, the building

2. base shears obtained from interpretation of recorded data are larger, sometimes substantially, than the base shears they have been apparently designed for. With the exception of the Van Nuys 7-story hotel, however, these buildings behaved remarkably well given the magnitude of forces they were subjected to. One could suggest that all these buildings performed much better than what would have been expected by routine design analysis techniques. Design procedures need to be modified to take advantage of this excess capacity which is not ordinarily addressed in design analysis schemes.

3. The ratio of the base shears experienced to design code base shears does not correlate very well with the extent of damage observed. The overall drift ratio, however, does correlate rather well. This statement, however, needs further clarification through system identification studies since it is not clear at this time whether the large drifts are contributing to damage or are caused by it.

4. Given the level of forces the building experienced, the overall drift ratios experienced are significantly less than what would have been expected from ordinary design analysis techniques.

5. While structural damage was generally less than what would have been expected, the content damage was generally extensive and usually the dollar value of the content damage and loss of occupancy far out-weighed the cost of structural repair.

6. In seismic response of majority of the buildings, different modes became predominant at different times during the response. In many cases, particularly for taller buildings such as the downtown skyscrapers, the Sherman Oaks 13-story office building, and the North Hollywood 20-story hotel, 2nd and or 3rd modes had more contribution to the overall response than the fundamental mode. In such cases application of the lateral story force profiles as suggested by static lateral force procedures may grossly underestimate the demand on the middle floors of the building. This can be further illustrated by examining the story force diagrams at the time of maximum base shear which indicate that except for the shorter buildings, the story force profile at the instant of maximum base shear is radically different from that recommended by static lateral force procedures. Lateral force distribution over the height of the building as suggested by static lateral force procedures is generally based on the static deflected shape of the building. Evaluation of the deformed shape at the time of maximum lateral displacement shows that the lateral deformation at this instant almost always follows a shape similar to the first mode of vibration.

7. Our studies indicate, however, that in most cases maximum forces and maximum displacements are not concurrent. In most cases the maximum force response occurs first and the maximum displacement response occurs many seconds later. Current edition of the UBC code requires dynamic (i.e., response spectrum) distribution of forces for irregular structures. In light of observations presented here it might be prudent to require dynamic distribution of forces for buildings exceeding a certain height (65 feet for example) and limiting the application of static lateral force distribution to the regular buildings of less height.

8. Except for the case of the 6-story Sylmar County hospital, behavior of mounted mechanical equipment was not a strong function of the severity of the ground motions but rather the quality of design and construction. (see for example photos of equipment mounted at the roof of the 3-story commercial building or the Van Nuys 7-story hotel in the Information System developed as a part of this project).

9. Except for buildings with observed structural damage, the period of the building as interpreted from the recorded data did not elongate significantly and when elongation occurred the period came back to the vicinity of the initial value towards the end of the ground motion. The period of damaged buildings however did decidedly elongate.

10. For several buildings, torsion contributed significantly to the seismic response. In one of these cases (Van Nuys 7-story hotel) the building experienced major damage.

4. Conclusions

An interactive information system was developed which contains all of the gathered information, inspection results, recorded data, performance analysis results, and analytical tools utilized for evaluation of twenty extensively-instrumented buildings which were subjected to significant ground shaking during the January 17, 1994 Northridge earthquake (see Naeim, 1996). This CD-ROM based interactive information system can be a very valuable tool in teaching and learning earthquake engineering and seismic response principles as well as a tool for further research on response of buildings to strong earthquake ground motions. A brief overview of the information generated regarding the seismic performance of these buildings were presented. The interested reader is referred to Naeim (1996) report to California Strong Motion Instrumentation Program and its companion CD-ROM based information system for more details.

5. Acknowledgments

Funding for this project was provided in part by State of California, Department of Conservation,Division of Mines and Geology, Strong Motion Instrumentation Program (SMIP) under Contract Number 1093-557. Significant supplemental funding was provided by John A. Martin and Associates, Inc.

The opinions expressed in this paper are those of the author and do not necessarily reflect the views of the California Strong Motion Instrumentation Program or John A. Martin and Associates, Inc.

6. References

1. California Strong Motion Instrumentation Program (CSMIP), 1994, *Processed Data for Burbank 6-story Commercial Building from the Northridge Earthquake of 17 January 1994*, Report No. OSMS 94-11A.
2. California Strong Motion Instrumentation Program (CSMIP), 1994, *Processed Data for Sylmar 6-story County Hospital from the Northridge Earthquake of 17 January 1994*, Report No. OSMS 94-11C.
3. California Strong Motion Instrumentation Program (CSMIP), 1994, *Processed Data for Van Nuys 7-story Hotel from the Northridge Earthquake of 17 January 1994*, Report No. OSMS 94-11D.
4. California Strong Motion Instrumentation Program (CSMIP), 1994, *Processed Data for Los Angeles 7-story UCLA Math-Science Building from the Northridge Earthquake of 17 January 1994*, Report No. OSMS 94-11J.
5. California Strong Motion Instrumentation Program (CSMIP), 1994, *Processed Data for Los Angeles 7-story University Hospital from the Northridge Earthquake of 17 January 1994*, Report No. OSMS 94-11E.
6. California Strong Motion Instrumentation Program (CSMIP), 1994, *Processed Data for Los Angeles Hollyood Stroage Building Grounds from the Northridge Earthquake of 17 January 1994*, Report No. OSMS 94-11K.
7. California Strong Motion Instrumentation Program (CSMIP), 1994, *Processed Data for Los Angeles Hollywood Storage Building Grounds from the Northridge Earthquake of 17 January 1994*, Report No. OSMS 94-11K.
8. California Strong Motion Instrumentation Program (CSMIP), 1994, *Processed Data for Los Angeles Hollywood Storage Building from the Northridge Earthquake of 17 January 1994*,Report No.OSMS94-11K.California Strong Motion Instrumentation Program (CSMIP), 1994, *Processed Data for North Hollywood 20-story Hotel from the Northridge Earthquake of 17 January 1994*, Report
9. No.OSMS94-11L.California Strong Motion Instrumentation Program (CSMIP), 1994, *Processed Data for Sherman Oaks 13-story Commercial Building from the Northridge Earthquake of 17 January 1994*, Report No. OSMS 94-11B.
10. California Strong Motion Instrumentation Program (CSMIP), 1995, *Processed Data for Los ANgeles 17-story Residential Building from the Northridge Earthquake of 17 January 1994*, Report No. OSMS 95-01D.
11. California Strong Motion Instrumentation Program (CSMIP), 1995, *Processed Data for Los Angeles City Terrace from the Northridge Earthquake of 17 January 1994*, Report No. OSMS 95-01A.
12. California Strong Motion Instrumentation Program (CSMIP), 1995, *Processed Data for Los Angeles 2-story Fire Command Control Building from the Northridge Earthquake of 17 January 1994*, Report No. OSMS 95-01A.
13. California Strong Motion Instrumentation Program (CSMIP), 1995, *Processed Data for Los Angeles 3-story Commercial Building from the Northridge Earthquake of 17 January 1994*, Report No. OSMS 95-01C.
14. California Strong Motion Instrumentation Program (CSMIP), 1995, *Processed Data for Los Angeles 5-story Warehouse from the Northridge Earthquake of 17 January 1994*, Report No. OSMS 95-01B.
15. California Strong Motion Instrumentation Program (CSMIP), 1995, *Processed Data for Los Angeles 52-story Office Building from the Northridge Earthquake of 17 January 1994*, Report No. OSMS 95-01E.
16. California Strong Motion Instrumentation Program (CSMIP), 1995, *Processed Data for Los Angeles 54-story Office*

Building from the Northridge Earthquake of 17 January 1994, Report No. OSMS 95-01G.

17. California Strong Motion Instrumentation Program (CSMIP), 1995, *Processed Data for Los Angeles 6-story Office Building from the Northridge Earthquake of 17 January 1994*, Report No. OSMS 95-01R.
18. California Strong Motion Instrumentation Program (CSMIP), 1995, *Processed Data for Los Angeles 6-story Parking Structure from the Northridge Earthquake of 17 January 1994*, ReportNoOSMS95-11F.California Strong Motion Instrumentation Program (CSMIP), 1995, *Processed Data for Los Angeles 9-story Office Building from the Northridge Earthquake of 17 January 1994*, Report No. OSMS 95-01F.
19. International Conference of Building Officials (1994 and previous editions), *Uniform Building Code*, Whittier, California.
20. Naeim, F., 1995, "On Seismic Design Implications of the 1994 Northridge Earthquake Records", *Earthquake Spectra*, EERI, Vol. 11, No. 1.
21. Naeim, F., 1996, *Performance of Instrumented Buildings During the January 17, 1994 Northridge Earthquake -- An Interactive Information System --*, Draft report CSMIP, February, 1996.
22. Shakal, A., Huang, M, Darragh, R., Cao, T, Sherburne, R., Malhorta, P., Cramer, C., Sydnor, R., Graizer, V., Maldonado, G., Peterson, C., and Wampole, J.(1994), *CSMIP Strong Motion Records from the Northridge, California Earthquake of 17 January 1994*, Report No. OSMS 94-07, CSMIP, Sacramento, February.
23. United States Geological Survey (1994), Northridge Strong Ground Motion FTP Site on the Internet.

Table 1. Base Shear, Drift and Overall Damage Summary

Building Acronym	Maximum Direct Base Shear Interpreted (% Total Weight)[a]	Maximum Differential Base Shear Interpreted (% Total Weight)[a]	Design Code Strength Level Base Shear (% Total Weight)[b]	UBC-94 Strength Level Base Shear (% Total Weight)[c]	Maximum Overall Drift Index Interpreted (in./in.)[a]	Overall Structural Damage[d]	Overall Non-strucutral and/or Equipment Damage[d]
BURBANK 10	34	14	14	20	0.0023	Insignificant	Insignificant
BURBANK 6	22	7	10	7	0.0039	Insignificant	Moderate
LARES 17	17	16	18	18	0.0022	Insignificant	Insignificant
LAOFFI 19	34	22	8	12	0.0039	Moderate	Insignificant
LACC 2	7	11	–[e]	–[e]	–	None	Moderate
LACOMM 3	49	27	18	27	0.0111	None	Insignificant
LAWH 5	17	10	8	17	0.0020	Insignificant	Insignificant
LAOFFI 52	9	4	–[e]	–[e]	0.0011	None	None
LAOFFI 54	4	4	–[e]	–[e]	0.0008	None	None
LAOFFI 6	–[f]	–[f]	–[f]	–[f]	–	None	None
LAPARK 6	27	13	12.5	18	0.0016	Insignificant	Insignificant
UCLA 7	27	21	–[f]	–[f]	0.0047	Moderate	Insignificant
LAHOSP 7	–[g]	–[g]	–[e]	–[e]	–	None	None

Table 1. (Continued)

Building Acronym	Maximum Direct Base Shear Interpreted (% Total Weight)[1]	Maximum Differential Base Shear Interpreted (% Total Weight)[1]	Design Code Strength Level Base Shear (% Total Weight)[2]	UBC-94 Strength Level Base Shear (% Total Weight)[3]	Maximum Overall Drift Index Interpreted (in./in.) [a]	Overall Structural Damage	Overall Non-strucutral and/or Equipment Damage
LAOFFI 9	20	18	–[f]	–[f]	0.0018	Insignificant	Insignificant
HWSTOR	10	10	–[f]	–[f]	0.0013	Insignificant	Insignificant
NHHOTEL 20	12	9	6	6	0.0036	Insignificant	Moderate
SHERMAN 13	13	3	6	6	0.0067	Moderate	Insignificant
SYLMAR	97	63	–[e]	–[e]	0.0022	Insignificant	High
VAN NUYS 7	33	41	7	21	0.0117	High	Moderate

NOTES:

a) For analytical assumptions see (Naeim, 1996).
b) Estimate of the code WSD value times 1.4 at the time of building design.
c) Estimate of the UBC-94 code WSD value times 1.4.
d) ATC-38 Definitions of overall damage states are used.
e) Design was not based on code static lateral force procedures.
f) Sufficient information not available for compiling this value.
g) Approximate methods used are not applicable to this case.

Table 2. Vibration Periods (seconds)

Building Acronym	Interpreted N-S & E-W Fundament al Periods [a]	Interpreted Predominan t Response Periods[a]	Significant Period Elongation? [a]	Design Code Period Estimates[a]	UBC-94 Period Estimates[a]
BURBANK 10	0.57 - 0.62	Same	No	0.30	0.58
BURBANK 6	1.28 - 1.28	Same	No	0.60	0.95
LARES 17	0.80 - 1.20	Same	Moderate	--[b]	--[b]
LAOFFI 19	2.60 - 3.41	Same	No	0.76 - 1.90	1.24 - 2.33
LACC 2	1.28 - 0.2 to 1.14	Same	Yes in E-W Dir.	--[b]	--[b]
LACOMM 3	0.55 - 0.51	Same	No	0.16	0.40
LAWH 5	1.46 - 1.37	Same	No	0.60	0.73
LAOFFI 52	6.0 - 6.0	1.6 to 2.0	No	--[b]	--[b]
LAOFFI 54	6.0 - 6.0	1.0 to 2.0	No	--[b]	--[b]
LAOFFI 6	0.85 - 0.85	Same	No	0.56	0.56
LAPARK 6	0.35 - 0.40	Same	No	0.18	0.44
UCLA 7	0.66 - 1.02	Same	No	--[b]	--[b]
LAHOSP 7	0.64 to 1.5 - 0.79 to 1.5	Same	Yes	--[b]	--[b]
LAOFFI 9	1.21 - 1.71	Same	No	--[f]	--[f]
HWSTOR	--[b]	--[b]	--[b]	--[b]	--[b]
NHHOTEL 20	2.20 - 2.50	Same & 0.70	Moderate	1.20	1.60
SHERMAN 13	2.6 - 2.9	Same & 1.0	Yes	1.27	1.60
SYLMAR	0.46 - 0.46	Same	No	--[b]	--[b]
VAN NUYS 7	1.1 to 1.8 - to2.2	Same	Yes	0.70	0.70

a) for analytical assumptions, methods and procedures see (Naeim, 1996)

b) Either sufficient information not available for compiling this value or value not applicable.

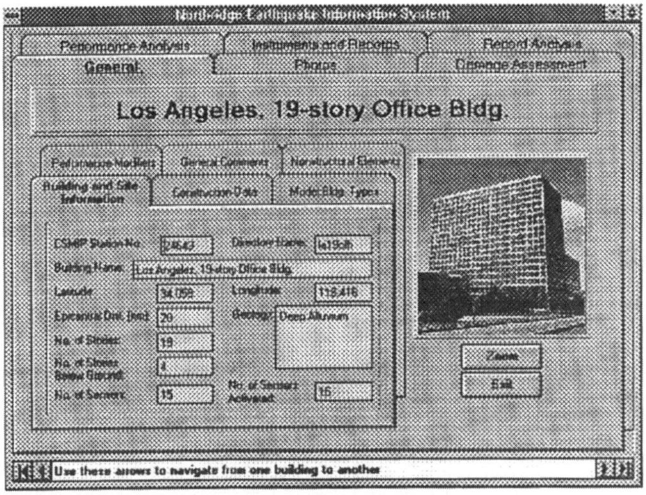

Figure 1. The Information System main folder LAOFFI 19 building.

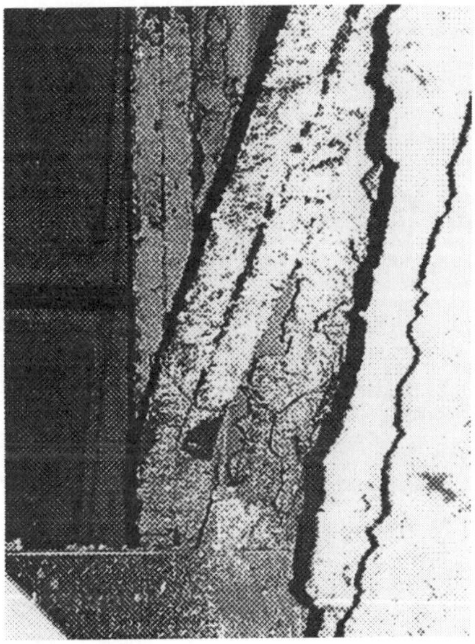

Figure 2. Buckled brace at the penthouse of the LAOFFI 19 building.

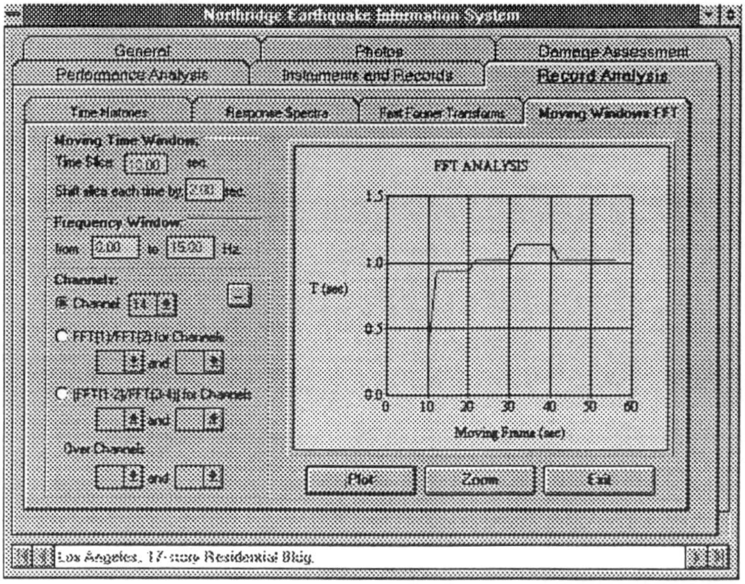

Figure 3. Moving windows FFT analysis for the N-S response of the LARES 17 bldg.

Figure 4. Tiles over isolation pit of the LACC 2 bldg. after the earthquake (photo courtesy of Robert Bachman).

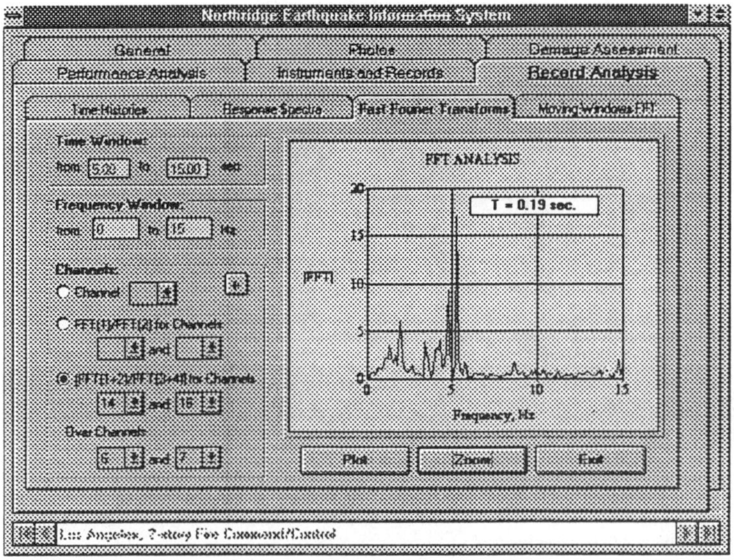

(a) time = 0 to 15 seconds

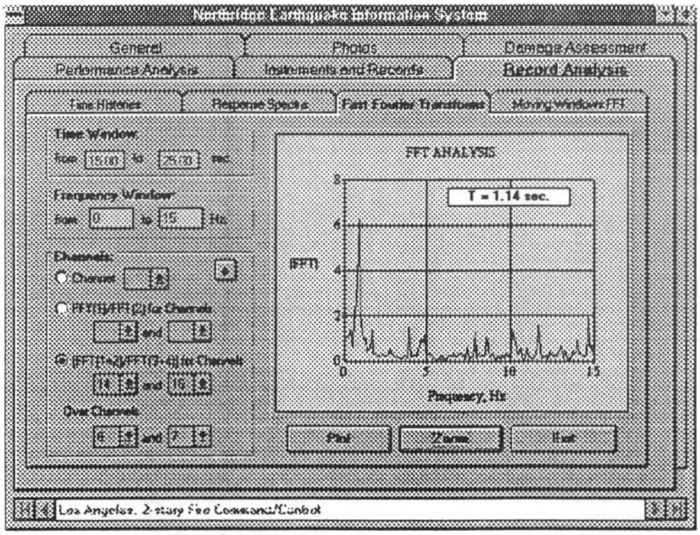

(b) time = 15 to 25 seconds

Figure 5. **Fast Fourier Transform of E-W response of LACC 2 building shows that the pit separation shown in Figure 4 has permitted the building to behave as an isolated system after 15 seconds into the ground motion.**

Figure 6. Mechanical equipment damage at the roof of BURBANK 6 building.

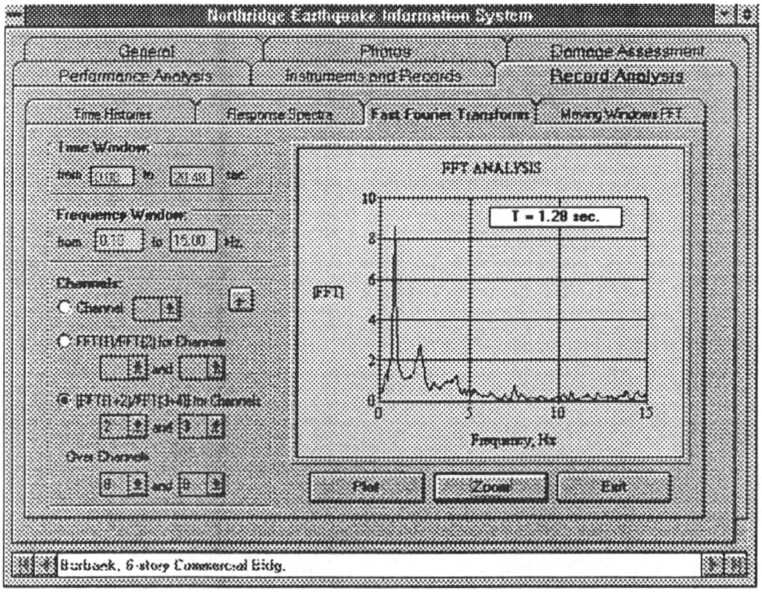

Figure 7. FFT analysis depicting fundamental N-S period of BURBANK 6 building.

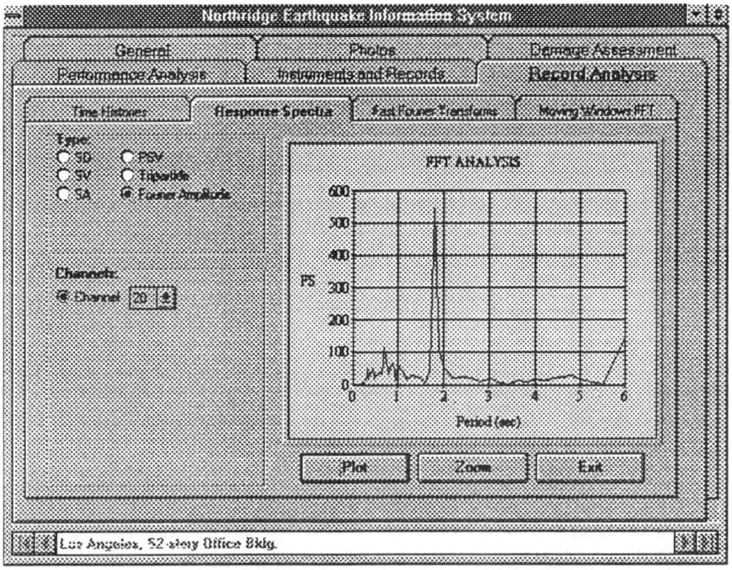

Figure 8. Fourier spectrum indicating a fundamental period of 6.0 sec. For LAOFFI 52 building. Notice that the predominant period however is slightly below 2 seconds.

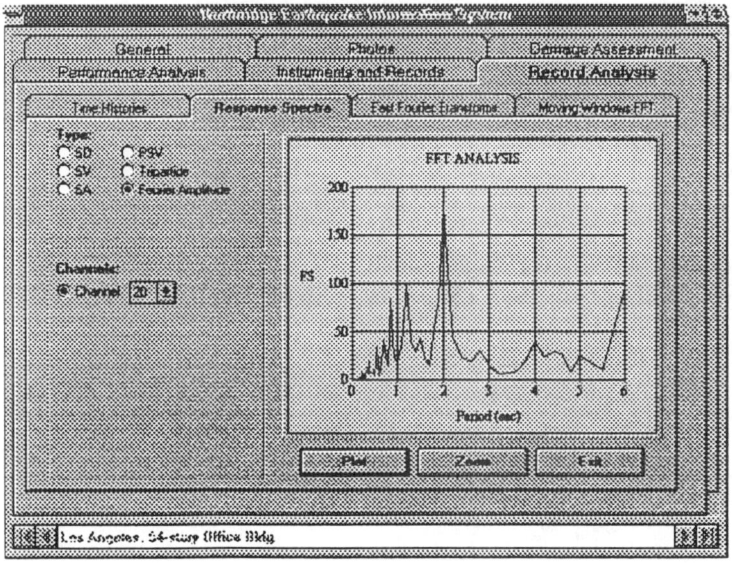

Figure 9. Fourier spectrum indicating a fundamental period of about 6.0 sec. For LAOFFI 54 building. Notice that the predominant period however is at about 2 seconds.

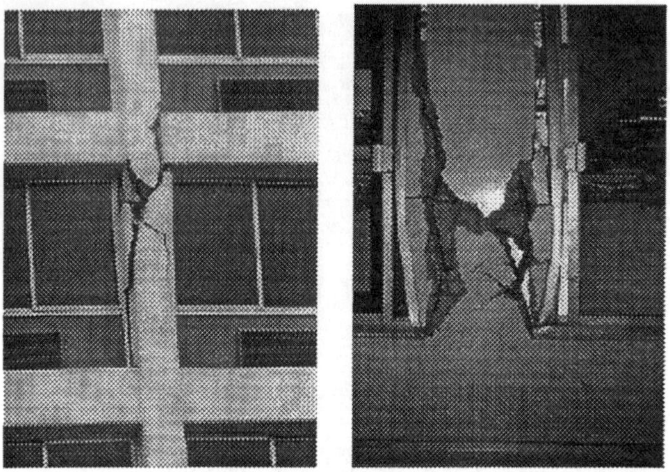

Figure 10. Shear failure of column at the fourth floor of the VAN NUYS 7 building (views of the same column from outside and inside).

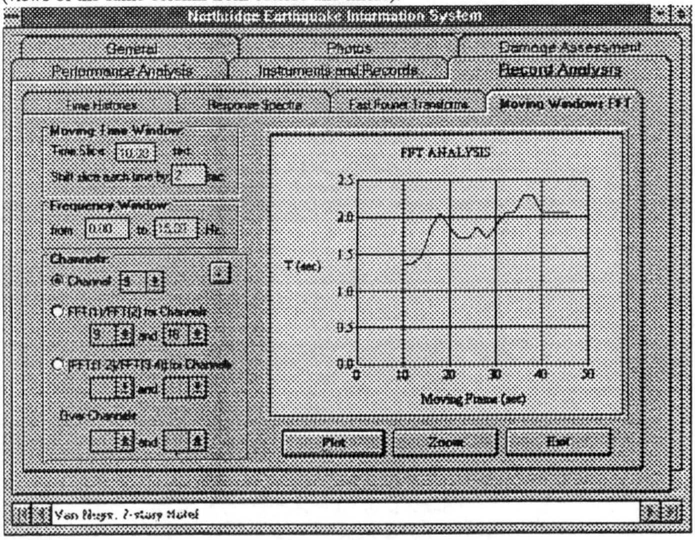

Figure 11. Moving windows FFT analysis showing significant softening of the VAN NUYS 7 building largely due to damage at the fourth floor.

NEW DEVELOPMENTS IN HEALTH MONITORING FOR CIVIL STRUCTURES

ROBERT NIGBOR[1], JOHN FORT[2]
[1] *University of Southern California, Los Angeles, California*
[2] *Kinemetrics Inc., Pasadena, California*
e-mail:rnigbor@agbabian.com

Abstract

Practical examples of health monitoring of civil structures now exist, and recent or coming improvements in technology will likely add more examples. This paper discusses several health monitoring projects, both commercial and research. Some new technologies in instrumentation, communications, and analysis are then briefly listed. However, technology alone cannot improve the "market" for structural health monitoring. The decision process for monitoring must be studied, and developers of monitoring technology must realistically compare the costs and benefits. Barriers to wider adoption and implementation are discussed and some needs expressed.

1. Introduction

A vision for structural health monitoring, as applied to civil structures, is one in which a structure can sense its condition, express the information both internally and externally, and possibly use that information as the basis for some corrective action. The time scale for this monitoring can vary from immediate (for extreme events such as earthquake) or on a scale of years for long-term degradation due to corrosion or fatigue.

At the present time, technologies are available to perform real-time monitoring of various physical quantities such as acceleration, displacement, strain, force, temperature, etc. Technologies are also available to perform rapid, even real-time analysis on the measured data and to communicate both data and analysis results to a remote location. Many research examples, and several practical examples, exist. Three examples are discussed below: the Rama IX Bridge, the Kingston Bridge, and Winooski One Dam. An attempt at standardization, the Remote Bridge Monitoring System (RBMS), is also discussed.

The difficulty with such monitoring projects, when applied to practical situations, is that the analysis technology is not yet ready to robustly tackle the damage detection problem. This is a difficult problem that may never be completely solved. Until somewhat reliable and robust condition assessment technology is available, the cost-benefit relationship for structural health

M. Erdik et al. (eds.), Strong Motion Instrumentation for Civil Engineering Structures, 53–59.

monitoring will be weighted toward the cost side, preventing practical implementation in all but a few extreme cases.

2. Health Monitoring Examples

In the following section, three examples of practical structural health monitoring systems are briefly described. A commercial monitoring system specification, RBMS, is also discussed.

Rama IX Bridge

The Rama IX Bridge in Bangkok, Thailand is a large single-plane cable-stayed bridge crossing the Chao Phraya River. It measures 782 meters in total length including the 450m main span and two 166m back spans. Concrete approach viaducts extend the span to three kilometers. This is a single-plane symmetric cable-stayed bridge with steel deck and two support pylons rising 78m from the bridge deck.

Wind tunnel tests during design forecasted unacceptable vibration of the bridge deck due to wind-induced oscillations, so eight tuned-mass dampers were installed near the bridge deck center. However, soon after construction in 1987 perceptible vibrations were and are observed in the bridge deck. These vibrations are traffic-induced, with a peak acceleration (measured later) of about 0.1g.

Concerns about possible long-term effects of the deck vibration prompted installation of a simple monitoring system immediately after construction. In June, 1995, a more complex monitoring system was installed. This was an OASIS (Online Alerting of Structural Integrity and Safety) system developed by Agbabian Associates and Kinemetrics (Nigbor, 1996).

This monitoring system consists initially of 12 accelerometers (high dynamic range) plus 3D wind and temperature sensors, an on-bridge data acquisition and recording system, a real-time data link (initially hardwired), and a monitoring computer at the bridge headquarters about 3km from the bridge. Data are continuously monitored both by the on-bridge triggered recorder and the remote real-time monitoring computer. Data are distilled into a graphical display, with green/yellow/red alarms when threshold levels are exceeded. The primary focus of the real-time monitoring system is occupant safety (environmental and vibration). Structural health is monitored through periodic recordings of ambient vibrations. Figure 1 shows an example of these data.

In the near future this health monitoring system will be expanded and used as an integral part of a thorough structural inspection of the Rama IX Bridge.

Kingston Bridge

The Kingston Bridge in Glasgow, Scotland is a twin three-span structure carrying traffic over the River Clyde. Built in 1970, the continuous superstructures are cast-in-place post-tensioned concrete three-cell box girders constructed segmentally as balanced cantilevers. In 1990, an assessment of the bridge revealed serious problems with the decks and piers. Safety was a concern in several extreme traffic loading scenarios.

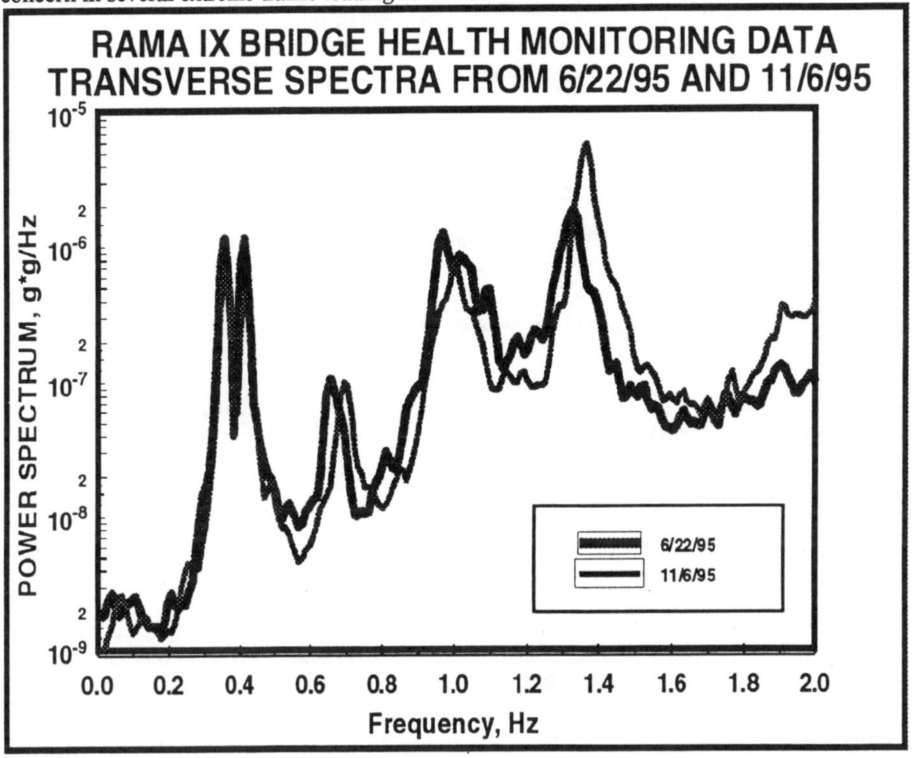

Figure 1: Sample data from the Rama IX Bridge OASIS System

To solve this structural problem, traffic control was implemented and repairs were started. During the multiple-year repair program the bridge could not be shut down. So, an extensive bridge monitoring system was installed to assist engineers in preventing collapse (Robison, 1996). This extensive monitoring system consisted of displacement, stress, strain, temperature, wind, and meteorological sensors. Quasi-static data from the sensors were collected on the bridge and transmitted off-site for analysis. One feature of the monitoring system was real-time alarms available to alert users to abnormal or critical conditions.

Winooski One Dam

The previous two examples of structural monitoring were both based upon practical, safety-related problems. There are numerous examples of monitoring systems installed for research purposes to advance the development of monitoring technology. One such example is the Winooski One Dam in Vermont. During construction, this concrete dam was extensively instrumented with fiber optic strain sensors and the data made available in real-time through the internet (Fuhr et al, 1998).

Physical placement of the fiber optic sensors was based upon a detailed structural analysis. Key structural areas were chosen for structural health monitoring. At some locations, crack detection was of prime importance. At others, vibration analysis or intrusion detection were important. The resultant sensor "web" embedded into the powerhouse was composed of 47 sensors.

Radio telemetry was used to collect all data at one location, where these data were made available through the internet.

This research project, typical of many, demonstrated the available technologies and provided data for development of analysis technologies.

Remote Bridge Monitoring System

Each of the above three examples is unique, and the monitoring systems are unique. Some standardization of monitoring systems is perhaps desirable for future developments.

An attempt at standardization has been made by the New York Department of Transportation. A Remote Bridge Monitoring System (RBMS) has been developed based upon commercial off-the-shelf (COTS) components by Alampalli et al (1994). This system uses dynamic measurements and modal-parameter-based analysis to look for long-term degredation or damage. A standardized commercial specification was developed for a generic RBMS and is in use by the State of New York.

3. New Developments

Through both commercial developments and academic research, there continue to be new developments which improve the structural health monitoring capabilities for civil structures. Often, these developments are in other fields, such as electronics or aerospace. This brief paper is not a venue for detailed discussion of this subject. Following is a list of significant new developments in instrumentation, communications, and analysis:

Instrumentation:

- ñ High dynamic range (>16 bits) digitizers which allow both ambient vibration and extreme event monitoring without gain change.
- ñ High dynamic range accelerometers which allow measurement of the full amplitude range of structural vibrations.
- ñ Intelligent dataloggers, which contain microprocessors or digital signal processors. These can perform some of the required calculations in the field, providing distributed intelligence within a monitoring system.

Communications:

- ñ Fiber optic data communications allows fast, noise-free communication both within a structure and from the structure to an external computer.
- ñ Internet communications (such as in the Winooski One system) hold promise for remote communications.
- ñ Wireless digital point-to-point communications now provide relatively fast data transfer over intermediate distances.
- ñ Low Earth Orbit satellite communications systems may offer future improvements in the robustness of monitoring system communications.

Analysis:

- ñ New probabilistic analysis approaches based upon statistical pattern recognition show promise at answering the question "is there damage?".
- ñ Real-time analysis and information display is simplified with the faster computers now available.

4. Barriers To Wider Implementation

There are very few practical examples of structural health monitoring. This is because there exist both technical and economic barriers to wider implementation.

Technical Barriers

In the laboratory it is relatively easy to infer structural damage or degradation using structural monitoring. However, in the field things are not so simple. One reason is that changing environmental conditions, including structural loading, produce a variability in a structure's behavior. Figure 1 shows two spectra from the RAMA IX Bridge taken at two different times. Other than the first two modes, frequencies are different, probable due to different loading conditions (traffic). Figure 2 shows similar data from another large bridge, the Vincent Thomas Bridge in Los Angeles. Here, data are from three different times in the same day.

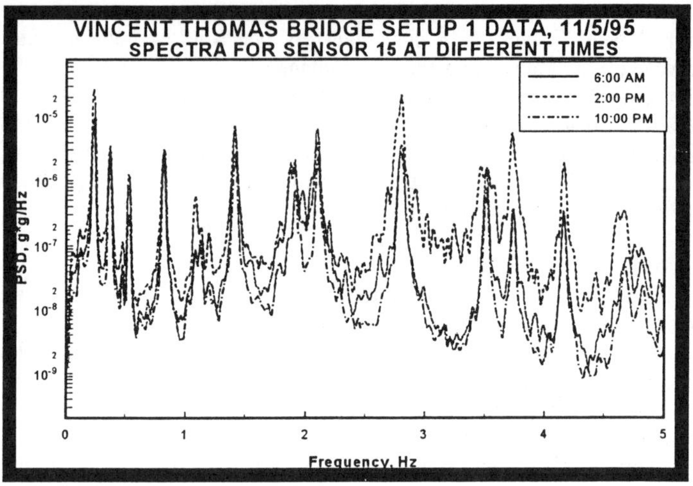

Figure 2: Ambient Vibration Data from the Vincent Thomas Bridge (*Nigbor, 1995*)

This variability in modal and other parameters calculated from real structural data make the detection of all but large damage difficult.

There are other technical barriers, including:

- Data transmission limitations
- Sensor dynamic range and frequency response limitations
- Issues due to high background noise for forced vibration
- Issues due to multiple inputs for ambient vibrations
- Small effect of local damage on global response of the structure

Economic Barriers

Cost is always an issue. Installation of a structural monitoring system on a large civil structure is not cheap because of the large distances and numerous sensor locations. Hardware costs range from US$2000-10000 per channel, and typical large structural monitoring systems cost several US$100k. In addition, there are training, maintenance, and operating costs; about 10% of the initial cost per year is required for maintenance and operation of a typical monitoring system.

Health monitoring of a civil structure can have benefits to the structure's management, can assist in the diagnosis or characterization of structural problems, can enhance the safety of occupants or contents, and can provide data for research. However, none of these benefits are easily quantifiable. This is a significant barrier for the future of structural health monitoring. Unless a problem is known (as in the case for the Rama IX and Kingston Bridges), it is doubly difficult to argue that monitoring has a direct benefit to the structure and its owners or occupants.

So, we have an expensive technology with mostly indirect benefits. When the costs and benefits are weighed by management, it is often difficult to justify implementation of a structural health monitoring system.

5. Needs

In light of the comments in the last section, it is the author's opinion that we must address and quantify the benefits of structural health monitoring in order to encourage wider implementation. This can be done technically by developing more reliable and robust damage detection algorithms and procedures that can be used with relatively sparse sensor distributions. New instrumentation concepts that reduce cost will help also, but the major problem with the economics is on the benefit side.

Finally, most practical monitoring systems have been implemented on large structures where a problem is suspected or known. To expand structural health monitoring beyond these few pathological instances, we must develop simple, robust health monitoring schemes for "ordinary" civil structures such as mid-rise buildings or medium0size bridges, with costs low enough to be comparable to frequent visual inspection.

6. References

1. Alampalli, S. and G. Fu (1994), "Remote Bridge Monitoring Systems for Bridge Condition," Report 70, Engineering Research and Development Bureau, New York State Department of Transportation, Albany, New York, August 1994.
2. Fuhr, P. et al (1998), "An Internet Observatory: Remote Monitoring of Instrumented Civil Structures Using the Information Superhighway," J. Smart Materials and Structures, 1998.
3. Nigbor, R. et al (1995), "Preliminary Report on the Vincent Thomas Bridge Monitoring Test", Report M9510, Dept. of Civil Engineering, University of So. Cal., December, 1995.
4. Nigbor, R. (1996), "State-of-the-Practice in Structural Monitoring," Proc. 2nd Int'l Workshop on Structural Control, Nong Kong, Dec. 18-21, 1996.
5. Robison, R. (1996), "Saving Scotland's Busiest Bridge," Civil Engineering, January 1996.

EXPERIMENTAL DYNAMIC ANALYSIS OF HISTORIC MONUMENTS AND BUILDINGS

DARIO RINALDIS,
ENEA, AMB-CAT-GET, C.R. Casaccia, Via Anguillarese 301,
00060 S. M. Galeria (Rome) Italy.
e-mail:utente@sunrin.casaccia.enea.it

1.Introduction

The earthquake effect on monumental heritage is a critical issue in areas of ancient civilisation. Monumental buildings require specific strengthening techniques, related to a clear understanding of all factors affecting its vulnerability.
Moreover, it is important to outline the singularity of the environmental built in Italy: the originality of historical centres (in each village as well as in the most important cities) and the peculiarity of the urban planning, suggests to involve in the matter, scientists specialised on topics like history of the art and architectures, to take advantage of collection of reliable historic information about the building and the constructive techniques. The evaluation of the seismic vulnerability is the evaluation of the degree of damage that a structure with a specified typology would suffer, when subjected to a seismic action with specified characteristics. The evaluation of seismic vulnerability is linked to the knowledge of the dynamic behaviour of buildings during seismic events.

2.The Methodology

There are two main approaches to evaluate, experimentally, evaluate seismic behaviour and performance of structural systems. One requires a laboratory in which subsystems, components, or (if the facility is large enough) prototypes or large, scaled models of complete systems are tested under static, quasistatic, or dynamic loading. This approach does not necessarily demand a time-dependent testing scheme, such as a shaking table; however, testing of structural systems under controlled simulated dynamic environments is required if a mathematical model (generally numerical) of the structures is required. The second approach, to evaluate behaviour and performance of structural, by observing and studying damage to structures from earthquakes. By determining why specific designs lack earthquake resistance and then by using extensive laboratory testing of modified designs, if significant progress in improved designs can be achieved. Nevertheless, the historical environmental built, might not wait for a future seismic event to be rehabilitate as the recent Umbria-Marche earthquake effects clearly stated. Thus is essential that, to characterise the dynamic behaviour of monumental buildings, a campaign of ambient-vibration tests is performed. Generally speaking, the importance of ambient-vibration analysis is that it gives

M. Erdik et al. (eds.), Strong Motion Instrumentation for Civil Engineering Structures, 61–79.

the elastic properties of the structure: during strong earthquakes the structure may experience non-linear behaviour; thus, if the parameters of linear behaviour are known beforehand, it is easier to analyse the non-linear behaviour. Speaking about, old, slightly or greatly damaged, masonry or monumental buildings, is quite difficult to study their elastic properties: even for very small excitation a non-linear behaviour is experienced by the structure: we can approximate their behaviour to a pseudo-linear trend. To evaluate its dynamic behaviour an experimental approach is necessary by means of deploying velocimeter (ambient vibrations) and accelerometers (earthquakes) in optimal configurations[1-3].

3. Case-History

Researchers at ENEA Laboratories, studied during the last decade the dynamic behaviour of monumental buildings in Rome (such as «Tempio della Minerva Medica», Colosseo, Traiana and Antonina Columns, «Tempio Rotondo al Foro Boario», etc.) and in others Italian cities (such as the bell-tower of San Giorgio church in Trignano, near by Reggio Emilia or the Duomo of Orvieto). Test and data analysis of the last two structures is presented in this paper[4-14].

3.1. THE BELL TOWER OF S. GIORGIO CHURCH IN TRIGNANO

The bell tower of San Giorgio church in Trignano, damaged by the 15 October 1996 Novellara (RE) earthquake (Epicentre 44° 48' 00''N 10° 42' 00''E Ml=4.8), has been surveyed to characterise its dynamic properties by ambient noise measurements and by temporary installation of strong motion accelerometers suited to record aftershocks.

A preliminary analysis of the records (noise and earthquakes), plots of the time histories and of the Fourier amplitude spectra, made possible the following consideration:

1. both ambient noise and earthquakes enhance resonance of the bell tower close to 3 Hz and between 7 and 10 Hz;
2. the bell tower shows the main resonance close each to the other and close to 3 Hz in the two main directions;
3. significant frequency content in the range 1÷5 Hz for the earthquakes records at the basement of the tower suggest a possible effect of the local soil condition on the seismic wavefield;

Examples of noise and earthquake recordings are illustrated in the following figure (fig. 1).

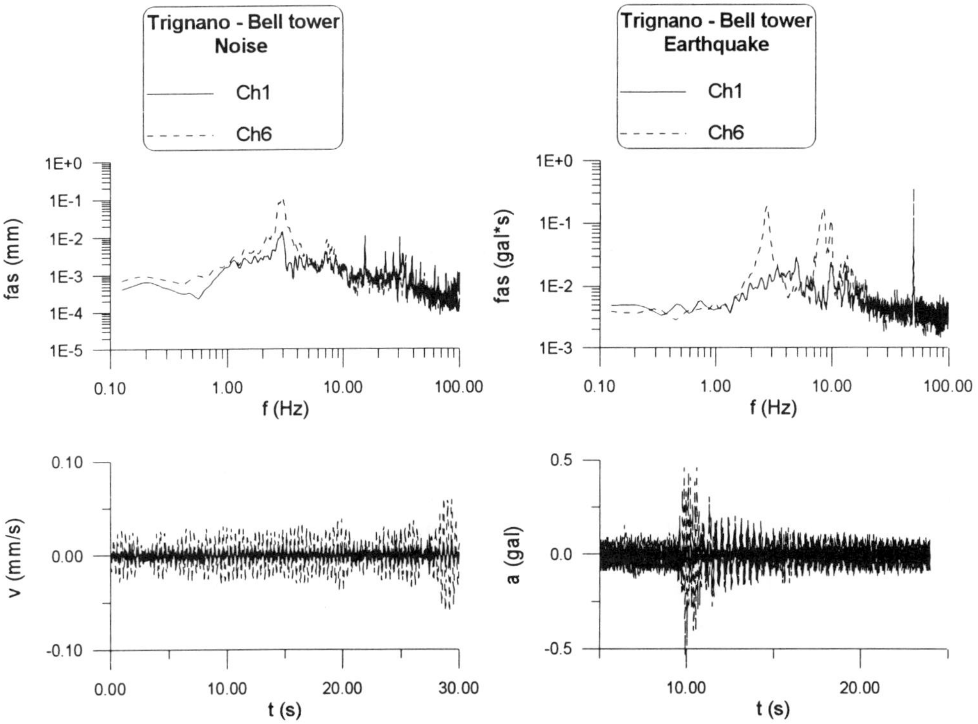

Figure 1 Noise and earthquake recordings

3.1.1. Description of the structure.

The medieval Bell Tower of S. Giorgio Church in Trignano is 18.5 m tall and 3.35*3.00 m at the basement. The original structure withstood several changes and additions in the past centuries. Four masonry pillars compose the main structure, about 40 cm thick, at the corners. Very poor masonry walls, whose connections with the pillars are not effective, fill the spaces between the pillars. The first three floors are made of timber, the forth one was substituted by a two brick little vault floor, supported by a central steel I beam. The stairs were composed by wooden and steel flights. The tower is connected to the structure of the church and to other buildings on three sides, up to the height between 6 and 7 m. All the others are masonry structures too.

The effects of the October 15, 1996 earthquake was the opening of a near horizontal crack in the freely rising part, above the roofs of the adjacent buildings, which interests three of the four sides of the tower. A 3-cm offset was also apparent between the upper and lower part of east wall, due to a clockwise rotation of the upper part with respect to the lower one.

3.1.2.The ambient-vibration tests.

3.1.2.1.Configuration and tests

Nineteen tests lasting 32 sec were performed, deploying the seismometers in two configurations, shown in figure 2. These differ only for the position of sensor Ch3. In the 1-st configuration Ch1, Ch2and Ch3 were on the ground level; Ch4 and Ch5 were just under the cracks; Ch6, Ch7 and Ch8were at the top.In the 2-nd configuration sensor Ch3 was moved to the same level of Ch4 and Ch5. Both ambient and forced vibrations were considered, by recording the vibrations due to the noise only and to the effect of a mass dropped on the ground near the tower.

3.1.2.2.Data analyses

The recorded data were studied in the frequency domain by means of cross-spectral analysis. For each sensor location, the records obtained from each test in the same configuration were considered as subsets of a longer record, so that the analysis was actually performed over very long record. Of course, before doing that, each record was half cosine tapered. In this way we obtained the average cross spectrum estimate, with the corresponding phase factor and the coherence function. The spectral window used was Hanning passed twice. The motion in terms of modal shapes was examined by means of the PSD amplitudes.

3.1.2.3.Results

In figure 3 the power spectral densities of records relative to the 1-st configuration are shown. A resonant frequency of 2.7 Hz is apparent in the spectra relative to sensors in the N-S direction (CH3, CH5, CH7, and CH8) if exception is made of the sensor CH2, probably due to a poor basement-structure connection reducing the signal as in free-field situation. Peaks at 2.9 Hz are instead evident in the spectra of records in the W-E direction (CH1, CH4, and CH6). The cross analysis allowed to point out that two modal shapes, with predominant displacements in the N-S and W-E directions, are associated to the above mentioned frequencies (sensors in the same direction always in phase at these frequencies). Analysis of the cross CH6-CH7 and CH6-CH8 do not evidence torsional modes associated to those frequencies. A peak at 6.9 Hz is evident in all the PSD. The cross analysis between records at CH7 and CH8 indicated that it is associated to a torsional mode shape.
The most significant cross-spectral densities are shown in figure 4 with the relative phase factor and coherence function.

The records of sensor CH5 show a smaller contribute of the frequency components in the range[0, 4[Hz, while the spectral amplitudes are much higher in the range [6, 8] Hz, if comparison is made with records from other channels. Therefore location CH5 does not take part in the first and second vibration modes, but it has a very important contribution in the torsional mode. This particular behaviour is related to the presence of the other structures, which influence the dynamic characteristics of the tower very much.

The records of the 2-nd configuration allowed analysing in more detail the movements of the floor under the crack. The cross-spectral densities of the couples Ch3-Ch4, Ch4-Ch5 and Ch3-Ch5 are shown in figure 5. The presence of the already mentioned resonant frequencies is confirmed. At the frequencies 2.7 and 2.9 Hz the phase factors of the cross spectrum of records at Ch3 and Ch5 are not always significant. The same happens for the spectrum of records Ch3 and Ch4.

3.1.3.Earthquake Measurements.

The 6 accelerometers of the mentioned configuration recorded 67 aftershocks in a period of two months (October-December 1996). The events were classified on the basis of an estimate of the energy content (Arias Intensity) [13]. A frequency domain analysis was performed for all the events and the main three resonance frequencies obtained. From the frequency analysis versus the earthquake event is evident the changes of the resonant frequency values [13]. Plotting the Arias Intensity for each of the recorded event versus the obtained resonant frequencies, gave the following results [13]:

- The energy range can be subdivided into two intervals: in the first one resonant frequency are randomly distributed to the energy increase; in the second one a linear regression between frequencies and the Log(E) seems to fit the data.
- This is valid for all the 3 resonant frequencies.

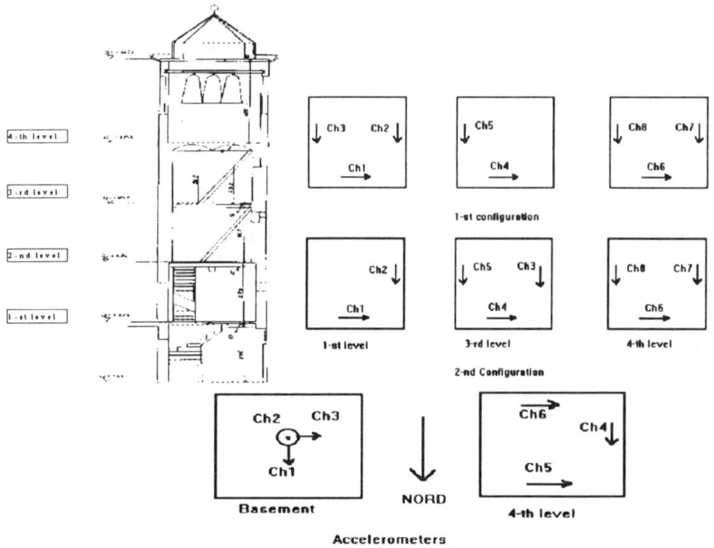

Fig 2. Velocimeters and accelerometers sensors layout

The coherence is always quite good. This occurrence could be related not only to the presence of the cracks, which determine the non linear behaviour of the structure, but also to the very low contribute of Ch5 at these frequencies. The phase between records Ch3 and Ch5 is not significant in the range [4, 8] Hz and so is the phase between Ch4 and Ch5.

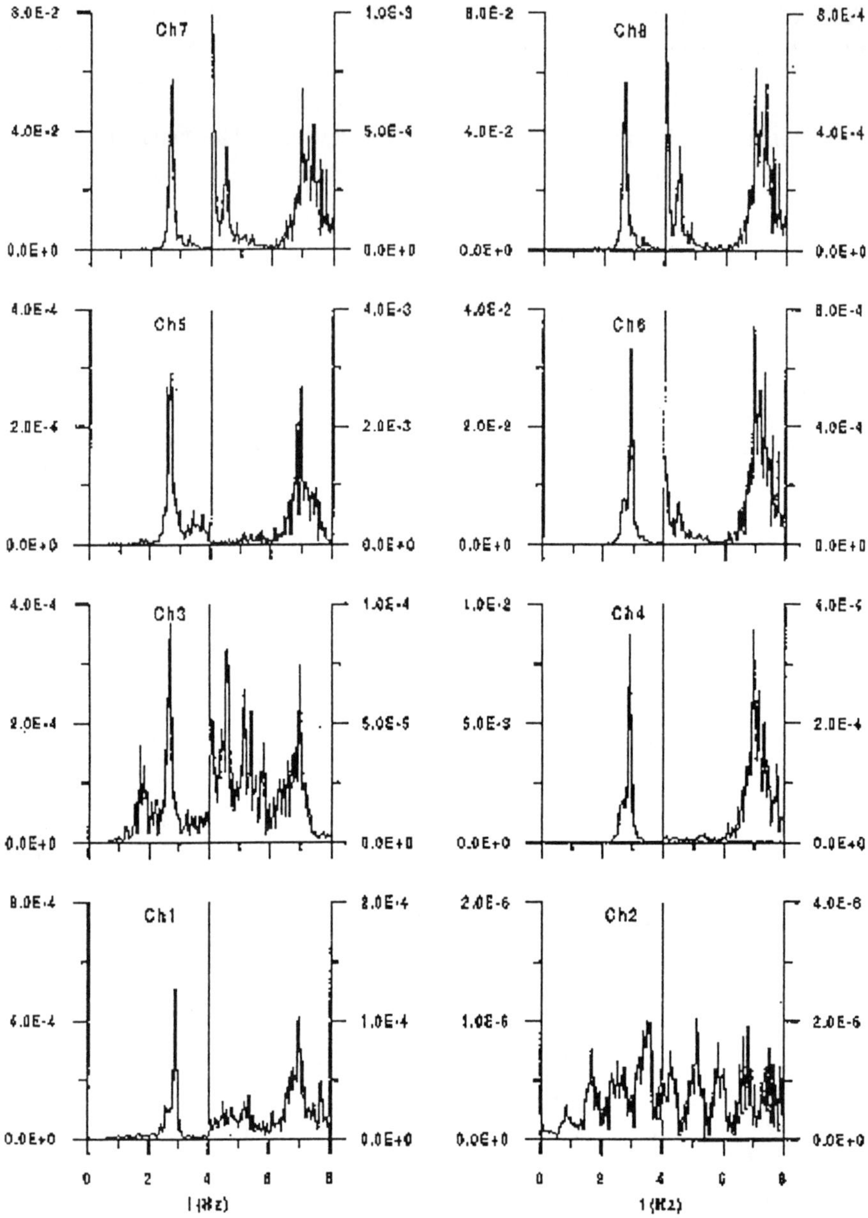

Fig. 3 Bell Tower of S. Giorgio Church. 1-st Configuration: PSD

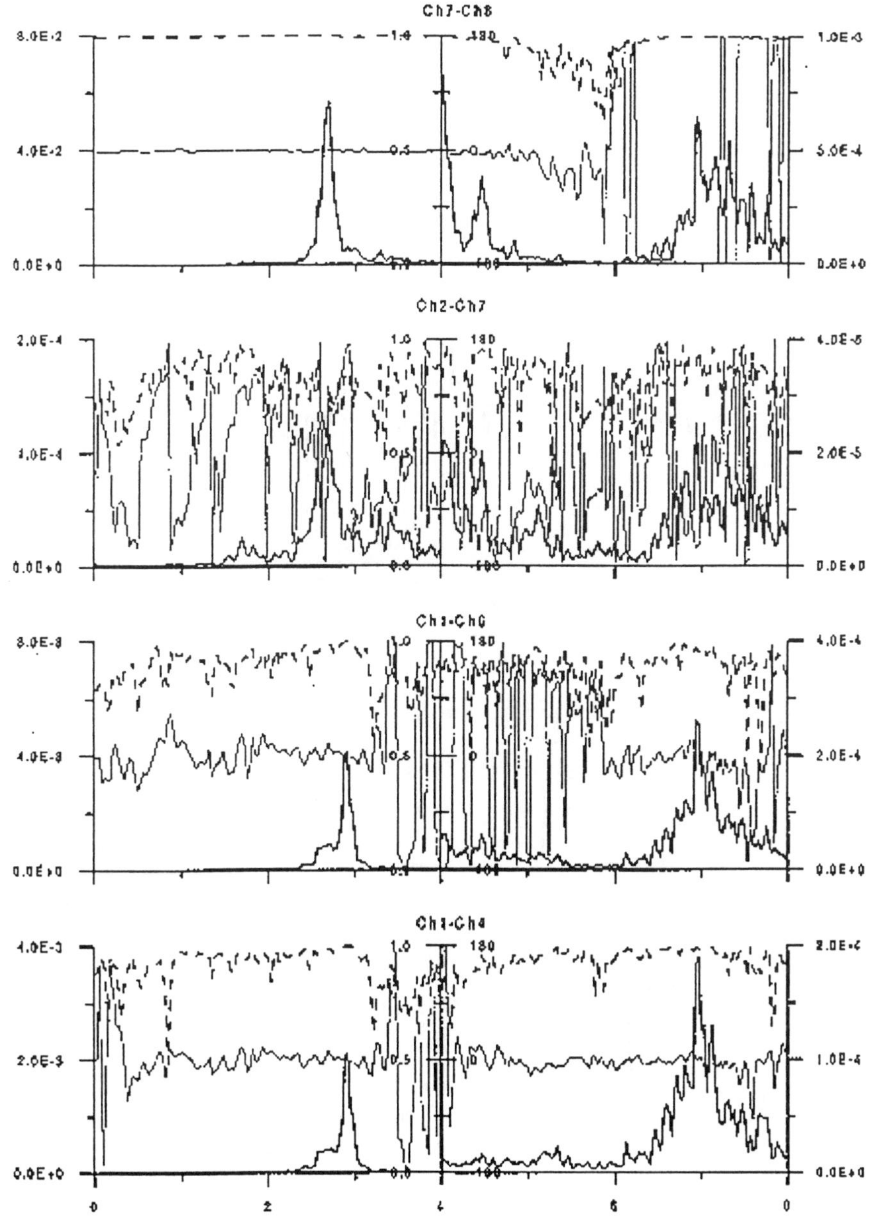

Fig. 4 Bell Tower of S. Giorgio Church. 1-st Configuration: CSD

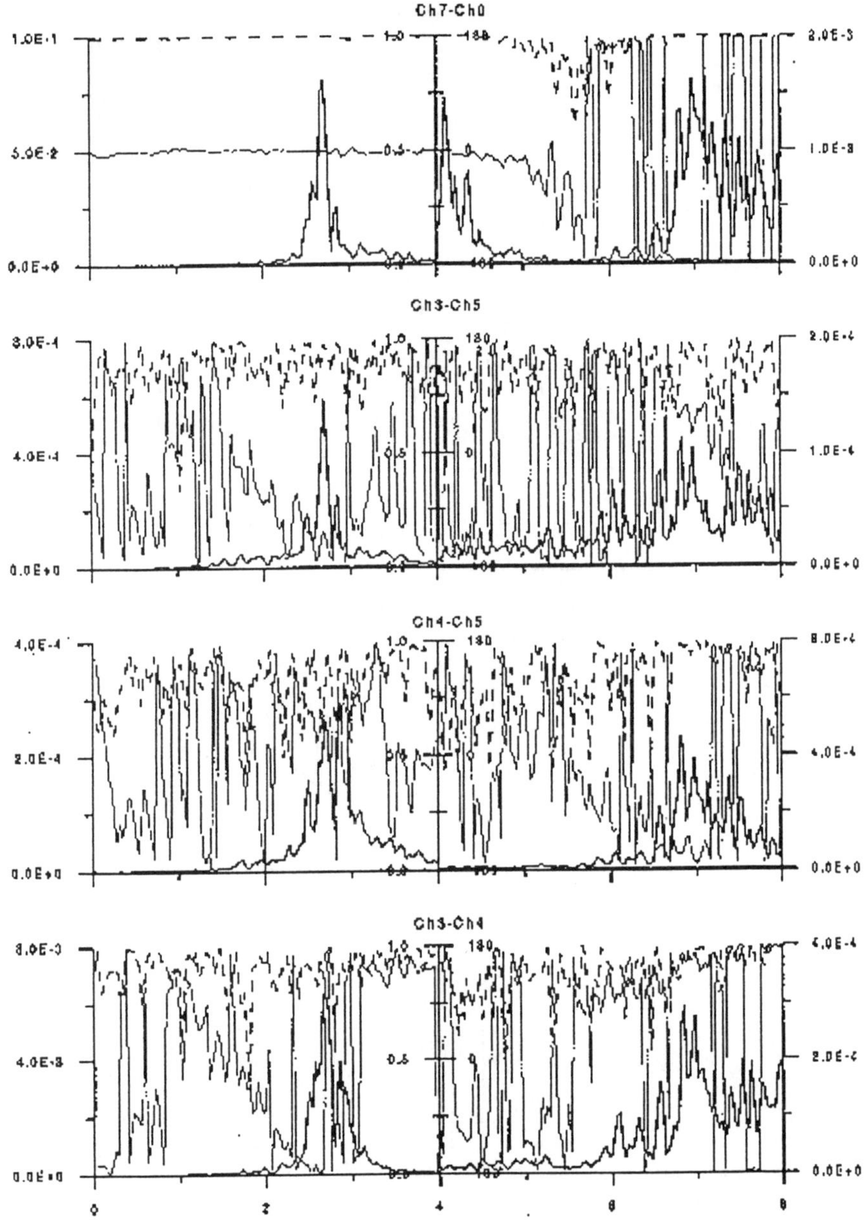

Fig. 5 Bell Tower of S. Giorgio Church. 2-nd Configuration: CSD

3.2. THE CORPORAL'S CHAPEL OF THE DUOMO OF ORVIETO

The monumental heritage of the Central Italy suffered very much because of the Umbro-Marchigiano Earthquake of September-October 1997. The first shock happened in September 26th at 2.33 a.m., Italian time, with magnitude ML=5.5. The epicentre was localised in Colfiorito (Lat. 43.0°N, Lon. 12.9°E). Two shocks, with magnitude 5.8 and 4.7 respectively, happened at 11.40 and 11.46 a.m.. Several shocks followed the first ones, most of them characterised by a magnitude higher than 3.0 [15].

The seismic motion had effects also in Orvieto, a small town 70 Km far the epicentre area, where the first shocks caused the opening of cracks in the structures of the very famous Duomo, especially in the vaults of the Corporale's Chapel.

The Environment Department of ENEA was involved in the experimental dynamic analysis of the structural behaviour, in order to evaluate the health status of the building and to locate any damage. The vaults of the Corporale's Chapel were particularly studied.

3.2.1. Description of the structure.

The Duomo of Orvieto is a Gothic-Romanesque style church, built between 1290 and 1320. Its very famous facade is a mix of marble and mosaics. It has a 17-m wide nave and two 8.5-m wide aisles. A wooden truss roof, supported by masonry walls covers the nave, whose length is 59 m. Each wall is supported by six arches whose springing start from circular masonry pillars, which separated the nave from each aisle (Fig. 6).

The roof of the transept is composed by three cross vaults, the central one larger than the other two. From the transept you can get to the Corporale's Chapel and to the S. Brizio's Chapel. The covering of both them is composed by two cross vaults.

A square shaped choir substituted the original apse. Also the Corporale's Chapel and to the S. Brizio's Chapel were built later in correspondence of existing flying buttresses. The basement of the transept and the choir is formed by an old masonry structure.

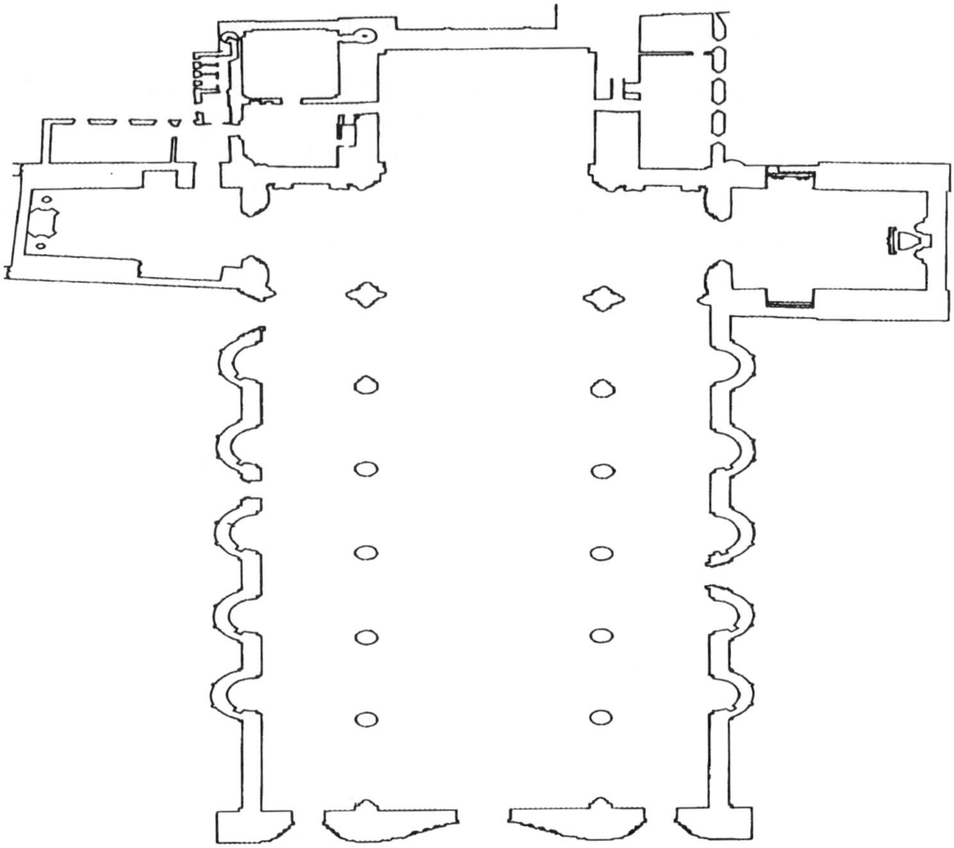

Figure 6 Plan of the Duomo of Orvieto.

3.2.2.Corprale's Chapel Configurations

Preserving notation of the site-tests report [14] configurations L to P were relative to the two vaults of the Corporale's Chapel. In the first one (Configuration L) three sensors (S1, S3 and S7) were on the ground level, the other five were on the vaults. Configurations M and N were relative to the left vault, while Configuration O and P were relative to the right vault. In all these configuration, seven sensors were deployed on the vaults, while S1 was on the ground level in the vertical direction.

3.2.3.Results

The analysis of data shows that peaks at 1.8 and 2.5 Hz are present in the spectra of the records of the Corporale's Chapel (fig. 8), in which also resonance frequencies between 3.0 and 3.5 Hz are apparent. Those are associated to local modes, in which the Corporale's Chapel moves in the longitudinal direction. Cross-spectra of records obtained during ambient vibration tests of the external vault of the Corporal Chapel (configuration M of figure) are plotted in fig. 9. The 11.5 Hz sharp peak is probably related to a modal shape of the vault. The phase factor is not always significant because of the non-linear behaviour of the vault, which shows relevant cracks induced (or may be just evidenced) by the Umbria-Marche earthquake. The spectra shown on figure 10 are from records obtained in the configuration O. The peaks at low frequencies are associated to Chapel modes already described in the spectra of fig. 8. A higher resonance frequency at 12.5 Hz is evident, which seems to be associated to a mode of the vault. The phase factor is, again, often not significant as well as the coherence function, denoting non-linearity in the structural behaviour, confirmed by comparison with forced vibrations, showing different features.

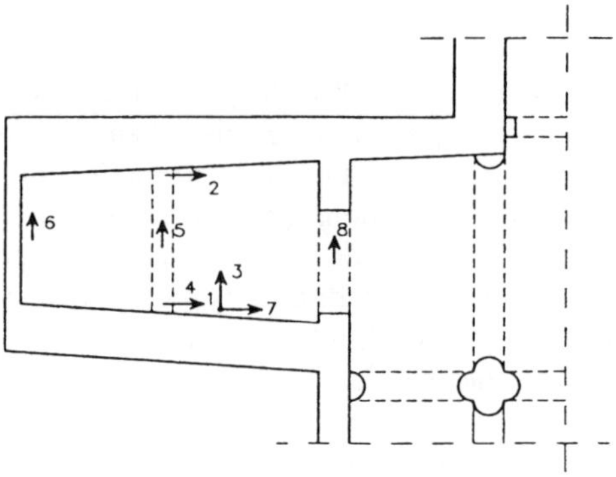

l) Configuration L

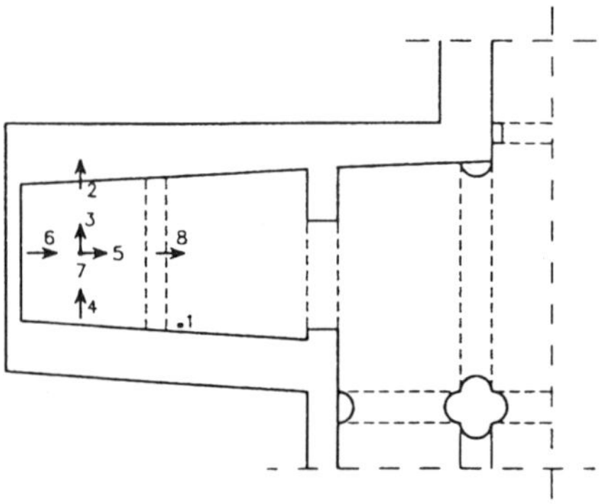

m) Configuration M

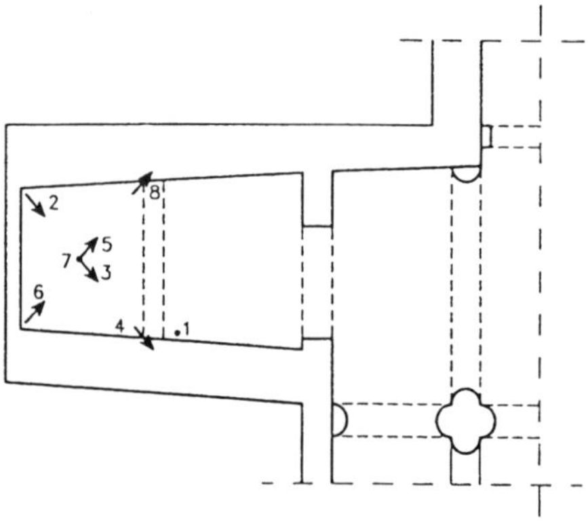

n) Configuration N

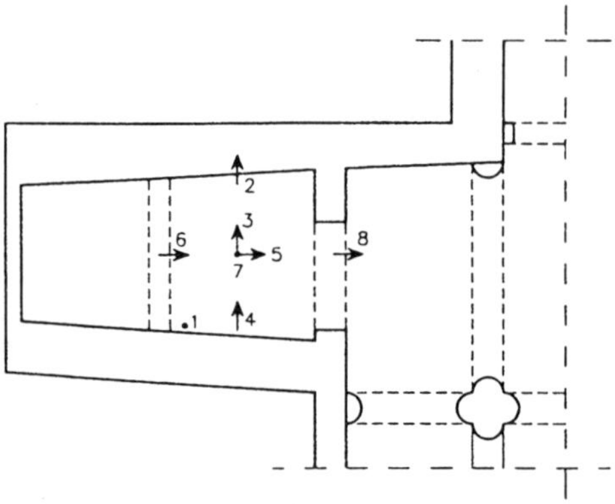

o) Configuration O

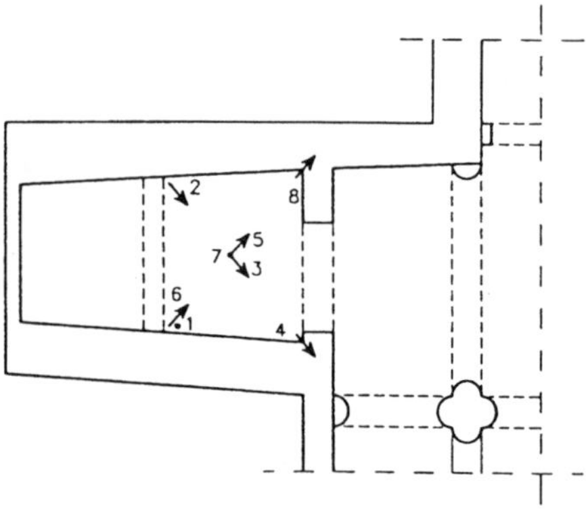

p) Configuration P

Fig. 7 l), m), n), o), p)

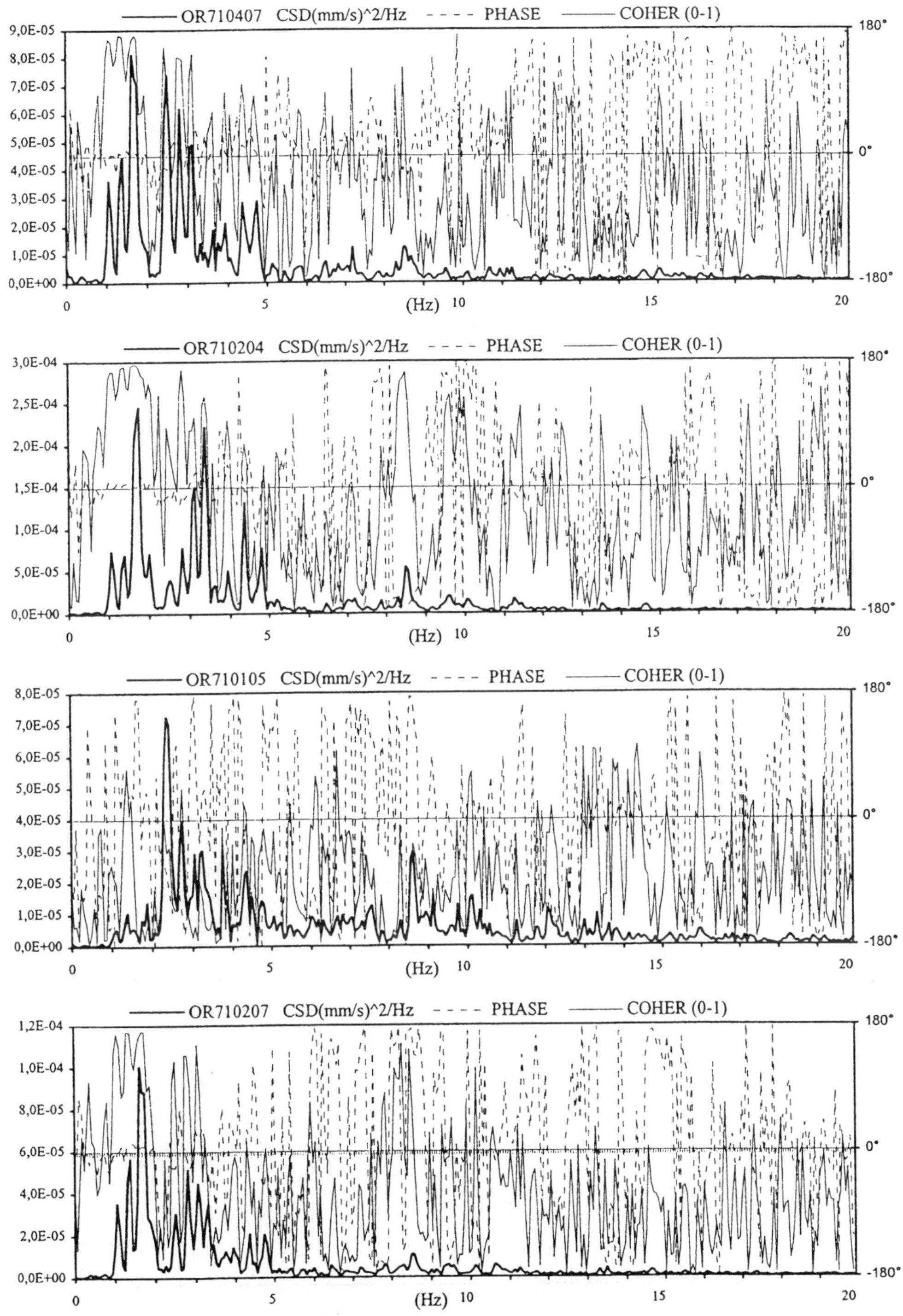

Fig. 8 Phase Factor, Coherence / A

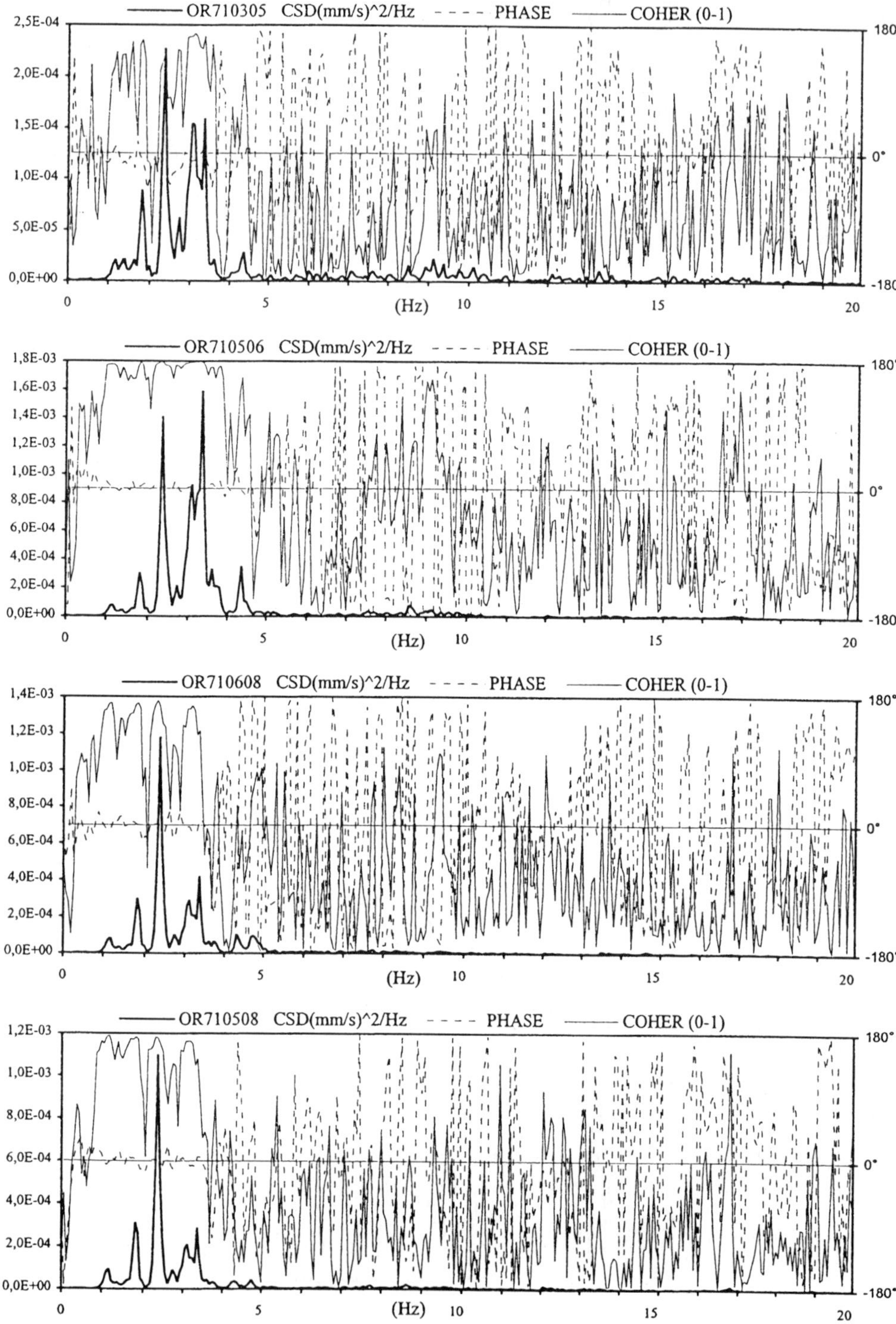

Fig. 8 Phase Factor, Coherence / B

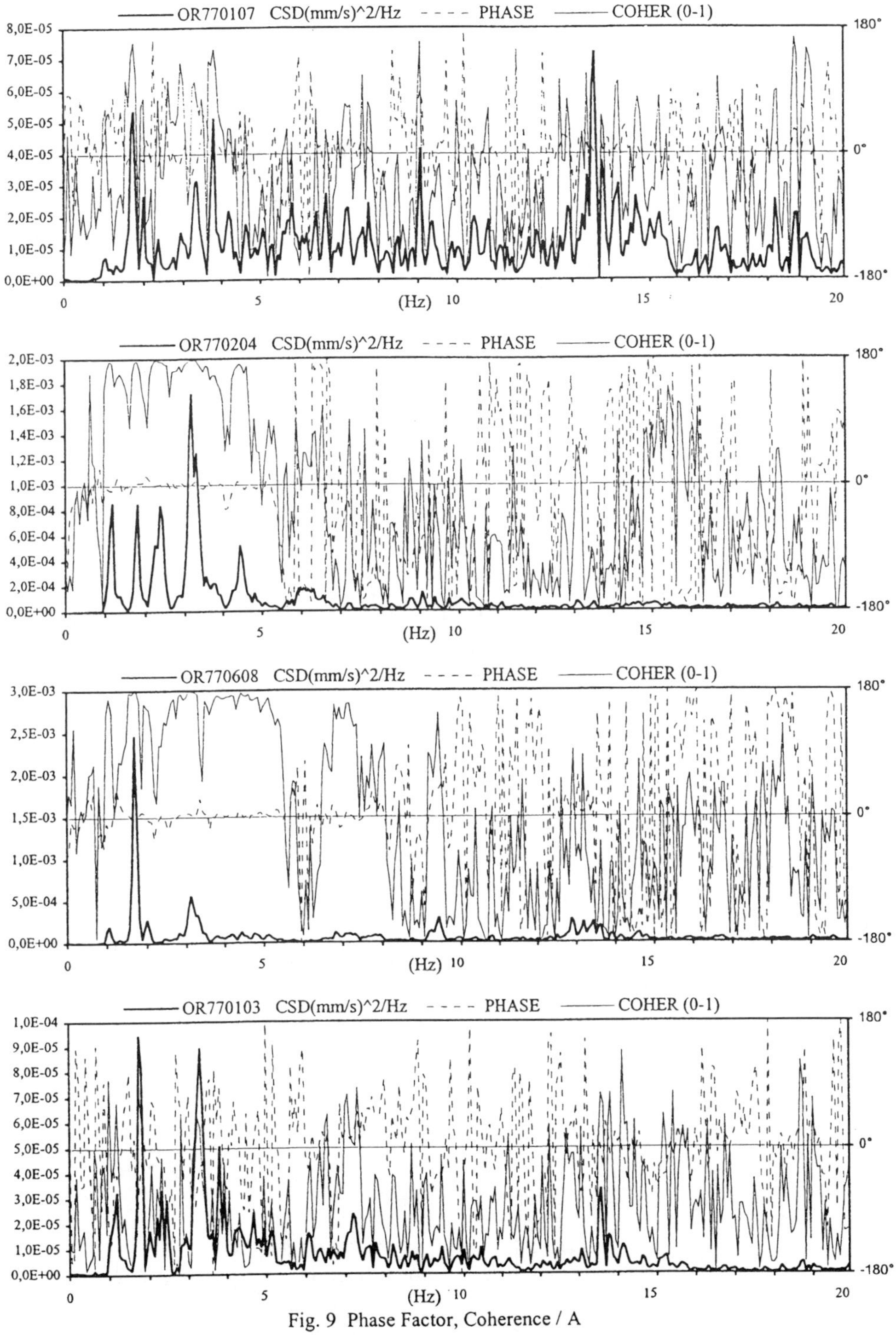

Fig. 9 Phase Factor, Coherence / A

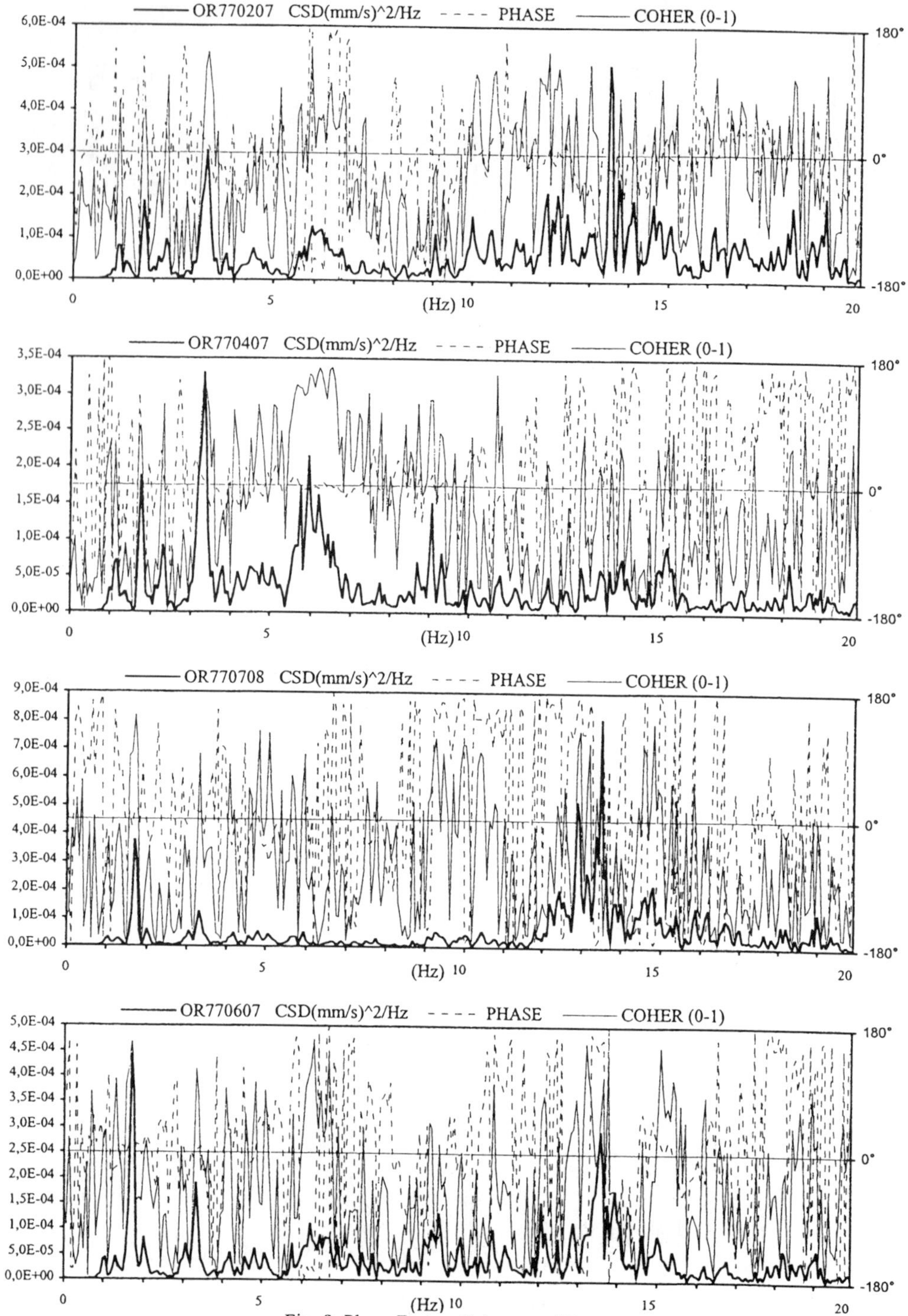

Fig. 9 Phase Factor, Coherence / B

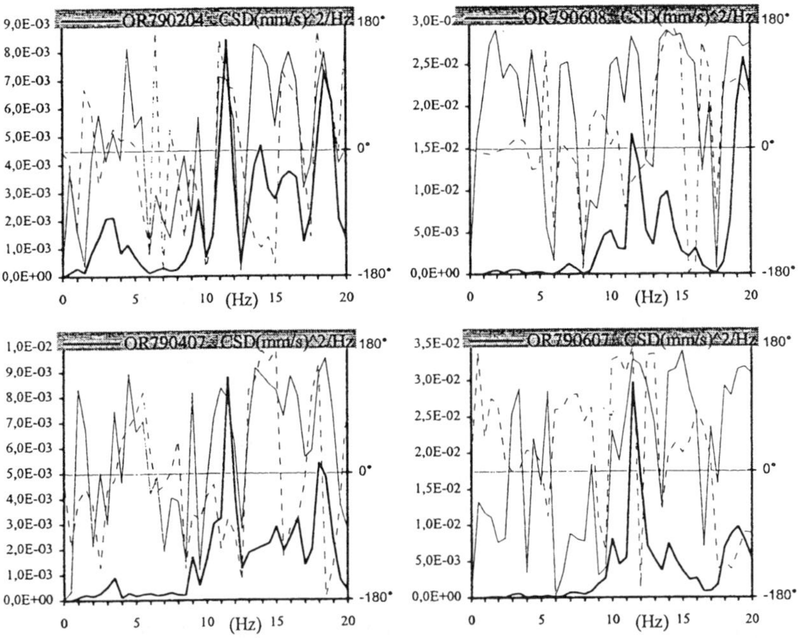

Figure 10a - Vault of the Corporale's Chapel (5b). CSD S2-S4, S6-S8, S4-S7 and S6-S7

...... Phase ___ Coherence [0,1]

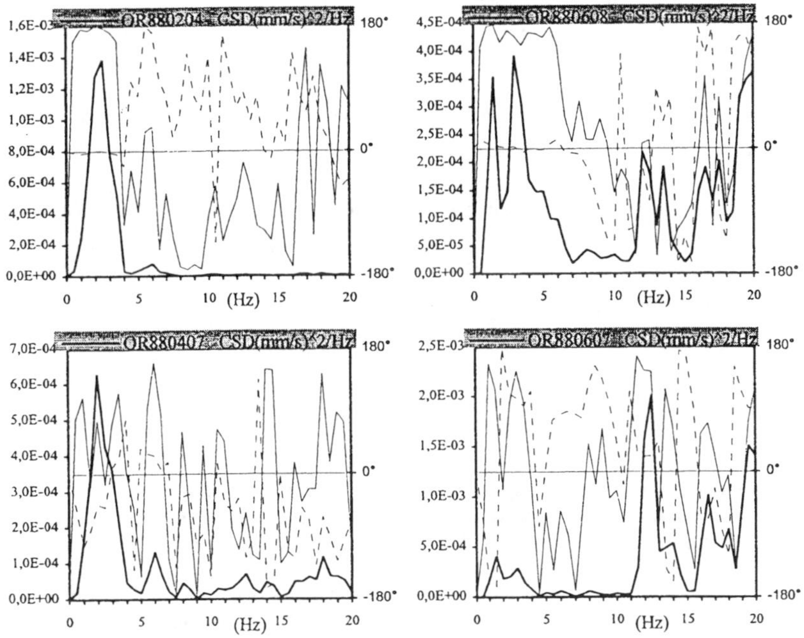

Figure 10b - Vault of the Corporale's Chapel (5c). CSD S2-S4, S6-S8, S4-S7 and S6-S7

4. Conclusions

The bell tower of the S. Giorgio in Trignano church, stroke by an earthquake, showed damages consisting in near horizontal cracks, just above the roof level of the connected buildings. The experimental analysis allowed to identify the influence of the damage on the structural behaviour and to state the absence of other important cracks. The dynamic interaction between the bell tower and the connected buildings certainly played a fundamental role in the crack formation and so did the bad static conditions of the structure.

The results of the dynamic test of the Duomo of Orvieto showed a good performance of the structure both under ambient vibration and forced tests. The structure behaves almost linearly in the tested range. The analysis of the recorded data relative to ambient vibration tests allowed identifying the resonance frequencies of the structure and of the vaults too. The main structure of the nave shows a good performance, even though there is no rigid connection between the longitudinal walls. In fact horizontal constrains between the wooden roof and the masonry walls are missed. Several structural resonance frequencies related to different modal shapes were pointed out.

Forced tests of the vaults showed resonance frequencies only slightly different from those pointed out by ambient vibration tests. More significant differences were observed in the records obtained on the vaults of the choir and that of the transept during earthquakes. This occurrence was due to the mechanical non-linearity of masonry. The presence of some cracks in the vaults could also played an important role. Even vibration of small amplitude could determine the deterioration of the mortar and its detachment from the stones.

Numerical modelling of the structure, associated with another experimental campaign based on a higher number of sensors, would be advisable in order to perform the system identification.

Due to the great importance of such a monument, a monitoring system should be installed or the testing of the structure at specific time interval during its life should be carried out, to analyse any changes in either the static and the dynamic behaviour and therefore in its integrity and service performance.

References

1. Bendat J.S. & Piersol A.G. (1980). Engineering application of correlation and spectral analysis. John Wiley and Sons.
2. Chopra A.K. (1995). Dynamics of Structures. Prentice Hall, Englewood Cliffs, New Jersey.
3. Ewins D. J. (1984). Modal testing: theory and practice. Research Study Press.
4. Clemente P., Bongiovanni G. (1993). "Ambient Vibration Effects on the Colosseum." Proc., IABSE Symposium "Structural Preservation of the Architectural Heritage" (Rome, September), IABSE, Zurich, 107-114.
5. Clemente P., Rinaldis D., Bongiovanni G. (1994). "Dynamic characterization of the Tempio della Minerva Medica." Proc., 10th European Conference on Earthquake Engineering (Vienna, August), Balkema, Rotterdam, Vol. 2, 981-986.

6. Rinaldis D., Clemente P., Donzelli R. (1995). "Analisi e controllo delle vibrazioni della Farnesina." Atti del Convegno "Terremoti in Italia: previsione e prevenzione dei danni" (Roma, Dicembre 1994), Accademia Nazionale dei Lincei, Roma, 69-77.
7. Buffarini G., Clemente P., Rinaldis D. (1997). Tempio Rotondo al Foro Boario: Rilievo delle vibrazioni ambientali (1a fase: Struttura non consolidata).ENEA RT/AMB/97/11, ENEA, Roma.
8. Buffarini G., Clemente P., Rinaldis D. (1997). Tempio Rotondo al Foro Boario: Rilievo delle vibrazioni ambientali. 2a Fase. ENEA RT/AMB/97/12, ENEA, Roma.
9. Clemente P., Bongiovanni G., Marzi C. (1988). "La colonna Antonina in Roma: valutazione degli effetti delle vibrazioni ambientali." Atti del III Convegno Nazionale ASS.I.R.C.CO. (Catania, Novembre), ASS.I.R.C.CO., Roma, 207-217.
10. Bongiovanni G., Celebi M., Clemente P. (1990). "The Flaminio Obelisk in Rome: vibrational characteristics as part of preservation efforts." Int. J. of Earthquake Engineering and Structural Dynamics, Vol. 19, John Wiley & Sons, 107-118, January.
11. Buffarini G., Clemente P., Rinaldis D. (1996). "Vibration Test of an old Masonry Building." Proc., 3rd European Conference on Structural Dynamics, eurodyn'96, (Firenze, June), Balkema, Rotterdam, Vol. 2, 825-832.
12. Clemente P., Rinaldis D. (1997). "Protection of an Ancient Building from Traffic-Induced Vibrations." Volume of Extended Abstract, Eight International Conference on Soil Dynamics and Earthquake Engineering, SDEE'97 (Istanbul, July), 316-317.
13. Clemente P., Bongiovanni G., Buffarini G., Dynamic Behaviour of a masonry tower subject to successive earthquakes, submitted to SDEE , Bergen, Norway, (1999)
14. Buffarini, G., Clemente P., Experimental dynamic analysis of the Duomo of Orvieto, ENEA, Rome, (1998).
15. Rinaldis D., Bongiovanni G., Verrubbi V., Le registrazioni accelerometriche delle reti ENEA in Umbria ottenute durante la crisi sismica del settembre-ottobre 1997 nell'Appennino umbro-marchigiano, ENEA, Rome, (1998)

UTILIZATION OF RAPID POST-EARTHQUAKE DATA BY UTILITIES

WILLIAM U. SAVAGE
Geosciences Department
Pacific Gas and Electric Company P. O. Box 770000
San Francisco, CA 94177
e-mail: wus1@pge.com

1. Abstract

When a potentially damaging earthquake occurs, utilities (electricity, gas, water, and telecommunications) have an urgent need for information about the effects of the event so that they can make optimal decisions regarding safety and maintaining and restoring utility functionality. Modern earthquake instrumentation systems, including strong-motion recorders and regional seismic networks, can collect data and provide information products that can greatly improve this decision-making and action-taking process. Four areas of utility response to earthquakes illustrate the utilization of these data: (1) Rapidly available network and strong-motion data can provide an earthquake alert that will make utility personnel aware that an earthquake is occurring, what area of the utility's service territory is affected, and the likely extent of damaging ground motions. This alert will focus the earthquake response attention of the utility and may permit quick operational and life-safety actions. (2) Within 10 to 30 minutes after the earthquake, analysis of strong-motion data from key utility sites can provide assessments of the likelihood of damage that can be used to prioritize deployment of field personnel and guide the initial operational control and recovery plans. (3) In the same time frame, similar strong-motion-based damage assessments of transportation routes (e. g. freeways, bridges, and overpasses) along with reported damage and disruption can help the utilities plan how to get inspection and repair crews to key facilities. In addition, damage likelihood assessments of commercial, industrial, and residential buildings can indicate where utility service connections may need rapid responses to safety and secondary damage threats. (4) Within a few hours of the earthquake, pre-arranged building inspectors can use building response strong-motion measurements to help evaluate the safety of continued occupancy of structures housing critical post-earthquake response functions.

M. Erdik et al. (eds.), Strong Motion Instrumentation for Civil Engineering Structures, 81–89.

2. Introduction

Modern society is increasingly dependent on safe and reliable services provided by the utility industries, including electric power, natural gas, telecommunications, and water and wastewater. The utilities consist of networks of transportation conduits (primarily pipelines, conductors, and telecommunication cables) and myriad ancillary facilities that connect the sources and destinations of utility products. Significant earthquakes in recent years have demonstrated that seismic vulnerabilities in utility networks can lead to interruptions in customer service that are dangerous to life and property, disruptive and costly to commercial enterprises, and produce long-term hardships on affected communities and people. While such long-term measures as improved seismic design practices and seismic retrofit programs can reduce the likelihood of damage, many utilities systems contain older components that are more vulnerable than their modern versions. Thus, in the coming years, utility system damage and service disruptions need to be anticipated, and utilities need to prepare for effective responses.

Accompanying the increased dependence on utility services in the Information Age have come major new developments in earthquake data acquisition, data analysis and processing, and communication of data and resultant information products (Kanamori and others, 1997). Digital strong-motion recorders, broadband seismometers with digital data loggers, high-speed and reliable telecommunications, and powerful computers and user-friendly software combine to make possible the rapid notification of earthquake occurrence and effects to emergency responders. This paper focuses on the increasingly important role that modern strong-motion data, along with other earthquake data, can play in effective and rapid earthquake response by utilities.

3. What Utilities Need

When a potentially damaging earthquake occurs, personnel in a utility who are responsible for operations, maintenance, and emergency response want to know:

- What happened: Was it an earthquake, an explosion, a plane crash? In a large utility, key personnel may be too far away from the earthquake to feel it, yet they still need to know what has happened.
- Where is the affected area: The locality affected is critical information to start the response process.

How much damage and disruption is there: This is the most important information to utility personnel. The level of damage not only affects their response effort, but may have personal impact because of potential threats to their families and homes.

Utility personnel are practiced in gathering this emergency response information from such traditional sources as fire and police reports, field reports from utility personnel and customers, and media announcements. But the utility response can be significantly improved by additional rapidly provided data and information to speed up the decision-making processes of utilities (and other emergency responders). What is now possible is to use modern scientific and engineering information based on real-time and near-real-time earthquake data to rapidly develop an accurate description of the likely state of damage of utility facilities, and other structures of interest such as transportation corridors and customer buildings. From this knowledge, utility decision-makers can take optimal action to address the damage conditions their organization is facing.

Basic earthquake data (such as recordings of strong ground shaking and of regional and broadband seismographic networks) are used to derive earthquake information such as location coordinates, focal depth, local magnitude, focal mechanism, moment tensor and moment magnitude, aftershock locations and rates, tectonic association, and others. Strong-motion data analysis adds peak acceleration, velocity, and displacement and response spectral ordinates at each instrument site; contour maps of ground motion parameters; event-specific attenuation, and others. These information products can be directly used or transformed into additional products that directly address what the earthquake did to utility facilities and other parts of the built environment. Figure 1 illustrates the relationship between earthquake data and information products and the sequence of utility responses that will be discussed in the next section.

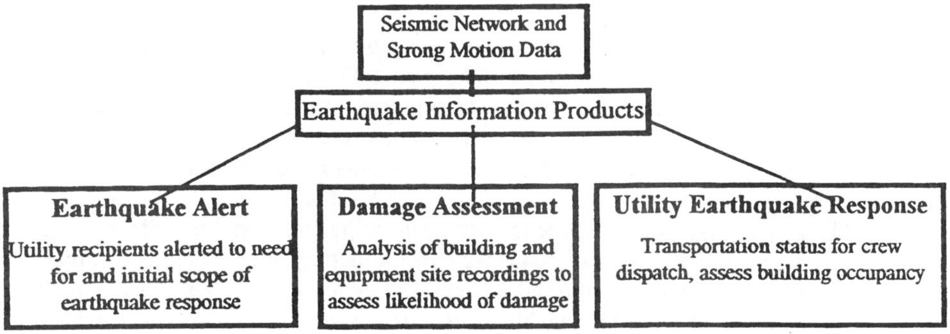

Figure 1. Rapid earthquake data used in earthquake response sequence by utility decision-makers.

4. Rapid Data Utilization By Utilities

4.1 EARTHQUAKE ALERT

In a modern-instrumented urban region, as the Los Angeles basin or the San Francisco Bay Area are becoming, an alert that an earthquake is occurring and information about its location (geographic coordinates) and size (magnitude) can be provided to utility recipients in a minute or two. This notification has immediate value to the recipients:

- Utility personnel know that an earthquake has just happened, rather than some other kind of disruptive or damaging event.

- The region affected by the earthquake is known. Current systems incorporate paged or e-mailed notification information into utility computers or GIS map servers, and distribute the combined earthquake and facilities location information within the utility via its intranet. This provides a vivid picture of where the earthquake occurred with respect to utility facilities.

- From the earthquake location and magnitude, computations using attenuation relations can give an initial map of the areal distribution and severity of ground shaking. The utility personnel can gain a graphic understanding of the extent of the region affected by the earthquake.

The initial rapid notification information serves to immediately initiate the utility organization's response to the consequences of the event. The CUBE and USGS/UC Berkeley notification systems are excellent examples of providing increasingly rapid information. Because the utility response is people- and decision-intensive, the faster they know the area within which a serious event has taken place, the more accurate their response can be.

It is desirable from the utility perspective to have faster initial information made available on earthquake location and size, instead of delaying while getting highly accurate values. A quick location that is accurate within 5 miles is close enough for the initial utility response to begin. It is particularly important to quickly distinguish between earthquakes with significant damage potential (M6½+) and those that may be locally damaging but that are not large enough to affect a widespread area (M5½ to 6½). It would be useful to develop a fast, approximate measure of earthquake size, even if the size were described by words (e.g. "moderate, big, very big") or magnitude ranges (e.g. M5±, M6±, or M5½ to M6½).

If the earthquake alert is prepared and distributed sufficiently quickly, it becomes an early warning of strong earthquake shaking for locations relatively far from large earthquakes. Possible benefits of such early warnings are enabling employees to get out of potentially dangerous working situations, or alerting remote backup operational facilities to take over for possibly threatened primary facilities. However, utilities need to practice receiving earthquake alerts, and to gain experience and confidence in using such information before the implementation of substantial early warning actions will be acceptable. Utility operators are reluctant to take significant preemptive operational actions that result in customer services being interrupted. There are likely to be major liability issues if electric power is cut off in a region that is threatened by strong shaking, for example. In general, any responsive actions taken as a result of receipt of an earthquake early warning will need to be preceded by extensive training and practice by employees, and the benefits of such actions will need to be clearly established.

4.2 RAPID DAMAGE LIKELIHOOD ASSESSMENT FOR FACILITIES

As noted previously, the information that utility personnel want to have immediately after an earthquake is the state of damage of utility equipment (substations, pump facilities, etc.) and of office buildings and buildings housing repair equipment and spare parts. Since many facilities and buildings are not routinely occupied, particularly at night or on weekends, on-site personnel may not be available to conduct immediate inspections. If personnel are on-site, they may be initially occupied with injuries or other immediate safety matters. Strong-motion instruments located at the facilities and buildings provide a means to perform a remote damage likelihood assessment. Using a free-field recording taken at the facility, the ground motion can be compared with previously determined fragility curves for the equipment and structures at a site. The comparison would take about 15 minutes after the earthquake begins, allowing time for data transmission from 10 or so key sites in the affected region. The fragility curves indicate probability of damage as a function of ground motion parameter for specified damage states (e.g., porcelain break for a transformer, limited entry for a building). While this is presently an approximate procedure due to the uncertainties in available fragility curves, it is a useful first assessment of what the damage is at a utility site, prior to conducting field inspections, which may occur several hours later.

Detailed contoured ground shaking maps can be constructed that include strong-motion data provided by many public and private sources, including utilities. For utility sites that do not have strong-motion instruments, ground motion estimates can be obtained from these maps. These estimates can them be used in fragility comparisons to assess likely damage. This approach depends upon having maps with sufficient input ground-motion observations to provide a reasonably accurate estimate of site ground shaking.

Installation of strong-motion instruments at utility facilities provides a valuable additional benefit in that ground-motion data from earthquakes can be used to refine the fragility functions for the equipment and buildings exposed to strong shaking from future earthquakes. Along with the site ground-motion recording, it is critical to accurately compile characteristics of the facility or equipment damage that occurs along with the lack of damage.

4.3 DAMAGE LIKELIHOOD ASSESSMENT OF TRANSPORTATION SYSTEM AND CUSTOMER BUILDINGS

The response of a utility to a damaging earthquake is also dependent upon the utility having information about the status of other parts of the built environment. Of particular importance is the operational status of the freeways, roads, and streets in the earthquake-affected area immediately after the event. Emergency access is needed to shut off downed power lines, or shut in leaking gas lines or water mains. The California Department of Transportation has instrumented a number of bridges and overpasses; just as for utility facilities, rapid analysis of these data can result in remote assessments of the likely damage to the structures and the prudence of continuing to use them or not. When combined with on-site reports of damage, useful early information about traffic access can be provided to utilities.

It is also valuable information for the initial utility response effort to have a preliminary picture of the localities in the earthquake-affected area where there are concentrations of building damage. These are places where there is likely to be extensive damage to customer utility connections, threatening life and property, and requiring immediate utility attention.

4.4 POST-EARTHQUAKE BUILDING INSPECTION FOR OCCUPANCY

Utilities, medical providers, and other organizations with critical post-earthquake emergency response functions have recognized that certain occupied buildings or buildings containing vital information or equipment may be closed by authorized building inspectors, or may be evacuated by frightened employees, some of who have critical post-earthquake response roles. To address this problem, some organizations have established Post-Earthquake Inspection Programs to facilitate the access to and use of essential buildings (Matsuda and others, 1995). Qualified engineers are contracted by the owner and pre-assigned to automatically inspect key facilities given certain earthquake location and magnitude criteria. They will have the benefit of pre-earthquake evaluations and post-earthquake inspection manuals, and will be able to tag the facilities regarding occupancy (red/yellow/green) under authorization from the local jurisdiction.

To assist the inspecting engineers, certain buildings can be instrumented to provide time history and response spectral data on the actual behavior of the building. The engineer can thus assess the potential for concealed structural damage that may make the building too dangerous for further occupancy, or that may be sufficiently minor to permit necessary emergency access or even full normal occupancy. The data could be available in the same time frame that the inspector arrives at the facility, within an hour after the earthquake.

5. Additional Comments

5.1 FALSE ALARMS

The providers of rapid earthquake information are sincerely concerned about the occurrence of false alarms—issuing an announcement of a large earthquake that has not occurred—or failing to report the occurrence of a large event. For the foreseeable future, utilities will most likely rely on their practical experience with their own status and information notification systems, and expect occasional problems and errors coming from the earthquake notifiers . It is simply not prudent for a utility to initiate aggressive or expensive initial response efforts on the basis of an earthquake alert that is not consistent with or verified by other available information. It is incumbent on the organizations issuing rapid notifications to have in place reasonable systems to detect false alarms and to rapidly correct them.

The problem of not issuing an accurate rapid report for a major damaging event that actually occurs has been the case for the 1989, 1992, and 1994 events in California. Recognizing this unfortunate history, the various seismological organizations have taken extensive efforts to build "earthquake-proof" systems. While these efforts are commendable, it is still possible for failures to occur. As the current notification systems continue to be installed and operated, additional opportunities to improve their reliability will no doubt be found and instituted. It should be the responsibility of the utilities and other recipients of rapid earthquake information to continue to use the notifications to gain experience in how to best apply them, and to let reliance on notifications grow with that experience.

5.2 RAPID EARTHQUAKE INFORMATION AS A SERVICE

Rapid earthquake information capabilities have been largely developed in the research seismology community. For some years, these capabilities have been provided to a limited number of recipients on a somewhat experimental basis. These arrangements have been helpful in learning how to improve essentially all aspects of the notification systems. With the development of the FEMA-supported TriNet project in Southern California, it is time to begin to view the provision of rapid earthquake information, from initial alerts, to shake maps and site-specific ground motions, to final archived data sets, as a public service, not an add-on to a research program. Instituting this distinction will clarify many of the uncertainties that currently exist in defining the responsibilities of the issuers and the recipients of information, in determining the sources of financial support for the service versus related research, and other issues. Shifting to a service-oriented perspective should result in a more professional and customer-oriented relationship between issuers and recipients. Federal/state agency partnerships with research/educational organizations appears to be developing into a good model for establishing an "earthquake service" that draws on the best of both academic and agency capabilities.

5.3 UTILITY RESPONSIBILITIES

Utility users of earthquake information must take responsibility for effectively understanding and applying the data and information they receive. This includes training and exercises in simulating the use of the rapid notification and damage likelihood assessments. The more frequent occurrence of small events help keep both the notification providers and notification users practiced.

6. Conclusions

Utilities and the communities they serve are at the dawn of a new era in earthquake response, as they begin to use the earthquake information products from modern seismic instrumentation. Through the application of such key technologies as digital strong-motion instruments, alternative reliable telecommunication pathways, powerful computers and software, and digital maps for displaying earthquake information, utility personnel will receive faster and more accurate information on earthquake occurrence and damage to be used for improved utility decision-making. This new capability is consistent with the increasing societal value placed on utility safety and reliability.

7. Acknowledgments

Discussions and critical reviews by Marcia McLaren and Norm Abrahamson are gratefully appreciated. This paper is based on numerous and ongoing interactions with many colleagues in both the seismological and utility communities regarding utility applications of earthquake information products. In particular, enthusiasm and support from the Geosciences Department members, key leaders in the operating utility groups, and the management of PG&E are acknowledged. This paper was originally presented at the SMIP98 Seminar on Utilization of Strong-Motion Data, Oakland, California, September 15, 1998; and a modified version was published in the proceedings of that seminar, which were published by the Strong Motion Instrumentation Program, California Division of Mines and Geology, Sacramento, California.

8. References

1. Kanamori, H., E. Hauksson, and T. Heaton (1997). Real-time seismology and earthquake hazard mitigation, *Nature*, Vol. 390, p. 461-464.
2. Matsuda, E., L. Cluff, C. Poland, and W. Savage (1995). Post-earthquake inspection of utility buildings, in *Lifeline Earthquake Engineering, Proceedings of the Fourth U. S. Conference*, M. J. O'Rourke, ed., Technical Council on Lifeline Earthquake Engineering Monograph No. 6, American Society of Civil Engineers, New York, p. 787-794.

ANALYSIS OF EARTHQUAKE RECORDS FROM STRUCTURES: AN OVERVIEW

ERDAL ŞAFAK
U.S Geological Survey, Golden, Colorado, USA
e-mail: safak@ usgs.gov

Abstract

The paper presents an overview of the methods that are available to analyze seismic records from structures. A typical analysis involves data processing, system identification, and damage detection. Data processing aims to minimize the ambient and instrument noise in the data, as well as possible low-frequency drifts, outliers, and other unwanted signals. System identification deals with determining the dynamic characteristics of a structure from its recorded response. There are a large number of methods available in the literature for system identification, varying from simple Fourier analysis to stochastic adaptive filtering. Unless data require otherwise, simple methods should be preferred for identification, because they are more robust and results are easier to interpret. Modal identification is the most widely used form of system identification. An alternative is the discrete-time filters, which provide a convenient model for identification of linear as well as nonlinear structures. Special techniques can be developed to identify a particular component of response, such as torsion, soil-structure interaction, and inter-story drift. Damage detection is a subject that is closely related to nonlinear system identification. Since a damaged structure almost always behaves in a nonlinear fashion, the problem of damage detection becomes equivalent to identification of the nonlinear behavior in the structure. The standard method for damage detection has been to observe the changes in the frequencies of the structure. However, unless it is a major damage, frequencies are not very sensitive to damage, particularly to localized damage. More reliable methods for damage detection can be developed by using time-frequency analyses and wave propagation techniques.

1. Introduction

Rapid developments in instrumentation and computer technologies have resulted in a large number of seismic monitoring systems installed in structures. Consequently, the amount of earthquake response data from structures has increased enormously in recent years. Data analyses, however, have not kept up with data accumulation, and a large number of structural data sets are yet to be analyzed properly Data analysis requires utilization of digital signal processing and system identification techniques. Structural and earthquake engineers are generally not trained on such subjects.

In this paper, we present an overview of the methods that are available to analyze seismic records from structures. The analysis involves three basic steps: data processing, system identification, and damage detection. These steps are discussed in the following paragraphs.

M. Erdik et al. (eds.), Strong Motion Instrumentation for Civil Engineering Structures, 91–107.

2. Data Processing

In general, recordings in their original (i.e., raw) form do not provide a good identification of the structure because of the ambient and instrument noise in the data, as well as possible low-frequency drifts, outliers (i.e., erroneous isolated peaks), and other flaws in the sensors, installations, and recordings. To minimize the adverse effects of such factors on the identification, we should first process the data. Processing involves the removal of mean, low-frequency drifts, outliers, and other unwanted frequencies by filtering, base-line corrections, instrument corrections, and decimation. For filtering, we use band-pass filters to eliminate the components of the record below and above a specified frequency band. Typically, we chose a low-frequency corner that will remove the mean (i.e., DC or static component) and other low-frequency errors, and a high-frequency corner that will remove the high frequencies that are dominated by noise.The filter bandwidth should be selected such that all the significant modes of the structure are within the bandwidth. It is important that we do not distort the phase of the signal during filtering. To ensure this, we either use band-pass filters with zero phase (e.g., Ormsby filter), or apply the filter twice (e,g., Butterworth filters), first forward and then backward. The band-pass filtering might not completely remove the low-frequency errors in the record. Low-frequency errors will show in the form of linear trends in the calculated velocity and displacement time histories. To eliminate linear trends we first fit a straight line to the accelerations and velocities , and then subtract this line from the records (for detail, see Hudson, 1979).

If somehow we can estimate the Fourier spectrum of the noise in the records, a more efficient filtering of the noise can be done by using Wiener (i.e., optimal) filters. Wiener filters remove the noise optimally over the entire frequency band, whereas the band-pass filters remove not the noise but the components of the record dominated by noise. The detail on Wiener filters can be found in Press et al. (1992).

The natural frequency of a typical strong-motion accelerometer is around 25 Hz, which records the true accelerations up to about 20 Hz. Beyond that the recorded accelerations gradually become smaller than the true accelerations because the amplitudes of the accelerometer's frequency response function are no longer one, but less than one. If we are interested in data at frequencies higher than the instrument's frequency, we should apply an instrument correction to the record. Instrument correction involves dividing the Fourier spectrum of the record with the frequency response function of the instrument, and can be implemented either in the time-domain, or in the frequency-domain, by using various algorithms available in the literature (e.g., Hudson, 1979; Press et al., 1992).

In a record, the highest frequency that we can extract information on is $1/(2\Delta)$ Hz, where Δ is the sampling interval. This frequency is called the Nyquist frequency. In most cases, the maximum frequency that we are interested is much smaller than the Nyquist frequency. If this is the case, we re-sample the record at a lower sampling rate such that the new Nyquist frequency is only slightly higher than the highest frequency in which we are interested. This is known as decimation. The procedure for decimation involves first low-pass filtering the record below the new Nyquist frequency, and then resampling the signal by skipping appropriate number of data points. Commercial signal processing software packages, such as the Signal Processing Toolbox of MATLAB (MathWorks, 1999a), all provide tools for decimation. Increasing sampling interval helps to improve the accuracy and the convergence of system identification algorithms. However, if the records are to be used in numerical analysis (e.g., input to a finite element model), we require sampling intervals that are much smaller than those needed for identification in order to prevent numerical instability. A rough guide is to have at least 10 sampling points for the shortest period that we want to consider in the analysis.

3. System Identification

System identification involves determining the dynamic characteristics of a structure from its recorded response. For earthquakes, the recorded response is typically the accelerations at various locations of' the structure. System identification involves three major steps: (1) selection of a mathematical model for the structure, (2) identification of model parameters from the recorded response, and (3) evaluation of the match of the model's response to the recorded response. There are two general approaches to identify model parameters. The first approach is a deterministic one, in which we determine model parameters by minimizing the difference between the model output and the recorded output. This is know as the prediction-error approach, and is based on the least-squares method for error minimization. The second approach is a probabilistic one, where we determine the model parameters by maximizing the probability that the model output will take the values of the recorded output. This is known as the maximum likelihood, or Bayesian identification. The advantage of the deterministic approach is its simplicity, whereas the advantage of the probabilistic approach is its independence from the model type. It can be shown that for records with Gaussian properties, the two approaches are identical (Ljung and Söderström, 1983). In general, the more complex the model is the better the match becomes. However, there is a trade off between the model complexity and the model reliability Complex models tend to match the noise as well as the structure. The goal is to determine the simplest model that gives the best match.

System identification is a subject that has applications in a wide range of fields, varying from biology to econometrics. Consequently, there are literally hundreds of methods and ad-hoc techniques available in the literature.

The fact that we can always develop another variation of an existing system identification method to solve our problem makes the subject of system identification an art as well as science. There are a large number of excellent textbooks on system identification for beginners (e.g., Mendel, 1973; Sinha and Kuszta, 1983; Graupe, 1984; Ljung, 1987; Bendat and Piersol, 1993). The texts by Pilkey and Cohen (1972), Newland (1989), and Norton (1989) are specifically on vibration analysis. More theoretically oriented treatments are given in Jaswinski (1970), Eykoff (1974), Goodwin and Payne (1977), Ljung and Söderström (1983), Young (1984), and Söderström and Stoica (1987). The text by Schoukens and Pintelon (1991) presents the frequency-domain methods for system identification. Publications that present applications in structural dynamics and earthquake engineering include Beck (1978), Natke (1982), Şafak (1988), and Ghanem et al. (1991).

There are also a large number of papers (too many to list here) appeared in structural dynamics and earthquake engineering journals.

In what follows, we will present an overview of the basic techniques available for system identification of structures from their measured seismic response. Parts of the paper are taken from the state of the art report in structural identification that is being prepared by the ASCE Technical Committee on System Identification and Health Monitoring of Constructed Facilities (ASCE, 2000). The author is a member of that committee and one of the co-authors of the report.

3.1. SIMPLE IDENTIFICATION TECHNIQUES

It is always a good practice to start with simple system identification techniques before applying complex identification methods to data. Simple techniques provide a quick assessment of the expected ranges of structural parameters, as well as giving insight which may help in interpreting results from more sophisticated methods. The tools for simple system identification are standard Fourier transforms, correlation analyses, and band-pass filtering techniques.

Fourier amplitude spectra (FAS) provide the simplest means to see the dominant frequencies of a signal. For records with large signal-to-noise ratios, such as those from the upper stories of a building, FAS can usually detect the fundamental frequency (the resonance frequency of the first mode) of the structure. Once the fundamental frequency is determined, we can approximate some of the key response values (e.g., maximum displacement, base shear, inter-story drift) from a single record. As an example we will present the results for a building that has the roof acceleration records only (Jennings, 1997). If the peak roof acceleration of a building is a_{max}, the peak roof displacement in the same direction is approximately

$$d_{max} \approx \left(\frac{1}{2\pi f_0}\right)^2 \cdot a_{max} \tag{1}$$

where *fo* is the fundamental frequency of the building. The base shear is the sum of the mass times accelerations of the floors:

$$V = \sum_{i=1}^{n} m_i a_i \tag{2}$$

where m_i and a_i denote the mass and the acceleration of the i'th floor, and n is the number of floors. If M is the total mass of the building and the floor masses are approximately equal, then $m \approx M/n$. Also, if $\varnothing(x)$ is the fundamental mode shape with a unit amplitude at the top, we have $a_i = a_{max}\,\varnothing(x_i)$ These lead to

$$V = M a_{max} \sum_{i=1}^{n} \frac{\phi(x_i)}{n} \tag{3}$$

The sum on the right-hand side gives the average ordinate of the mode shape. If the fundamental mode shape is assumed to be a straight line, Eq. 3 results in V = $0.5M\ a_{max}$ If the mode shape is equal to that of a shear beam, the base shear becomes V = $2/\pi\ M\ a_{max}$. The expressions for the inter-story drifts can be derived by using similar assumptions. They are given in Jennings (1997).

The response time histories at the resonance frequencies are obtained by simply band-pass filtering the records around a narrow-band filter centered at the resonance frequency. As mentioned earlier, to ensure that the phase spectrum of the signal is not distorted during filtering, we either use zero-phase filters or filter the signal twice, first forward and then backward. The filtered records are all sinusoidal-like, corresponding to the motion of the structure at that resonance frequency. We determine the mode shapes by plotting the resonance displacements for one cycle of vibration, and observing the relative directions and amplitudes of the displacements at each of the observed degrees of freedom.

Modal damping values are calculated from the rate of decay of resonance displacement amplitudes towards the end of the record, which represents the free vibration part of the response after the earthquake has ended (Clough and Penzien, 1975). Modal displacement time histories can be used to check if the energy dissipation is well-modeled as classical linear viscous damping (for example, the damping distribution in the structure is proportional to the mass and/or stiffness distribution). This is an assumption regularly made in structural analysis and design. If the damping has a classical form (see Caughey and O'Kelly, 1965), then the modal displacements from all sensors should reach their maxima and minima at the same time. It is observed in practice that this often does not happen (e.g., Şafak, 1993). Indeed, experience with system identification applied to seismic response records shows that the linear viscous damping model is not a very good model of the energy dissipation in structures at strong-motion amplitude levels.

If a structure is subjected to SSI (soil-structure interaction), its resonance frequency represents not the frequency of the fixed-based structure but the frequency of the soil-structure system. The resonance frequency with SSI is always smaller than the fixed-based frequency. For buildings, the presence of SSI and the fixed-based frequency are determined from the foundation-to-roof transfer function. The peak of the foundation-to-roof transfer function gives the fixed-based (i.e., no SSI) resonance frequency of the building (Şafak, 1995). If this frequency is identical to the dominant frequency of the records from the upper stories (e.g., the roof) we conclude that there is no SSI. If the transfer function frequency is larger, the building is subjected to SSI. The foundation-to-roof transfer function can be estimated by using one of the following two equations:

$$H(f) = \frac{\sum_{j=1}^{2k+1} W(j) \cdot |Y(f-k-1+j)|}{\sum_{j=1}^{2k+1} W(j) \cdot |X(f-k-1+j)|} \tag{4.a}$$

or

$$H(f) = \frac{\sum_{j=1}^{2k+1} W(j) \cdot Y(f-k-1+j) \cdot X^*(f-k-1+j)|}{\sum_{j=1}^{2k+1} W(j) \cdot |X(f-k-1+j)|^2} \tag{4.b}$$

where $X(f)$ and $Y(f)$ are the complex Fourier transforms of foundation and roof records, respectively, $W(j)$ *is* the symmetric smoothing window with $2k + 1$ points, f is the cyclic frequency, and the superscript * denotes the complex conjugate. The smoothing aims to reduce the effects of noise that is always present in the records. The ratio of the FAS of two noisy records is very sensitive to noise and would have unrealistically high amplitudes if no smoothing were performed on the FAS prior to taking the ratio. It can be shown that for noisy records, the second definition gives a more accurate estimate of the transfer function (Safak, 1997*a*). There are no clear-cut rules for the selection of smoothing windows. However, once the shape of the window is selected (e.g., Hanning, Gaussian, triangular, etc.), its optimal length can be determined by observing the decay of FAS amplitudes with increasing window length (Şafak, 1997*a*). More on the detection and identification of SSI can be found in Şafak (1995).

There are simple techniques to identify a specific response component, such as torsional and rocking motions, and inter-story drifts. The time histories for torsional vibrations are calculated by taking the differences of horizontal records. The FAS of the difference highlights the torsional mode frequencies, at least for the lower modes. We can separate the torsional and translational components of the motion by band-pass filtering the recorded horizontal motions around the torsional and horizontal natural frequencies, provided these are reasonably well separated. Similarly, we can identify the modes with significant rocking motions by taking the differences of two vertical records from the foundation. The FAS of the difference signal highlights the frequency of the modes with significant rocking. The rocking component of these motions is separated by band-pass filtering the records around each rocking mode frequency. For noisy records, taking the differences of the signals can amplify the noise to unacceptable levels, making the identification of torsion and rocking difficult. In such cases, more advanced techniques (e.g., cross-spectrum techniques, adaptive noise canceling filters, or multi-input, multi-output modal identification) might be needed. Identifications of torsion by using cross-spectrum techniques and adaptive noise canceling filters are given in Şafak and Çelebi (1990*a*, 1990*b*).

In general, instrumented buildings do not have sensors at every floor. Therefore, we cannot determine inter-story drifts accurately for all the stories. Mau and Aruna (1994) give a methodology to estimate inter-story drift in buildings from a limited number of records. The method is based on the assumption that the displaced configuration of a building is a combination of those of a shear beam and a bending beam.

The simple techniques outlined above give fairly accurate identification of the modal properties if the modes have well-separated frequencies so that modal interference is not severe. In some cases structures exhibit coupled modes where the mode involves translation as well as torsion. The presence of coupled modes can be detected by plotting the displaced configuration of the structure (or a section of the structure, e.g., a floor slab) for one full cycle of vibration at that model frequency. In other cases, structures might exhibit closely-spaced modes such as a torsional mode with a frequency close to a translational mode (e.g. separated by less than the half-power bandwidth of each mode), or two closely-spaced translational modes. Such modes can appear to be a single mode with a mode shape which is some combination of the contributing modes. In this case, the multi-input, multi-output modal identification method described in the next section can help to resolve the properties of the closely-spaced modes.

3.2. MODAL IDENTIFICATION

Modal identification is the most popular form of system identification in structural dynamics where modal parameters of a structure are estimated from its vibration records. There are various techniques for modal identification, such as direct identification (e.g., Beck, 1978; Beck and Jennings, 1980), frequency domain identification (e.g., Schoukens and Pintelon, 1991), and subspace-based identification (e.g., Overschee and Moor, 1996). We will briefly outline the direct identification method, which was introduced by Beck (1978) and widely used to analyze earthquake data from structures.

For a n-DOF (n degrees-of-freedom) linear structure subjected to seismic loads, the relative displacement $y_i(t)$ of node i can be written as a superposition of n modal responses:

$$y_i(t) = \sum_{j=1}^{n} x_{ij}(t) \tag{5}$$

where $x_{ij}\ (t)$ denotes the amplitude of j'th modal response at node i. The equation for $x_{ij}(t)$ is

$$\ddot{x}_{ij}(t) + 2\xi_j\omega_j\dot{x}_{ij}(t) + \omega_j^2 x_{ij}(t) = -p_{ij}\ddot{x}_g(t) \tag{6}$$

where ω_j and ξ_j are the modal frequency (in radians) and damping ratio for the j'th mode, p_{ij} is the participation factor of the j'th mode at node i, and $\ddot{x}_g(t)$ is the ground acceleration. We can determine $\ddot{x}_{ij}(t)$ by solving Eq. 7 in terms of ω_j, ξ_j ,p_{ij} and the initxal conditions $\dot{x}_{ij}(t)$

and $x_{ij}(0)$. They are the unknowns that we like to identify. If we denote the unknowns for the j'th mode by θ:

$$\theta_j = [\omega_j, \xi_j, p_{ij}, \dot{x}_{ij}(0)\, x_{ij}(0)] \tag{7}$$

we can write

$$\ddot{x}_{ij}(t) = F_j(\theta_j) \tag{8}$$

where $F_i(\theta_i)$ is a nonlinear function of θ_j. By calculating F for all the modes, i.e., $j = 1,\ldots, n$, and inserting in Eq. 5, we obtain $\ddot{y}_i(t)$ in terms of modal unknowns; that is

$$\ddot{y}_i(t) = \sum_{j=1}^{n} F_j(\theta_j) \tag{9}$$

We determine θ_j by minimizing the difference between the recorded and calculated accelerations at node i. If recorded acceleration at node i is $\ddot{y}_{ri}(t)$, the difference is defined as

$$J = \int_{t=0}^{t_d} [\ddot{y}_{ri}(t) - \ddot{y}_i(t)]^2\, dt \tag{10}$$

where t_d denotes the duration of earthquake. Since J is a nonlinear function of modal parameters, its minimization is done numerically by applying iterative optimization algorithms. It can be shown that J has a global minimum, such that modal parameters can be identified uniquely (Beck, 1978). The detail on the minimization of J, as well as its convergence and uniqueness properties are given in Beck (1978).

The formulation given above is for single-input and single-output only. The procedure can be extended to multi- input, multi-output cases, where all the records from the structure are considered simultaneously for identification (e.g., Werner et al, 1987). The same methodology can also be implemented in the frequency domain by matching the complex Fourier transforms of the recorded and model outputs (McVerry, 1980).

3.3. IDENTIFICATION USING DISCRETE-TIME FILTERS

The response of a structure during an earthquake is similar to that of a filter. The structure filters the ground motions entering through the foundation (the input of the filter) into those appearing in the superstructure (the output of the filter). Therefore, we can model the structure as a filter, and the identification of the structure becomes equivalent to determining the parameters of that filter. Since the response is recorded at discrete times, it is natural to use discrete-time (or digital) filters to model the structure. For linear behavior, these filters are expressed in terms of linear difference equations, where the present output of the filter is a linear combination of past outputs, and the present and past inputs. The following equation represents a linear filter

$$y(t) + a_1 y(t-\Delta) + \cdots + a_m y(t-m\Delta) = b_0 x(t) + b_1 x(t-\Delta) + \cdots + b_n x(t-n\Delta) \tag{11}$$

where *x(t)* and *y(t)* are the input and the output of the filter, respectively, Δ denotes the sampling interval, and a_j and b_j are the parameters of the filter By using the delay operator q^{-1}, defined as $q^{-j}x(t) = x(t - j\Delta)$, we can write the same equation in a more compact form as

$$y(t) = \frac{B(q)}{A(q)} \cdot x(t) \tag{12}$$

where *A(q)* and *B(q)* are polynomials in the delay operator q^{-1}

$$A(q) = 1 + a_1 q^{-1} + \cdots + a_m q^{-m} \qquad \text{and} \qquad B(q) = b_0 + b_1 q^{-1} + \cdots + b_n q^{-n} \tag{13}$$

In its most general form, the relationship between the input, x(t), and the output, *y(t),* of a structure is represented by the following equation:

$$A(q)y(t) = \frac{B(q)}{F(q)} x(t - n_d) + \frac{C(q)}{D(q)} e(t) \tag{14}$$

where *A(q),*..... , *F(q)* are all polynomials in the delay operator q^{-1}, n_d is the number of delays from input to output , and *e(t)* denotes a white-noise sequence which models uncertainties in the input-output relationship. This general form is known as the (linear) Black-Box model. In a typical application, we do not need all the polynomials in the Black-Box model. The special forms of the Black-Box model obtained by eliminating some of the polynomials in Eq. (14) are summarized in Table 1.

For SISO system identification, a popular model is the ARX model (see Table 1), which is

$$y(t) + a_1 y(t-\Delta) + \cdots + a_m y(t-m\Delta) = b_0 x(t) + b_1 x(t-\Delta) + \cdots + b_n x(t-n\Delta) + e(t) \tag{15}$$

The system identification of an ARX model involves determining the coefficients of a_j,b_k, where $j = 1,\ldots,m$ and $k = 0, \ldots, n$, for given *x(t)* and *y(t)*. To accomplish this, we first write the expected value, $\hat{y}(t)$, of $y(t)$. Since $e(t)$ is a zero-mean white noise process (i.e., its expected value is zero), we have

$$\hat{y}(t) = -a_1 y(t-\Delta) - \cdots - a_m y(t-m\Delta) + b_0 x(t) + b_1 x(t-\Delta) + \cdots + b_n x(t-n\Delta) \tag{16}$$

By defining the following colum vectors

$$\theta = [a_1, a_2, \cdots, a_m, b_0, b_1, \cdots, b_n]^T \tag{17.a}$$

$$\phi(t) = [-y(t-1), \cdots, -y(t-m), x(t), x(t-1), \cdots, x(t-n)]^T \tag{17.b}$$

we can write the expected value of the output as

$$\hat{y}(t) = \phi(t)^T \theta \tag{18}$$

Total error between the model output $\hat{y}(t)$ and the recorded output $y_r(t)$ is

$$V = \sum_{t=1}^{N} [y_r(t) - \phi(t)^T \theta]^2 \tag{19}$$

To determine the unknown coefficient vector θ, we minimize V with respect to θ, which gives

$$\theta = \left[\sum_{t=1}^{N} \phi(t)\, \phi(t)^T\right]^{-1} \left[\sum_{t=1}^{N} \phi(t) y(t)\right] \tag{20}$$

Note that the first term on the right-hand side is a $(m + n + 1) \times (m + n + 1)$ matrix, whereas the second term is a $m + n + 1$ dimensional column vector .

There is a one-to-one correspondence between the filter coefficients and the modal parameters that were discussed in the previous section. To show the relationship, we apply the partial fraction expansion to the filter transfer function $H(q)$, which becomes

$$H(q) = \frac{b_0 + b_1 q^{-1} + \cdot + b_n q^{-n}}{1 + a_1 q^{-1} + \cdot + a_m q^{-m}} = \sum_{k=1}^{m} \frac{r_k}{1 - p_k q^{-1}} \tag{21}$$

where r_k and p_k are called the residues and poles, respectively, of the transfer function. The poles are the complex roots of the denominator polynomial $A(q) = 0$. The theory of the partial fractions expansion can be found elsewhere (e.g., Press et al., 1992). For stable systems (e.g., structures with no negative damping) the poles are all in complex conjugate pairs with modulus less than one. By combining the pairs of terms corresponding to complex-conjugate poles, we obtain

$$H(q) = \sum_{j=1}^{m/2} H_j(q), \qquad \text{where} \quad H_j(q) = \frac{2\Re(r_j) - 2\Re(r_j p_j^*) q^{-1}}{1 - 2\Re(p_j) q^{-1} + |p_j|^2 q^{-2}} \tag{22}$$

where R denotes the real part. Each $H_j(q)$ is equivalent to a SDOF damped oscillator, and represents a modal transfer function of the structure. Note that a m' th order A(q) polynomial results in m / 2 modes. The modal requency f_j and the damping ratio ξ_j are calculated from the coefficients of $H_j(q)$ by the following equations:

$$f_j = \frac{\ln(1/|p_j|)}{2\pi\xi_j\Delta} \qquad \text{and} \qquad \xi_j = \frac{\ln(1/|p_j|)}{\left[\tan^{-1}[\Im(p_j)/\Re(p_j)] + \ln(1/|p_j|)\right]^{1/2}} \tag{23}$$

where $\Im$ deriotes the imaginary part. The derivations of above equations are given in Şafak (1988, 1991).

For multi-input, multi-output structures, the form of discrete-time filter models is similar to that given by equation (14), except $x(t)$ and $y(t)$ now represent column vectors with each element corresponding to a different input and output. $A(q)$,...... , $F(q)$ are matrices with appropriate dimensions, where each element in the matrices is a polynomial in the delay operator q^{-1} . For an nx-input, ny-output ARX model, for example, $A(q)$ is a ny x ny matrix. The polynomial corresponding to element j, k of matrix $A(q)$ describes how the old values of output j ifluences the present value of output k. Similarly, $B(q)$ is a ny x nx matrix, whose element j, k describes hcw the present and past values of input j influences the present value of output k.

The methodology outlined above can also be used for the output-only identification, where we do not know the input signal. Identification from ambient vibration data is a typical example of output-only identification. In the case of unknown input, the ARX model becomes an AR model. The equations for output-only identification are obtained by simply deleting all x (or, b_j) terms in the above formulation, and assuming that the input has a wide-band spectrum such that it can excite all the modes of the structure (i.e., the input is similar to a white-noise sequence). The frequencies and damping ratios are calculated from the same equation, Eq. 23, since they are not influenced by the input. An example of identification from ambient vibration data can be found in Şafak (1988).

An important advantage of the discrete-time identification over the modal identification is that there are well- established algorithms and software tools to identify discrete-time filters. Detailed discussions of these algorithms or different model structures, including rigorous stability and convergence analysis, can be found in Ljung and Söderström (1983) and Söderström and Stoica (1989). The System Identification Toolbox of the numerical analysis Software package MATLAB (MathWorks, 1999b) includes most of the algorithms. The manual of the Toolbox is also a good introduction to system identification using discrete-time filters. The MATLAB-based software package MACEC (Branden et al., 1999) provides interactive tools for output-only identification. An overview of the methods and their applications in structural and earthquake engineering are presented in Safak (1988). Examples for the applications to real structural data are given in Şafak and Çelebi (1991, 1992) and in Şafak (1991, 1993).

4. Nonlinear Identification And Damage Detection

During large earthquakes, deformations and stresses in the structure and the foundation soil might exceed elastic limits causing nonlinear vibrations of the structure. Also, any structure that suffered damage during an earthquake will have experienced nonlinear vibrations. Nonlinear system identification is a subject with a wide range of tools and techniques, some of which are still being developed. In this section, we will briefly outline the basics and provide some references, where more detail can be found.

Any nonlinear model can be represented as a linear model with time-varying characteristics (Young, 1984). Therefore, nonlinear behavior can be investigated by applying system identification techniques to linear models in a time-varying fashion. For simple identification and modal identification, this involves dividing the records into segments where the dynamic behavior is approximately time-invariant, and applying the linear model techniques to each segment separately. For discrete-time filters, the time variations of the model can be captured by allowing the filter parameters to change with time. This is known as recursive, adaptive, or time-varying system identification. The two most widely used recursive identification techniques are the Extended Kalman filters (Kalman and Bucy, 1961) and the recursive prediction-error filters. The theory of Kalman filters and recursive identification are given in texts available in the literature (e.g., Jaswinski, 1970; Kailath, 1981; Ljung and Söderström, 1981; and Gelb, 1992). An application of Extended Kalman filters in structural identification can be found in Hoshiya and Saito (1984), Whereas the applications of recursive prediction-error filters are given in Şafak (1989). Several other methods were also used to track time variations in vibration signals. They include those that utilize neural networks (Wu et al, 1992; Masri et al, 1993; Chen et al., 1995), orthogonal decomposition (Udwadia and Jerath, 1980; Benedettini, 1995; Zeldin and Spanos, 1998), and wavelets (Kitada, 1998).

Another approach to nonlinear system identification is to assume an appropriate nonlinear model for the expected behavior of the structure, and determine the parameters of the model by matching the model output with the recorded output. This approach is feasible only if we have a very good idea on the type of nonlinear behavior that is likely to occur, and a nonlinear model that is flexible enough to account for all the variations of the expected nonlinearity. The identification is achieved numerically by using iterative minimization techniques. The examples of model-based nonlinear identification in structural dynamics are given in Cifuentes (1984), Sues at al. (1988), and Kunnath et al. (1997).

A subject closely related to nonlinear system identification is damage detection. Since a damaged structure almost always behaves in a nonlinear fashion, the problem of damage detection becomes equivalent to identification of the nonlinear behavior in the structure. The simplest and most straightforward technique to detect damage is to search for the time variations in the frequency content of the structure's vibrations. Any significant change in the modal firequencies is a strong indication for damage. The subject of studying time-varying characteristics of signals is known as time-frequency analysis (Cohen, 1995). The methods that have been used for time-frequency analysis include the spectrogram (Oppenheim and Schafer, 1989), wavelet analysis (Strang and Nguyen, 1996), singular-value decomposition (Cohen, 1995), and evolutionary spectrum (Priestley, 1981). A new method, called Hilbert-Huang transforms (HHT), has been shown to have better frequency tracking abilities than the other methods (Huang et al., 1998). Recently, Şafak (1997b) and Pines (1997) have shown that for some structures the wave propagation analysis provides a superior alternative to the time-frequency analysis for damage detection. In the wave propagation analysis, the damage is detected by observing the changes in the wave travel times, and the wave reflection and transmission coefficients, instead of frequencies. More specialized techniques for damage detection in civil engineering structures are also available in the literature (e.g., Nielsen et al., 1995; Köylüoglu et al., 1998).

5. Summary And Conclusions

Analyses of earthquake data from structures involve data processing, system identification, and damage detection. Data processing aims to minimize the effects of noise and other unwanted signals on the records, and is performed by utilizing band-pass filters, base-line and instrument corrections, and decimation. System identification deals with the estimation of dynamic characteristics of a structure from its recorded response. There are a large number of methods available in the literature for system identification, varying from standard Fourier analysis to stochastic adaptive filtering. If they work, simple identification techniques should be preferred to more complex methods. The tools for simple identification include Fourier transforms, correlation analyses, and band-pass filters. Modal identification is the most widely used form of system identification. An alternative to modal identification is the identification using discrete-time filters. Discrete-time filters provide a convenient model for the identification of linear as well as nonlinear structures. Special techniques are available to identify a particular component of response, such as torsion, soil-structure interaction, and inter-story drift. Damage detection requires the applications of nonlinear system identification techniques. Since a damaged structure almost always behaves in a nonlinear fashion, the problem of damage detection is equivalent to identification of the beginning of a nonlinear behavior in the structure. The standard method for damage detection has been to search for the changes in the frequencies of the structure. However, unless it is a major damage, frequencies are not very sensitive to damage, particularly to localized damage. More reliable methods for damage detection can be developed by using the tools of time-frequency analyses and wave propagation techniques.

TABLE 1. Special forms of Black-Box model

Abbreviations:
AR– Autoregressive, MA– Moving average, X– Auxiliary input, FIR– Finite impulse response.

1. Black-Box Model	$A \cdot y = \frac{B}{F} \cdot x + \frac{C}{D}$
2. Box-Jenkins Model :	$y = \frac{B}{F} \cdot x + \frac{C}{D} \cdot e$
3. Output-Error Model :	$y = \frac{B}{F} \cdot x + e$
4. ARARMAX Model	$A \cdot y = B \cdot x + \frac{C}{D} \cdot e$
5. ARARX Model	$A \cdot y = B \cdot x + \frac{1}{D} \cdot e$
6. ARMAX Model	$A \cdot y = B \cdot x + C \cdot e$
7. ARMA Model	$A \cdot y = C \cdot e$
8. ARX Model	$A \cdot y = B \cdot x + e$
9. AR Model	$A \cdot y = e$
10. FIR Model	$y = B \cdot x + e$

References

1. ASCE (2000) *The State of the Art Report in Structural Identification,* Technical Committee on System Identification and Health Monitoring of Constructed Facilities, ASCE, Reston, VA (in preparation).
2. Beck J.L. (1978). Determining Models of Structures form Earthquake Records, *EERL Report No. 78-01,* California Institute of Technology, Pasadena
3. Beck J.L. and P.C. Jennings (1980). Structural identification using linear models and earthquake records. *Earthq. Engng. aud Struct.* Dyn., 8, 145-160.
4. Bendat J.S. and Piersol, A.G. (1993). *Engineering Application of Correlation and Spectral Analysis* (2nd Ed.), John Wiley & Sons, Inc. New York, NY.
5. Benedettini, F., D. Capecchi and F. Vestroni (1995). Identification of hysteretic oscillators under earthquake loading by nonparametric models, *J. Eng. Mech., ASCE,* 121, 606-612.
6. Branden B.V., Peeters, B. and Roeck, G.D. (1999). *Introduction to MACEC (v 2.0): Modal Analysis on Civil Engineering Construction,* Dept. of Civil Eng., ,Katholieke Iniversiteit Leuven, Belgium.
7. Caughey, T.K. and M.E.J. O'Kelly (1965). Classical normal modes in damped linear dynamic systems. *J. Applied Mech., ASME,* 32, 583-588.
8. Chen H.M., G.Z. Qi, J.C.S. Yang, and F. Amini (1995). Neural networks for structural dynamic model identification. *J. Eng. Mech.; ASCE,* 121, 1377-1382.
9. Cifuentes, A.O. (1984). *System Identification of Hysteretic Structures,* EERL Repoıt No. 84-04, California Institute of Technology, Pasadena.
10. Clough R.W and Penzien, J. (1975). *Dynamics of Strucıures,* McGraw-Hill, New York, NY.
11. Cohen, L. (1995). *Time-Frequency Analysis,* Prentice Hall PTR, Upper Saddle River, NJ.
12. Eykhoff, P (1974). *System Identification: Parameter and State Estimation.* Wiley, New York.
13. Gelb, A. (Ed.) (1992). *Applied Optimal Esrimation,* The M.I.T. Press, Cambridge, MA.
14. Ghanem, R., Shinozuka, M. and Gavin, H. (1991). Expeıimental verification of a number of system identification algoıithms, *NCEER Tech. Rep. 91-0024,* Nat. Ctr for Earthquake Engrg. Res., Buffalo, NY.
15. Goodwin, G.C. and R.L. Payne (1977). *Dynamic System Identification. Experiment Design and Data Analysis.* Academic Press, New York.
16. Graupe, D. (1984). *Time Series Analysis, ldentification and Adaptive Filtering,* Robert E. Krieger Publishing Co., Malabar, FL.
17. Hoshiya, M. and E. Saito (1984). Structural identification by extended Kalman filter. *J. Eng. Mech., ASCE,* 110, 1757-1770.
18. Huang, N.E., Shen, Z., Long, S.R., Wu M.C., Shih, H.H., Zheng, Q., Yen, N: C., Tung, C.C., and Liu, H.H. (1998). The empirical mode decomposition and the Hilbert spectrum for nonlinear and non-stationary time series analysıs. *Pıoc. R. Soc. Lond.,* 454, 903-995.
19. Hudson, D.E. (1979). *Reading and Interpreting Strong Motion Accelerograms,* Earthquake Engineering Reserach Institute, Oakland, CA.
20. Jayakumar, P. and J.L. Beck (1988). System identification using nonlinear structural models. In: *Pmc. Int. Workshop on Structural Safety Evaluation based on System Identification Approaches* (H.G. Natke and J.T.P Yao, Eds), Vieweg and Sons, Wiesbaden.
21. Jazwinski, A.H. (1970). *Stochastic Processes and Filtering Theory.* Academic Press, New York.
22. Jennings P. (1997). Use of strong motion data in earthquake resistant design, in *Proc. SMIP97 Seminar on Utilization of Sırong-Motion Data* (Ed. by MJ. Huang), California Division of Mines and Geology, Sacramento, CA.
23. Kailath, T. (1981). *Lectures on Wiener and Kalman Filtering,* Springer-Verlag, New York, NY
24. Kalman, R.E. and R.S. Bucy (1961). New results in linear filtering theory. *J. Basic Eng., ASME,* 83, 95-108.
25. Kitada, Y. (1998). Identification of nonlinear structural dynamic systems using wavelets, *J. Eng. Mech., ASCE, 124,* 1059-1066.
26. Köylüoğlu, U.H., Nielsen S.R.K., Abbott, J., and Çakmak, A. Ş. (1998). Local and modal damage indicators for RC frames subject to earthquakes, *J. Eng. Mech., ASCE, 124,* 1371-1379.
27. Kunnath, S., J.B. Mander, and L. Fang (1997). Parameter identification for degrading and pinched hysteretic structural concrete systems, *Engineering Structures,* 19, 224-232.
28. Ljung, L. and T. Söderström (1983). *Theory and Practice of Recursive ldentification,* MIT Press, Cambridge, MA.
29. Ljung, L. (1987). *System Identification: Theory for the User.* Prentice-Hall, New Jersey.
30. Masri, S.F, .A.G. Chassiakos and T.K. Caughey (1993). Identification of nonlinear dynamic systems using neural networks. *J. Applied Mech., ASME,* 60, 123-133.

31. Mathworks (1999*a*). *Signal Processing Toolbox for MATLAB,* Mathworks, Inc., Nattick, MA.
32. Mathworks (1999*b*). *System Ident,fication Toolbox for MATLAB,* Mathworks, Inc., Nattick, MA.
33. Mau, S.T. and V Aruma (1994). Story drift, shear and OTM estimation from building seismic records. *J. Struct. Eng., ASCE,* 120, 3366-3385.
34. McVerry, G.H. (1980). Structural identification in the frequency domain from earthquake records. *Earthq. Engng. and Struct.* Dyn., 8, 161-180.
35. Natke, H.G. (Ed.) (1982). *Identification of Vibrating Structures.* Springer-Verlag, New York.
36. Newland, D.E. (1989). *Mechanical Vibrations Analysis and Computation,* Longman Scientific & Technical, Essex, England.
37. Nielsen , S.R.K, Skjarbek, P.S., Köylüoğlu U.H., and Çakmak, A. Ş (1995). Preditiction of global damage and reliability based upon sequential identification and updatıng of RC stuructures subject to earthquakes, *Soil Dynamics & Earthquake* Engineering *VII* (Çakmak, A.Ş. and Brebbia, A., Eds.), Computational mEchanics Pubs., Southampton, England, 361-369.
38. Norton, M.P. (1989). *Fundamentals of* Noise *and Vibration Analysis for Engineers,* Cambridge University Press, Cambridge, England.
39. Oppenheim, A.V and Schafer, R.W. (1975). *Digital Signal Processing,* Prentice-Hall, Englewood Cliffs. NJ.
40. Overschee, P.V. and Moor B.D. (1996). *Subspace İdentification of Linear Systems:Theory - Implemention - Applicasion.* Kluwer Academic Publishers.
41. Pilkey.WD. and R.Cohen (Eds) (1972). *System İdentification of Vibrating* Structures. ASME. N.Y
42. Pines DJ. (1997) The use of wave propagation models for structural damage identification, in *Structural Health Monitoring: Current* Status *and Perspectives (Ed. By F .- K* Chang). Technomic Publishing Co. Inc., Lancaster, PA.
43. Press W H., Tekolsky, S.A., Vetterling, W.T., and Flannery, B.P. (1992). *Numerical Recipes in Fortran* (2nd Ed.), Cambridge University Press, Cambridge, England.
44. Priestley M.B. (1981). *Spectral* Analysis *and Time Series,* Academic Press, New York.
45. Şafak, E. (1988) Analysis of recordings in stuructural engineering: Adaptive filtering, prediction, and control, *Open-File Report 88.647, U.* S. Geological Survey, Menlo Park, California.
46. Şafak, E. (1989).Adaptive modeling, identification, and control W dynamic structural systems: Part II-Applications, *J. Eng. Mech., ASCE,* 115, 2406-2426
47. Şafak, E. (1991). Identification of linear stuructures using discrete-time filters, *J. Struct. Eng., ASCE,* 117, 3046-3085.
48. Şafak, E. (1993). Response of a 42-story steel-frame building to the Ms=7.1 Loma Prieta earthquake, *Engineering* Structures, 15, 403-421.
49. Şafak, E. (1995). Detection and identification of soil-sturucture interaction in buildings from vibration recordings, *J.Struct.* Eng., ASCE, 121,
50. Şafak, E. (1997a). Models and methods to charactarize site amplification from a pair of records, *Earthquake Spectra, EERI,* 13, pp,97-129.
51. Şafak, E. (1997*b*) New directions in seismic monitoring of multi-story buildings, in *Structural Health Monitoring: Current Status and Perspectives* (Ed. by F. - K. Chang). Technomic Publising Co. Inc., Lancaster, PA., 418-429.
52. Şafak, E. and M. Çelebi (1990*a*). "Method to estimate center of rigidity of a building using vibration recoordings", J. Struct. Eng., ASCE. 116,85 97
53. Şafak, E. and M. Celebi (1990b). New techniques in record analyses: Torsional vibrations, *Proc.* 4th *U. S. Nat. Conf on Earthq. Eng., EERİ, 2,*411-420
54. Şafak,E and M. Celebi (1991). Seismic response of Transamerica Building:Part II - System identification, *J.Struct.* Eng., ASCE, 117, 2405-2425.
55. Şafak, E.and M. Çelebi (1992). Recorded seismic response of Pasific Park Plaza: Part II - System identification, *J.Struct. Eng., ASCE,* 118, 1566-1589.
56. Schoukens, J. And Pintelon R. (1991). *Identification of Linear Sytlems: A Practical* Guideline *for Acccurate Modeling.* Pergamon Press, London England.
57. Sinha, N.K. and Kuszta, B. (I993). *Modeling and Identification of Dynamic Systems.* Von Nostrand Reinhold Co. Ioc., New York, NY
58. Söderström. T. and Stoica (1989). *System ldentification.* Prentice-Hall, New York, NY .
59. Strang, G. md Nguyen, T. (1996). *Wavelets amd Filter* Banks, Wellesley-Cambridge Press. Wellesley, MA
60. Sues, R.H., S.T. Mau, and Y.- K. Wen (1998). System identification of degrading hysteretic restoring forces, J. *Eng. Mech.,* ASCE, 114, 833-846.
61. Udwadia, F.E.and N. Jerath (1980).Time variations *of* structural properties during strong ground shaking. *J.*

Eng. Mech. ASCE, 106, 111-121.

62. Werner, S. D., J.L. Beck and M, B. Levine (1987) Seismic ıesponse evaluation *of* Meloland Road Overpass using 1979 Imperial Valley earthquake records. *Earthq.* Engng and *Struct.* Dyn., 15. 749.274.
63. Wu, X, J. Ghaboussi, and J.H.Garrett (1992). Use of neural networks in detection of structural damage, *Computer & Structures,* 42, 649-659.
64. Y, P.C. (1984). *Recursive* Estimation *and Time-Series Analysis.* Sprınger-Verlag, New York, NY

Zeldin, B.A. and P.D. Spanos (1998). Spectral identification of nonlinear structural systems, J. Eng. *Mech.,* ASCE, 124, 728-733.

STRUCTURAL MONITORING OF BRIDGES - AN OVERVIEW

R.T. SEVERN
Earthquake Engineering Research Centre
University of Bristol, Bristol, BS8 lTR, UK

Abstract
The author's experience is used to describe two kinds of monitoring of suspension and cable-stayed bridges. These include the Humber and Bosporus suspension bridges and the Second Severn cable-stayed bridge. Special attention is paid to a very detailed study at Humber of correlation between wind speed/direction and response. Use of a new video vision system for measuring dynamic displacement is described. The importance of wind speed/direction and rain is noted for cable-stayed bridges, as well as cable/deck interaction. Mention is made of monitoring studies on highway bridges.

1. Introduction

In the design process of structures for static loads, it can be argued that analysis does not now need the support of observed behaviour, either of prototypes or models. Methods are well established, and proven software allows parameters to be modified so easily that the best solution can be selected from a large number of options. The situation becomes quite different when dynamic loads are involved. Here, there continues to be a mutual dependency between calculation and measurement In fact, dynamic response calculations cannot proceed without knowledge of the damping characteristics of the structure, and these can only be obtained by measurement. It is also useful to have experimentally obtained values of natural frequencies and modal shapes to compare with computed values, because it is dynamic material properties which are being used to obtain the calculated values, and these are usually found by small samples tests on individual structural components, not on the as-built structure.

In the ideal case of dynamic design, calculation and measurement would go forward together, and this is possible in several fields of engineering (eg: aircraft, automobiles, switchgear) where early versions of the product can be designed, tested and modified in an iterative way, before the final version is manufactured in quantity. But this is not so for the great majority of civil engineering structures. Here, the contribution from measurement has been achieved previously by tests on similar structures and careful interpretation of the results for the dynamic design of the new structure. The importance of such interpretation cannot be over-emphasised. If the experiments are made at full-scale, it is unlikely that the necessary input forces can be applied having the correct distribution, whereas if models are to be used, scaling laws relating to material properties and other parameters cannot be fully satisfied, and here again the correct pattern of input forces can rarely be applied. This last problem becomes even more uncertain when the behaviour of the model becomes non-linear - a usual requirement in earthquake engineering. In collecting and processing data

M. Erdik et al. (eds.), Strong Motion Instrumentation for Civil Engineering Structures, 109–130.

from dynamic tests there are many opportunities for procedures leading to false conclusions, even when the more obvious of them are avoided.

This paper is concerned with structural monitoring of the dynamic response of bridges of various types , and the valuable lessons which can be learned from such monitoring. Because wind and traffic are more reliable exciting agencies than earthquakes, they are used as input in many studies, but the natural frequencies, modal shapes and damping which can be obtained, are, of course, essential ingredients in any seismic design process. In addition to these structural characteristics, the paper describes new methods of measuring displacement response of structures, which becomes of increasing value now that displacement-based approaches to design are being actively promoted. To a large extent the illustrations used are taken from the author's own experience but many references are given of similar studies made by other researchers.

2. Suspension Bridges

2.1 INTRODUCTION

During the last 20 years increase in computer power has made possible theoretical studies in earthquake engineering for asynchronous and multiple-support ground input (1), and it is likely that this is the reason why Eurocode8 now requires asynchronous ground input to be considered for highway bridges greater than 200m in length where there are discontinuities in the ground, and for all bridges over 600m notwithstanding ground conditions. The term asynchronous is properly reserved for structural ground support points receiving the same input - but with a time difference, whereas multiple support input refers to completely different inputs at the ground support points. This is the more usual practical occurrence since earthquake motion is modified by the ground through which it passes. It is clear that long-span bridges are structures in which the effects could be important, because at a speed of 250 m/s, say, the input would take almost 6 sec. to travel the 1410m distance between the two towers of the Humber bridge (Fig. 1), which must be considered in relation to its fundamental natural period in the vertical direction which is about 9 sec.

Increased computer power has also made possible studies aimed at producing stochastic parameters for long-span bridges (2), but both these and multiple-support studies require not only a very large number of degrees of freedom in the finite element modelling, but also require several major assumptions to be made, particularly in the modelling of the box-deck. To validate such calculations scale model studies are not feasible for such complex structures, and recourse must be had to full-scale testing, which brings with it knowledge of real damping values, even though these may not be completely valid for actual seismic input forces.

2.2 HUMBER BRIDGE: ACCELERATION MEASUREMENTS (3)

At the time of testing (1986) Humber had the longest main-span length of all suspension bridges (1410m) and two unequal side-spans of 530m and 280m. It has a steel box deck, concrete towers and inclined hangers which are articulated about the lateral (x) axis in all spans; additionally, for a considerable distance either side of the mid-point, there is articulation about the longitudinal (z) axis. Response to forces produced by a set of 4 eccentric-mass exciters was considered, but these produced too little force at the low frequencies being studied (0.1-0.3Hz), and so the ambient forces of wind and traffic were used.

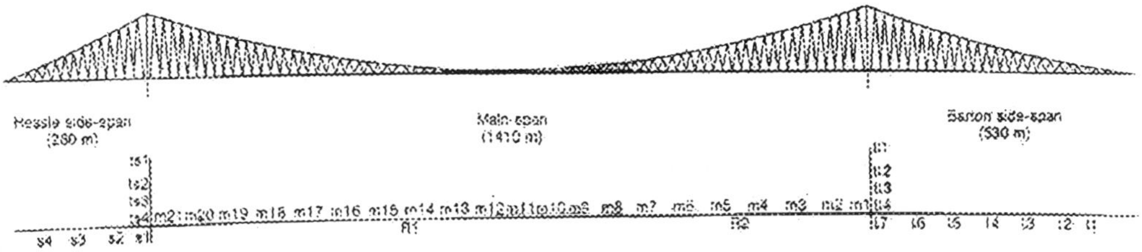

Figure 1. Humber Bridge,showing measurement station

It is relevant to note that three different finite element models were produced. First, one in the vertical (yz) plane having y and z modal displacement and rotation about the x- axis. Second, one in the horizontal (zx) plane having z and x nodal displacements and rotation about the y-axis. Neither of these models is able to produce torsional modes, and for this purpose a third, fully 3-dimensional, model was produced having 6 nodal variables. Two results of importance came out of the many calculations. First, convergence tests indicated that at least 40 modes must be considered because tower and deck frequencies are well-spaced. Second, modes in both the vertical and horizontal planes have an appreciable longitudinal component.

The measurement stations at which accelerometers were placed are shown in Fig.1, enabling full mode-shape mapping to be obtained for main span, side-span decks, and towers. Stationary reference accelerometers were located at Rl and R2, whilst at other stations a number of accelerometers were located in turn, a sufficient number to measure all translational motions and principal rotation. For example, in the deck, if vertical acceleration was being measured, the average from two accelerometers at the extreme edges gave vertical acceleration, whilst the difference gave torsion. The measurement period at each station was around 2 hours, the whole monitoring programme taking 10 days, requiring many months of data processing. It was possible to isolate a large number of modes (3); for example, in the main span 25 predominantly vertical, and 23 lateral modes were found, 7 in each case having a substantial torsional component. Fig. 2 gives the first 8 predominantly vertical modes, both measured and calculated, together with modal damping valves. The design conditions allow movement in the longitudinal direction at the deck bearings, and if the calculations permit this, the first vertical mode is antisymmetrical in shape, with a frequency of 0.108Hz, compared to the first measured mode which is symmetrical at 0.1165 Hz.

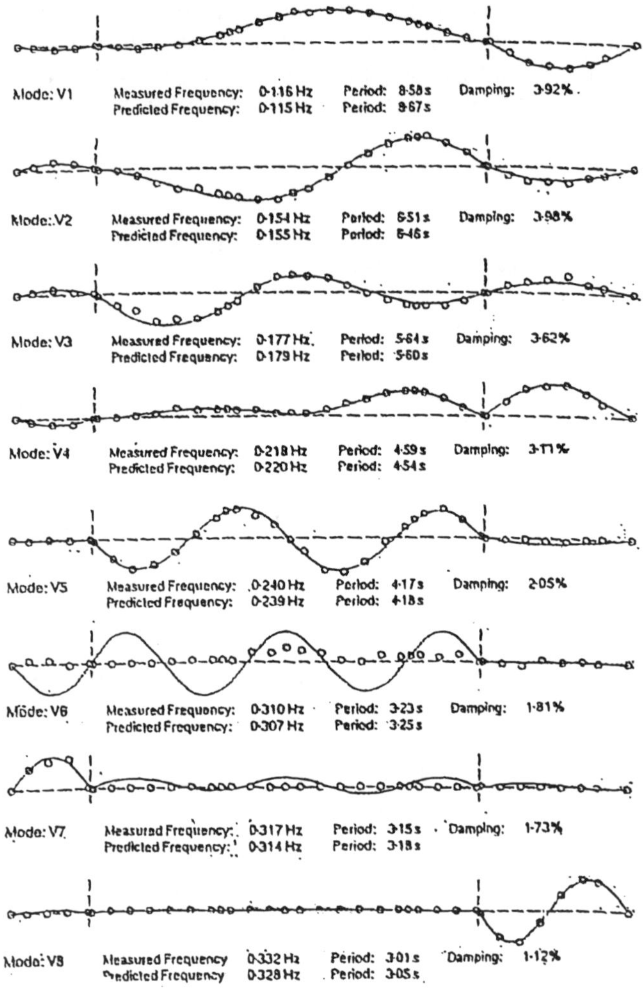

Figure 2. Humber Bridge:First 8 measured and predicted vertical modes

However, if the longitudinal movement at one bearing is restrained in the calculation, leading to a hinged condition there, the first vertical mode becomes symmetrical at 0.115 Hz, which is in excellent agreement with measurements. These are the end conditions assumed in Fig.2, it being assummed that the light loading during the test conditions was insufficient to overcome stiction at one bearing. Regarding damping, 1- 4% covered the 26 vertical models, 1-9% the lateral modes and all torsional modes had a value close to 1%.
damping, 1-4% covered the 26 vertical models, 1-9% the lateral modes and all torsional modes had a value close to 1%.

2.3 BOSPORUS BRIDGE: ACCELERATION MEASUREMENTS (4)

This was chosen as the second bridge to test because although also of `modern' design, it is different from Humber in several respects. Humber has concrete towers, a relatively narrow but deep box deck of high torsional stiffness; both side spans are carried by the cable but one is twice as long as the other; near centre span the hangers are doubly articulated. Bosporus has steel towers, a wide shallow deck, side spans nearly equal in length but which are supported on columns rather than by the cables, and singly articulated hangers throughout. A further, non-structural, but important, difference from the testing aspect is that whereas traffic at Humber is very modest, Bosporus was saturated by more than 140,000 vehicles per day, many being heavy lorries, resulting in the live load being a not insignificant proportion of the dead load.

These Bosporus tests throw further light on the dependence of the fundamental vertical mode on the precise deck support conditions. In calculation it is antisymmetrical if both ends are free to slide, but symmetric if one end is hinged. This difference appears to result from inclined hangers, because it disappears if vertical hangers are used. At Bosporus, the first measured vertical mode was antisymmetric, indicating sliding bearings at both ends, caused physically by the very heavy traffic loading. During testing, both bearings could be seen to be in frequent jarring motion of around 50mm in amplitude. Regarding damping, at Bosporus there was an increase with frequency, an opposite observation to that at Humber.

2.4 SECOND BOSPORUS BRIDGE (FATİH SULTAN MEHMET): ACCELERATION MEASUREMENTS (5).

The interesting features of this bridge, compared to the previous two, is that it has vertical hangers, no side spans and a wide shallow deck carrying 8 traffic lanes. An unusual arrangement of rocker and expansion bearings at the check-tower connection is intended to permit longitudinal motion of the deck, and rotation about a lateral axis, but to restrict other translations and rotations.

The objectives of the tests, as for Humber and Bosporus, was to build up a picture of vertical and lateral mode shapes of the deck, and the longitudinal and lateral mode shapes of the towers, as well as damping ratios. From a 10-day period of measurement it was concluded that the vertical modes matched those predicted in a model in which the deck was fixed at one end and sliding at the other. This was so at Humber, but not at Bosporus, a possible cause being lack of traffic at Fatih during the tests. The vertical modes of the deck are clearly defined; the extreme width of the deck is a possible cause of this. It is noted that Brownjohn (23) has used the results from Humber and the two Bosporus bridges to investigate non-linear dynamic characteristics of such bridges.

Most of the published studies of measurements on actual suspension bridges have been those of the more `classical' type, reference to which is made in Ref. 5 of this paper. Somewhat different in nature is the study by Higashihara et al (6) concerning the use of microtremors in ambient vibration tests of a suspension bridge anchorage, so slender as to offer the possibility of resonance between it and the neighbouring tower. One valuable outcome of these tests was that the elastic modulus appeared to be 5 times that obtained from loading tests in boreholes. It is noted that recently completed long-span suspension bridges in Hong Kong (Tsing Ma) and Kobe are fully instrumented and should produce valuable information in the near future.

3. Computer Vision System For Prototype Displacement Measurements.

3.1 INTRODUCTION

The direct measurement of displacements of long span bridges presents practical difficulties, particular at centre span where vertical and lateral displacements are large, often more than lm. Such movements are a combination of dynamic modal response and quasistatic, or sub-modal response due to slowly varying loads of traffic and wind. The lack of suitable stable locations for use as datum precludes the use of contacting displacement transducers, such as pull-wire potentiometric devises. But a knowledge of response displacements is crucial for many aspects of the design process, and this is certain to increase in the near future as displacement-based, rather than force-based, methods of design are introduced.

A limited amount of displacement information can be derived from acceleration measurements as indicated in Fig. 17. Vertical (y) and lateral (x) displacements can be obtained by double integration of time or frequency domain information, the latter being equivalent to simply dividing amplitude spectral valves by the square of the angular frequency (w). For that part of the signal corresponding to low frequencies (w→o) this can result in unrealistically large values. For time histories, digital high-pass filtering is necessary to remove low frequency information and prevent instabilities occurring in the baseline correction algorithm. This is a particular problem for lateral motion, which has significant components at very low frequencies.

Conventional surveying techniques do of course provide a method of displacement measurement, but only for sparsely distributed static valves. For dynamic measurements alternative approaches are necessary, and a number have been tried. For example, at the Tagus Suspension bridge (7) telescopes controlled by servo-motors were used to track infra-red lamps on the main span of the bridge, while at the CN tower in Toronto (8) a servo-controlled tracking carriage at an upper level was used to track a laser beam pointing veitically upwards from ground level. A system used at the Forth bridge (9) located the brightest part of a video image which was arranged to be a light source. At Humber, a global positioning system (GPS) employing navigational satellites has been used (10). In principle the system has an accuracy of ± lmm horizontally and ± 3 mm vertically, but in the reported study the GPS sensor was placed at the top of a pole, the wind-induced vibration of which was itself of the order 1-2 mm, so that the overall accuracy was reduced to this level. All these systems are specific in their application, requiring separate instruments for each object being tracked, whereas a system (11) developed by ISMES (Bergamo, Italy) uses a telescope sighted on a linear image of high contrast (Fig. 3) together with image intensity

Figure 3. Humber Bridge: Vision sysrtem targets for ISMES and EERC instruments

thresholding software to determine a single displacement coordinate. Laser vibrometers are commercially available but they have safety considerations, are restricted in range, and work in the longitudinal axis only, making them of limited use for vertical and lateral movements of bridges.

3.2 THE EERC BRISTOL COMPUTER VISION SYSTEM (12,13,14)

The EERC vision system (Fig.4) uses a UK standard video camera to produce pictures at the rate of 25 per sec., digitising each one into a rectangular grid of picture elements (pixels). Each pixel carries a digital value of its greyness, from 0 for black and 255 for white. A specific object in the frame will exhibit its own particular pattern of pixel values, and if this pattern has been pre-set in the system, computer algorithms can be used to locate where it is in the frame; that is, the two-dimensional coordinates of its centre can be obtained. An advantage of the EERC system is that several targets can have their pixel pattern pre-set and their displacements monitored so long as they appear in the field of view. The focal length of the lens used in the video camera depends on the application; for measurements on long-span bridges it is typically 1.6m, but in the laboratory measurement of models, lenses of 6 mm and 75 mm have been used.

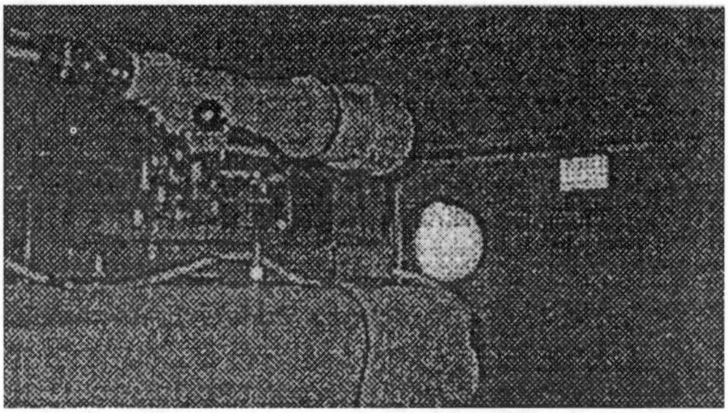

Figure 4. The EERCvision system

The system allows selection from a variety of algorithms for the various tasks which it performs; in general, there is a trade-off between speed and accuracy. With a limited time to perform calculations between frames, complex calculations may take too long and will result in the following frame being missed. This gives a reduced frequency of samples or possibly loss of tracking. Simpler calculations will reduce accuracy, but will cause fewer frames to be missed and reduce the chance of losing track. If tracking is lost, the system stops giving an output and starts a search from the last location of the target, spiralling outwards at increasingly coarse intervals until the target is re-located. Because the whole frame occupies 256 x 256 pixels and is refreshed every 0.04 sec, the precise manner and speed with which the target template is located is a key issue. It clearly does not make sense to scan the whole frame each time because the motion being tracked can be expected to follow a reasonably smooth trajectory, and having been initially located, its succeeding positions are capable of some measure of prediction. For long-span bridges, polynomial prediction based upon the previous few locations has been found satisfactory.

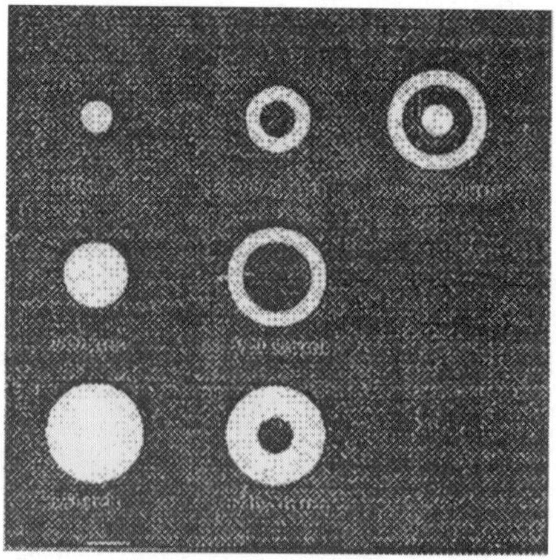

Figure 5. The EERC vision system: a selection of possible targets

For the targets, circular patterns have been used, with contrasting black and white rings as indicated in Fig. 5. The accuracy of measurement depends on the number of pixels on boundaries because they are sensitive to small displacements, whereas pixels within large uniform areas are not. But if too many rings are used, black and white areas can become confused with each other. In practice a single ring has normally been used, but for higher velocities a solid circle has been found more reliable.

Current research development of the EERC vision systems includes stereoscopic viewing of a target by two cameras, tracking of several targets in each view, sub-pixel interpolation of target positions, and re-location of lost targets.

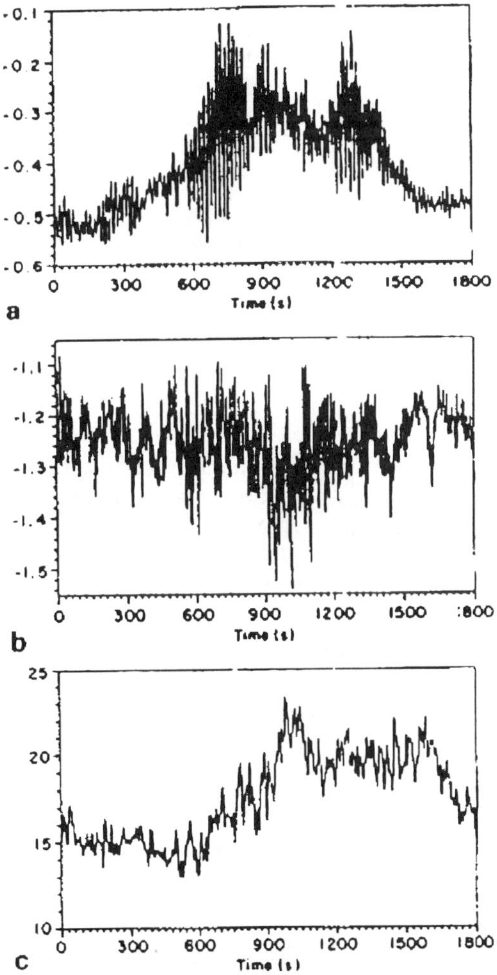

*Figure 6.*The EERC vision system:Humber bridge mid-span Displacements (a) lateral (b) vertical (c)normal component of wind speed

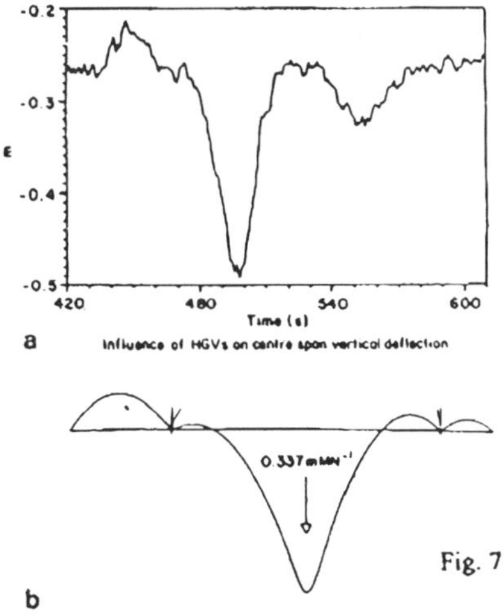

Fig. 7

Figure 7:Humber bridge:Influence lines for mid-span vertical deflection (a)measured (b)calculated gives a sample of the data obtained at the mid-span of the deck, (a) and (b) being lateral and vertical displacements, respectively, whilst (c) is the normal component of wind at mid-span. It is seen that the displacement of the deck has a major oscillatory component superimposed on a low frequency drift, and from such data it is possible to abstract the lateral and vertical natural frequencies given in Table 1 and to compare them with corresponding values obtained from the modal survey using accelerometers. Because of the predictable, low-frequency nature of the recorded motion it was possible to reduce the tracking and computational tasks by working only on every sixth frame, which means a sampling rate of 4.17 Hz and a corresponding Nyquist frequency of 2.08 Hz which, being much higher than any of the relevant frequencies of the bridge, causes no aliasing problems.

Table 1. Humber bridge deck vibration modes: comparison of values obtained from the vision system and the modal survey.

	Frequency from displacment spectrum (vision system) (Hz)	Vibration mode (from modal survey[19]) (Hz)
Lateral	0.054 0.247 0.309	L1 0.056 L6 0.260 T1 0.311
Vertical	0.115 0.158 0.176 0.309	V1 0.116 V2 0.154 V3 0.177 V6 0.310 T1 0.311

All displacement measurements were made at centre-span and Fig. 7a gives the variation of the vertical component as a single heavy vehicle crosses the bridge whilst the wind was very light. By definition, an influence line is a deflection at a point in a structure as a unit load passes across it, and Fig. 7b is such for the Humber bridge; the correspondence with the measured influence line is pleasing, even though the latter does contain some contributions from the wind. Fig. 8 presents information which has a very important design significance since it can be used to provide measuced lift (Cc) and drag (C_D) coefficients for comparison with those used in design or from wind-tunnel model studies. Presented there are lateral and vertical displacements plotted against wind velocity normal to the span. The coefficients are defined by

$$C_L = \frac{2K_y}{\rho B}\left(\frac{y}{V^2}\right) \quad \text{and } C_D = \frac{2K_x}{\rho B}\left(\frac{x}{V^2}\right)$$

where B is the deck width, ρ the air density, and K_y and K_X are stiffness coefficients obtained from the finite element model of the bridge. The bracketed terms are obtained from curves fitted to the data of Fig. 8. From this combination of calculated and measured values $C_L = 0.35$ and $C_D = 0.067$, which can be compared witti corresponding wind-tunnel values of 0.26 and 0.078, respectively. The rotation about the x-axis at mid-span can also be plotted against Vcosθ (15), and again using the fuiite element model to obtain the rotational stiffness (K_Φ), the aerodynamic moment coefficient can be obtained as $C_M = 0.6$. For these displacement measurements at Humber the vision system used a 1.6m focal length lens, measuring from the rigid base of the Hessle pier to mid-space, a distance of 705m. A comparative analysis of similar measurements using the ISMES system indicated that an accuracy of around 2mm was obtained by the EERC system over this distance using sub-pixel interpolation .

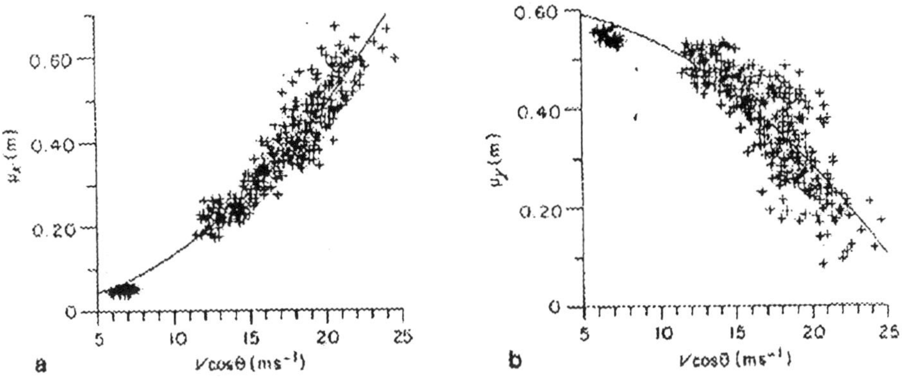

*Figure 8.*Correlation of 64s average displacements with wind speed
(a) μ lateral displacement $\mu x=0.0018\ (v \cos\theta)^2$
(b) μ vertical displacement $\mu y= -\ 0.00081\ (v \cos\theta)^2$

The practical value of the Humber displacement measurements was taken a stage further by Larose et al (16), not only for corroboration of wind tunnel tests but also as a means of validating the design procedure known as Assumption (QSA). The actual deck cross-section used was not that of Humber but was sufficiently similar to make the comparisons the Humber data, in which r.m.s. response, normalised by $Bw^2{}_j$ is shown to be proportional to the reduced velocity to the power 2 in vertical direction and 2.36 laterally. In Fig. 10 the Humber vertical measurements are compared with wind-tunnel studies for four different turbulence intensity (I) values.Going a stage further,the QSA predicts that if the normalising factor include I, the scatter of fig.10 will be removed; Fig. 11 shows that this is so. Equally good corroboration is obtained for the lateral direction.

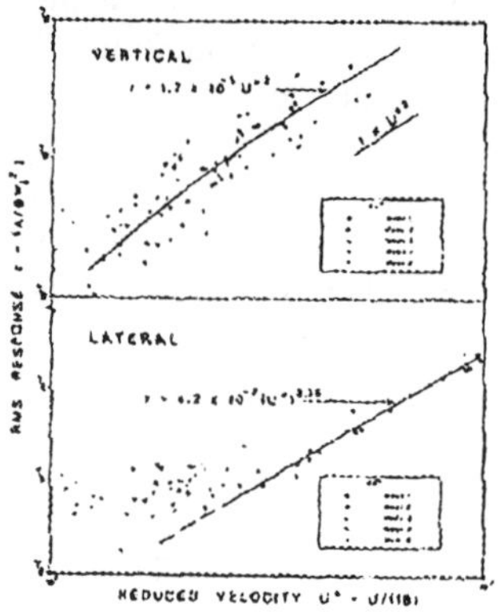

Figure 9. Humber Bridge:variation of modal responses with mean wind speeed

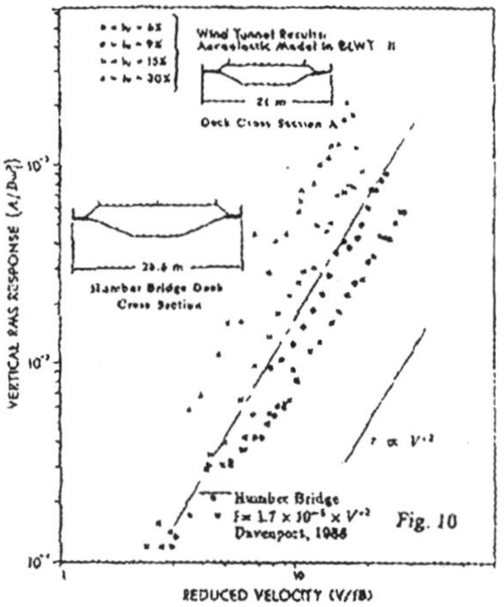

Humber bridge: measured vertical displacement response compar

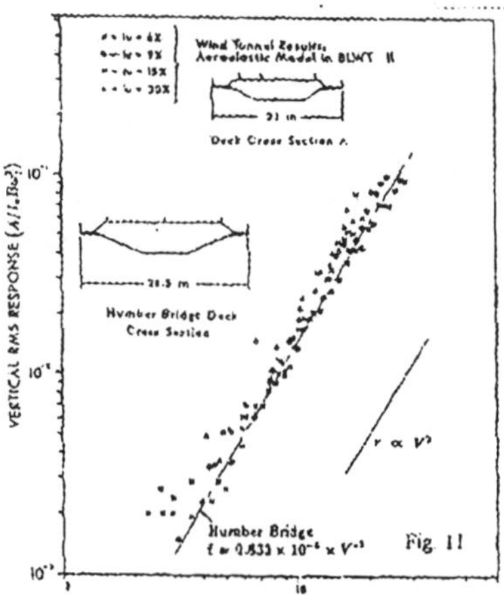

Figure 11. Humber Bridge :the data of *Figure 10.* normalised by turbulence intensity

3.3 HUMBER BRIDGE: STRETTO DI MESSINA STUDIES

The EERC programme just described attracted the attention of Stretto di Messina Spa, who employed ISMES and the Politecnico di Milano to collaborate with EERC in a detailed and extensive measurement programme aimed at supporting design and laboratory studies for the proposed Straits of Messina crossing. Fig. 12 shows that more than 70 transducers of various types were deployed, including 13 anemometers to measure spatial, as well as temporal, distribution of wind. The total programme covered 4400 hours of monitoring and a data segment length of 512s was chosen for analysis. This is longer than any period of oscillation, or the time taken for a vehicle to cross the bridge, and long enough to provide good resolution in the frequency domain.

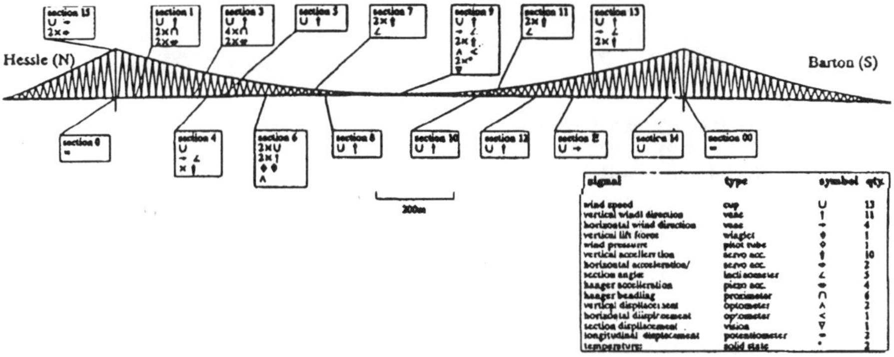

*Figure12.*Humber Bridge:instrumentation used by EERC and Politecnico di Milano

Many results of great practical value were obtained (15), only two of which will be discussed here. First, is the effect of temperature on mid-span deflection. Fig. 13a is a scatter diagram of mid-span lateral deck deflection and E. main cable mid-span temperature for a period of 24 days. The relationship is not clear because of the variable effect of the wind, but if the data is confined only to those for a wind speed of 3-5 m/s, Fig. 13b results, showing a clear linear relationship most likely due to unequal solar heating of the two sides of the deck. A linear temperature/displacement coefficient was derived from this plot, similarly for vertical and longitudinal responses.

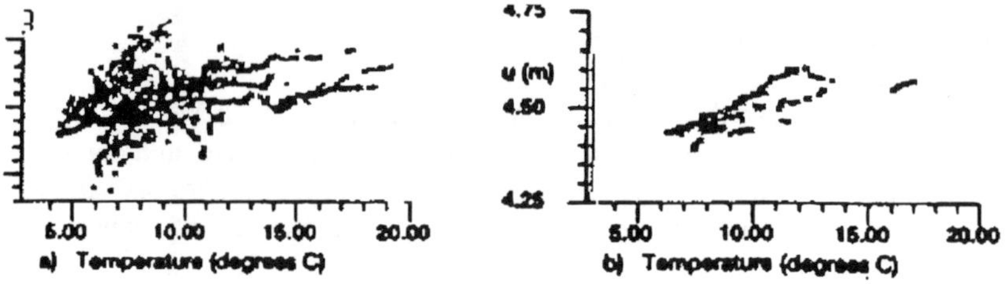

*Figure 13.*Humber bridge:main span lateral deflection against temperature;
(a) all wind speed (b) low wind speed only

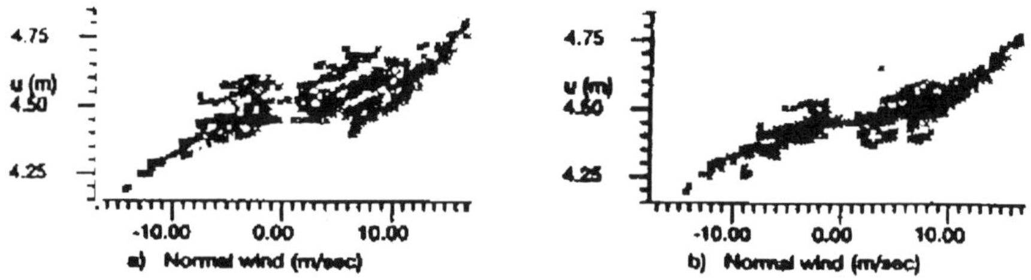

*Figure 14.*Humber bridge:main span lateral deflection (a) without and
(b) with compensation for temperature effects

Fig. 14a shows the mid-span lateral deflection and normal wind component (Vcosθ) for the same 24-day period. Using the coefficient from Fig. 13b, and knowing the temperature corresponding to data points in Fig. 14a the plotted data can be adjusted to a common (datum) temperature; this is shown in Fig. 14b. The second sample of results derived from this very extensive monitoring programme is shown in Fig. 15. For The test procedure used at SSC followed that established at Humber, having a combination of accelerometers and the video computer displacement tracking system, the latter embodying the results of research since 1991, particularly the multiplexing of alternate image fields from two cameras, stereoscopic three-dimensional tracking and full synchronisation with the rest of the acquisition system (Fig. 16). The cameras used lenses of 800 mm focal length, giving an accuracy of 0.5 mm at a distance of 220m. Regarding modal characteristics, modes up to 5 Hz could be measured and identified, but only those below 1Hz contributed significantly to displacements. The frequency of the first mode was always within 25% of 0.3Hz during construction, due to mass and stiffness increasing at much the same rates. Table 2 gives a summary of these measurements. Fig. 17 indicates the kind of agreement obtainable between direct displacement measurements and those obtained by integration of acceleration with suitable baseline correction. At the SSC good agreement could be obtained for periods up to 10 sec.

dynamic response the correlation of modal rms, frequency and amplitude with wind are derived from the FFT of the acceleration response, whilst damping and resonance frequency are estimated by fitting a single degree of freedom oscillator response curve to the data. In Fig. 15a the amplitude of the first vertical mode is plotted against normal wind speed, and Fig. 15b shows the corresponding damping. Fig. 15c gives the first torsional frequency against normal wind speed.

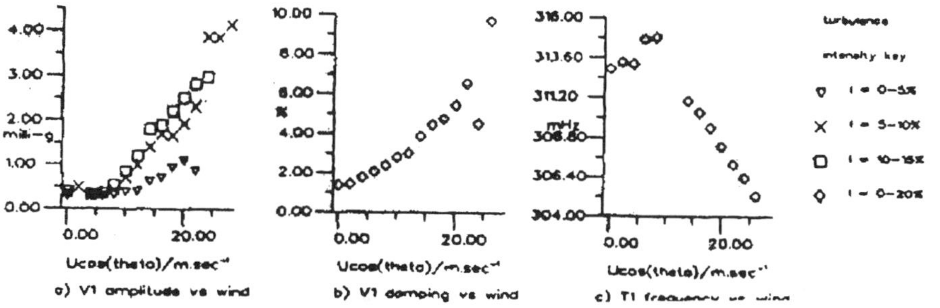

*Figure. 15.*Humber bridge: modal parameters versus wind speed for the lowest vertical (vi) and torsional modes.

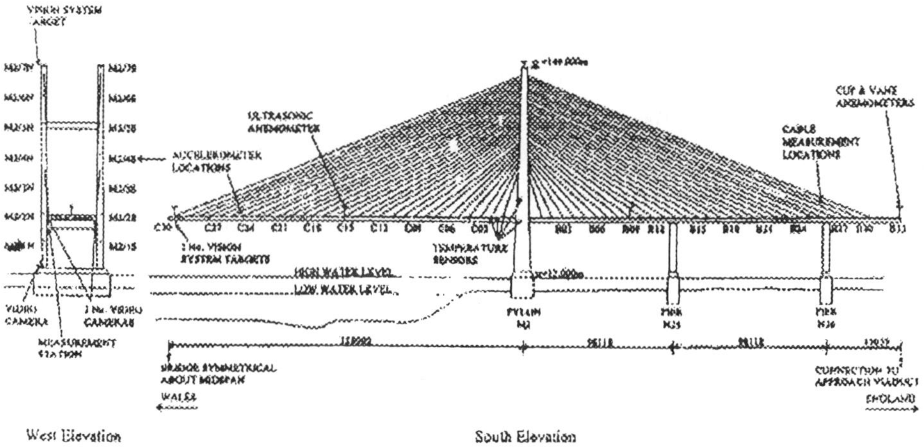

*Figure 16.*Second Severn crossing: arrangement of instrumentation on English side cantilever

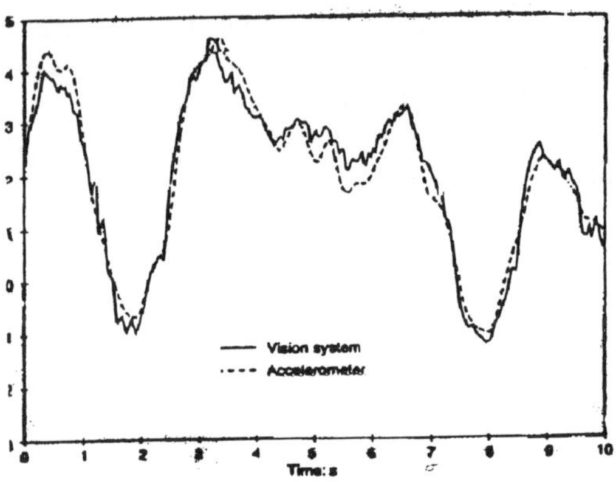

Figure 17. Second Severn Crossing: Vertical deck displacement from computer vision system and from inteerated acceleration

Table 2. Second Severn crossing:comparison of natural frequencies obtained by the different approaches

Mode no.	Principal component	Other components	Measured frequency: Hz	Finite element model		Wind tunnel tests	
				Frequency: Hz	% error	Frequency: Hz	% error
1	Deck lateral	Deck torsional Pylon lateral	0·34	0·34	0	0·313	-8
2	Deck vertical	Pylon longitudinal	0·35	0·38	+9	0·373	+7
3	Pylon torsional	—	0·45	0·59	+31	—	—
4	Deck vertical	—	0·52	0·54	+4	0·395	+14
5	Deck lateral	Pylon torsional	0·56	0·45	-20	—	—
6a	Deck torsional		0·56	0·59	+5	0·512	-9
6b		Pylon lateral				0·477	-15
7	Deck vertical	Pylon longitudinal	0·61	0·66	+8	0·906	+49

4. Cable Stayed Bridges

4.1 SECOND SEVERN CROSSING (13)

The Second Severn Crossing (SSC) near Bristol UK comprises two viaducts, each more than 2 km in length, and a central cable-stayed bridge of 948m, with a main span of 456m. The construction sequence was governed by time to construct the viaducts, and so the bridge was constructed in two halves sequentially. The English half (Fig. 16) therefore existed as a lone cantilever for several months in 1995, giving an unusual opportunity for monitoring a cable-stayed structure in this potentiatly more vulnerable state, but monitoring was also carried out on the completed bridge, and is, in fact, an ongoing activity for the next few years.

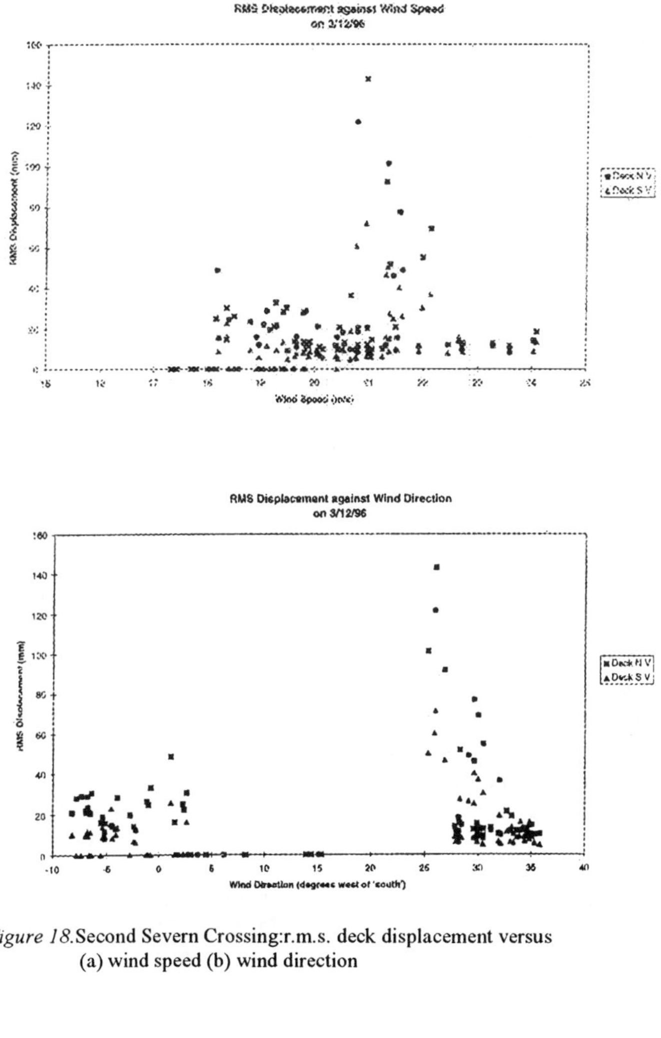

*Figure 18.*Second Severn Crossing:r.m.s. deck displacement versus (a) wind speed (b) wind direction

It was observed that large amplitude (lm) vibrations in the vertical plane occurred in many of the cables for wind speeds above 13m/s at a direction approximately 45° to the bridge normal, but only when wind was combined with rain. Similar behaviour has been observed at several other cable-stayed bridges (Fig.17). The natural frequencies of these individual cables and of the complete structure are in the same range, so that interaction between them is significant, and it was concluded that the principal excitation of the deck arose from this interaction, rather than from direct wind excitation of the deck. Figs 18(a) and (b) indicate the dependence of response on wind speed and direction (combined with rain), the former recording speed and the latter direction, for both the North (approximately) and South sides of the deck. It is important to note that although wind-tunnel tests gave adequate predictions for responses of the deck and pylons alone, the interactive responses between these and the cables could not be obtained because of the impossibility of satisfying all similitude requirements. The large cable displacements was a consequence of lower than expected damping values in the cables, measured both without and with the corrosion prevention wax; secondary cables, approximately at right-angles to the main cables have therefore been introduced connecting the main cables together. In two respects therefore, the monitoring studies on the SSC have been an example of where design has been modified during construction as a result of the measurements made, and a second phased wind tunnel tests were made to corroborate prototype findings.

4.2 KESSOCK BRIDGE (18)

The 240 m main span of this cable-stayed bridge, near Inverness, Scotland, experienced large oscillations in 1982, shortly after opening, in wind speeds of 22 m/s. Tuned mass dampers were installed, but further large oscillations occurred in 1988, and inspection showed that the dampers had experienced vibration beyond their capacity. The large oscillations had been predicted by wind tunnel tests as being due to vortex shedding.

The EERC became involved in 1991, installing accelerometers and anemometers for telemetric monitoring from Bristol over a period of 8 months. The results were not unexpected. Ttere were several periods when the wind speed reacted 22 m/s, but on only 3 occasions did large amplitude responses occur. In the first, in an E wind (I = 6%) the response was concentrated in the first vertical mode (0.5Hz), but it was so large as to be outside the range of the accelerometers and the peak-to-peak displacement of the deck could only be estimated at 200 mm. The second and third periods of large response occurred in a W. wind, the former (I = 10%) was again largely in the first vertical mode, but the latter (I = 14%) was mainly the first torsional mode but with a significant contribution from the second vertical mode. This severe coupled motion caused traffic restrictions to be enforced. Vortex excitation of the first torsional mode would occur at about 33 m/s, and this speed was certainly reached but occurred for only a short time. However, classical flutter seems unlikely because the two natural frequencies are well separated and wind tunnel tests predicted flutter at 61.5 m/s, a wind speed far higher than any measured. In view of the importance of rain accompanying wind, discovered later at the Second Severn Crossing, it was unfortunate that the incidence of rain was not recorded at Kessock.

4.3 OTHER CABLE-STAYED BRIDGE STUDIES

Valuable results relating to the Higashi-Kobe bridge have been reported by Ganev et al (19). This 485m main span cable-stayed bridge was instrumented by 3 accelerometers at different heights in the towers and 2 more buried in the ground close to the tower at depths of 1.5m and 35m; all were activated by the Kobe (1995) earthquake. It was found that the response of the soil-structure system was strongly influenced by pore pressure increase in the saturated surface layers. Non-linear effective stress analysis combining the Ramberg-Osgood stress-strain relation with a pore pressure model produced excellent agreement with observed behaviour.

Other recent measurements on cable stayed bridges have been carried out by Casas (20) on the very unusual Alamllo bridge in Seville, Spain; by Gentile and Cabrera (21) on a repaired bridge in Italy, and by Wilson and Lui (22) on the Quincy Bayview bridge in Illinois.

5. Hıghway Brıdges

5.1 INTRODUCTION

There can be no dispute that the most valuable form of dynamic monitoring occurs when an instrumented structure is subjected to strong winds or earthquakes. For suspension and cable-stayed bridges there have until recently been very few that have experienced on earthquake and so wind has produced the bulk of our information. For highway bridges, however, instrumentation programmes have been commonplace during the last 30 years in seismic regions and many valuable results have already been obtained. It is argued that bridges built prior to the middle of this century suffered damage in earthquakes due to sub-structure failure, but in the 1971 San Fernando earthquake, for example, where 62 bridges in the epicentral region were damaged, structural vibration was the principal cause of failure, and that veıtical motion was an important factor.

As far as this paper is concerned, the monitoring of highway bridges introduces no new methods or principles. The aim is to obtain natural frequency and damping values together with modal shapes. Where displacements are required, they are usually relative, rather than absolute, values, as, for example, between a bridge beam and its seating, and these can be measured with many simple devices. The literature contains many reports of measured seismic response characteristics being correlated with damage, and also with predictions at the design stage. A useful general summary of the overall problem is by Paultre et al (24).

5.2 OBSERVED SEISMIC RESPONSE

Modern developments probably start with the paper by Wilson (25) describing the observed seismic response of an irregular 6-span highway bridge in California which was designed with minimal seismic resistance, but which experienced the moderate 1979 coyote Lake earthquake without damage, probably because it responded with locked expansion joints.

Higher intensity motion would probably have released the joints, resulting in larger response and damage.

During the 1979 Imperial Valley earthquake it was fortunate that an array of strong- motion accelerometers produced records in the Meloland Road Overpass, a 2-span RC bridge located only O.5 km from the fault. These records were used by Werner et al (26) with a system identification methodology to obtain the response characteristics of the bridge. Three valuable results came from this study. First, linear models provided an excellent fit to the measured motions, despite the O.5g maximum measured acceleration. Second, the transverse response is controlled by abutment motions, with no significant amplification in the deck. Third, the vertical response of the bridge deck at the two mid-spans is dominated by a single vertical mode, whereas above the central pier the vertical response is governed by the vertical response of the base of the pier coupled with deck torsion. The effect of the same bridge abutments on the overall response was also considered by Wilson and Tan (27).

The use of recorded motions, coupled with system identification methodology, to obtain seismic response characteristics is the basis of a more recent paper by Loh and Lee (29). It refers to the New-Lian River Bridge in Taiwan, a 5-span continuous prestressed box-girder box, 700m in length which is almost symmetrical (regular). Records were obtained for both a weak motion earthquake having maximum acceleration of 0.083g, and a strong motion event where the corresponding value was 0.689g. It is particularly noted that records were obtained by a total of 24 accelerometers disposed at pier bases and pier heads, the former allowing multiple- support system identification procedures to be used. Three significant conclusions were made from this study. First, agreeing with the Meloland Overpass (26) results, linear models provide an excellent fit to the measured motions for both seismic events, even though the two events produce significant differences in response. Second, the rigid- body rocking of the bridge pier during strong shaking is significant and cannot be ignored during system identification. Third, the transverse motion at mid-span is controlled by the quasi-static response induced by the boundary system, which is quite significant during strong motion. This last conclusion gives support to the Eurocode8 requirement that bridges over 600m in length should consider the effects of multiple- support input. It is an important conclusion arising, as it does, from a field monitoring study. Laboratory studies (30) have so far proved inconclusive, and theoretical research (31) has claimed to show that in the special case of non-synchronous (ie: inputs differing only in time) inputs, there is little effect in comparison with the single input case at all support points.

A large volume of valuable measurements were of course made during the 1994 Northridge earthquake, and again during the 1995 Kobe event. For the former, many reports were written immediately after the earthquake, but a more valuable study has recently (1998) been published by Basoz and Kiremidjian (32), in which observed damage is correlated with local ground motions and bridge structural characteristics.

5.3 OTHER STUDIES

Artificial excitation of large bridges is possibly by the use of hydraulic actuators, but their low fundamental frequencies mean that the generated input forces at resonance are small. Richardson and Douglas (33) solved the problem by using 5 hydraulic rams, applying forces at 45°to the vertical at chosen points on the Dominion Road overpass bridge in New Zealand. A quick-release mechanism resulted in free vibration of the bridge, from which well-defined lateral, longitudinal, vertical and torsional modes were identified for comparison with computed values for this curved box-girder (277m) bridge.

It is often suggested that dynamic response measurements on structures can be used to assess deterioration and/or damage if they are repeated at suitable intervals. Work by the author's group on full-size bridge beams (34) suggested that serious damage was necessary before appreciable changes in natural frequency occurred. A more recent contribution to this issue has been made by Sohn et al (35) who studied the effect of temperature on the modal parameters of the Alamosa Canyon bridge, a highway bridge of 7 independent spans carried by steel beams and a concrete deck. They observe that changes in frequency are linearly correlated with changes in temperature, but definitive conclusions were marred by heavy rain which considerably increased the mass of the bridge. In summary, environmental effects can mask changes caused by deterioration and damage.

References

NB:In the following references the abbreviation EESD has been used for the Journal Earthquake Engineering and Structural Dynamics

1. Dumanoglu, A.A and Sevem RT. (1990) Seismic Response of Modern Suspension Bridges to Asynchronous Longitudinal and Lateral Ground Motion. Proc. Instn. Civ. Engrs. Pad 2, 1989, 87, 73-86.
2. Dumanogtu, AA and Severn, R.T. (1990) Stochastic Response of Suspension Bridges to Euthquake Forces, EESD, 19, 133-152.
3. Brownjohn, J.M.W., Dumanoglu , AA, Severn, RT. and Taylor, C.A (1987) Ambient Vibration Measurements ofthe Humber Suspension Bridge and Comparison with Calculated Characteristics, Proc. Iustn. Civ. Engrs. Pt2, 83, 561-600.
4. Brownjohn , J.M.W., Dumanoglu, A.A., Severn, RT., and Blakeborough A. (1989) Ambient Vibration Survey ofthe Bosporus Suspension Bridge, EESD, 18, 263-283.
5. Brownjohn, J.M.W., Dumanoglu, A.A. and Severn , RT. (1992) Ambient Vibration Survey of the Fatih Sustan Mehmet (Second Bosphorus) Bridge, EESD, 21, 907-924.
6. Higashirbara, H. and Mariga, J. (1987) Ambient Vibration Test of an Anchorage of South Bisan- Seto Suspension Bridge, EESD, 15, 6790695.
7. Marecos, J, Castanheta, M. and Trigo, J.T. (1969) Field observations of Tagııs River Suspension Bridge, J. Stnıck Div. ASCE 95 (ST4~ 555-583.
8. Isyumor, N., Davenpoıt, A.G. and Monbaliu. J. (1984) CN Tower: Model and Full-scale Response to Wind. IABSE 12s' Congess, Vancouver, 3-7.
9. I,eitch, J.G., Thompson. A and Sloan, T.D., (1989) A Novel Dynamic Deflection Measurement System for Large Structures. Civ. Comp. Press, Edinburgh, Vol.2, 301-306.
10. Ashkenazi, V., Roberts, G.W. (1997) Experimental Monitoring of tbe Humber Bridge, Prof. Instn. Civ. Engrs. Civ. Engrg. 120, 177-182.
11. Zasos, A., Vergani, M., Bocciolone, M. and Evans, R (1993) Use of a Newly Designed Optometric Instrument for Long-term, Long-Distance Monitoring of Structures. 2"~ Int. Conf. On Bridge Management, Univ. of Suırey, UK, 18-21.

12. Stephen, G.A, Brownjohn, J.M.W. and Taylor F.A (1993) Measurements of Static and Dynamic Displacement &om Visual Monitoring of the Humber Bridge, Engrg. Struct. 15, 197-208.
13. Macdonald, J.H.G., Dagless, E.L., Thomas, B.T., and Taylor, C.A, (1997) Dynamic Measurements ofthe Second Sevetn Crossing, Proc. Instn. Civ. Engrs. Transp. 123, 241-248.
14. Thomas, B.T. and Dagless, E.L (1998) Real-time Monitoring of Dynamic Displacements by Computer Vision VI SECED Conf, Oxford, UK, Balkema, Rotterdam, 389-396.
15. Brownjohn, J.M.W., Zasso A, Stephen , G.A. and Sevem , RT. (1995) Analysis of Eı~erimenta! Data &om Wind-induced Response of a Long Span Bridge, J.Wing. Eng and IndDyn. 54/, 13-24.
16. Larose, G.L., Davenpoıt, AG. and King, J.P.C., (1992) Wind Effects on Long Span Bridges: Consistency of Wind Tumıel Results. J. Wind. Eng. And 2"" Dyn. 41-44, 1191-1202.
17. Svensson , H.S. and Jordet, E . (1996) The Concrete Cable-Stayed Helgeland Bridge in Norway. Proc. Iıstn. Civ. Engrs (Civ. Eng), 114, 2, 54-63.
18. Owen, J.S., Vann, A.M>, Davis, J.P. and Blakeborough, A. (1996). The Prototype Testing of Kessock Bridge. J. Wind Eng. and Ind. Aerodyn. 60, Nos. 1-3, 91-108
19. Ganev, T., Yamazaki, F., Ishizaki, H. and Kitazawa, M. (1998) Response Analysis ofthe Higashi-Kobe Bridge and Surrounding Soil in the 1994 Hyogoken-Nanbu Esıthquake, EESD, 27, 557-576.
20. P. Casas, J.R.C. (1995) Full-scale Dynamic Testing of the Alamillo CableStayed Bridge in Sevilla (Spain), EESD, 24, 3S-S1.
21. Gentile, C. and Cabrera, F.M.Y. (199~ Dynamic Investigation of a Repaired Cable-stayed Bridge, EESD, 26, 41-59.
22. Wilson, J.C. and Liu, T. (1991) Ambient Vibration Measurements of a Cable-stayed Bridge, EESD, 20, 723-747.
23. Brownjohn, J.M.W. (1994) Observations on Non-Linear Dynamic Characteristics of Suspension Bridges. EESD, 23, 1351-1367.
24. Paultre, P., Challal O. and Proulx, J. (1992) Bridge Dynamics and Dynamic Amplification Factors - A Review of Analytical and Experimental Findings. Can. J. of Civ. Eng. 19(2), 260-278.
25. Wilson, J.C. (1986) Analysis ofthe Observed Vibration Measurements of a Cable-stayed Bridge, EESD, 20, 723-747.
26. Werner , S.D., Beck , J.L. and Levin, M.B . (1987) Seismic Response Evaluation of Memoland Road Overpass Using 1979 Lnperial Valley Earthquake Records, EESD, 15, 249-274.
27. Wilson, .I~C., and Tan, B.S. (1989) Bridge Abutments: Assessing Their Influence on the EarUıquake Response of Meloland Overpass, ASCE (EM) June.
28. Loh C.H. and Lee Z.K . (1997). Seismic Monitoring of Bridge: Assessing Dynamic Characteristics from Both Weak and Strong Ground Excitation, EESD, 26, 269-288
29. Casirati M. and Franchioni, G. (1994). Seismic Tests On A 1:8 Model of An Irregular bridge On Three Shaking Tables. Proc. IOs' Europ. Conf. On E'quake Eng., Vienna
30. Monti, G., Pinto, P.E. and Nuti, C. (1995) Response of Conventional and Isolated Bridges under Non-synchronous Seismic Motion. European Seismic Deaign Practice-Research and Application, Editor tIS. Elnashi, A.A. Balkema, Rotterdam/Brookfield
31. Basoz, N., and Kiremidjian , A. (1998) Evaluation of Bridg~ Damage Data from the Loma Prieta and Northridge California Earthquakes. Report 98-0004, MCEER Buffalo, NY, USA.
32. Richardson, J.A. and Douglas, B.M. (1993) Results from Field Testing G~ırved Box-Girder Bridge using Simulated Earthquake Loads, EESD, 22, 905-922
33. Maguire, J.R. and Severn , R.T., (1987) Assessing the Dynamic Properties of Prototype Structures by Hammer Testing, Proc. Instn. Civ. Engrs., Pt2, 83, 769-784.

RECORDING AND INTERPRETING EARTHQUAKE RESPONSE OF FULL-SCALE STRUCTURES

M.D. TRIFUNAC, M.I. TODOROVSKA
University of Southern California
Civil Engineering Department, Los Angeles, CA 90089-2531, USA
e-mail: trifunac@mizar.usc.edu

Abstract

Earthquake resistant design (or retrofit) of structures requires realistic and accurate physical and theoretical models. Validation and further improvement of these models can be done only by comparison with full-scale, in situ measurements of the response to earthquake excitation. This paper presents (a) a review of the principles this validation process is based on, (b) discussion of selected examples of past contributions to modeling of structures, and (c) an outline of some of the current research needs. It is concluded that, in the education of future engineers, the art of modeling full-scale structures, and breadth of knowledge in classical mechanics have been neglected, and that this trend should be reversed.

1. Introduction

To proceed with design of earthquake resistant structures, or to analyze the existing structures, the engineer begins by creating a model representation of the prototype. This model will form the basis for all subsequent analyses. It will be used to write the governing equations, to describe the structure by discrete or continuous parameters, and to compute the quantities required for design. Thus, this model must have the properties which will describe the response of the prototype as completely and as accurately as possible. The search for *a good model* is therefore the ırst and the most important step preceding the analysis and design processes. After the model has been specified, the engineer will analyze, design and test only that part of the real world which the adopted model is capable of representing.

The difficult task of selecting the model is compounded by the lack of attention the modeling process receives in our educational system. Too often we approach the modeling task backwards, choosing the model configuration and its properties so that we can accommodate the methods of analysis. We then complete the loop of appearances by experimental verification procedures which involve measurement of the most elementary (often incomplete) aspects of the response. Having accepted the simplified experimental program, which usually does not produce adequate constraints, we finally ignore the nonuniqueness of our solution. Of course, the critique will arrive eventually in the form of a

M. Erdik et al. (eds.), Strong Motion Instrumentation for Civil Engineering Structures, 131–155.

Key Words: Structural response, full-scale testing, modeling, soil-structure interaction
careful and detailed analysis of a full-scale experiment, but often too late to correct the original design.

Is detailed modeling necessary for many engineering design tasks? After all, many design methods continue to rely on empirical and ad hoc procedures, many of which have been shown to be practical and adequate. The answers to these questions require considerations of the criteria defining the occupant safety and economic consequences, which both change with time. For example, thirty to forty years ago, analyses of earthquake losses focussed mainly on the replacement value of structures [43]. Today, and more so in future, with the growth and intricate inter-dependence of technologically advanced societies, which will live in dense metropolitan areas, reduction of indirect losses will govern in the selection of the required levels of safety [24, 25]. These new priorities will require development of advanced design principles and criteria which will be aimed towards reducing losses from business intemıptions, and will provide answers to the above questions. To prepare for the required changes in the design codes, we must initiate systematic development of advanced modeling techniques. This is feasible only through the use of the full-scale experimental measurements, which will guide us on how to single out the important phenomena and how to develop representative modeling and analyses.

The availability of representative dynamic models, and of carefully tested methods of analysis, will set the stage for realistic consideration of base-isolation methods, and of passive control techniques. Structural health monitoring will also become feasible, because the availability of representative models will enable selection of proper variables for real time monitoring. To implement the current and future advanced analyses, design, and control techniques, we must first solve the modeling problem, and to do that we must learn how to interpret better and document the actual response of structures to strong earthquakes. This can be accomplished by analyses of past recordings, and by continued but more advanced and detailed monitoring of response during fiıture strong shaking.

The first key step towards creation of representative dynamic models of structures is to include the effects of soil-structure interaction. Soil-structure interaction has a profound inlluence on the form and on the solution of the governing equations of motion, and on the wealth of physical phenomena that accompany transfer of energy &om the ground to the structure, and back to the ground. It offers advantages in the form of absorption and scattering of incident wave energy, before this energy enters the structure. These advantages must be brought to the attention of designers, and exploited fully before other more expensive and complex layers of protection, such as base-isolation or passive control are considered. Neither base-isolation, nor various ideas on structural control can be designed or implemented without including the soil-structure interaction effects in the controlling motions [46, 52-54].

In the following, we discuss the iteration steps in the learning process for development of representative models of structures. We illustrate these steps by examples from our past and current work on full-scale experimental studies. Finally, we summarize our view of priorities for future work. The mentioned full-scale experimental studies are only for illustration purposes, and do not represent a general and comprehensive review of literature on this subject.

2. Selected Modeling Considerations

The steps of the learning process (with the feed-back loops) which describe how to select representative models of engineering structures for dynamic response analyses are summarized in Fig. 1. The forward analyses include only the first three steps: (I) idealization of real structures, (2) mathematical representation and modeling and (3) analysis of response. The iterative learning process involves two additional tasks: (4) full-scale experimental verification and (5) revision of the previously adopted models. In the following, we discuss selected aspects of these steps and illustrate their role by reviewing examples from our previous work.

2.1 IDEALIZATION OF REAL STRUCTURES

This step constitutes the most important and the most difficult first step in any dynamic response analysis. Here the aim is not to overlook the importance of maintaining clarity and simplicity, while developing the model on the basis of deep understanding of the physical problem [4].

2.2 SIMPLE VERSUS DETAILED MODELS.

A tall structure, responding to long period (long wave) dynamic loads mainly in its first mode of vibration, could be represented by a single-degree-of freedom (SDOF) system (Fig. 2a). The same structure, excited by high frequency (short wave) dynamic loads, will have to be represented by a detailed (continuous or discrete, two- or three-dimensional) model which is capable of representing relative deformations and the associated forces within the structure (Fig. 2b). A structure supported on "rigid" base (which deforms due to passage of earthquake waves, but does not deform due to forces caused by the structure), can be represented by a one-, two- or three-dimensional continuous shear beam model (Fig. 3a,b). Such representation can describe relative deformations within the structure and the way these depend on the internal distribution of rigidities [47-50]. In another example, the NS motions of a nine-storey reinforced concrete building, vibrating in its first mode during forced vibration tests (the effects of soil-structure interaction were included), could be represented by a bending beam model (Fig. 4a). Forced vibration response of the same building in EW direction leads to large deformations of the foundation and floor slabs, and must be modeled in detail by the finite element method [31]. Structures with rigid floor slabs and flexible columns can be modeled via lumped mass models, where the mass of the floor structure and bottom half of the columns above, and top half of the columns below, are lumped at each floor level. The lateral shear stiffness can be represented by massless springs representing the total stiffness of all columns between adjacent floors (e.g. Fig. 3c) [69]. Such models are very common in both theoretical [15, 16] and forward engineering analyses [27].

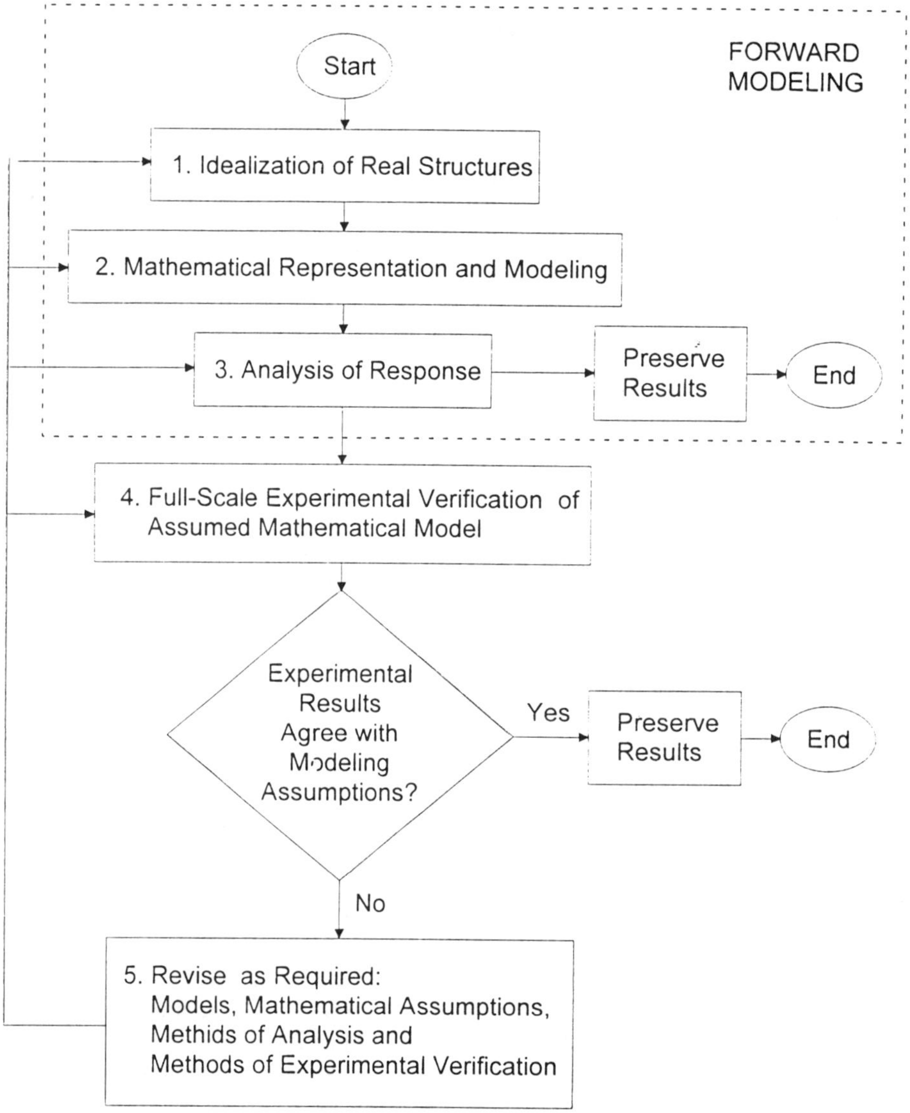

Figure 1. Steps in the learning process of selecting representative models of engineering structures for dynamic response analyses.

In the example shown in Figure 4a, the deformation of the model (structure) can be specified by only one coordinate and can be described by one-dimensional wave equation. For structures with smaller height-to-width ratio, and with nonuniform distribution of stiffness (e.g. Fig. 3a and b), the use of two- or three-dimensional models and wave propagation analysis may offer advantages over vibrational description of the response [55].

2.3 REDUCTION OF THE DEGREES-OF FREEDOM.

The number of the degrees of freedom, which must be considered in the analysis, will depend on the type of the model (e.g. model in Fig. 2a versus model in Fig. 2b), and on the band-widths of the excitation. For a given model (e.g. in Fig. 3c), it may be desirable to further reduce the number of degrees-of freedom. For example, two adjacent iloors could be lumped into oae mass, thus reducing the order of the linear system of differential equations by a factor of two. Such condensation of the degrees of freedom will lead to a reduction of computer time and may give adequate results as long as the outcome depends only on the low order modes of vibration (long wave lengths). Relative to the original model, the intermediate modes of vibration will have only approximate frequencies and mode shapes, while the higher modes and frequencies will be absent [9].

2.4 LINEAR VERSUS NONLINEAR ANALYSES

In addition to the geometrical description of the model, the specification of the maximum amplitudes of response, and of material properties, the following needs to be defined:

(i)Force displacement rslationships. These can be modeled, e.g. as linear elastic (Force, F, and dis lacement, x, are related by $F = k_o x$, where k_o is a constant), nonlinear elastic (e.g. $F = k_o x + k_1 x^3$, where k_1 is a positive or negative constant, this leads to Duffing equation), nonlinear hysteretic (e.g. F is defined by bi-linear hysteric spring) and so on.

(ii) Amplitudes of response. For large deflections, differential equations of dynamic equilibrium will involve nonlinear transcendental functions (e.g. sines and cosines).

(ü) Stability. For large amplitudes of response, simultaneous action of vertical and of horizontal inertial forces, in the presence of gravity, will lead to dynamic equilibrium equations which will result in time dependent excitation term influencing the stiffness term in the differential equations. The effective stiffness then may oscillate between positive and negative values, and for response amplitudes beyond the static stable deflections, the structure will enter the dynamic instability region and may collapse. This feature of the models is rarely considered in earthquake engineering studies [19], but should be included in every analysis dealing with large deflections and nonlinear response. The contribution of the soil-structure interaction to this dynamic instability is rarely investigated, even though it may be very important for "stiff' structures on "soft" soils.

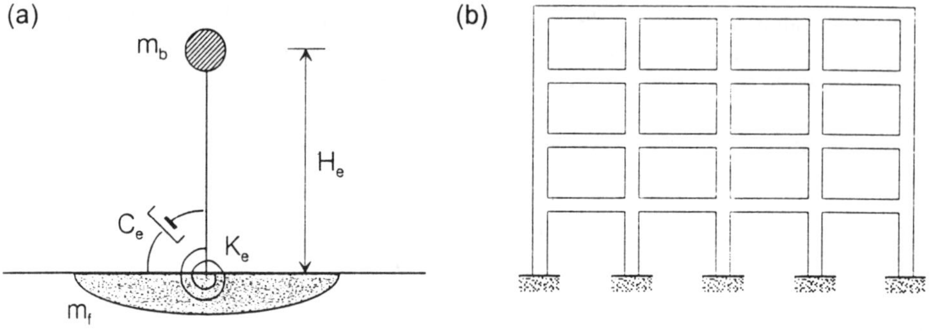

Figure 2. (a) Equivalent SDOF system (with damping C_e, rotational stiffness K_e, equivalent mass m_b, equivalent height K_e, on a rigid foundation with mass m_f) supported by flexible soil. (b) Multi-degree of-freedom system (e.g., moment resistant frame) supported by independent spread footings on elastic soil.

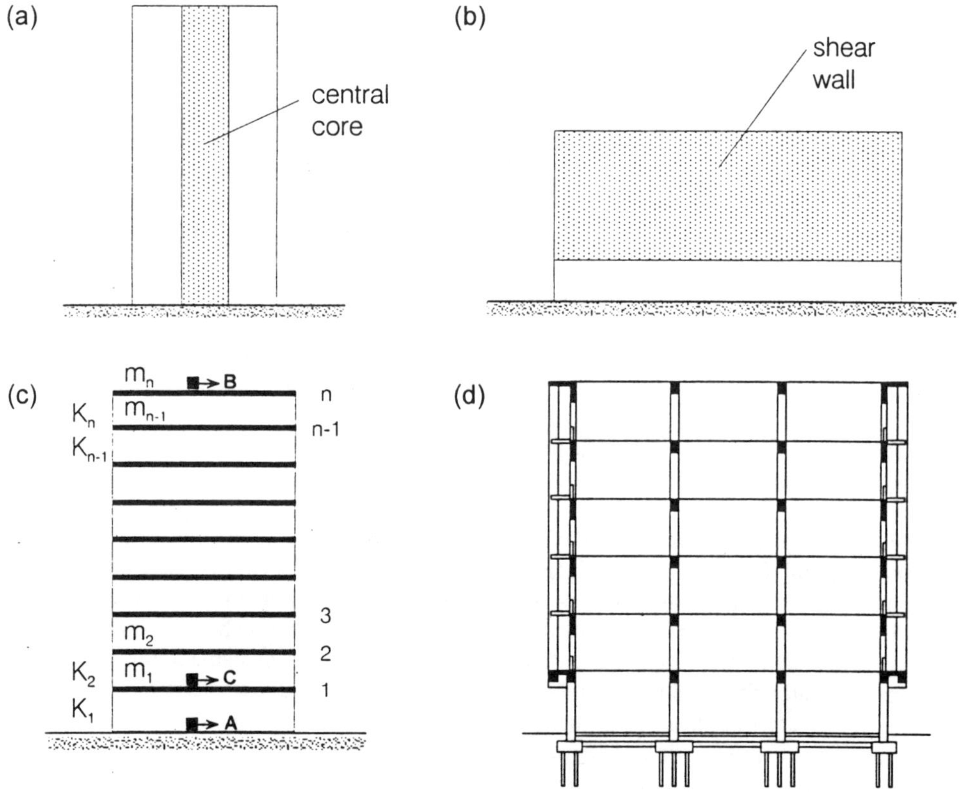

Figure 3. (a) Continuous model representation of a tall building with a stiff central core [55]. (b) Continuous model representation of a building with a soft first storey, on flexible soil [55]. (c) Discrete MDOF system of an n-storey building with rigid floors (m_i-floor mass, K_i-floor stiffness, A, B and C-strong motion instruments). (d) A section through a six-storey reinforced concrete structure supported by a foundation on piles [27].

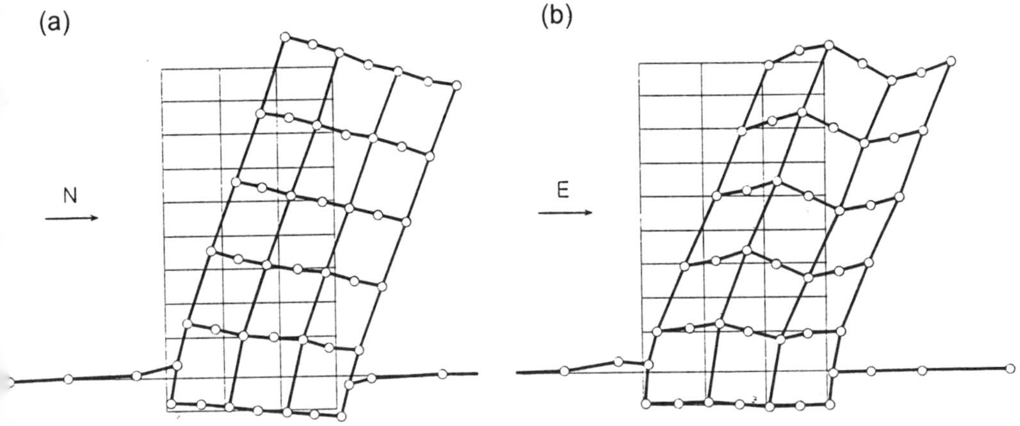

Figure 4. Deformation of a nine-storey reinforced concrete building excited at the roof by a shaker with two counter rotating masses [11] (a) along the west shear wall, and (b) along a section through the elevator core.

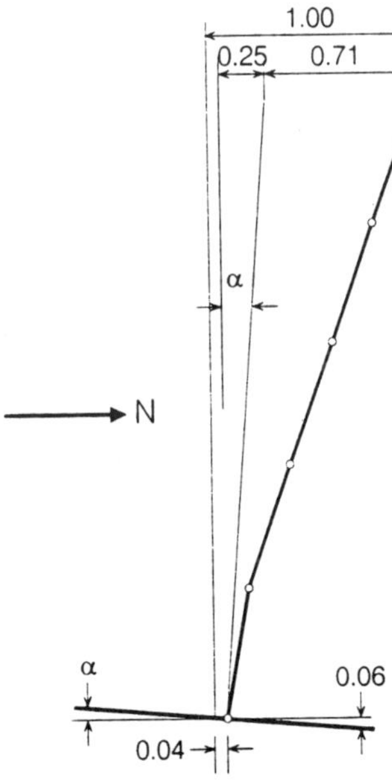

Figure 5. Contribution of foundation translation and rocking to the roof motion for N-S shaking of the nine-storey reinforced concrete building.

2.5 VARIOUS FORMS OF DYNAMIC COUPLING.

Many models of structures are symmetric and, for linear response, this makes the superposition principle applicable, and excitation in two orthogonal horizontal directions can be analyzed separately. For nonlinear response, the yielding members will cause sudden migration of the centers of stiffness, this will lead to strong coupling of the two horizontal translations with the torsional response, and the superposition principle does not hold. For nonlinear response of foundations, it is also usually assumed that the nonlinear response of symmetrically arranged piles, for example, does not lead to significant migration of the center of rigidity of the soil- structure interface. When this does occur, it can cause strong coupling of the torsional and the translational components of excitation [66].

2.6 SOIL-STRUCTURE INTERACTION

This describes a complex family of phenomena which are caused by the flexibility of the foundation soils. Analysis of soil-structure interaction requires consideration of additional degrees-of freedom and, depending on the model, it may call for methods of solutions based on wave propagation. In general terms, the soil-structure interaction will lengthen the apparent period of the system, will increase the relative contribution of rocking excitation of ground motion to the total response, and will usually reduce the maximum base shear [53]. The advantages of including soil-structure interaction in the design of structural systems result from the scattering of incident waves from the foundation, and from additional radiation of structural vibration energy into the soil. When the soil surrounding the foundation experiences small to modest levels of nonlinear response, the soil-structure interaction will lead to signi ıcant loss of the available input energy. Since this energy loss occurs outside the structure, if will be one of the important challenges for future design of safe structures to quantify this loss and to exploit it in design.

The simplest way to consider soil-structure interaction effects is to assume that the building is supported by a rigid foundation. This results in minimum number of additional degrees-of freedom (three translations and three rotations), but may lead to restrictive and too simple representation. Studies which model flexible foundations are rare [21, 30] and difficult to evaluate in absence of strong motion records. As far as we know, there exists no strong motion program to document distortions and warping of foundations of structures during passage of seismic waves [66].

The extent to which soil-structure interaction alters the apparent frequencies of the system response and changes the nature of the time functions of absolute and relative displacements, rotations, shear forces and bending moments in the structure ranges from negligible to profound, and depends mainly on the relative stiffness of the soil and the structure [53]. Recordings of strong motion in structures show that destructive shaking is usually accompanied by nonlinear response of the foundation soils [31, 64, 66], and that the time dependent changes of the apparent frequencies of response are usually accompanied by significant contribution of soil structure interaction [68]. Since the success of base-isolation systems, control of structural response, and of health monitoring depends on accurate representation of the anticipated and of recorded motions, it is clear that the nonlinear soil-structure interaction phenomena must be included in the analysis.

Experimantel studies of soil-structure interaction are best conducted in full-scale, in actuel buildings during microtermos [57-59], forced vibrations [5-18], and earthquake excitation [32]. It is difficult to coduct soil-structure interaction tests in laboratories, not only because of the constraints imposed by the need to satisfy to similarity laws, but mainly because it is almost impossible to model the half space boundary conditions for the soils.

2.7 DISSIPATION MECHANISMS

Modeling the dissipation of energy of a vibrating structure is constrained by the mathematical methods of analysis, and by the lack of comprehensive measurements which would show the physical nature of this dissipation [36]. Many linear response analyses use normal mode representation and, to maintain the advantages of working with decoupled equations, approximate the damping matrix by a linear combination $\alpha[m]+\beta[k]$ of the mass and stiffness matrices, [m] and [k], where α and β are constants. For an n degree-of freedom system, this allows one to choose the damping only for two modal frequencies, ω_i; and ω_j, and the remaining n-2 modes then have equivalent damping ratios $\varsigma_k=0.5(\alpha/\omega_k+(\beta\omega_k)$, which are not realistic. A common practice is to use constant damping ratios for all mode-shapes in the response analyses. This, of course, ignores the fact that the solution then violates the original differential equation.

In the presence of soil-structure interaction, the system damping depends on the damping in the building and in the soil, and on the scattering of wave energy from and through the foundation [31]. Design of foundations to scatter efficiently high frequency (short) waves can increase the apparent system damping and can reduce the amplification of the system response near the first mode of vibration [53].

In nonlinear response, the energy dissipation (damage) results from the work required to break structural components, to create plastic hinges and in nonlinear deformation of the soil supporting the foundation. For each structure and excitation, this energy can be quantified by computing the work dissipated by the equivalent hysteretic forces, but it will be different for each case.

3. Mathematical Representation And Modeling

Mathematical modeling of full-scale structures, for the purpose of performing response analyses, aims to find those representations which will satisfy all the modeling requirements and the constraints of the available analytical tools. For analytical representation of the incident waves, and for linear response analyses, the foundation soil (or rock) has to be geometrically simple. With the analytical approach, it is practical to consider only simple topographic irregularities [28, 29] and simple soil and alluvium layers [33, 34]. More complicated surface topography and layering must be represented by finite element or inite difference models.

Simple symmetric structures can be modeled by analytical methods and can be analyzed via wave propagation approach. Geometrically irregular structures and those which are expected to experience nonlinear response must be modeled by lumped masses, inite elements or by other discrete representations [72]. Simple surface or embedded foundations can be approximated by rigid slabs, when the soil is relatively soft and when the foundation and the structure are expected to experience only small relative deformations. However, it is difficult to predict intuitively how realistic is such an assumption and the decision should be guided by data and experience &om full-scale tests on similar structures. Soft (flexible) foundations can be represented by discrete lumped mass interconnected foundations [74], and this can be combined with finite element representation of the structure.

4. Analysis Of Response

The method which will be used for estimation of response will depend on whether the response will be linear or nonlinear. For linear response, the analysis can be formulated in terms of transfer functions, which express the quantity of interest (e.g. displacement of a point, bending moment, stress) in the frequency space. When response time functions are required, inverse transforms can be used to obtain the results, and when only the estimates of the peaks of response are required, spectrum superposition methods, coupled with probabilistic estimation of peak values, will give the final results [14-16J.

For nonlinear response, numerical integration of the system of governing equations must be carried out. This used to lead to lengthy calculations, but with modern computers it is becoming easier and faster [75].

5. Full-Scale Experimental Verification

For forward modeling, the above described three steps will produce the end result (Fig. 1). To verify the accuracy of this modeling, fiill-scale experiments can be carried out. The type of these experiments, the scope of measurements and the analysis of measured data will depend on the available sources of excitation, instrumentation and the specific purpose of the experiment. Components of full-scale structures may be tested in the laboratory, but evaluation of complete structures should be performed in full-scale. This limits the experiments to completed or to similar existing structures, and cannot be perfomied before the structure is built. However, systematic testing of the existing full-scale structures and careful interpretation and documentation of the results can go a long way towards creating a body of intuitive understanding, deep physical insight, and experience on how to extrapolate, and which essential steps to take leading to the core of complex new problems.

It should be noted that the laboratory tests can be very useful, but can never be as complete as full-scale experiments. Even the most carefully and completely planned laboratory work will represent only those aspects of the problem which the experiment designer chose to study and had incorporated into the model. That is, the best and the most complete laboratory tests can be used to verify and quantify mainly those aspects of the problem which the investigator knows. Except when fortunate accidents occur, we do not know how to model what we are not aware of and what we do not understand. The full-scale tests present a completely different set of practical problems, and the as built environment contains all the physical properties of reality. We only have to find ingenuous ways to discover, record and interpret the reality.

Another point to be made is that the physical completeness and the reality of the full-scale structures is necessary but not suffcient to guarantee correct end results. The discovery and understanding of the true nature of response tend to be born by the difficult labor involving reconciliation between our imperfect theories, modeling and analyses, with often incomplete data from measurements. Experienced experimentalists know that the first test rarely produces results, as we inevitably forget to measure something, or what we measure does not turn out to be useful. Thus, iterations are almost a rule, in both experiments and in the analyses.

Often, the difficulty lies in nonunique features of our starting models and assumptions. For example, the transfer functions of horizontal roof displacement of a fixed-base building, and of the same building on llexible soil, have very similar appearance near the peak of the first fixed- base frequency, or near the apparent frequency of the soil-structure system. Using a simple identification technique, it is easy to estimate the "frequency" and the associated "fraction of critical damping" from full-scale measurements during an earthquake, but it is not easy to identify the factors which contribute to the formation of these peaks. Separation of the fixed- base frequency of the structure from the rocking and translation &equencies associated with soil-structure interaction is less straight forward and can be performed only if special purpose instrumentation and records are available. The list of investigators who overlook this nonuniqueness is so long that it seems that this problem is ignored in most published work. Along the same lines, it is common to find papers presenting analyses of nonlinear response of structural components with discussions of structural ductility and how it relates to the observed changes in the response period, without including in their analyses the fact that, shortly after the earthquake, the apparent period of the soil-structure system was back at or near its pre- earthquake value, indicating that the main source of nonlinearity was not in the structure, but in the soil supporting it.

Most of the above mentioned nonuniqueness can be eliminated by placing additional instruments to measure the rotation of the building foundation [31, 36], but so far this is not possible with most of the data recorded in the buildings in California. It is interesting to note that in spite of the fact that transducers which recorded rotational acceleration and velocity have been constructed and tested [39, 42, 73], essentially no buildings are equipped with such instruments, and so far the earthquake engineers do not seem to request such data.

Rotation of the foundation about a horizontal axis (rocking) can be calculated from the difference of recorded vertical motions at two points on a line perpendicular to the axis of rotation. The result represents the average rotations between the two points. To evaluate the actual point rotations, it is necessary to map the pattern of deformations of the building foundation, associated with the apparent frequency of the system prior to and following the earthquake (provided no damage occurred), using forced vibration tests, for example. This requires detailed full-scale testing and is not available for most buildings.

The experience shows that full-scale testing of structures, before and after significant earthquake shaking, must be very detailed if one is to find the location and the extent of changes (failures) in the material properties, and to identify differences with respect to the previous models, assumptions and analyses. The next generation of improved rnodels and theory will be developed through reconciliation of differences with the current experimental results. This is clearly a difficult and time consuming work, and it is therefore not surprising that it is rarely carried out. Nevertheless, this is the only way we can learn how to measure what, where and how much to record. Before this work is completed, it is difficult (1) to see how to improve the full-scale observational programs, and (2) impossible to implement (correctly) various new ideas involving base-isolation, structural control, and structural health monitoring.

6. Examples Of Full-Scale Testing And Analyses

In the following, by way of examples, we illustrate the nonuniqueness associated with forward modeling, and some results on measurements of ilexibility of foundations and iterations in modeling.

6.1 FORWARD MODELING WITHOUT MODEL REVISION

Empirical evaluation of fundamental vibration periods of buildings is used in building codes to calculate lateral earthquake design loads. For a building on flexible ground, the period of vibration of the soil-structure system [23] is

$$T = [\ T_t^2 + \ T_r^2 + \ T_{fs}^2], \qquad (1)$$

where T_t is the period of horizontal translatory vibration of the rigid building, T_r, is the period of rocking motion of the rigid building, and T_{fs} the period of the first mode of vibration of an elastic building fixed at its base. Thus, the fundamental period of the system, T, compared to the period of the same structure on rigid base, *Tfs,* is longer.

Many empirical studies of T analyze the functional form of the dependence of T_{fs}, in terms of the type of lateral resisting system (e.g. moment resistant frames, shear walls), but ignore the contributions of T_r, and T_t. For example Housner and Brady [17] refer to "foundation yielding" as a "second order effect", while Goel and Chopra [13] ignore it [61].

An experimental study of the effects of soil-structure interaction in 1966 and 1967 concluded that the rigid body motion of the superstructure due to compliance of the soil contributed less than 1 percent to the total roof motion [20]. Repeated tests in the same building, in 1973 and in 1974 [11, 32), found that the rigid body motion contributes about 30 percent to the total roof motion (Fig. 5). While errors in data reduction cannot be ruled out [32], it has been argued [10] that this change is real and that it corresponds to degradation of the foundation system as a result of San Fernando Earthquake in 1971. That apparent periods of response, T, can change by factors approaching 2, during shaking by strong earthquakes, has been reported by Udwadia and Trifunac [68J, but further systematic studies identifying the primary sources of these changes have not been carried out. Papers have been written describing these changes, and often ascribing them to the nonlinear response of structural components [20], but field evidence and later tests suggest that these changes are associated mainly with nonlinear soil response.

The above illustrates that verification and revision of the models of analysis followed by improvement of the original models and governing equations are more of an exception rather than customary in comprehensive and complete analyses. The consequences for the future research are serious. With so many past studies opting for simplified and only forward modeling, it will be increasingly difficult in future to convince researchers who work with base- isolation, identification, control and health monitoring that to test these ideas and to make them work in real buildings, all aspects of soil-structure interaction must be included in the analysis.

6.2 NONUNIQUE REPRESENTATION

A weakness of many past and present instrumentation programs, which contribute data on actual earthquake response of full-scale structures, has to do with lack of adequate number of transducers, and lack of minimum data required to describe motion and deformation of foundations of buildings. Even in most carefully instrumented structures, the transducers have been installed on the basis of intuitive expectation rather than detailed measurements and analysis. A classical example of counter intuitive outcome is provided by the problem of optimal placing of two horizontal transducers into a one-dimensional lumped mass model of a tall building with n stories, on fixed ground and deforming in shear (Fig. 3c). Intuitively, with only two recording instruments, one would choose ground level (A) and roof (B) to record future shaking, as we have done to instrument many tall buildings during the past 65 years. This however leads to n! possible models which would be consistent with recorded motions at A as input and at B as output. A unique solution (i.e. one unique set of masses and springs) is possible, but the second instrument must be on the first floor (C), not on the n-th [67].

T'he lack of ideal physical access in buildings to the centers of mass, rigidity, and torsion results in placement of the recording instruments at some distance from these "ideal" locations. This results in recording a mixture of the desired components of motion. From the point of view of performing simplified identification analyses, such "contaminated" data may rule out the possibility for any meaningful interpretation. It is possible, however, to incorporate actual recording eccentricities for each position of the transducers into a linear representation of the

recorded time series [36], and in this way separate translations from rocking and from torsion. Unfortunately, the recorded data is usually not accompanied by information on the exact location of the instruments (e.g. within the plan of each floor of the building). Information on the centers of rigidity and of mass requires analyses and interpretation, but could be verified via fiill-scale experiments prior to strong motion instrumentation in a building. The lack of reports with such information forces researchers to work with simpli ied models and assumptions. This limits the quality and quantity of possible inferences and often results in problems associated with nonunique representation.

The above incomplete information is further complicated by the lack of documentation on the accuracy of the transducer orientations, accuracy of relative timing of digitized data from individual transducers, and on transducer characteristics. It is, of course, difficult and may not be practical to place the sensitivity vectors of all transducers exactly along specified directions. It is possible, however, to describe their relative orientations with sufficient accuracy and to perform the required corrections [45, 56]. Data can be processed with very accurate relative time for different components [65]. Sufficient detail on measured transducer constants (sensitivity, damping, natural frequency, and electrical frequency and damping parameters for force balance accelerometers, for example) could be provided, so that advanced instrument correction algorithms can be used. Most modeling and identification analyses will benefit immensely from data which incorporates all these corrections and instrument parameters [37,38].

At present, the digitized data is distributed without information on how the origin time of the traces has been identified, and it is not known how accurate the time coordinates are, and whether corrections for nonuniform film speed have been performed. It is further not clear what transducer corrections have been used. All this creates obstacles for knowledgeable investigators and limits the scope of possible inferences which otherwise they could extract from the recorded data.

Organizational efforts associated with maintenance of instrument arrays, data retrieval, data archiving, digitization, processing and dissemination are considerable. Therefore, it is difficult to suggest that instruments should be recalibrated at certain time intervals (and certainly following every strong shaking), and that important data should be digitized and processed again using advanced methods, followed by rigorous quality control [65], especially when unknown, but abundant recorded data has never been digitized and processed. The Strong Motion Instrumentation Program of the California Division of Mines and Geology (CDMG), the United States Geological Survey (USGS) and several private organizations have large volumes of excellent data which has not been processed (particularly data recorded in buildings). T'hese data already contains invaluable information for improvement and modification of the earthquake design codes, and can be used to teach us how and what to instrument and record in the future.

6.3 AN EXAMPLE OF ENGINEERING (FORWARD) MODELING ANALYSES

The Imperial County Services Building in El Centro (Fig. 3d) almost collapsed during the 1979 earthquake in Imperial Valley, California. We have studied this building and the data recorded in it to find how successful a typical engineering analysis would be in predicting such response [27]. We found that, in the first iteration, we were not able to predict the actual response, and that the reasons for this were complex and in some instances beyond the state of the art at that time. Perusal of more recent attempts to do the same today, will show that we would probably have the same difficulties at present. It is still difficult to model foundations on piles, and to describe realistic distribution of stiffness in the soil-pile-foundation-structure systems. Almost 20 years later, little has changed. It is still an art to model structures by simplified commercially available computer programs, and we still do not have adequate number of such studies to have gained good understanding of how to improve our ability to model.

6.4 RIGID VERSUS FLEXIBLE FOUNDATION MODEL

When soil-structure interaction is considered in the dynamic analysis of soil-structure systems, it is convenient to assume that the foundation is rigid. This assumption simplifies the analysis and reduces the required number of additional degrees-of freedom to model soil-structure interaction, and thereby the number of simultaneous equations which must be solved. Whether such assumption can be made must be carefully investigated, and the outcome does not depend only on the relative rigidity of the foundation and of the soil, but can be influenced also by the overall rigidity of the structure, its lateral load resisting system and its orientation. As we already noted, this can be illustrated by comparison of NS versus EW vibrations of a nine-storey reinforced concrete structure which was studies by Luco et al. [31]. Even though the foundation system of that building is relatively flexible, for NS vibrations, two symmetric shear walls at each end (east and west) of the building act to stiffen the foundation slab, and this allows one to proceed with "rigid" foundation representation (Fig. 4a). For EW vibrations, the building carries lateral loads by an elevator core, which deforms the foundation slab in the middle, while the shear walls act as membranes providing axial constraints, but little bending stiffness (Fig. 4b). For EW vibrations, the foundation slab cannot be approximated by a "rigid" foundation model. These three-dimensional deformation shapes, which showed how this structure defoms while vibrating in NS and EW fundamental modes of vibration, were measured during forced vibration tests, and were essential for this interpretation.

Recent ambient vibration tests in a seven storey, reinforced concrete, moment resistant frame building have shown that the foundation supported by piles deforms during passage of microtremor waves. It was then inferred that the same happens during passage of much larger strong motion waves. Detailed ambient vibration survey of this symmetric structure, on symmetric pile foundations, showed that the center of torsion for this structure is outside the building plan, close to its south-eastern corner. Subsequent reexamination of the strong motion records in this building has shown that this eccentricity may have been present in all post 1971 excitations and that it is associated with some asymmetry in the soil-pile system, since the date of its construction, in 1966, or that it was caused by some partial damage during the 1971 San Fernando earthquake [66].

Differential motion of building foundations [60] may reduce the translational response at the upper floors, but leads to large additional shear forces and bending moments in the columns of the first floor. The response spectrum method can be modified to include the consequences of such differential motions [62], but it is necessary to document this via full-scale measurements during future strong earthquakes, and to correlate the theory with observations.

The assumption that foundations can be represented by rigid slabs seems to be implicit in most fiill-scale instrumentation programs for the buildings where strong motion was recorded so far. Technically, it should be easy to supplement the existing instrumentation to provide data on differential motion of building foundations. Ideally this should be done first in instrumented buildings where strong motion has already been recorded during many past earthquakes, so that additional value can be added to the existing data, interpretation and analyses.

6.5 AN EXAMPLE OF ITERATION IN MODELING SOIL-STRUCTURE INTERACTION

Following many ambient, forced vibration, and earthquake recording experiments in a nine storey reinforced concrete building, and apparent inconsistencies in the data and its interpretation, we decided in mid 1970's to develop a comprehensive model, which includes soil-structure interaction, so that we could use it in interpretation of all the recorded data [31]. When this model was completed, comparison of the theoretical predictions with the recorded motions showed that the theory for computation of compliances available at that time (for rigid surface foundations) was not adequate to interpret the results. Our analysis and writing of the report were intemipted, and we started to work on refinement of compliance functions, so that the embedment could be considered explicitly. After new compliance equations were developed and tested, the original full-scale tests of the building could be explained, now resulting in excellent agreement between the theory and the measurements. Finally, the report could be finished, almost ten years after it was started.

Not every iteration of an experiment verification will take ten years to complete. With more focus and effort, our work could have been completed earlier, but it should be understood that complicated subjects take more time to understand and to master. The nature is fiill of fascinating examples showing that the time from conception to complete delivery is proportional to the complexity of the product. Clearly, there exists an upper bound on the time rate of creation and realization of technologically advanced ideas.

Finally it must be remembered that many valuable discoveries tend to be accidental, and so their production rate can be measured only in probabilistic terms. We may approach more advanced levels sooner by designing more ingenious experiments, processing and analyzing more recorded data in many different full-scale structures, all these over a longer period of time and focussing our attention on how to explain even innocent and small inconsistencies between the data and the theory. Focused initiatives and organized five or ten year programs may be counter productive, because the education of new experts and the time they need to produce significant new results are increasing with the complexity of the new challenges, and the time to delivery may be significantly longer than five to ten years.

7. Summary, General Comments And Selected Priorities For Future Research

Successful and lasting contributions to advanced modeling require: (1) comprehensive gathering and preparation of representative databases, (2) education of young generations in modeling and analysis, and (3) development of general nonlinear theory to guide future data gathering, interpretation, modeling and education. The following discussion and presentation of selected priorities for future work may help initiate some efforts towards these goals.

7.1 DIGITIZATION, PROCESSING AND DISSEMINATION OF RESPONSE DATA IN BUILDINGS AND OTHER STRUCTURES

The ultimate tests of earthquake resistant design provisions came from recorded response and from observation of the performance of full-scale structures during severe earthquake shaking. Following the introduction of the first strong motion accelerographs to record ground motion in early 1930's, gradually instruments were placed also in buildings, dams and bridges to record their response. Following many organizational changes, today many cities in California still have local provisions requiring one or three accelerographs to be installed in buildings over six stories high (when the floor area is greater than 60,000 feetz) or over 10 stories high. At least several hundred buildings in the Los Angeles metropolitan area are believed to have recorded the 1994 Northridge earthquake. A &action of these records is processed and distributed to researchers, but most of these data eventually end up in various archives, never to be seen or used in engineering analyses. Also, the United States Geological Survey (USGS), the California Division of Mines and Geology (CDMG), and several private and public organizations maintain recording stations in selected buildings. Following the Northridge earthquake, USGS collected records from over 30 high-rise buildings, 7 hospitals, 12 dams, 6 fire stations and 7 water and power distribution facilities [40]. CDMG recorded motion in 77 structures, including 57 buildings, 12 dams, 5 freeway interchanges, a toll bridge, an airport tower and a power plant [41].

In the former Soviet Union, various strong motion transducers and recorders were developed, mainly by the laboratories associated with the Institute of Physics of the Earth, in Moscow [12]. The lack of modern digitization systems for processing of analog records, and the emphasis on direct recording rather than computing the derived quantities (e.g., computing velocity and displacement from recorded acceleration) resulted in a series of transducer-galvanometer combinations which could be paired to measure directly acceleration, velocity and displacement [12, 37]. Buildings were instrumented systematically and in detail with many channels of simultaneous acceleration, velocity and displacement recorders (e.g. [3]). Most of the recorded data is still archived in various laboratories of the newly formed independent countries (Armenia, Uzbekistan, Tajikistan, Kazakhstan), has never been processed, and soon may be lost due to the lack of support or local wars.

Our goal should be to digitize and process these data, and then to analyze it so that we can be guided by the results, in formulation of future instrumentation and recording needs.

7.2 ANALYSIS OF DATA RECORDED IN BUILDINGS AND OTHER STRUCTURES

All the data recorded so far should be analyzed and interpreted, so that systematic work on linear and nonlinear response and modeling can be initiated without delay. This analysis will also provide sound basis for future selection of recording techniques and for better coverage of recorded motions through design of more advanced recording systems. Finally, it will help by leading towards wiser deployment of the available transducers.

For many years, typical building instrumentation consisted of two (basement and roof) or three (basement, roof and an intermediate level) self contained tri-axial accelerographs interconnected for simultaneous triggering. The early studies of recorded motion [31] noted that such instrumentation cannot provide information on rocking of building foundations, and that this information is essential for identification of the degree to which soil-structure interaction contributes to the total response. Beginning in the late 1970's, new instrumentation was introduced, based on central recording of individual, one component transducers (usually force balance accelerometers). This instrumentation provided greater ilexibility to adapt the recording systems to the needs of different structures, but budget limitations and lack of understanding of how different structures would deform during earthquake response, often resulted in recording incomplete information [27]. The outcome is that the recorded data is used rarely in advanced engineering research, and usually only to provide general reference for the analyses.

7.3 FULL-SCALE TESTING AND UPGRADING OF THE EXISTING INSTRUMENTATION

In addition to the analyses of earthquake records in buildings, it will be necessary to document further the three-dimensional deformations accompanying the principal modes of vibration of instrumented buildings. This will facilitate the work on advanced modeling tasks.

Full-scale forced vibration and ambient vibration tests will help demonstrate how important it is to consider soil-structure interaction. Simplified soil-structure interaction provisions are included in NEHRP [6) and Applied Technology Council [2] codes, but these provisions are voluntary and are often neglected in design. At present, there is a common perception that ignoring soil-structure interaction results in conservative design. The factors which have contributed to this situation could be related, in part, to a common belief, prevailing up to the early 1970's, that the contribution of soil-structure interaction to the observed response is small. By late 1970's, it became clear that soil-structure interaction effects cannot be ignored, but neither code design provisions nor the principles governing the decisions on how to instrument the buildings seem to have recognized this. A simple and useful "standard" approach would have first performed detailed three-dimensional ambient vibration tests in the buildings [22, 31, 57-59]. Next, based on the three-dimensional defomiatio~ and mode shapes from these tests, a knowledgeable committee could have selected the optimum number, location and orientation of sensors for future recordings. Wtien necessary, subsequent repetition of ambient vibration surveys could have compared the results with the previous "finger-print" of the structural system, thus documenting the state of the structure preceding the earttiquake and the subsequent changes. This approach would have created numerous possibilities for new studies and better understanding of the response of actual structures, and would have acted in a positive way towards quality control for all forward calculations of response [27].

The Northridge earthquake of 1994 provided, so far, the most valuable and abundant data on strong ground motion and on response of structures. However, it seems that little will change, in spite of the fact that many experts have started to recognize the discrepancies between what we perceive and what the nature does. In contrast, looking back more than 60 years ago, it is remarkable with how much foresight and how energetically the researchers of that time set out to measure and understand the phenomena associated with the response of full-scale structures. In the introduction to a report entitled "Vibration Observations" (by Carder [7]), referring to instrumentation and testing of full-scale structures, L. Jacobsen wrote: "The work .... done by .... the United States Coast and Geodetic Survey is of such fundamental importance to the engineering profession that it makes a distinct milestone in our quest for better knowledge of how to build earthquake resistant structures." Then he continues to say that "without these and similar tasks, we should be in the dark as to whether or not certain proposed theories have experimental confirmation." In the same issue of "Earthquake Investigations in California, 1934-1935," after introducing another article (by Blume [5]) which describes an early model of a "centrifugal force agitator," Jacobsen states: "An experimental study of periods of vibration of buildings may be made by observing free transitory vibrations set up in the structures by impulsive external agencies, as for instance, wind gusts. Under these conditions a building will vibrate in a number of its natural modes ...." It will take forty years for this idea to be revived and fully affirmed as a useful tool for full-scale in situ testing of engineering structures [8, 59, 71].

7.4 INSTRUMENTATION DEVELOPMENT AND DEPLOYMENT

Transducers which measure rotation have been developed and used to record the response of soils and structures to strong shaking, but the measurements of angular accelerations and velocities did not gain popularity so far. Except for a few isolated cases, such instrumentation is generally not available and there are no significant databases to study strong motion rotations [39, 42, 73].

The traditional view has been that strong motion instruments should be designed to have high dynamic range, proper sensitivity and recording accuracy in as broad frequency band as possible. What we tend to overlook is that many problems in engineering can be understood and interpreted only by capturing the spatial variations of response, and that this calls for dense array type measurements, with small inter-station separation, to be able to record "short" wave lengths. Simple estimates call for deployment of many small aperture arrays, with each array containing hundreds to thousands of transducers, all working together. In the framework of the currently available instrumentation costs and the budgets which can be expected for such projects, the initial purchase and maintenance costs, immediately rule out such experiments. We must take advantage of the new technology to develop such systems, deploy such dense arrays, and maintain them in operation for 30-50 years. understanding of the response of actual structures, and would have acted in a positive way towards quality control for all forward calculations of response [27].

7.5 EDUCATION

It seems that, at present, the earthquake engineers are not properly trained in mechanics. Paraphrasing Biot [4], both scientists and engineers working in the field may have "become victims of narrow specialization... Many are almost totally ignorant in classical mechanics and are not able to understand the formulation of even simple problems unless it can be reduced to the solution o£.." a differential equation of single-degree-of freedom viscously damped oscillator.

Perhaps it is dif ıcult to find another topic in earthquake engineering which has grown and matured so well, and for which so much has been accomplished as has been the case for soil- structure-interaction. Furthermore, there is hardly any other topic which is more central and fundamental to the response analysis, and which offers so many powerful possibilities for implementation of passive energy dissipation, and which is at the same time less used, less understood and less considered in engineering design. There is no doubt that the principal reasons for this lie in our educational programs. Most universities just do not offer adequate number of lectures on soil-structure interaction, and so cannot build up the students understanding to the required mathematical level, which would allow professors to teach the intricacies of the physical nature of this problem. Timoshenko wrote "...a student should be taught, of course, *how* to use the method, but this is not enough. He should also understand why it works..." [44]. Since all advanced concepts of response control, such as base-isolation, active and passive control, and use of smart materials, all depend on our ability to write correct and representative differential equations describing the response, it is essential to initiate systematic and rigorous education of the future researchers and engineers in soil-structure interaction.

7.6 FULL-SCALE TESTS AND MODELING STUDIES

Following a brief and productive period, from about 1965 to 1975, when many useful force vibration and ambient vibration experiments were conducted, the earthquake profession seems to have converged toward small-scale laboratory experiments. From 1984 to 1992, perusal of papers published in the proceedings of World Conferences on Earthquake Engineering, will reveal about 100 to 150 papers (per conference) describing various experimental investigations. Only ~6 percent of these papers in 1984 (8WCEE, San Francisco), ~8 percent in 1988 (9WCEE, Tokyo-Kyoto) and ~3 percent in 1992 (10 WCEE, Madrid) were related to experiments with fiill-scale structures. In 1996, "Earthquake Spectra" [1] published a theme issue entitled "Experimental Methods" which contains nine papers. These papers neither discuss nor reference full-scale tests of structures. An uninitiated reader might conclude that we are dealing with an anachronic subject, but the time will show that systematic gathering and processing of strong motion data recorded in buildings, combined with comprehensive full-scale tests of the same structures, is the best and the only real laboratory where scaling and similarity laws do not pose problems, and where the boundary conditions are never approximated. By contrasting and combining the full-scale ambient vibration tests of a building with what the available records of earthquake response can show, we may iniluence those who decide where and how many transducers to use in instrumenting the full-scale structures, and what and how many records to digitize for engineering research applications. With this paper, we wish to contribute towards initiating a new phase of rational full-scale measurements for future earthquake engineering research and design.

Analyses and interpretation of experimental measurements are difficult because the recording resolution may not be adequate or the location or type of the instruments are not suitable for the structure being tested. Most successful experiments require a series of tests where the preceding tests and their interpretation serve as a guide on how to perform the next test. This may require far more work than what is feasible for most commercial analyses, but the results are always invaluable and always teach us how to model the structure for computer simulation [22, 26, 35, 70].

8. References

1. Abrams, D.P. (editor) (1996). Experimental methods, *Earthquake Spectra,* 12(1), 1-80.
2. Applied Technology Council (1978). Tentative provisions for the development of seismic regulations for buildings: a cooperative effort with the design profession, building code interests, and the research community, *Report No. ATC 3-06,* U.S. Dept. of Commerce, National Bureau of Standards.
3. Begiev, B.B., V.A. Nechaev, V.A. Tokmakov and D.A. Harin (1975). Opit raboti Inzenerno-seismometriceskoi sluzbi G. Dushanbe, Voprosi Inzenemoi Seismologü, Vipusk 17, Akademia Nauk SSSR, Inst. Fiziki Zemli, Izdatelstvo Nauka, Moscow.
4. Biot, M.A. (1963). Are we drowning in complexity? *Mechanical Engrg,* 85(2), 26-27.
5. Blume, J.A. (1936). The building and ground vibrator, Chapter 7 in Earthquake in Califomia 1934-1935, U.S. Dept. of Commerce, Coast and Geoletic Survey, *Special Publication No.* 20/, Washington, D.C.
6. Building Seismic Safety Council, BSSC (1995). NEHRP recommended provisions for seismic regulations for new buildings, Part 1, Provisions and Part 2, Commentary, *Repori No. FEMA 222A,* Federal Emergency Management Agency, Washington D.C.
7. Carder, D.S. (1936). Vibration observations, Chapter 5 in earthquake investigations in California 1934-1935, U.S. Dept. of Commerce, Coast and Geologic Survey, *Special Publicaiion No. 201,* Washington, D.C.
8. Crawford, R. and H.S. Ward (1964). Determination of the natural periods of building, *Bull. Seism. Soc. Amer., 54,* 1743-1756.
9. Duncan, W.J. (1952). A critical examination of the representation of massive and elastic bodies by systems of rigid masses elastically connected, *Quart. Journ. Mech. Appl. Math.,* 5(1), 97-108.
10. Foutch, D.A. and P.C. Jennings (1978). A study of the apparent change in the foundation response of a nine-story reinforced concrete building, *Bull. Seism. Soc. Amer.,* 68, 219-229.
11. Foutch, D.A., J.E. Luco, M.D. Trifunac and F.E. Udwadia (1975). Full-scale, three-dimensional tests of structural deformation during fotced excitation of a nine-story reinforced concrete building, *Proc. U.S. National Conference on Earthquake Engrg,* Ann arbor, Michigan, 206-215.
12. Fremd, V.M. (1978). Instrumentalnie Sredstva i Metodi Registraciji Silnih Zemljetresenü, *Izdatelstvo Nauka,* Moscow.
13. Goel, R.K. and A.K. Chopra (1998). Period formulas for concrete shear wall buildings, *J. of Structural Eng.,* ASCE, 124(4), 426-433.
14. Gupta, I.D. and M.D. Trifimac (198'7). Statistical analysis of response spectra method in earthquake engineering, Dept. of Civil Eng., *Report No. 87-03,* Univ. of Southem Califomia, Los Angeles, Califomia.
15. Gupta, V.K. and M.D. Trifimac (1989). Investigation of building response to translational and rotational earthquake excitations, Dept. of Civil Engrg, *Report No. 89-02,* Univ. of Southern Califomia, Los Angeles, California.
16. Gupta, I.D. and M.D. Trifunac (1996). Investigation of nonstationarity in stochastic seismic response of structures, Dept. of Civil Engrg, *Report No. 96-Ol,* Univ. of Southem Califomia, Los Angeles, California
17. Housner, G.W. and A.G. Brady (1963). Natural periods of vibration of buildings, *J. ojEngrg Mech.* Div., ASCE, 89(EM4), 31-65.
18. Hudson, D.E. (1970). Dynamic tests of full-scale structures, Chapter 7 in Earthquake Engineering, edited by R. Wiegel, *Prentice Hall,* New Jersey.

19. Husid, R. (1967). Gravity effects on the earthquake response of yielding structures, *PIı.D. Thesis,* Calif. Inst. of Tech., Pasadena, Califomia.
20. Iemura, H. and P.C. Jennings (1974). Hysteretic response of a nine-story reinforced concrete building, *Earthquake Engrg and Struct. Dynam., 3,* 183-201.
21. Iguchi, M. and Luco, J.E. (t982). Vibration of flexible plate on visoelastic medium, *J. ojEngng. Mecl., ASCE,* 108(6), t 103-I 120.
22. Ivanovic, S. and M.D. Trifunac (1995). Ambient vibrations survey of full-scale stnictures using personal computers (with examples in Kaprielian Hall), Dept. of Civil Engrg, *Report No. 95-OS,* Univ. of Southern Califomia, Los Angeles, Califomia.
23. Jacobsen, L.S. and R.S. Ayre (I958). Engineering vibcations, *McGraw Hill, N.* York.
24. Jordanovski, L.R., M.I. Todorovska and M.D. Trifunac (1992). The total loss in a building exposed to earthquake hazard, Part I: the model, *European Earıhquake Engrg,* Vol. VI-n.3, 14-25.
25. Jordanovski, L.R., M.I. Todorovska and M.D. Trifunac (1992). The total loss in a building exposed to earthquake haıard, Part II: a hypothetical example, *European Earthquake Engrg,* Vol. VI-n.3, 26-32.
26. Kadakal, U. and Ö, Yüzügüllü (1996). A comparative study of the identification methods, for the autoregressive modeling from the ambient vibration records, *Soil Dynam. and Earthquake Engrg,* 15(1), 45-49.
27. Kojic, S., M.D. Trifiınac and J.C. Anderson (1984). A post earthquake response analysis of the Imperial County Services Building, Dept. of Civit Engrg, *Report No. 84-01,* Univ. of Southern California, Los Angeles, Califomia.
28. Kojic, S. and M.D. Trifunac (1991). Earthquake stresses in Arch Dams: I - theory and antiplane excitation, *J. of Engrg Mech.,* ASCE, 117(3), 532-332.
29. Kojic, S. and M.D. Trifunac (1991). Earthquake stresses in Arch Dams: II - excitation by SV, P and Rayleigh Waves, *J. ofEngrgMech.,* ASCE, 117(3), 553-574.
30. Liou, G: S. and Huang, P.H. (1994). Effects of tlexibility on impedarice functions for circular foundations, *J. of Engng. Mech.,* ASCE, 120(7), 1429-1446.
31. Luco, J.E., H.L. Wong, and M.D. Trifunac (1986). Soil-structure interaction effects on forced vibration tests, Dept. of Civil Engrg, *Report No. 86-O5,* University of Southern Califomia, Los Angeles, California.
32. Luco, J.E., M.D. Trifunac and H.L. Wong (1987). On the apparent change in dynamic behaviour of a nine-story reinforced concrete building, *Bull. Seism. Soc. Amer., 77(6),* 1961-1983.
33. Moeen-Vaziri, N. and M.D. Trifiınac (1988). Scattering and diffraction of plane SH-waves by two-dimensional inhomogeneities, *Soil Dynam. and Earthquake Engrg, 7(4),* 179-188.
34. Moeen-Vaziri, N. and M.D. Trifiınac (1988). Scattering and diffraction of plane P and SV waves by two-dimensional inhomogeneities, *Soil Dynam. and Earthquake Engrg, 7(4),* 189-200.
35. Morassi, A. and N. Rovere (1997). Localizing a notch in steel frame from frequency measurements, *J. Engrg Mech.,* ASCE, 123(5), 422-432.
36. Moslem, K. and M.D. Trifunac (1986). Effects of soil structure interaction on the response of buildings during strong earthquake ground motions, Dept. of Civil Engrg, *Report No.* 86-04, Univ. of Southern California, Los Angeles, Califomia.
37. Novikova, E.I. and M.D. Trifunac (1991). Instrument correction for the coupled transducer-Galvanometer systems, Dept. of Civil Engrg, *Report No. 91-02,* Univ. of Southern Califomia, Los Angeles, California.
38. Novikova, E.I. and M.D. Trifiınac (1992). Digital Instrument response correction for the force balance accelerometer, *Earthquake Spectra,* 8(3), 429-442.
39. Phinney, R., R. Gilman, and F. Press (1962). Progress report on a short period rotational seismometer, *J. Ceophys. Res.,* 67(9).
40. Porcella, R.L., E.C. Etheredge, R.P. Maley, and A.V. Acosta (1994). Accelerograms recorded by USGS national Strong Motion Network Stations during M, = 6.6 Northridge Califomia Earthquake of January 17, 1994, *Open File Report 94-741,* Dept. of the Interior, U.S. Geological Survey.
41. Shakal, A., M. Huang, R. Darragh, T. Cao R. Sherbume, P. Malhotra, C. Cramer, R. Synov, V. Graizer, G. Maldonado, C. Peterson and J. Wimpole (1994). CSMIP strong motion records from the Northridge, Califomia, Earthquake of 17 January 1994, *Report No. OSMS 94-07,* Calif. Dept. of Conservation, Div. of Miner and Geology, Sacramento, California.
42. Shibata, H, Shigeta, T. and Sone, A. (1976). A note on some results of observation of torsional ground motions and their respon5e analysis, *Bull Eartlıquake Resistant Struct. Research Center,* 10, 43-47.
43. Steinbrugge, K.V. (1982). Earthquakes, Volcanoes and Tsunamis, an anatomy of hazards, Scandia America Group, N.Y. N.Y.
44. Timoshenko, S.P. (1968). As I remember, *Pan Nostrand,* Princeton, New Jersey.

45. Todorovska, M.I. (1998). Cross-axis sensitivity of accelerographs with pendulum like transducers - mathematical model and the inverse problem, *Earthquake Engrg and Struct. Dynam.*, 27(10), 1031-1051.
46. Todorovska, M.I. (1999). Base isolation by a soft first storey with inclined columns, *J. Engrg Mech.*, ASCE, *125(4)*, 448-457.
47. Todorovska, M.I., and M.D. Trifunac (1989). Antiplane earthquake waves in long structures, *J. Engrg Mech.*, ASCE, 115(12), 2687-2708.
48. Todorovska, M.I., and M.D. Trifunac (1990). A note on the propagation of earthquake waves in buildings with soft first ftoor, *J. Engrg Mech.*, ASCE, 116(4), 892-900.
49. Todorovska, M.I., and M.D. Trifunac (1990). A note on excitation of long structures by ground waves, *J. Engrg Mech.*, ASCE, 116(4), 952-964.
50. Todorovska, M.I., and M.D. Trifunac (1990). Analytical model for in-plane building-foundation-soil interaction: incident P-, SV- and Rayleigh waves, Dept. of Civil Engrg, *Report No. 90-01,* Univ. of Southem California, Los Angeles, Califomia.
51. Todorovska, M.I. and M.D. Trifımac (1991). Radiation damping during two-dimensional building-soil interaction, Dept. of Civil Engrg, *Report No. CE 9l-01,* Univ. of Southem Califomia, Los Angeles, Califomia.
52. Todorovska, M.I. and M.D. Trifunac (1992). Effects of the base input rocking on the relative response of long buildings on embedded foundations, *European Earthquake Engrg,* Vol. VI-n.l, 36-46.
53. Todorovska, M.I. and M.D. Trifunac (1992). The system damping, the system frequency and the system response peak amplitudes during in-plane building-soil interaction, *Earthquake Engrg and Struct. Dynam., 21(2),* 127-144.
54. Todorovska, M.I. and M.D. Trifunac (1993). The effects of the wave passage on the response of base-isolated buildings on rigid embedded foundations, Dept. of Civil Engrg, *Report No. CE 93-10,* Univ. of Southern Califomia, Los Angeles, California.
55. Todorovska, M.I., M.D. Trifunac and V.W. Lee (1988). Investigation of earthquake response of long buildings, Dept. of Civil Engrg, *Report No. CE 88-02,* Univ. of Southern California, Los Angeles, California.
56. Todorovska, M.I., E.I. Novikova, M.D. Trifunac and S.S. Ivanovic (1998). Advanced sensitivity calibration of the Los Angeles Strong Motion Array, *Earthquake Engrg and Struct. Dynam., 27,* 1053-1068.
57. Trifunac, M.D. (1970). Wind and microtremor induced vibrations of a 22-story steel frame building, Earthquake Engrg Res. Lab., *Report EERL 70-Ol,* Calif. Inst. of Tech., Pasadena, California.
58. Trifunac, M.D. (1970). Ambient vibration test of a 39-story steel frame building, Eaıthq. Engrg Res. Lab., *Report EERL 70-01,* Calif. Inst. of Tech., Pasadena, California.
59. Trifunac, M.D. (1972). Comparison between ambient and forced vibration experiments, *Earthq. Engrg and Struct. Dynam., l,* 133-150.
60. Trifunac, M.D. (1997). Differential earthquake motion of building foundations, *J. of Structural Engrg,* ASCE, 123(4), 414-422.
61. Trifunac, M.D. (1999). Period formulas for concrete shear buildings, Discussion of paper No. 16087 by Goel R. and A.K. Chopra, *J. ojStructural Engrg,* ASCE (in press).
62. Trifunac, M.D. and M.I. Todorovska (1997). Response spectra for differential motion of columns, *Earthquake Engrg and Struct. Dynam.*, 26(2), 251-268.
63. Trifımac, M.D. and M.I. Todorovska (1998). AmpliFıcation of ground motion and damage patterns during the 1994 Northridge, California, Earthquake, Proc. ASCE Specialty Conf. on Geotechnical Earthquake Eng. and Soil Dynamics, Seattle, Washington, *Geotech. Special Publ. No.* 75, ASCE, Vol. I, 714-725.
64. Trifımac, M.D. and M.I. Todorovska (1998). Nonlinear soil response as a natural passive isolation mechanism - the 1994 Northridge, California, earthquake, *Soil Dynam. and Earthquake Engrg,* 17(1), 41-51.
65. Trifunac, M.D., V.W. Lee and M.I. Todorovska (1999). Common problems with automatic digitization of strong motion accelerograms, (submitted for publication).
66. Trifımac, M.D., S.S. Ivanovic, M.I. Todorovska, E.I. Novikova and A.A. Gladkov (1999). Experimental evidence for flexibility of a building foundation supported by concrete friction piles, *Soil Dynam. and Earthquake Engrg,* 18(3), 169-187.
67. Udwadia, F.E. (1977). Uniqueness problems in structural identification from strong motion records, *Proc. Sixth World Conf. On Earthquake Eng.,* Sarita Prakasan, Meerut, India, Vol. II, 1010-1015.
68. Udwadia, F.W. and M.D. Trifunac (1974). Time and amplitude dependent response of structures, *Earthquake Engrg and Stnıct. Dynam.*, 2, 359-378.
69. Witson, E.L., J.P. Hollings and H.H. Dovey (1975). Three dimensional analysis of building systems, Earthquake Engrg Research Center, *Report No. 75-l3,* Univ. of California, Berkeley.

70. Wang, D. and A. Haldar (1997). System identification with limited observations and without input, *J. Engrg Mech.*, ASCE, 123(5), 504-511.
71. Ward, H.S. and R. Crawford (1966). Wind induced vibrations and building modes, *Bull. Seism. Soc. Amer.*, *56*, 793-813.
72. Wemer, S.D., L.C. Lee, N.L. Wong and M.D. Trifunac (1977). An evaluation of the effects of traveling seismic waves on the three-dimensional response of structures, *Report R07720-4514,* Agbabian Assoc., El Segundo, Califomia.
73. Whitcomb, J. (1969). Detecting of SH-type seismic shear waves by means of angular accelerometers, *Geol. Survey Professional Paper No. 599-D.*
74. Wong, H.L., J.E. Luco and M.D. Trifunac (1977). Contact stresses and ground motion generated by soil-structure interaction, *Earthquake Engrg and Struct. Dynam., S,* 67-79.
75. Wunderlich, W., E. Stein and K .J. Bathe (editors) (1981). Nonlinear finite element analysis in structural mechanics, *Proc. Europe-U.S. Workshop,* Springer-Verlag, Berlin.

THE U.S. ARMY CORPS OF ENGINEERS SEISMIC STRONG-MOTION INSTRUMENTATION PROGRAM

ROBERT F. BALLARD, JR.
Geophysicist, Geotechnical Laboratory,
U.S. Army Engineer Waterways Experiment Station,
Vicksburg, MS 39180-6199
Email: ballard@mail.wes.army.mil

Abstract

The U.S. Army Corps of Engineers (USACE) currently operates a seismic Strong-Motion Instrumentation Program (SMIP) throughout the United States to provide a measure of project performance, provide insight into the safety of USACE projects, and establish a data base for earthquake research. Strong-motion instruments used for SMIP consist of digital and analog accelerographs, peak acceleration recorders, and seismic alarm devices. These instruments have been placed at earth, rock, earth and rock, arch, and gravity dams owned and operated by the USACE. At present, 123 projects located in 32 states and Puerto Rico are monitored with more than 500 strong-motion instruments. The USACE network is second in size to that operated by the California Division, of Mines and Geology.

The purpose of this paper is to present various aspects of the USACE SMIP including criteria for design of installations, recording equipment, operation, maintenance, performance to date, upgrades, future goals and the importance of interagency cooperation. Particular attention is focused on economics and advantages associated with ultimate conversion to remotely accessed digital instrumentation.

1. Evolution Of Smip

1.1 PURPOSE

During the 1970s, the U.S. Army Corps of Engineers (USACE) embarked upon an undertaking which has since been termed the Strong-Motion Instrumentation Program (SMIP). SMIP was designed to allow observation and analysis of seismic waves produced by earthquakes and explosions to examine the effect of these motions on USACE projects. The objectives of SMIP are threefold:

M. Erdik et al. (eds.), Strong Motion Instrumentation for Civil Engineering Structures, 157–166.

(a) to provide insight into the safety of and to act as inspection guide for existing USACE projects,
(b) to provide a measure of project performance, and
(c) to act as a data base for performance predictions and earthquake research.

The existence of performance data in the engineering profession is unequivocally beneficial. For instance, D'Appolonia (1990) describes the value of field performance data for geotechnical engineering. He summarizes:

"Data from long-term monitoring should be integrated into the design process [to] provide a basis for future decisions and maintain a facility in a functional state consistent with its intended purpose. A planned approach to decision making over time that draws on long-term field measurements for input, with planned analysis of the measurements and appropriate contingent actions, is sought. A monitored-decision process provides a means to gain knowledge, be innovative, and mitigate adverse relationships between parties involved in . the ownership, construction, and operation of a facility."

1.2.BACKGROUND

As described by Ballard et al. (1990), the SMIP was formalized in 1973 by issuance of Engineer Regulation (ER) 1110-2-103 (Department of the Army, 1973) which essentially required instrumentation of all USACE dams that exist within seismic risk zones 2, 3, and 4 (after Algermissen, 1969). After careful planning and deliberation with experts in the field of earthquake measurements and analysis, a revised Engineering Manual (EM) 1110-2-1908 (Department of the Army, 1976) was published for use by USACE district offices. This EM provides guidance and information concerning the selection of instruments for measuring movements and earth pressures in earth and rock-fill dams and describes techniques for collecting and analyzing data.

In 1973, the USACE also entered into agreements with the U.S. Geological Survey (USGS) to assist in SMIP. Specifically, the US GS was to:

(a) provide guidance in the selection of instruments,
(b) review installations for conformance with network specifications and provide suggestions for proper mounting from weather and other elements,
(c) act as recipient (thus assuring adherence to specifications) for new instruments and to calibrate them for installation, and
(d) install and maintain those instruments at regular intervals throughout the federal fiscal year (October through September).

At that time, estimates of less than 200 installed instruments were provided to the USGS as the number that would ultimately make up the USACE strong -motion network. Since a large number of these installations would be east of the Rocky Mountains, the USGS viewed them as a necessary extension of an envisioned national network which at that time had most recording stations located in the state of California.

In 1977, the USGS was assigned additional missions and funding without commensurate increases in personnel allotments. The U.S. Army Engineer Waterways Experiment Station (WES) subsequently proposed alternative plans for programs which would enable the USACE to absorb these additional requirements. After a very thorough investigation of alternatives, it was decided that the US GS would continue to provide installation, service, and data collection for instruments in the Western U.S. WES was to phase itself into installation and maintenance of approximately one-half of the USACE instruments located in the central and eastern part of the U.S. Additionally, on a reciprocal arrangement, WES was to service certain USGS and Veterans Administration (VA) instruments. This transition occurred over a two-year period. Since 1978, WES Instrumentation Services Division (ISD) has assumed responsibility for in-house maintenance of more than fifty percent of the USACE instruments with USGS providing service for the remainder. Personnel of WES Earthquake Engineering and Geosciences Division (EEGD), Geotechnical Laboratory (GL), provide overall project management of SMIP and analyze recorded data.

Functionally, the SMIP has been structured so that a USACE agency can design its own program for strong-motion instrumentation with guidance from the Engineering Manual. Upon completion of a plan for instrumenting a specific structure, the individual agency then forwards its plan to WES for approval. Once it is determined that all criteria have been taken into consideration for that particular project, WES then approves the installation. By directive, WES is also responsible for:

(a) maintaining records of instrument servicing and location,
(b) reviewing instrument locations and type to assure conformance with USACE policy,
(c) processing and analyzing records obtained,
(d) furnishing copies of obtained records to the USACE district offices concerned,
(e) coordinating with USGS and the USACE district offices to establish schedules for inspection visits,
(f) billing USACE district offices for services provided,
(g) reimbursing USGS for expenses incurred,
(h) providing personnel for installation and maintenance of USACE instruments not serviced by USGS,
(i) preparing an Engineer Circular on the status of the program for Corps-wide distribution.

In addition to its heavy involvement with the USGS, the USACE has established a working arrangement whereby data are exchanged and coordination established with the state of California strong-motion network and with the U.S. Department of the Interior, Bureau of Reclamation (BuRec) In actuality, the state of California operates the largest network of strong-motion instrumentation. The strong-motion instrumentation program instituted by the BuRec is rapidly expanding. It is intended that close ties remain in effect between all of these agencies.

1.3.USACE MANDATE FOR STRONG-MOTION INSTRUMENTATION

According to an updated ER 1110-2-103 (Department of the Army, 1981) issued by USACE, all dams in seismic risk zones 2, 3, and 4 of the seismic risk maps should be instrumented for strong-motion earthquake measurement. As previously mentioned, guidance on details concerning instrumentation, location, and selection is currently given in EM 11102-1908. These documents are adequate for most situations; however, numerous questions have since been raised regarding instrumentation of dams in seismic risk zone 2. A popular viewpoint was that the low probability of obtaining meaningful data does not offset the cost of installing and maintaining instruments. In an effort to supplement the above documents (particularly for zone 2) so that sound decisions can be made regarding dams with uncertainties about the need for instrumentation, additional guidance was advocated. Considerations below are listed in order of relative importance.

1. <u>Nature of Foundation</u>. If foundation materials underlying the dam are composed of sands or silty sands that might be subject to liquefaction, the dam should be instrumented. If the foundation materials are rock or other materials that are not subject to liquefaction, the remaining factors below should be taken into account.

2. <u>Type of Construction</u>. Regardless of seismic risk zone, all hydraulic fill dams should be instrumented. Rolled earth fill or rock fill (being less susceptible to liquefaction) should be considered for instrumentation by other influencing factors.

3. <u>Height of Dam</u>. Most dams more than 33 m high should be instrumented.

4. <u>Presence of Known Capable Faults</u>. If the dam is located nearer than 40 km to a known capable fault, it should be instrumented.

5. <u>History of Seismic Activity at the Site</u>. If acceleration levels greater than 0.2 g have been recorded in the vicinity of the dam, it should be instrumented.

6. <u>Distance from Higher Risk Zone Boundaries</u>. If the dam is located less than 160 km from a higher risk zone boundary, it should be instrumented.

2. Locations Of Strong-Motion Instruments

As of September 1992, the USACE SMIP consisted of the following: 123 instrumented projects located in 32 states and 1 commonwealth. The locations of these projects are shown on a map which also incorporates seismic zoning boundaries.

3. Descriptions Of Strong-Motion Instruments

The instruments used for SMIP range from seismic alarm devices to digital accelerographs. In all, 424 accelerographs, 45 peak acceleration recorders, and 35 seismic alarm devices are presently used for SMIP. Seismoscopes were eliminated from the network during Fiscal Year 92 because the value of seismoscopes as reliable devices for documenting sites subjected to earthquakes has been subject to question.

Accelerographs are the most versatile and widely used instruments by SMIP for recording strong motions. Accelerographs may be analog or digital devices. For new SMIP installations and upgrades of existing sites, WES is currently installing digital accelerographs manufactured by Kinemetrics Inc., of Pasadena, California (Model SSA-1, SSA-2 or Etna).

The Kinemetrics Model SMA-1 is an analog triaxial strong-motion accelerograph that photographically records strong motions on 70 mm film. It employs three flexure-type accelerometers (longitudinal, vertical and transverse) in a orthogonal arrangement and has a maximum recordable peak acceleration of a standard 1.0 g. A vertical acceleration-sensitive starter (pre-set at a level of 0.01 g for all instruments) senses the initial ground motion (P-wave), and actuates the SMA-1 in less than 50 msec (0.05 sec). The device continues to operate for a duration of 10 seconds after the vertical starter no longer senses motions above the preset trigger level. The film is recovered during semi-annual service trips or shortly after a known strong-motion event. An attached event counter provides the number of times the instrument was activated. This count is very important to service personnel. Although it is possible to develop film in the field, photographic laboratory developing is preferred. The Kinemetrics SMA-1 is the most widely used analog accelerograph for the SMIP (these are, however, being replaced as rapidly as possible with digital units).

The Kinemetrics Model SSA-1 is a solid-state, digital strong-motion accelerograph that records seismic events at 200 samples per second per channel. The SSA-1 is configured to record four external channels of data from Kinemetrics FBA-11 and FBA-13 force-balance accelerometers. SSA-1 trigger thresholds are determined by a software-based algorithm with a bandwidth of 0.1 to 12 Hz, preset for each of the three data channels. When signal amplitude exceeds a preset trigger threshold (normally 0.01g), the SSA-1 records and stores acceleration data in CMOS RAM. On-site data retrieval involves downloading to an IBM-compatible laptop personal computer by copying data files to a floppy diskette or hard disk. The Kinemetrics SSA-1 can be interrogated remotely via telephone modem and is considered one of the most technically advanced accelerographs available today.

Addressing the market need for a more economical digital accelerograph, Kinemetrics introduced the model SSA-2 which borrows heavily from the SSA-1. This unit retains all of the "necessary" features of the SSA-1 minus a few convenience items to appreciably reduce cost. The SSA-2is currently under evaluation by WES to determine its position within the SMIP.

Peak Acceleration Recorders (PAR) provide a low-cost method for detecting a strong-motion event. Terra Technology of Redmond, Washington, and Engdahl Enterprises of Costa Mesa, California, manufacture peak acceleration devices installed and maintained by the SMIP. The Terra Technology PRA-103 employs a spring-mass magnetic stylus on a magnetic tape and records a peak acceleration of 2.0 g. The tape is returned to the laboratory where it is dusted with ferric powder and analyzed. The PAR-650L from Engdahl records a peak acceleration level of up to 2.5 g. A diamond stylus rests on a soft metal plate that is etched when strong motions are detected. The metal plates are returned to the manufacturer for resurfacing after the recorded event is analyzed. The PAR-650L can have a local or remote annunciator for preset accelerations. Approximately 75 percent of peak acceleration devices serviced by WES are the model PRA-103 from Terra Technology.

Due to non-availability on the commercial market, the Seismic Alarm Device (SAD) was designed and is produced by WES. This alarm package contains a vertical accelerometer with ten individual threshold level relays. The latching relay bank stores accelerations greater than the preset threshold. A light-emitting diode (LED) indicates peak acceleration .on the main control board. The standard SAD is calibrated to display peak accelerations in steps of 0.05 g from 0.05 to 0.50 g. Accelerations at or greater than the threshold cause the LED to illuminate and sound an alarm indicating the instrument has triggered and should be inspected. This device has also been adapted to activate an automatic telephone dialer and remote annunciator.

4. Operation And Maintenance

ISD personnel, an electronics engineer and two electronic technicians, service SMIP instruments under the jurisdiction of VIES. Typically one electronics technician is involved in servicing that includes four routes looping through all project sites twice a year. Plenty of spare parts are at hand to make the service teams autonomous. A detailed inspection record for each device is completed on location and accompanies recorded data to WES for interpretation and cataloguing in a computer data base.

Once accelerometer recording media are delivered to WES, it is processed and baseline and instrument corrections are made. Kinemetrics developed (PC-compatible) software is used to process data received at VIES. For analog records, the traces are first optically digitized at a rate of 600 points per second of record. Computer algorithms written by US GS are used to integrate the variation of acceleration with time to obtain velocity and displacement records. Further processing includes plots of response spectra. Once all data reduction is completed, a WES seismologist analyzes results to determine natural periods and various amplification factors. Reports are published for larger events under sponsorship of the respective USACE district office (e.g., Chang, 1985).

In an effort to keep operating costs at a minimum while increasing reliability and overall effectiveness of the SMIP, it is periodically necessary to modify and/or upgrade various instruments as maintenance records dictate and technological advances are made. On an annual basis, WES prepares an Engineering Circular which serves as a status report for the USACE. Included therein are current estimated costs for installation and servicing instruments and a tabulation of strong-motion instrumentation both operational and planned. Charges for services are adjusted annually on the basis of actual cost experience.

5. Earthquake Data Retrieval

The National Earthquake Information Center (NEIC),located in Golden, Colorado, dispenses factual information about an earthquake within minutes of its occurrence. Such factors as location, magnitude, damage, casualties, and history of previous seismic events are sent to interested parties throughout the world. Access to the NEIC data base is available to anyone via computer and telephone modem. The system is called "Quick Epicenter Determinations" (QED) and there is no charge for logging onto the NEIC system. More information may be obtained by calling the USGS.

WES is linked to NEIC via teletype located in a USACE communications center in Vicksburg, Mississippi. When an earthquake occurs anywhere in the world, NEIC information is received at the communications center and relayed to designated WES personnel. If the earthquake exceeds magnitude 4.5 and its epicenter is located within 80 km of an instrumented USACE structure, WES contacts the responsible USACE district office to determine whether' or not instruments have triggered. If personnel on the site confirm instrument activation, special arrangements are made to service the instrument(s).

6. Significant Earthquakes Recorded To Date

Since inception of the SMIP, a number of significant earthquake records have been acquired. One of the most important recent events was an earthquake that occurred near Franklin Falls, New Hampshire, on 18 January 1982. This event was rated at 4.8 on the Richter Scale and triggered some 13 instruments in the New England area. These data are the most significant strong motions recorded in the New England area in over 40 years. Detailed analyses indicated that preconceived notions about attenuation factors, frequencies, and amplitudes should be revised for the New England area.

Other high quality records have been obtained at Coyote Dam, California, in March 1978, Mt. Borah, Idaho,in October 1983 (Chang 1985), and Whittier Narrows, California, in October 1987 and in March 1988. The Mt. Borah earthquake (Ms = 7.3; USGS, 1983) was recorded at Dworshak Dam (330 km from epicenter), Lucky Peak Dam (180 km from epicenter), and at Ririe Dam (179 km from epicenter) which was constructed and instrumented by USACE but is presently owned and operated by the BuRec. The 1987 Whittier Narrows earthquake was recorded at Brea, Carbon Canyon, Prado, San Antonio, Sepulveda, and Whittier Narrows Dams. No SMIP instruments were triggered during the October 1989 Loma Prieta, California, earthquake (M = 7.1). The nearest USACE project (New Hogan Dam, California) was located more than 130 km from the epicenter. More recently, the magnitude 5.8 "Sierra Madre" earthquake of 28 June 1991 in the Los Angeles area was recorded by USACE instruments at seven sites. On 22 April 1992, a magnitude 6.1 quake epicentered near Desert Hot Springs, California, was recorded by six USACE instruments. Numerous other less significant earthquakes were also recorded.

7. Future Goals

In the beginning, the SMIP was designed to provide insight into the safety of, and to act as an inspection guide for existing and future structures. It was devised to provide a measure of project performance and design performance comparisons, and to act as a data base for performance predictions and earthquake research. These goals are being achieved. Because of the relative short recorded history of seismic events in the United States, seismic risk maps are being continually updated but still can only give an approximation of the long-term hazard.

As more information is gained and technological advances made, both in terms of instrumentation and analytical seismic analysis techniques, more reliable assessments of USACE projects will be made. As confidence is gained, many of the very conservative, assumptions now being used to assess structure stability will undoubtedly be revised to more realistically approach the problem. By so doing, it is reasonable to assume that many structures which are thought to be borderline by today's standards can be conclusively assessed as safe by future standards thus eliminating or drastically modifying expensive remedial actions. These measures will invariably result in reduced costs.

A second goal, minimizing on-site inspections after earthquakes, has already proven feasible in areas of high earthquake activity through the development and installation of seismic alarm devices on USACE projects. Following an event, site personnel can readily determine if any of the preset threshold levels were reached. After enough data are obtained through the SMIP, an acceptable threshold level of acceleration can be safely established for each individual structure.

Inspections would then be required only if that threshold level is exceeded- thereby saving numerous operational man-hours. As an additional feature which is highly desirable for unmanned structures, an automatic telephone dialer can be incorporated to relay information to district or division offices.

In the normal course of technological improvements in instrumentation, it is expected that digital accelerographs will ultimately replace analog instruments. Performance and reliability evaluation of digital devices is an ongoing goal of the SMIP. As performance evaluations show more and more positive signs of reliability, recommendations will be made to incorporate specific models which meet SMIP criteria. A prime objective of the strong-motion instrumentation program is to automate data acquisition and instrument status, via computer modem or FM radio telemetry remote capability, for better response and control of the network.

Acknowledgments

The seismic instrumentation program described herein, unless otherwise noted, was conducted under the Strong-Motion Instrumentation Program of the United States Army Corps of Engineers by the Waterways Experiment Station. Permission was granted by the Chief of Engineers to publish this information.

Mrs. Tina H. Grau, GL, WES, prepared the figures for and edited this paper. Messrs. M. B. "Joe" Savage, Jr. and Lewis Smithhart, ISD, WES, assisted in providing descriptions of instrumentation systems, maintenance and operation procedures, and the data reduction process.

References

1. Algermissen, S. T. "Seismic Risk Studies in the United States." *Proceedings, 4th World Conference on Earthquake Engineering,* Santiago, Chile, 1969.
2. Ballard, R. F., Jr., Comes, G. D., and Sykora, D. W. "Overview of the U.S. Army Corps of Engineers Seismic Strong-Motion Instrumentation Program (SMIP)", *Proceedings, Symposium on Engineering Geology and Geotechnical Engineering,* Pocatello, Idaho, 1990.
3. Chang, F. K. "Analysis of Strong-Motion Data from the Mount Borah, Idaho, Earthquake of 28 October 1983," Miscellaneous Paper GL-85-12, U.S. Army Engineer, Waterways Experiment Station, Vicksburg, MS, 1985.
4. D'Appolonia, E. "Monitored Decisions," *Journal of Geotechnical Engineering,* AS CE, Vol 116, No. 1, pp 4-34, 1990.Department of the Army. "Instrumentation of Earth and 5-5-5-5-
5. Rock-Fill Dams," Part 2 of 2, EM 1110-2-1908, USACE, Washington, DC, 19 November 1976.
6. __. "Engineering and Design: Strong-Motion Instruments for Recording Earthquake Motions on Dams," ER 1110-2-103, USACE, Washington, DC, 10 December 1981.

Corps of Engineers Strong-Motion Instrumentation Program Status - September 1992

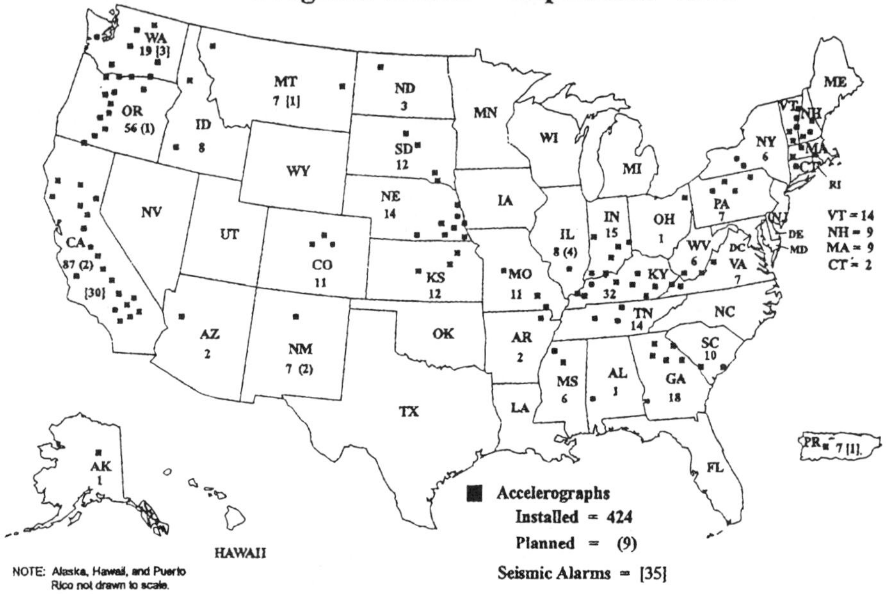

Figure 1. Location of strong-motion instrumentation.

INTEGRATED SURFACE AND BOREHOLE STRONG-MOTION, SOIL-RESPONSE ARRAYS IN SAN FRANCISCO, CALIFORNIA

R. D. BORCHERDT, H. P. LIU, R. E. WESTERLUND, C. DIETEL, J. F. GIBBS, R. E. WARRICK
United States Geological Survey
Menlo Park, CA 94025
e.mail:borcherdt@usgs.gov

1. Abstract

An integrated set of four borehole arrays and ten surface installations is installed in the city of San Francisco, California to measure the response of soft-soil deposits to strong earthquake ground motions. The borehole arrays extend through thick layers of soft water-saturated soils of Holocene age and older more consolidated soils of Pleistocene age into bedrock at depths up to 100 m. The surface installations are configured in pairs to provide simultaneous comparative surface measurements of soft soils and nearby rock. The rock locations also permit comparative measurements of rock as observed at the surface and in nearby boreholes. Complementary structural response arrays also are installed near each array.

The arrays are designed to address a wide variety of scientific and engineering issues, and especially the issue of anelastic and nonlinear soil response at high strain levels. Exact anelastic models have been developed to account for contrasts in anelastic properties at boundaries and the resultant inhomogeneity of propagating wave fields. These models predict that significant amounts of energy may be trapped in soil basins with resultant larger amplifications than can be predicted using conventional homogeneous wave-field models with damping. Results of these models are discussed.

2. Introduction

Structural damage and consequent loss of life from earthquakes is often concentrated in areas underlain by soft soils. Damage from modern earthquakes such as those affecting Mexico City (1985), Leninakan, Armenia (1988), San Francisco, California (Loma Prieta 1989), and Kobe, Japan (1995) imply that a major portion of the damage occurred in areas underlain by soft soils. These increased amounts of damage are due to amplification effects of soft soil deposits on certain frequencies of ground shaking and in some cases associated soil failures. Nevertheless, relatively scarce empirical information exists concerning the in-situ response of such deposits. Few strong-

M. Erdik et al. (eds.), Strong Motion Instrumentation for Civil Engineering Structures, 167–178.

motion recordings have been obtained in boreholes necessary to define the actual amplification characteristics as a function of input ground shaking.

In the city of San Francisco, damage from both the great California earthquake of 1906 and the Loma Prieta earthquake of 1989 was concentrated in areas underlain by soft soil deposits. Many of these areas are along the margins of San Francisco Bay and are densely urbanized. However, no in-situ borehole measurements of the response of these deposits have yet been obtained at damaging levels of motion. Such measurements of the next large damaging earthquake will be of special importance for mitigation of future earthquake losses.

Several important questions exist concerning the in-situ response of soft soil deposits at damaging levels of shaking. A question of particular interest concerns the role of non-linearity in modifying the in-situ amplification characteristics of soft soil deposits as a function of shaking amplitude. This is a major issue yet to be resolved with in-situ evidence. Other questions concerning changes in constitutive properties with duration, ground motion characteristics necessary for the onset of liquefaction, and the influence of basin geometry also are of considerable interest.

This paper describes a set of integrated borehole arrays and surface installations in the city of San Francisco designed to address these questions. It describes the design objectives of the arrays, their location, geologic setting, seismic and geologic information, and instrumentation configuration.

3. General Design Objectives

The integrated strong-motion soil response arrays in San Francisco operated by the US Geological Survey are designed to provide comprehensive data on the response of "soft soils" at four different locations in the city underlain by soft water saturated clays, sands, and silts. Major concentrations of damage occurred in these areas during both the 1906 and 1989 earthquakes. Evidence for liquefaction during the earthquakes exists at three of the four sites. Locations of the arrays are shown in Figure 1a. The arrays are located in the vicinity of Embarcadero Plaza (EMB) near lower market street, in Levi Strauss Plaza near the base of Telegraph Hill (LP), at Winfield Scott School in the Marina district (WSS), and at Bessie Charmichael School (BCS) in an area south of Market street. Surface installations with three-component velocity and three-component force-balance accelerometers are installed at 10 additional sites (see Figure 1a).

Measurements in these areas are particularly relevant for purposes of hazard mitigation, because a number of important facilities including high-rise buildings, apartment buildings, rapid transit facilities, bridges, and underground utilities are located in these areas. The arrays are designed to provide new *in-situ* data on the constitutive behavior of soils as a function of strain amplitude and liquefaction-induced failure. They are designed to complement strong-motion data recorded on nearby well-instrumented buildings together with data recorded at surface installations (see Figure 1b).

The arrays are designed to provide an integrated set of data pertinent to addressing a wide variety of scientific and engineering issues. Each array includes sensors located at depths ranging from the surface to below the bedrock interface. Sensors at three of the borehole locations (BCS, LP, EMB) are at four different levels. Sensors at the WSS location are at three different depths. The surface installations permit comparative studies of soil and and rock motions as observed at pairs of closely spaced soil-rock locations. The surface rock locations also permit comparative studies of motions as observed both at the surface and in nearby boreholes at the soil-rock interface. Surface and borehole motions will provide an improved definition of input ground motions for understanding the response of nearby high-rise buildings, bridges, harbor facilities, and a variety of other types of well instrumented structures in San Francisco.

The borehole instrumentation at three of the locations includes pairs of pore-pressure and force-balanced accelerometers (FBA) collocated at two depths. These pairs of sensors provide *in-situ* measurements of the temporal and spatial variations in pressure and ground displacements. Such collocated measurements using FBAs and dilatometers at other locations in California have been shown (Borcherdt, et al. 1988 and 1989) to provide important new information on *in-situ* material behavior as a function of time not provided by either measurement alone. These studies have shown that simultaneous collocated measurements provide *in-situ* constitutive properties of the surrounding medium as a function of strain amplitude and number of cycles. The extent to which these concepts can be used to help resolve important issues regarding non-linear dynamic soil behavior and its role in influencing ground and structural response is an important design objective of the arrays at three of the locations (BCS, LP, EMB). Each of these arrays also includes two additional FBAs at greater depths including one in bedrock. Installation of an additional velocity transducer is being planned for each site.

One of the arrays is designed with collocated sets of FBAs and velocity transducers (WSS). This array is intended to provide high resolution recordings ranging in amplitude from seismic background noise levels to damaging levels of shaking up to 2g. The array, initially installed with velocity transducers, has provided an extensive data set interpreted by Liu, et al. (1992).

4. Geologic Setting

The San Francisco Bay basin is a northwest-trending graben bounded on the northeast and southwest by the active Hayward and San Andreas faults. San Francisco Bay, located in the center of this graben, is bounded by marshlands, alluvial plains, and hills at distances of 1-15 km from the Coast Ranges. A wide variety of geologic units exist in close proximity throughout the region. They range from Mesozoic sedimentary and plutonic rocks in the hills, to Holocene alluvium and estuarine mud near the margins of the bay. Mesozoic and Tertiary rock units form the floor of the graben. This surface is overlain by a sedimentary sequence of which the lower part was initially designated Older bay sediments and the upper part Younger bay mud. Geologic characteristics, physical properties, and seismic shear-wave velocities for the various surficial units in the San Francisco Bay region are summarized in detail by Borcherdt and Glassmoyer (1994).

The Younger bay mud unit is of particular interest, because it yields the largest amplifications of ground motion (Borcherdt 1972, Borcherdt and Gibbs 1976). Thick sections of this unit exist at each of the borehole sites. In general, it is Holocene in age. It is composed of unconsolidated, water-saturated, dark plastic clay and silty clay rich in organic material with local lenses of well-sorted silt and sand and some beds of peat. It attains thickness of 37 meters beneath the bay and thins to vanishing thickness along the margins of the bay. It generally contains more than 50 percent water. It is overlain at each of the borehole array locations by fill, comprised of hydraulically emplaced sands and clays or earth and rock with varying degrees of compaction. Measurements of shear-wave velocity range from 55 to 100 m/s in the depth interval 0 to 10 m and from 100 to 115 m/s in the depth interval 10 to 20 m (Fumal, 1978).

The Older bay sediment unit is in general comprised of Pleistocene deposits ranging in composition from weakly to moderately consolidated, deeply weathered, poorly sorted, irregularly interbedded clay, silt, sand, and gravel to loose, well-sorted, fine- to medium-grained sand with subordinate silt (Helley and Lajoie 1979). This Older Bay Sediment unit reaches overall thickness generally less than 200 m in the northern bay region. Shear-wave velocities range from 150 to 219 m/s for dune sands near the surface to 418 to 749 m/s for coarse-grained, sandy gravels of late Pleistocene age.

The semi-consolidated and consolidated units in the hills of the San Francisco Bay region include plutonic, volcanic, and a wide range of sedimentary rocks. Shear-wave velocities in the top 30 m of these units range from about 425 m/s for soft sandstone to about 1,650 m/s for hard greenstone and fresh granite.

The depths for sensors were chosen to monitor behavior of potentially liquefiable granular layers near the surface, the underlying Holocene bay mud, the underlying Older bay sediment layer, and the underlying bedrock.

5. Geologic and seismic Logs

During the initial drilling a relatively complete set of geologic and seismic logs were obtained at each of the sites (Gibbs, 1994, Kayan, et al., 1990). A summary of the logs for each of the sites is shown in Figures 2, 3, 4, and 5. In addition, undisturbed samples were obtained and tested at the WSS site by Kayan et al. (1990).

The logs at each of the sites shows evidence of a thin compacted layer of fill at the surface overlying deposits of soft clay, silty clay, and silts with the lowest shear wave velocities in the top 20 meters being slightly greater than 100 m/s. In general, the shear velocities increase with depth to values near 350 to 400 m/s then show a significant increase to values from 780 to slightly more than 800 m/s for the underlying bedrock.

6. Instrumentation

The sensors deployed for direct measurement of acceleration are three component borehole force-balance accelerometers with full scale at $\pm$ 1 or $\pm$ 2 g ($\pm$ 5 volts), and frequency response from dc to 100 Hz, models FBA 13 and FBA 23 DH produced by Kinemetrics, Inc. The velocity transducers with a natural frequency of 2 Hz have been developed for borehole deployment (Liu, et al.,1991). The pore-pressure transducers are wide dynamic range (~140 dB) transducers with digital output proportional to pressure applied to a quartz crystal resonator, model 8DP depth sensor produced by Parascientific, Inc.

Signals from each of the sensors at each location are recorded on site using broadband digital recorders (General Earthquake Observation Systems, GEOS, Borcherdt, et al., 1985) with event derived parameters transmitted and recorded in near real time via GOES satellite (Mueller et al. 1993). Signals from the various sensors are recorded at various gain levels depending on the depth and type of sensor. Gains up to 84dB selectable in 6dB steps are available to permit signals over a dynamic range near 180 dB at frequencies near 1 Hz to be resolved from a variety of sensor types. In general, gains for similar sensors in a single borehole array vary between 6 and 18 dB.

Events are recorded in event triggered mode at 200 samples per second per channel using linear 16-bit linear analog to digital conversion. Absolute time and each of the various recording parameters are recorded simultaneously. Event and instrument parameters are transmitted at a selectable time interval usually chosen to be less than 10 minutes via GOES satellite. Parameters such as trigger time, peak amplitude, time of peak amplitude and duration together with a variety of instrument status parameters are transmitted. The satellite transmissions permit the retrieval of critical event and instrument data in near real time. These data are especially useful in maintaining the array and initial evaluations following a major event.

Sensor and recording system calibrations are recorded before and after each storage media change. Since installation of the array no earthquakes in the area have generated signals of sufficient strength to be recorded on the FBAs above the relatively high seismic background noise levels. Signals recorded on the velocity transducer array at the WSS site have been analyzed in detail by Liu, et al. (1992).

7. Exact Anelastic Models of Soil Response

Damping plays an important role in determining the response of soils. In general, it increases with frequency of shaking. Exact anelastic 2 and 3 dimensional wave- propagation models, which account for damping, show that wave fields, which propagate across anelastic boundaries, must, in general be inhomogeneous. In other words, as energy propagates across soil and soil-rock boundaries, contrasts in damping at the boundaries cause amplitudes to vary along surfaces of constant phase, that is the direction of phase propagation differs from that of maximum attenuation of the wave field. An example showing the directions of propagation and maximum attenuation for an inhomogeneous wave field incident on a soil basin is illustrated in Figure 6.

Inhomogeneity of the wave fields in a soil basin gives rise to a set of physical characteristics for the wave field distinct from those that would be predicted by incorrectly assuming homogeneous wave fields. In particular, exact inhomogeneous models imply two types of inhomogeneous S waves, one with linear particle motions and one with elliptical particle motions. They predict the wave speeds, damping, and other physical characteristics change as the angle of incidence changes. (Borcherdt, 1973, Borcherdt and Wennerberg, 1985, Borcherdt et al., 1986).

Numerical results have been derived using Mathematica for a single layer over a half space based on exact theoretical solutions. The material parameters used to characterize the constitutive characteristics of the material are the velocity ratio (V_R) and damping characterized by twice the reciprocal quality factor (Q^{-1}) for homogeneous waves. Inhomogeneity of the wave fields is denoted by gamma, the angle between the direction of propagation and maximum attenuation. The amplitude response of the layer is shown in Figure 7 for a homogeneous wave as a function of angle of incidence and normalized period. (The period of the wave is normalized by the fundamental period.) Figure 7 shows that as the angle of incidence increases, the amplitude response of the surface layer decreases. This result predicts that for conventional homogeneous wave fields propagating nearly parallel to the surface little or no energy is trapped by the soft surface layer, and hence no amplification occurs.

Results also are calculated for an inhomogeneous wave field incident on a surface layer (Figure 8). These results show that as the angle of incidence increases, the amplitude response of the surface layer may increase significantly. In particular this result predicts that for inhomogeneous wave fields propagating nearly parallel to the surface a significant amount of energy may be trapped by a soft surface layer, and hence give rise to a significant amount of amplification. The propagation of such a wave would correspond to a pseudo surface wave being generated at the margin of a soil basin. This theoretical prediction may help explain significant amplification sometimes observed near basin margins.

8. Conclusions

The borehole arrays are designed to provide *in-situ* data not previously available on the non-linear response of "soft soils" as a function of strain amplitude with and without liquefaction-induced failure. Each borehole array includes four three-component forced-balance accelerometers (FBA), two co-located wide dynamic-range pore-pressure transducers, and at least one velocity transducer. The co-located *in-situ* measurements of acceleration and pore pressure are designed to provide estimates of the constitutive properties as a function of time and strain amplitude that cannot be inferred from either measurement alone. A separate borehole array of velocity transducers collocated with FBAs installed at one of the sites (WSS) permits wide dynamic range signals useful for studies of non-linearity to be acquired. Complementary surface bedrock and structural response installations at several nearby sites permit a wide variety of additional scientific and engineering issues to be addressed. Sensor signals are recorded using broadband, wide dynamic range, on-site digital recorders (GEOS). Parameter information for each instrument and each recorded event is transmitted in near-real time via GOES satellite.

Exact theoretical models of anelastic soil response predict that significant amounts of body wave energy being converted to surface waves may be trapped by a layer of soil near the surface giving rise to significant amounts of amplification not predicted by conventional homogeneous wave propagation models. Measurements of strong earthquake ground motions on the integrated arrays will provide empirical evidence to evaluate these and other theoretical model predictions.

9. References

1. Borcherdt, R.D, 1970, Effects of local geology on ground motion near San Francisco Bay, *Bull. Seismol. Soc. Am.*, **60**, 29-61.
2. Borcherdt, R.D., 1973, Energy and plane waves in linear viscoelastic media, *Jour. Geoph. Res.*, **78**, 2442-2453.
3. Borcherdt, R.D, 1994, Estimates of site-dependent response spectra for design (Methodology and Justification), *Earthquake Spectra*, **10**, 617-653.
4. Borcherdt, R.D. and G. Glassmoyer, 1992, On the characteristics of local geology and their influence on ground motions generated by the Loma Prieta earthquake in the San Francisco Bay region, California. *Bull. Seismol. Soc. Am.*, **82**, 603-641.
5. Borcherdt, R.D. and G. Glassmoyer, 1994, Influences of local geology on strong and weak ground motions in the San Francisco Bay region, California and their implications for site-specific code provisions, *in* The Loma Prieta earthquake of October 17, 1989 -- strong ground motion, R. D. Borcherdt, *ed., U.S. Geol. Surv. Prof. Paper 1551-A*, 77-108
6. Borcherdt, R.D., Glassmoyer, G., and Wennerberg, G., 1986, Influence of welded boundaries in anelastic media on energy flow and charac-teristics of general P, S-I and S-II body waves: Observational evidence for inhomogeneous body waves in low-loss solids, *Jour. Geoph. Res.*, **91**, 11,503-11,518.
7. Borcherdt, R.D., and J. F. Gibbs, J.F., 1976, Effects of local geological conditions in the San Francisco Bay region on ground motions and the intensities of the 1906 earthquake, *Bull. Seismol. Soc. Am*, **66**, 467-500.
8. Borcherdt, R.D., Fletcher, J. B., Jensen, E. G., Maxwell, G. L., Van Schaack, J. R., Warrick, R. E., Cranswick, E., Johnston, M.J.S., and R. McClearn, 1985, A general earthquake observation system (GEOS). *Bull. Seismol. Soc. Am.*, **75**, 1783-1825.
9. Borcherdt, R.D, 1988, Volumetric strain and particle displacements for body and surface waves in a general viscoelastic half-space: *Geophys. J. Royal Astr. Soc.*, **93**, 215-228.
10. Borcherdt, R.D., Johnston, M.J.S., and G. Glassmoyer, 1989, On the use of volumetric strain meters to infer additional characteristics of short-period seismic radiation. *Bull. Seismol. Soc. Am*, **79**, 1006-1023.
11. Borcherdt, R.D., Fletcher, J.B., Jensen, E.G., Maxwell, G.L., Van Schaack, J.R., Warrick, R.E., Cranswick, E., Johnston, M.J.S., and McClearn, R., 1985, A general earthquake observation system (GEOS), *Bull. Seismol. Soc. Am.*, **75**, 1783-1825.
12. Borcherdt, R.D., and Wennerberg, L., 1985, General P, type-I S, and type-II S waves in anelastic solids: Inhomogeneous wave fields in low-loss solids, *Bull. Seismol. Soc. Am.*, **75**, 1729-1763.
13. Fumal, T.E, 1978, Correlations between seismic wave velocities and physical properties of geologic materials in the San Francisco Bay region, California, *U.S. Geol. Surv. Open-File Rept. 78-1067.*
14. Fumal, T.E., 1991, Shear-wave velocity estimates and site geology for strong-motion recordings sites of the Loma Prieta earthquake of October 17, 1989, *U.S. Geol. Surv. Open-File Rept. 91-311.*
15. Gibbs, J.F., Fumal, T.E., Borcherdt, R.D. Warrick, R.E., Liu, H.-P., and R.E., Westerlund, 1975, Seismic velocites and geologic logs from boreholes at three downhole arrays in San Francisco, California, *U.S. Geol. Surv. Open-File Rept. 94-706.*
16. Helley, E.J., and K.R. Lajoie, 1979, Flatland deposits of the San Francisco Bay region, California—their geology and engineering properties, and their importance to comprehensive planning, *U.S. Geol. Surv. Prof. Paper 943.*
17. Kayan , R.E., H.-P., Liu, Fumal, T.E., Westerlund, R.E., Warrick, R.E., Gibbs, J.F., and H.J. Lee., 1990, Engineering and seismic properties of the soil column at Winfield Scott School, San Francisco, California, *U. S. Geol., Surv., Open-File Rept. 90-253*, G1-G18.
18. Liu, H.-P., Warrick, R.E., Westerlund, and J.B. Fletcher, 1991, A three-component borehole seismometer for earthquiake seismology. *Bull. Seismol. Soc. Am.* **81**, *2458-2485.*
19. Liu, H.-P., Warrick, R.E., Westerlund, R.E., Sembera, E.D., and L. Wennerberg, 1992, Observation of local site effects at a downhole -and surface station in the Marina district of San Francisco, *Bull. Seismol. Soc. Am.*, **82**, 1563 - 1591.
20. Mueller, R.J., Lee, M., Johnston, M.J.S, Borcherdt, R.D., Glassmoyer, G., and Silverman, S., 1993, Near-real time monitoring of seismic events and status of portable digital recorders using satellite telemetry, *Bull. Seismol. Soc. Am.*, **85**, 640-645.

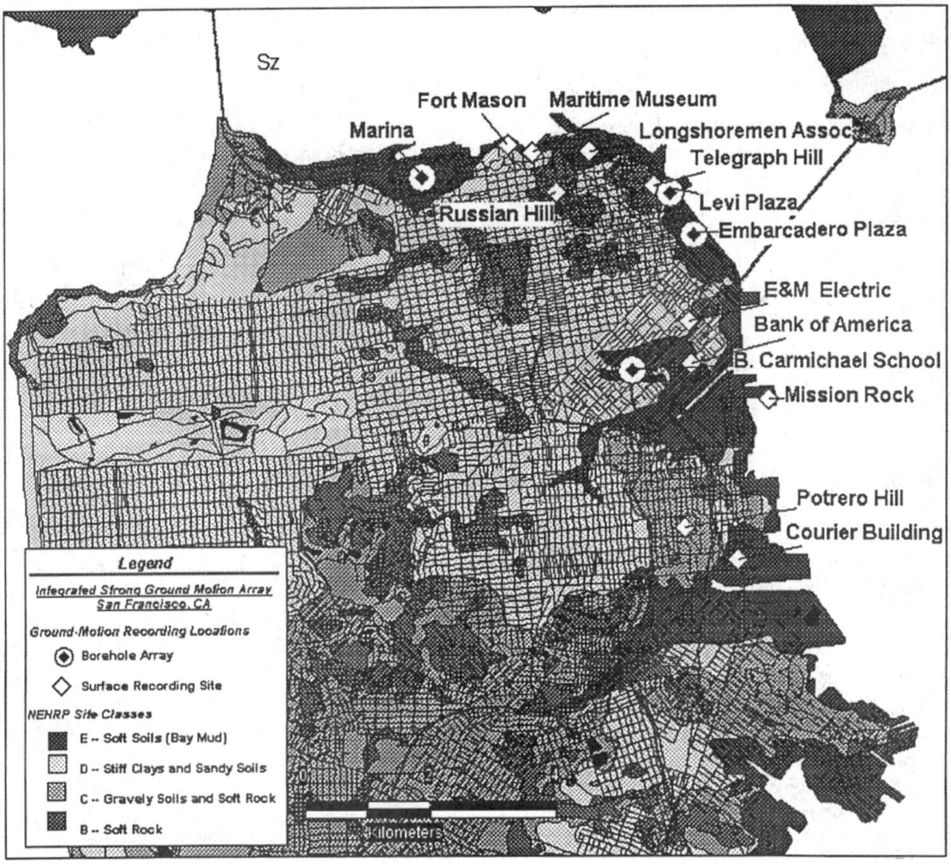

Figure 1a. Map showing location of borehole arrays and complementary surface installations superimposed on map of San Francisco, California showing streets and the NEHRP site classes recently adopted for recommended US building code provisions.

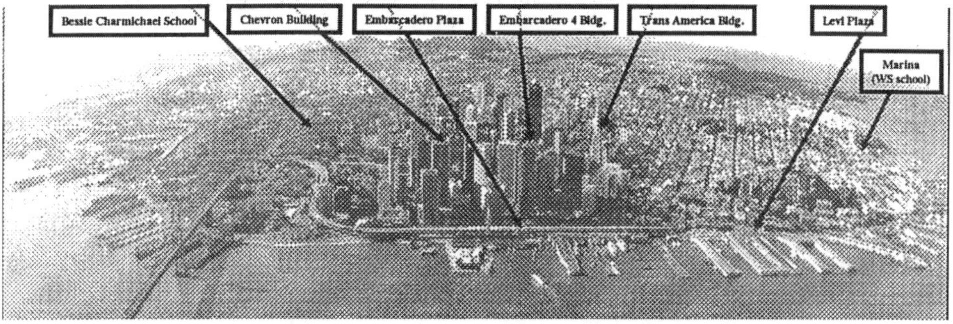

Figure 1b. Map showing location of borehole arrays and nearby structural response arrays superimposed on an areal view of San Francisco, California.

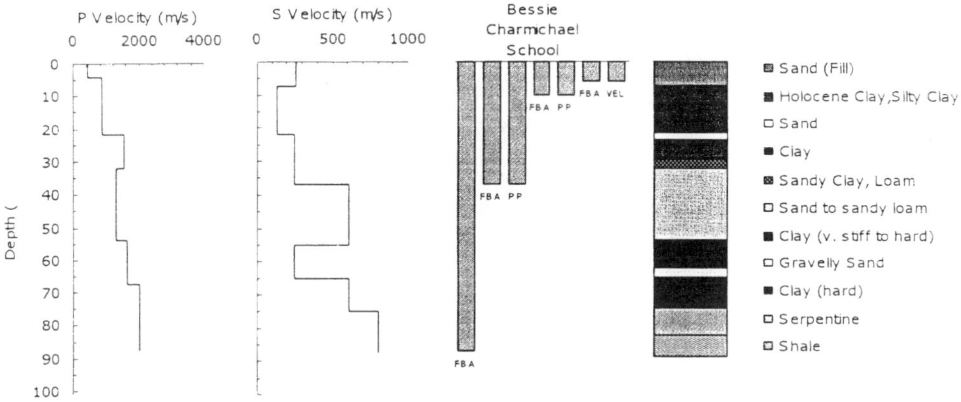

Figure 2. Seismic logs, geologic log, and sensor borehole configuration for the Bessie Charmichael School (BCS) site.

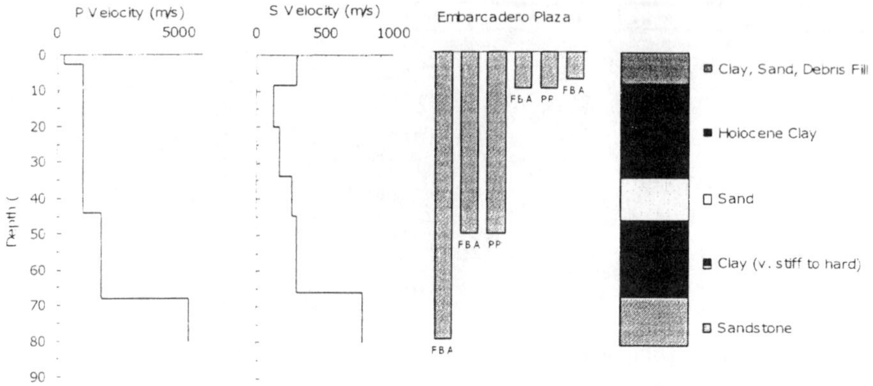

Figure 3. Seismic logs, geologic log, and sensor borehole configuration for the Embarcadero Plaza (EMB) site.

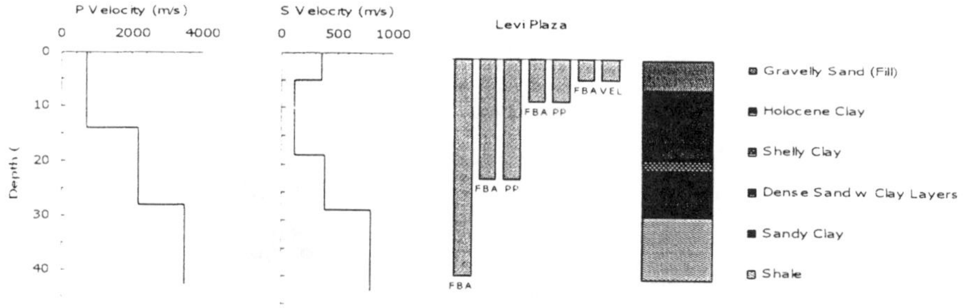

Figure 4. Seismic logs, geologic log, and sensor borehole configuration for the Levi Plaza (LP) site.

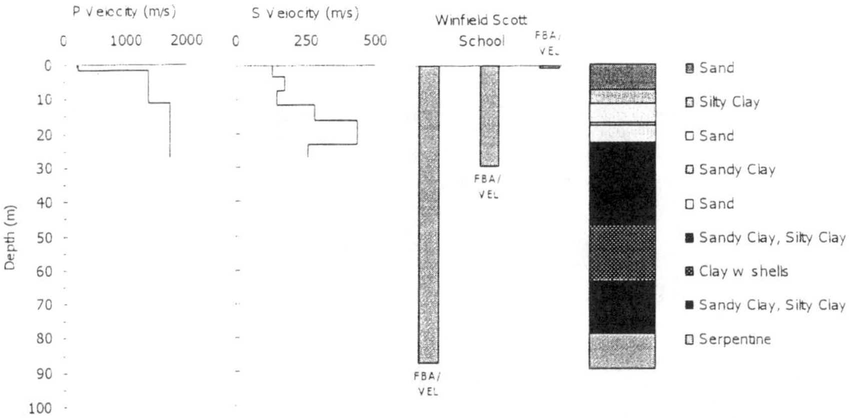

Figure 5. Seismic logs, geologic log, and sensor borehole configuration for the site at Winfield Scott School (WSS) site.

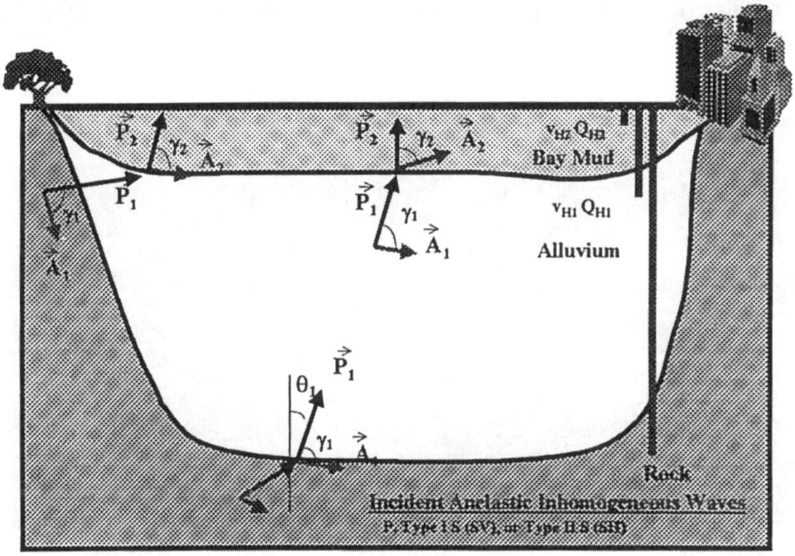

Figure 6. Illustration showing typical configuration for P and S waves in an anelastic soil-rock basin

Configuration illustrates direction of propagation (P) and maximum attenuation (A) for P, SV, and SH waves.

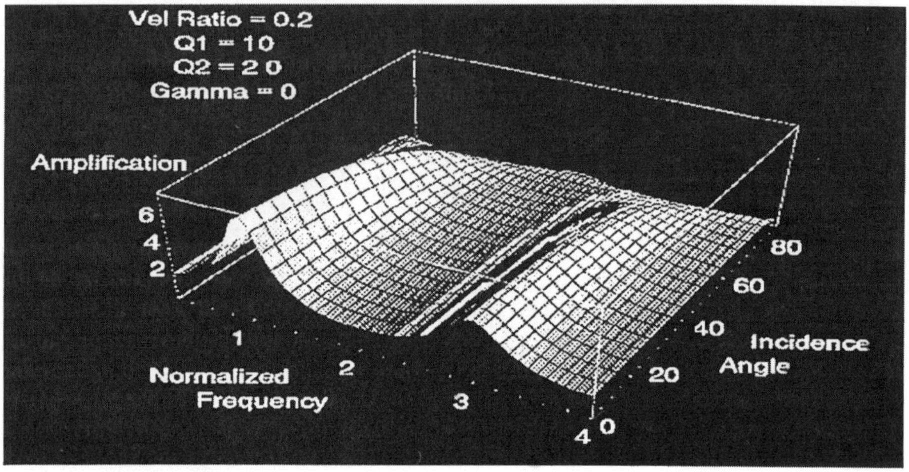

Figure 7 Anelastic response of a soft-soil layer over soft-rock to an incident homogeneous SH wave as a function of angle of incidence.

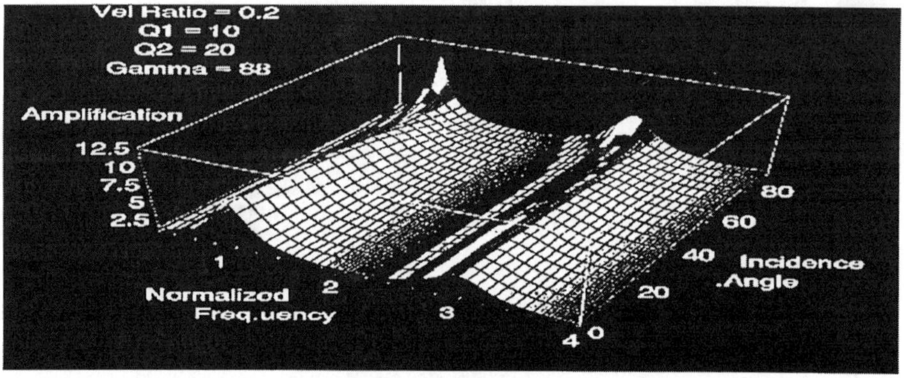

Figure 8. Anelastic response of a soft-soil layer over soft-rock to an incident inhomogeneeous SH wave as a function of angle of incidence.

CURRENT AND NEW TRENDS IN UTILIZATION OF DATA FROM INSTRUMENTED STRUCTURES

MEHMET ÇELEBI
U.S. Geological Survey,
345 Middlefield Rd. (MS977),
Menlo Park, CA 94025
e-mail:celebi@samoa.wr.usgs.gov

Abstract

The many uses of seismic response data from instrumented structures include assessment of design and analysis procedures, improvement of code provisions, and correlation of system response with damage. A preliminary list of applications of response data with sample references is provided. Three examples of recorded data and summary analyses are provided to illustrate uses of response data.

KEYWORDS: monitoring, instrumentation, utilization, strong-motion, building, structural response, frequency, displacement, acceleration

1. Introduction

Seismic monitoring of structural systems constitutes an integral part of the National Earthquake Hazard Reduction Program in the United States and similar programs in other countries. Recordings of the acceleration response of structures have served the scientific and engineering community well and have been useful in assessing design/analysis procedures, improving code provisions and in correlating the system response with damage. Table 1 summarizes some of the uses for the data from instrumented structures. Unfortunately, only a few damaged structures have been instrumented in advance to perform studies of the initation and progression of damage during strong shaking (e.g. Imperial County Services Building during the 1979 Imperial Valley earthquake, [Rojahn and Mork, 1981]). In the future, instrumentation programs should consider this deficiency. Jennings (1997) summarizes this view as follows: "As more records become available and understood, it seems inevitable that the process of earthquake resistant design will be increasingly, and quite appropriately, based more and more upon records and measured properties of materials, and less and less upon empiricism and qualitative assessments of earthquake performance. This process is well along now in the design of special structures".

M. Erdik et al. (eds.), Strong Motion Instrumentation for Civil Engineering Structures, 179–194.

The methods used in studying structural response records are quite diverse: (a) mathematical modeling (finite element models varying from crude to very detailed, subjected to time-history, response spectrum or modal analyses). The procedure requires the blueprints of the structures which may not be readily accessible; (b) system identification techniques: single input/single output or multi input/multi output. In these procedures, the parameters of a model are adjusted for consistency with input and output data (Ljung, 1987); (c) spectral analyses: response spectra, Fourier amplitude spectra, autospectra, S_x or S_y, cross-spectral amplitudes S_{xy}, and coherence functions (γ) [using the equation : $\gamma^2{}_{xy}(f) = S^2{}_{xy}(f) / S_x(f) S_y(f)$] and associated phase angles (Bendat and Piersol, 1980); and (d) simple procedures based on principles of structural dynamics (*e.g.* recently Jennings (1997) analyzed data from two buildings within close proximity (<20 km) to the Northridge epicenter, calculated the base shear from the records as 8 and 17 % of the weights of the buildings, drift ratios as 0.8 and 1.6 % (exceeding code limitations). Jennings (1997) states: "A difference between code design values and measured earthquake responses of this magnitude – approaching a factor of ten – is not a tenable situation."

Until recently, in general, only accelerometers (single, biaxial or triaxial) were used to instrument structures. However, observations of damages during the 1994 Northridge and 1995 Kobe earthquakes, have forced engineers and scientists to focus on performance based seismic design methods and to find new techniques to control drift and displacements. To verify these developments, sensors directly measuring displacements or relative displacements (transducers, laser devices and GPS units) are now being considered. A recent development in use of global positionaning systems (GPS) in dynamic monitoring of long-period structures is described by Çelebi and others (1999).

In general, accelerometer deployments, as depicted in Figure 1, fall into three categories: (a) minimal [a triaxial accelerograph only at the roof of a building or three triaxial accelerographs deployed at the roof, mid-floor and gound (or basement) levels – the later better known as the UBC recommended instrumentation]; (b) extensive [combinations of uniaxial, biaxial and triaxial accelerometers to record translational, torsional and rocking motions]; (c & d) special cases [additional accelerometers to detect deformations of in-plane motions of flexible diaphragms or relative vertical displacements of isolators of a base-isolated structure].

2. Case Studies

2.1. PACIFIC PARK PLAZA (EMERYVILLE, CA.)

The set of records from the 30-story Pacific Park Plaza (PPP) building is possibly the most studied building response data recorded during the M_s=7.1 Loma Prieta earthquake of October 17, 1989. The building is an equally spaced three-winged, cast-in-place, ductile, moment-resistant framed structure. Constructed in 1983 and instrumented in 1985, it is the tallest reinforced concrete building in northern California. A general view, a plan view, a three-dimensional schematic, and its instrumentation is shown in figure 2 (Çelebi, 1992, 1996). Twenty-one channels of synchronized uniaxial accelerometers are deployed throughout this structure. Three channels of accelerometers are located at the north free-field outside the building. All are connected to central recording systems. In addition, a triaxial strong-motion accelerograph is deployed at a free-field site on the south side of the building (SFF or EMV[1,2]).

The foundation of PPP is a 5-foot-thick concrete mat supported by 828 (14-inch-square) prestressed concrete friction piles, each 20-25 m in length, in a primarily soft-soil environment, with an average shear-wave velocity between 250 and 300 m/s and a depth of approximately 150 m to harder soil. The building, at 100 km from the epicenter of the earthquake, had considerably amplified input motions but was not damaged during the earthquake. The east-west components of acceleration recorded at the roof and the ground floor of the structure, at the associated free-field station (SFF in fig. 2) and, for comparison, the motion at Yerba Buena Island (YBI), the closest rock site with a peak acceleration of 0.06 *g*, are shown in figure 3. The response spectra also shown in figure 3 clearly demonstrate that the motions at EMV were amplified by as much as five times when compared with YBI. Amplification is also indicated by the amplitude of the peak accelerations (0.26 *g* for EMV and 0.06 *g* for YBI). The differences in peak acceleration at the free-field station (0.26 *g*) and at the ground floor of the building (0.21 *g*) (fig. 2a) suggest that there was soil-structure interaction (SSI).

In the design of the building, site-specific design response spectra (based on three probabilistic earthquakes based on expected levels of performance) were used: (a) the maximum probable earthquake (50 % probability of being exceeded in 50 years with 5 % damping) anchored at zero period acceleration (ZPA) of 0.32g., and two maximum credible earthquakes both with 10 % damping but 10 % probability of being exceeded in (b) 50 years [ZPA of 0.53 g] and (c) 100 years [ZPA of 0.63 g]. The design response spectra and the spectrum of the EW component of recorded motion at the SFF is shown in Figure 4. The ZPA of the recorded EW acceleration at SFF (0.26 g)

---

[1] In most studies, the site of south free-field (SFF) is referred to as the Emeryville site (EMV).
[2] In 1997, the analog recording instruments at Emeryville were upgrade to digital. A downhole accelerograph was installed at the same location as the surface free-field station, SFF.

(at 100 km from the epicenter) is close to that of the postulated maximum probable earthquake (0.32 g). Furthermore, the spectral accelerations of the EW component of SFF is considerably higher than the maximum probable earthquake for periods >0.6 seconds – that is, practically for the first three modes of the building. Therefore, one important conclusion derived from the records is that improvements are necessary in establishing site-specific design response spectra to account for realistic shaking at a specific site taking into account expected future closer earthquakes likely to produce larger peak accelerations.

Using different methods, the building has been studied in detail by Anderson and Bertero (1994), Anderson and others (1991), Kagawa and others (1993), Kagawa and Al-Khatib (1993), Aktan and others (1992), Kambhatla and others (1992) and Çelebi and Safak (1992). All investigators agree that the predominant three response modes of the building and the associated frequencies (periods) are 0.38 Hz (2.63 s), 0.95 Hz (1.05 s), and 1.95 Hz (0.51 s). These three modes of the building are torsionally-translationally coupled (Çelebi, 1996) and are depicted in the cross-spectra (S_{xy}) of the orthogonal records obtained from the roof, ground floor and SFF (the south free-field site) (figure 5) and the normalized cross-spectra of the orthogonal records (bottom right in figure 5). The frequency at 0.7 Hz (1.43 s) observed in the spectra is this site frequency (Çelebi, 1996).

System identification techniques, when applied to the records of this building, yielded unusually large damping ratios corresponding to the 0.38-Hz first-mode frequency [11.6 % (NS) and 15.5 % (EW)] [Table 2] (Çelebi, 1996a). Such unusually high damping ratios attributed to a conventionally designed/constructed building with its large mat foundation in a relatively soft geotechnical environment is due to radiation (or foundation) or material damping. This is one of two cases where large damping percentages implied by the recorded responses of buildings have been attributed to radiation damping; the other case is from the Olive View Hospital in Sylmar, Ca. – data from the Northridge earthquake (Çelebi,1997).

The dynamic characteristics determined from Loma Prieta response records of Pacific Park Plaza as well as those determined from modal analyses (Stephen and others, 1985) are summarized in Table 2. Also, it is noted in table 2 that although flexibility of the foundation was considered in the 1985 analyses, the structural frequency remained the same as the frequency determined with fixed base assumption. Clearly, the mathematical models developed at that time needed improvements. This conclusion could only be reached because we have recorded on scale motions. Most recent studies indicate that the frequencies from recorded motions can be matched when soil-structure interaction (SSI) is incorporated into the mathematical models (Kagawa and others, 1993; Aktan and others, 1992; Kambhatla and others, 1992). Furthermore, a study of the building for dynamic-pile-group interaction by (Kagawa and Al-Khatib, 1993; Kagawa and others, 1993) indicates that there is significant interaction. Their studies show that computed responses of the building using state-of-the-art techniques for dynamic-pile-group interaction compares well with the recorded responses. On the other hand, Anderson and others (1991) and Anderson and Bertero (1994) concluded that soil-structure interaction was insignificant for Pacific Park Plaza during the earthquake. They compared the design criteria, code requirements, and the elastic and nonlinear dynamic response of this building due to the earthquake using both simplified and detailed analytical models.

2.2 . TRANSAMERICA BUILDING ROCKS

In Figure 6, the instrumentation scheme of the 48-story Transamerica Building in San Francisco, California and the accelerations, recorded during the 1989 Loma Prieta earthquake, (and disploacements calculated from accelèrations) are shown. The most unique behavior of this building (with a 2.75 m thick basemat without piles) is rocking as determined from the coherency and phase angle calculations using the horizontal accelerations from the 21st floor and the vertical accelerations from the basemat (Figure 7). The figure shows that at 2 Hz, the pair of horizontal and vertical motions are coherent and are in phase. The fundamental frequency (period) of the building including the rocking mode of soil-structure interaction is 0.28 Hz (3.57). During the design/analyses process of this building, such interaction effects were not accounted for (Çelebi, 1998).

2.3. FLEXIBLE DIAPHRAGMS

Buildings with flexible diaphragms may exhibit an important performance defect during strong shaking that requires particular attention. In Figure 8, a schematic of West Valley College Gymnasium in Saratoga, California is shown. This building has a very flexible diaphragm. The acceleration records ontained during the 1984 Morgan Hill and 1989 Loma Prieta earthquakes (summarized in Table 3) from the building roof center are approximately three times that of the roof edges (Figure 9). The records from the building was studied in detail by Çelebi et al (1989) and by Bouwkamp, Hamburger, and Gillengerten (1991). From the recorded accelerations, the fundamental frequencies throughout the structure_ are calculated as 3.8 and 3.9 Hz in the NS and EW directions, respectively. All records, even those retrieved from the ground floor are affected by the shaking frequency of the flexible diaphragm. Therefore, the flexibility of the diaphragms should be accounted for in the design process. As a result of studies performed on these records, the SEAOC Blue Book and later Uniform Building Code provisions were changed.

The West Valley College Gymnasium was studied in detail by Çelebi et al (1989) using the 1984 MHE records and by Bouwkamp, Hamburger, and Gillengerten (1991). A schematic of the gymnasium and its instrumentation scheme is shown in figure 52 (Çelebi et al, 1989). Essentially, the records of the building are greatly influenced by the diaphragm frequency at approximately 4 Hz. Significant peak accelerations of this gymnasium are summarized in Table

Essentially, the records of the building are greatly influenced by the diaphragm frequency at approximately 4 Hz. Significant peak accelerations of this gymnasium are summarized in Table 3 for the two earthquakes (Çelebi 1998). The calculated or the recorded displacements do not compare well with those from either the ATC—7 recommended formulas or by calculations based on UBC seismic forces. In addition, the recorded or calculated first mode of the structure is that of the diaphragm only and the frequencies, recorded or calculated, do not match with the frequencies determined from standard period formulas for structures. This is recognized, since, in the design-analysis process, the standard code formulas for building periods are not used to predict the periods of buildings with large-span diaphragms or to predict forces and displacements. On the other hand, the calculation of the static displacement in the design process depends upon rational evaluation of the stiffness of the structure and not that of the diaphragm.

As a result of these studies, the design requirements for the restraints at the edges of the roof diaphragm were increased in UBC (1991) by 50 %.

3. Conclusions

In this paper, various uses of data from instrumented structures are summarized. A preliminary listing and classification of the different uses of data is provided with some sample references. A data set from an instrumented building is used to demonstrate extraction of dynamic characteriscs (modal frequencies) and other features (radiation damping). It is shown for this case that during future earthquakes that are closer to the building, the design response spectra will be significantly exceeded. On a generic note, it is noted that development of design response spectra particularly for longer periods should be improved and soil-structure interaction should be considered in design/analyses procedures.

References

1. Aktan, H., Kagawa, T., Kambhatla, A., and Çelebi, M., 1992, Measured and analytical response of a pile supported building, *in* Proceedings, Tenth World Conference on Earthquake Engineering: A.A. Balkema, Rotterdam, v. 3, p. 1791-1796.
2. Anderson, J.C., Miranda, E., and Bertero, V.V., and Kajima Project Research Team, 1991, Evaluation of the seismic performance of a thirty-story RC building: Earthquake Engineering Research Center, University of California, Berkeley, Report: UCB/EERC-91/16, 254 p.
3. Anderson, J.C., and Bertero, V.V., 1994, Lessons learned from an instrumented high rise building, *in* Proceedings, Fifth U.S. National Conference on Earthquake Engineering: Earthquake Engineering Research Institute, Oakland, Calif., v. II, p. 651-660.
4. Astaneh, A., Bonowitz, D., and Chen, C., 1991, Evaluating design provisions and actual performance of a modern high-rise steel structure, *in* Seminar on Seismological and Engineering Implications of Recent Strong-Motion Data: California Department of Conservation, Division of Mines and Geology, p. 5-1–5-10.

5. Bendat, J.S., and Piersol, A.G., 1980, Engineering applications of correlation and spectral analysis: John Wiley and Sons, 302 p.
6. Boroschek, R. L., Mahin, S. A., and Zeris, C., A., 1990, Seismic response and analytical modeling of three instrumented buildings, PROC., 4th U.S. National Conference on Earthquake Engineering, v.2, pp. 219-228, Palm Springs, Ca., May 20-24.
7. Boroschek, R. L., and Mahin, S., 1991, An Investigation of the Seismic Response of a Lightly-Damped Torsionally-Coupled Building, University of California, Berkeley, California, Earthquake Engineering Research Center Report 91/18, December, 291 p.
8. Bouwkamp, J. B., Hamburger, R. O., and Gillengerten, J. D., 1991, Degradation of Plywood Roof Diaphragms under multiple earthquake loading, Seminar on Seismological and Engineering Implications of Recent Strong-motion Data, Office of Strong Motion Studies, California Div. of Mines and Geology, Sacramento, May 1991, pp. 6-1 – 6-10.
9. Çelebi, M., Prescott, W., Stein, R., Hudnut, K., Behr, J., and Wilson, S.,1999, GPS monitoring of dynamic behavior of long-period structures: *Earthquake Spectra* (Journal of EERI), vol. 15, no. 1, pp. 55-66.
10. Çelebi, M., 1998, Performance of Building structures – A Summary, *in* The Loma Prieta, California, Earthquake of October 17, 1989 – Building Structures (M. Çelebi, *editor*), USGS Prof. Paper 1552-C, pp. c5-c76, January 1998
11. Çelebi, M., Prescott, W., Stein, R., Hudnut, K., Behr, J., and Wilson, S. GPS monitoring of dynamic behavior of long-period structures : *Earthquake Spectra* (Journal of EERI), Feb. 1999, vol. 15, no. 1, pp. 55-66.
12. Çelebi, M., Bongiovanni, G., Safak, E., and Brady, G., 1989, Seismic response of a large-span roof diaphragm: Earthquake Spectra, v. 5, no. 2, p. 337-350.
13. Çelebi, M., and Safak, E., 1991, Seismic response of Transamerica Building—I, data and preliminary analysis: Journal of Structural Engineering, v. 117, no. 8, p. 2389-2404.
14. Çelebi, M. and Safak, E., 1992, Seismic response of Pacific Park Plaza—I, data and preliminary analysis: Journal of Structural Engineering, v. 118, no. 6, p. 1547-1565
15. Çelebi, M., Phan, L. T., and Marshall, R. D., 1993, Dynamic characteristics of five tall buildings during strong and low-amplitude motions, Journal of Structural Design of Tall Buildings, J. Wiley, v. 2, pp. 1-15.
16. Çelebi, M., 1994, Response study of a flexible building using three earthquake records, Structures Congress XII: Proceedings of papers presented at the Structures Congress '94, Atlanta, GA, April 24-28, American Society of Civil Engineers, New York, Vol. 2, 1220- 1225.
17. Çelebi, M., 1996, Comparison of damping in buildings under low-amplitude and strong motions, Journal of Wind Engineering and Industrial Aerodynamics, Elsevier Science, v. 59, pp. 309-323.
18. Çelebi, M., Presscott, W., Stein, R., Hudnut, K., and Wilson, S., 1997, Application of GPS in Monitoring Tall Buildings in Seismic Areas, 1997a, Abstract, AGU Meeting, San Francisco, Ca., Dec.
19. Çelebi, M. and Liu, H-P., Before and After Retrofit – Response of a Building During Ambient and Strong-motions, 8USNational Conference on Wind Eng,, The John Hopkins Univ. June 5-7.
20. Çelebi, M., 1993, Seismic response of eccentrically braced tall building, Journal of Structural Engineering, v. 119, no. 4, p. 1188-1205.
21. Çelebi, M., 1996, Comparison of damping in buildings under low-amplitude and strong motions: Journal of Wind Engineering and Industrial Aerodynamics, Elsevier Science, v. 59, p. 309-323.
22. Çelebi, M., 1997, Response of Olive View Hospital to Northridge and Whittier earthquakes, American Society of Civil Engineers, *Journal of Structural Engineering*, April ,v.123, no. 4, p. 389-396.
23. Çelebi, M, GPS and/or Strong and Weak Motion Structural Response Measurements – Case Studies, 1998 , Structural Engineers World Congrass (invited paper), San Francisco, Ca. July 18-23, 1998.
24. Çelebi, M., 1995, Successful Performance of base-isolated hospital building during the 17 January 1994 Northridge earthquake, *Journal of the Structural Design of Tall Buildings*, v. 5, pp.95-109.
25. Crosby, P. , Kelly, J., and Singh, J. P., Utilizing visco-elastic dampers in the seismic retrofit of a thirteen story steel framed building, ASCE Structures Congress XII, Atlanta, Ga., 1994, v. 2, 1286-1291.
26. Hall, J. F., Heaton, T. H., Halling, M. W., and wald, D. J., 1996, Near-source ground motion and its effects on flexible buildings, Earthquake Spectra, v. 11, no.4, pp. 569-605.
27. Hamburger, R. O., 1997, FEMA-173 Seismic Rehabilitation Guidelines: The next step – Verification, in Proc.SMIP97 Seminar on Utilization of Strong-motion Data, California strong Motion Instrumentation Program, Div. of Mines and Geology, California Dept. of Conservation, Sacramento, Ca., 51-69.

28. Jennings, P.C., 1997, Use if strong-motion data in earthquake resistant design, in Proc.SMIP97 Seminar on Utilization of Strong-motion Data, California strong Motion Instrumentation Program, Div. of Mines and Geology, California Dept. of Conservation, Sacramento, Ca.,1-8.
29. Kagawa, T., Aktan, H., and Çelebi, M., 1993, Evaluation of soil and structure model using measured building response during the Loma Prieta earthquake: Department of Civil and Environmental Engineering, Wayne State University, Detroit, Michigan, 169 p.
30. Kagawa, T., and Al-Khatib, M.A., 1993, Earthquake response of 30-story building during the Loma Prieta earthquake, *in* Third International Conference on Case Histories in Geotechnical Engineering, June 1-4, University of Missouri-Rolla, v. I: p. 547-553.
31. Kelly, J., 1993, Seismic isolation, passive energy dissipation and active control, PROC. ATC 17-1 Seminar on State of the Art and State of the Practice of Base Isolation, vol. 1, 9-22.
32. Kelly, J.M., Aiken, I.D., and Clark, P.W., 1991, Response of base-isolated structures in recent California earthquakes, *in* Seminar on Seismological and Engineering Implications of Recent Strong-Motion Data, Preprints: California Division of Mines and Geology, Strong Motion Instrumentation Program, p. 12-1--12-10.
33. Ljung, L. ,1987, System identification – Theory for the User: Prentice-Hall, 519 p.
34. Porter, L.D., 1996, The influence of earthquake azimuth on structural response due to strong ground shaking, *in* Eleventh World Conference on Earthquake Engineering, Acapulco, Mexico (June), (No. 1623): Elsevier Science Ltd. (CD-ROM).
35. De La Llera, J., and Chopra, A., 1995, Understsanding of inelastic seismic behavior of symmetric-plan buildings, Earthquake Engineering and Structural Dynamics, 24, pp. 549-572.
36. Chopra, A., and Goel, R.K., 1991, Evaluation of torsional provisions of seismic codes, J. Struct. Eng. ASCE, 117, 12, 3762-3782.
37. Heo, G., Wang, M. L., and Satpathi, D., 1977, Optimal transducer placement for health monitoring, Soil Dynamics and Earthquake Engineering, 16, pp-496-502
38. Rojahn, C., and Mork, P.N., 1981, An analysis of strong-motion data from a severely damaged structure, the Imperial County Services Building, El Centro, California: U.S. Geological Survey Open-File Report 81-194.
39. Safak, E., and Çelebi, M.,1991, Seismic response of Transamerica Building; - II, System identification and preliminary analysis: Journal of Structural Engineering, v. 117, no. 8, p. 2405-2425.
40. Safak, E., and Çelebi, M., 1992, Recorded seismic response of Pacific Park Plaza; - II, System identification: Journal of Structural Engineering, v. 18, no. 6, p. 1566-1589
41. Straser, E., 1997, Toward wireless, modular monitoring systems for civil structures, in the John A. Blume Earthquake Engineering Center Newsltter, Issue No. 2.
42. Lin, B. C., and Papageorgiou, A. S., Demonstration of torsional coupling caused by closely spaced periods---1984 Morgan Hill Earthquake Response of the Santa Clara County Building, Earthquake Spectra, 1989, vol. 5, No. 3, pp. 539--556.
43. Marshall, R. D., Phan, L. T., and Çelebi, M., 1992, Measurement of structural response characteristics of full-scale buildings: Comparison of results from strong-motion and ambient vibration records, NISTIR REPORT 4884, National Institute of Standards and Technology, Gaithersburg, Maryland.
44. Uniform Building Code, International Conference of Building Officials, Whittier, CA, 1970, 1976, 1979, 1982, 1985, 1988, 1991, 1994 editions.

Table 1. A Preliminary List of Data Utilization & Sample References

GENERIC UTILIZATION
Verification of mathematical models (usually routinely performed) (*e.g.*Boroschek et al, 1990)
Comparison of design criteria vs. actual response (usually routinely performed)
Verification of new guidelines and code provisions (*e.g.*Hamburger, 1997)
Identification of structural characteristics (Period. Damping, Mode Shapes)
Verification of maximum drift ratio (*e.g.* Astaneh, 1991, Çelebi, 1993)
Torsional response/Accidental torsional response (*e.g.* Chopra, 1991, DeLalera, 1995)
Identification of repair & retrofit needs & techniques (Crosby, 1994)
SPECIFIC UTILIZATION
Identification of damage and/or inelastic behavior (*e.g.* Rojahn & Mork, 1981)
Soil-Structure Interaction Including Rocking and Radiation Damping (Çelebi, 1996, 1997)
Response of Unsymmetric Structures to Directivity of Ground Motions (*e.g.* Porter, 1996)
Responses of Structures with Emerging Technologies (base-isolation, visco-elastic Dampers, and combination (Kelly and Aiken, 1991, Kelly, 1993, Çelebi, 1995)
Structure specific behavior (*e.g.* diaphragm effects, Boroschek and Mahin,1991, Çelebi, 1994)
Development of new methods of instrumentation/hardware (Çelebi, 1997, Straser, 1997)
Improvement of site-specific design response spectra
Associated free-field records (if available) to assess site amplification, SSI and attenuation curves
Verification of Repair/Retrofit Methods (Crosby et al, 1994, Çelebi and Liu, 1997)
Identification of Site Frequency from Building Records (more work needed)
RECENT TRENDS TO ADVANCE UTILIZATION
Studies of response of structures to long period motions (e.g. Hall et al, 1996)
Need for new techniques to acquire/disseminate data (Straser, 1997, Çelebi, 1997, 1998)
Verification of Performance Based Design Criteria (future essential instrumentation work)
Near Fault Factor (more free-field stations associated with structures needed)
Comparison of strong vs weak response (Marshall, Long and Çelebi, 1992)
Functionality (Needs additional specific instrumentation planning)
Health Monitoring and other Special Purpose Verification (Heo et al, 1997)

Table 2. Summary of dynamic characteristics for Pacific Park Plaza

	Frequencies (Hz)			Damping (%)		
	Mode			Mode		
	1	2	3	1	2	3
1989 (LPE) STRONG-MOTION DATA (from Çelebi, 1996)						
N-S	0.38	0.95	1.95	11.6		
E-W	0.38	0.95	1.95	15.5		
MODAL ANALYSES (from Stephen and others, 1985)						
N-S	0.60	1.67	3.10			
E-W	0.60	1.67	3.10			
TORSION	0.57	1.70	3.25			

Table 3. Comparison of Peak Acceleration Responses from West Valley College (Saratoga) Gymnasium

		Earthquake	
Location	Direction	Morgan Hill M_s =6.1 April 24, 1984 Accel (g)	Loma Prieta M_s =7.1 October 17, 1989 Accel. (g)
Ground	NS	0.10	0.26
	EW	0.04	0.33
Roof Edge	NS	0.14	0.36
	EW	0.065	0.43
Roof Center	NS	0.42	0.77
	EW	0.20	0.87

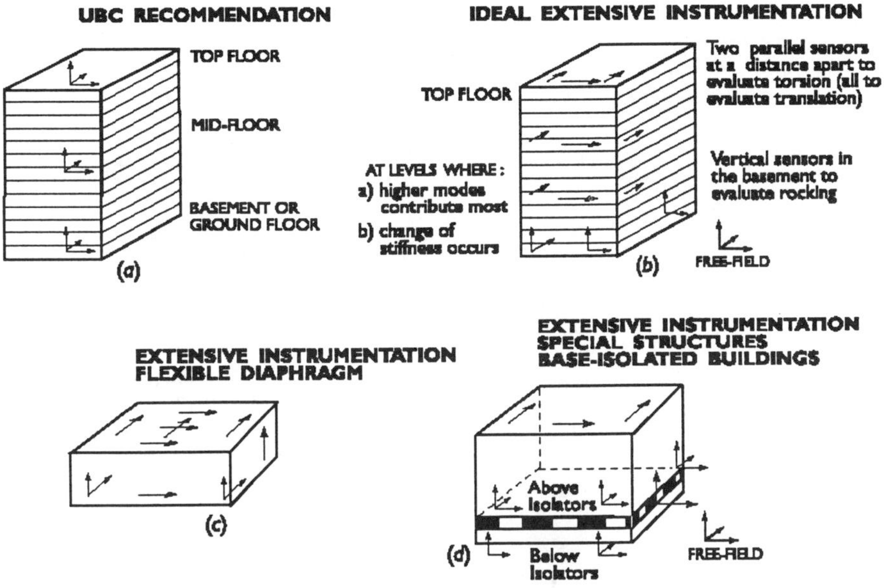

Figure 1. Schematic of commonly applied instrumentation schemes.

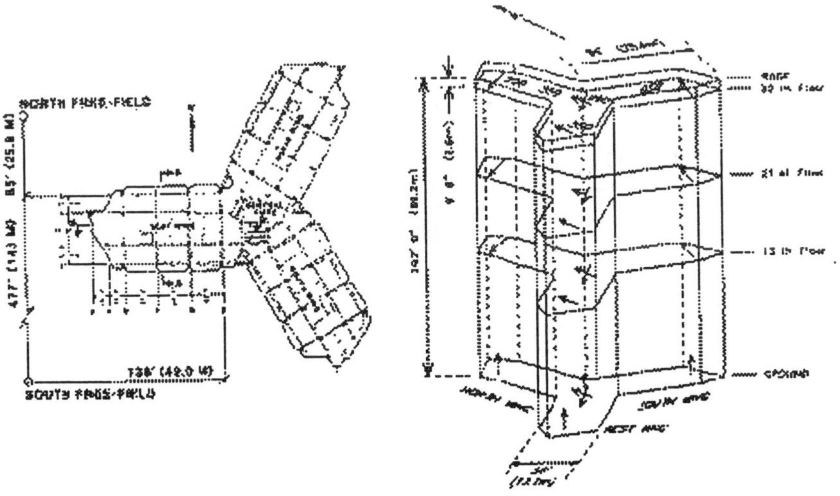

Figure 2. Plan layout and three-dimensional schematic and instrumentation scheme of Pacific Park Plaza (PPP), Emeryville, Ca.

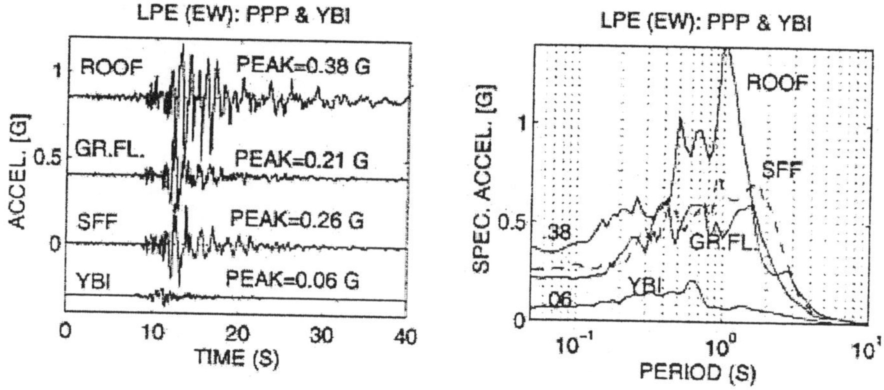

Figure 3. Recorded (EW components) of accelerations and corresponding response spectra at the free-field, ground floor and roof of Pacific Park Plaza (PPP), and at Yerba Buena Island (YBI), at approximately the same distance as PPP, depict the level of amplification.

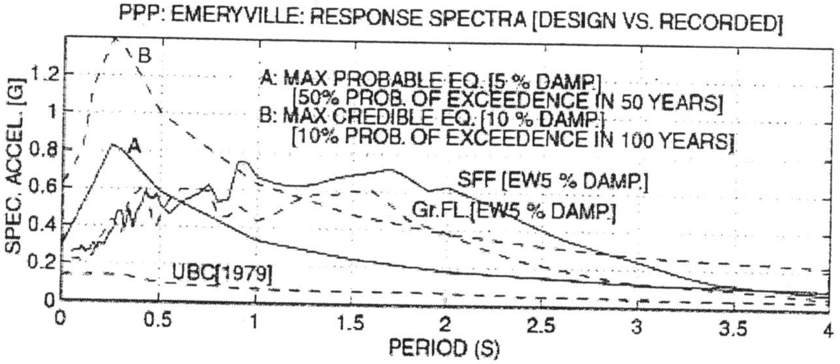

Figure 4. Design response spectra and response spectra of recorded motions at the ground floor and SFF of Pacific Park Plaza. Also shown is the 1979 UBC response spectrum for comparison.

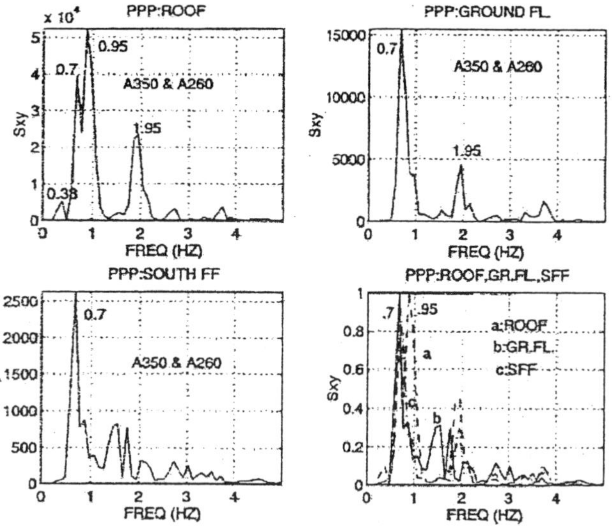

Figure 5. Cross-spectra of orthogonal accelerations (A350 & A260) at the roof, ground floor, free-field of PPP. Also shown (bottom right) is the normalized cross-spectrum depicting structural and site frequency peaks. (350 & 260 depict degrees clockwise from true north).

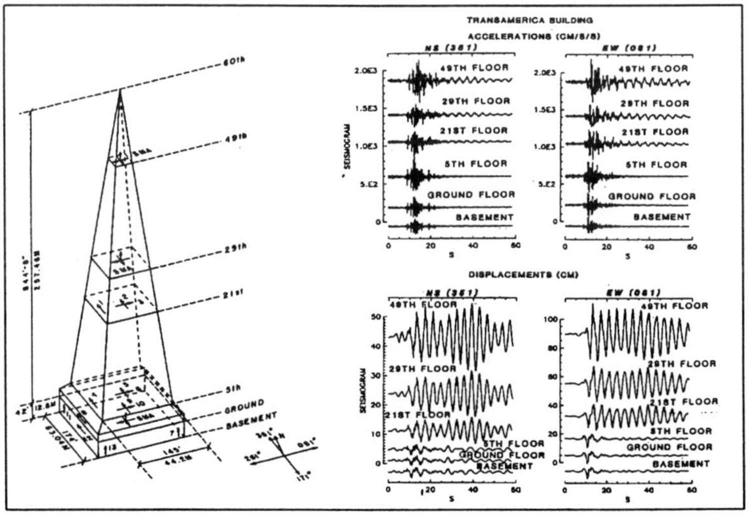

Figure 6. Instrumentation scheme of Transamerica Building and recorded Loma Prieta earthquake response.

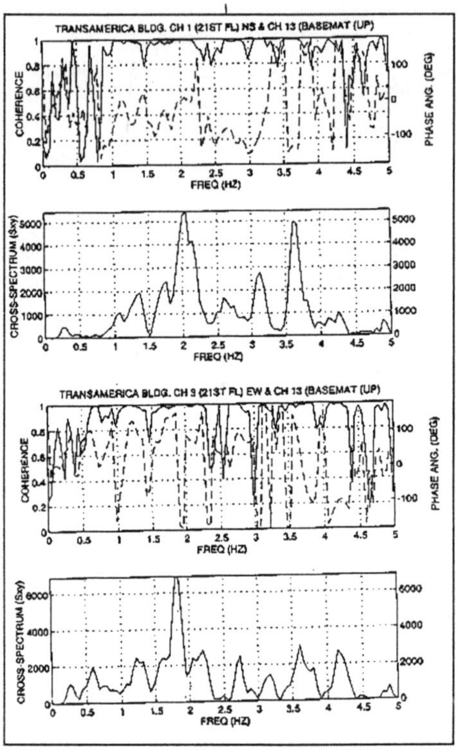

Figure 7. Coherency and phase angle plots to determine rocking at 2 Hz (Transamerica Building).

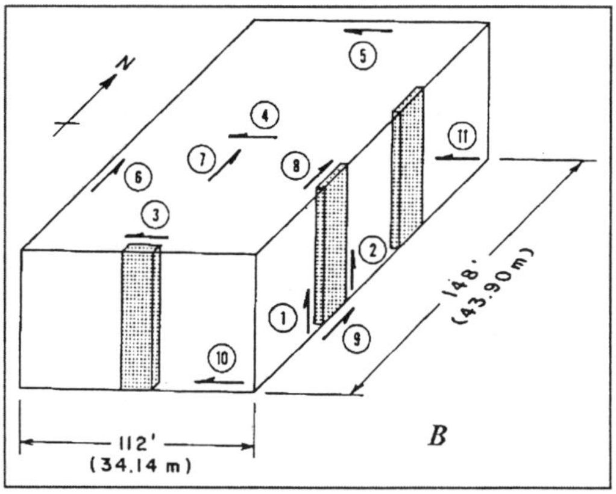

Figure 8. Schematic of West Valley College Gymnasium (Saratoga, Ca.)and its instrumentation

CDMG MORGAN HILL RECORD (30 km from the epicenter)

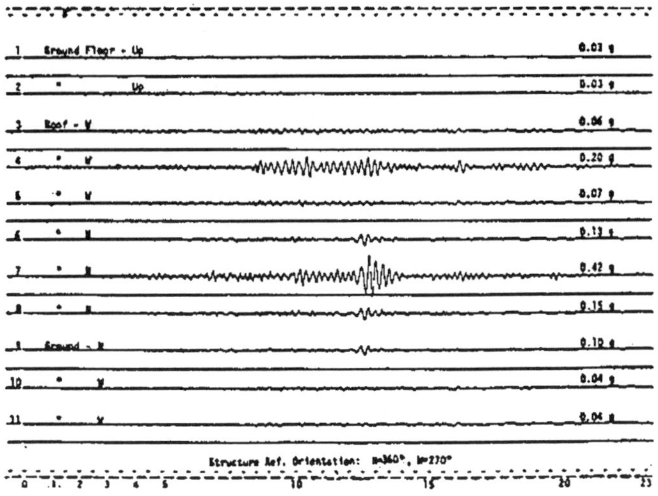

CDMG LOMA PRIETA RECORD (27 km from the epicenter)

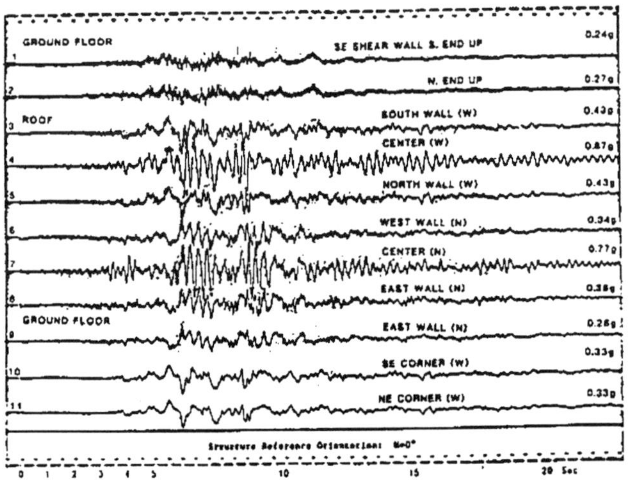

Figure 9. Recorded accelerations of West Valley College Gymnasium (Saratoga, Ca.).

INSTRUMENTATION OF DAM STRUCTURES IN SWITZERLAND

G.R. DARBRE
Federal Office for Water Management
Dam Safety
P.O.Box 957
CH-2501 Bienne, Switzerland
e-mail:Georges.darbre@bww.admin.ch

Abstract

A dam network was created in 1992 and 1993 in Switzerland. It encompasses 29 accelerographs placed in the dams of Mauvoisin (250.5 meter high arch dam, 12 instruments), Punt-dal-Gall (130 meter high arch dam , 7 instruments), Grande-Dixence (285 meter high gravity dam, 6 instruments) and Mattmark (120 meter high embankment dam, 4 instruments). The aim is to observe the free-field motions at the dam sites, the motions at the abutments and the dams' dynamic responses and characteristics. Array configurations and instrument specifications were developed based on the stated objectives. Earthquake events have been recorded at the dams.

1. Introduction

The validity of the excitations and of the models used in the earthquake analysis of structures can only be checked by way of comparison with relevant field observations. The latter are at the present incomplete and there is a need for field measurements during earthquakes in general and for strong-motion instrumentation in particular. The measurements needed relate to the free-field motions from which information on the earthquake recurrence and characteristics (including wave-propagation aspects) at specific sites can be obtained, and to the structural motions from which selected aspects of the earthquake behavior of different types of structures can be studied (high-rise buildings, industrial facilities, nuclear power plants, bridges, dams). The instrumentation of 4 large dams in Switzerland acknowledges this need [1].
The objectives of strong-motion instrumentation of dams and the corresponding array configurations and instrument specifications are reviewed in Section 2. The installations performed are presented in Section 3 and the records obtained are introduced in Section 4. Further details on this instrumentation are found in [2] to [5].

M. Erdik et al. (eds.), Strong Motion Instrumentation for Civil Engineering Structures, 195–208.

2. Review of instrumentation schemes

2.1 Observational needs

Free-field motions. The prediction of the behavior of large dams during future earthquakes requires the use of analytical and numerical techniques that utilize a temporal description of a design earthquake (synthetic accelerogram obtained directly or compatible with a design spectrum). The observational basis needed to characterize such earthquakes (amplitudes of motions, strong-motion duration, influence of local geological and soil conditions, attenuation laws and coefficients of wave propagation) is still insufficient. Such observations made in the free field also feed the data bases used when determining the seismic hazard at a site and permit to calibrate the design spectra used during preliminary studies, comparison studies and analyses of smaller dams.

Abutment motions. From the shape and the dimensions of the dam-foundation interface, various aspects of soil-structure interaction can contribute significantly to the earthquake response of a dam. First, the topography of the canyon and the inertial and energy dissipation properties of the foundation rock lead to a non-uniform motion at the interface (Figure 1a). This also holds true when it is assumed that the dam is not yet present and that the waves come from a single direction in a synchronized fashion. Second, assuming for a moment that the dam is massless, the motion along the interface is affected by the static resistance offered by the dam to any deformation of its support (kinematic interaction, Figure 1b). Third, the seismic excitation originates from waves arriving from several directions (incoherent excitation, Figure 1c). Four, the motions at the abutments are affected by the inertial, vibratory response of the dam (inertial interaction, Figure 1d). All these effects combine into the total effective input motion along the interface, whose observation is still largely missing. Its observation would contribute to gaining a better understanding of all the phenomena involved, permit calibration of the existing analytical models and the development of new models, and help in specifying design excitations.

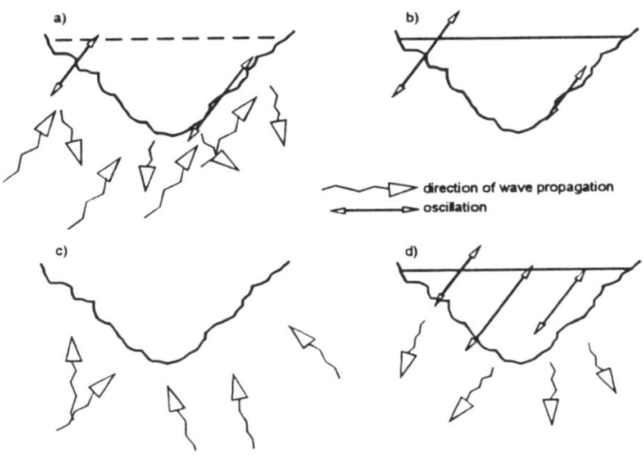

Figure 1. Foundation-structure interaction for arch dams:

a. Canyon effects;

b. Kinematic interaction;

c. Incoherent excitation;

d. Inertial interaction.

Dynamic response and characteristics of dams. The synthesis of the response of the individual modes of vibration of a dam to an earthquake determines its global response (linear range). The resonance frequencies and the associated modes of vibration and levels of energy dissipation can be established relatively easily by way of shaker tests. However, such tests are limited to very low levels of excitation and of response that are not representative of those encountered during large earthquakes. While the mode shapes and the resonance frequencies are not expected to vary much with the level of shaking (provided there is no major incursion in the inelastic range), this is not the case of the energy dissipation that needs to be better apprehended.

2.2 Array Configurations

General. The following array configurations for arch, gravity and embankment dams have been developed to address these observation needs. The use of three-component accelerometers is assumed. In the figures, a circle identifies an accelerograph station and the letter inside the circle refers to the instrumentation scheme.

Arch dams. The free-field instrument must be located far enough from the dam and from the appurtenant structures so as not to be affected by their presence and vibrations. At the same time, it must be close enough to measure the motions that are representative of those at the site. Considering the studies of [6], a distance equal to twice the dam height is appropriate for concrete dams. It can be reduced to once when the modulus of elasticity of the foundation is equal to or larger than the modulus of the dam concrete.
The total effective input motion at the abutments is measured by instruments located along the dam-foundation interface. The deformation of the interface can be observed in addition to the rigid-body motions (translations and rotations) when they are placed in sufficient number.

Wave propagation in the foundation can be apprehended by instruments installed in exploratory and grouting galleries.

The response of the dam can be assessed globally by taking advantage of modal decomposition (linear range). The instruments at the crest of the dam are located at points of maximal modal deflections (middle and quarter points for a dam that is essentially symmetrical). In this way, the responses of the lower modes of vibration are well captured.

The resulting instrumentation schemes are shown in Figure 2a. A minimum observation of the excitation and of the response is made through scheme A, complemented by schemes B and E for a more detailed observation. The effective input motion is captured by schemes D and G (complementary to schemes A and B). Galleries in the foundation may allow observation of wave propagation (scheme F) while scheme C provides for the free-field motion. The instruments are positioned with respect to a fictitious rectangular grid (dashed lines in the figure), thus allowing easier correlation between the recordings made at different stations.

Gravity dams. Gravity dams are generally designed and analyzed by assuming that the blocks (monoliths) behave independently from one another (bidimensional behavior). The instrumentation is consequently concentrated to the middle/highest block. The observational objectives of base input motion, tridimensional response and independent behavior of blocks are more particularly relevant to gravity dams, in addition to those of free-field and in-structure responses identified for arch dams.

The variation of the motion across the base can possibly be substantial in a large dam (base thickness comparable to a predominant wave length). The associated effective base input motion should be observed. A tridimensional response is expected due to longitudinal excitations, when the dam is built in a narrow canyon and due to transverse excitations in a dam with shear keys. The extent to which the blocks have an independent behavior can be appreciated from the comparison of their individual motions.

A minimal observation of the excitation and of the response of the dam is obtained through scheme A of Figure 2b, complemented by scheme G for a more detailed observation. The independent behavior of neighboring blocks is investigated through schemes C, D and F while the variation of the motion across the canyon is addressed by scheme E. The tridimensional response of the dam is captured by scheme H (crest instruments at the quarter points) while scheme B provides for the free-field motion.

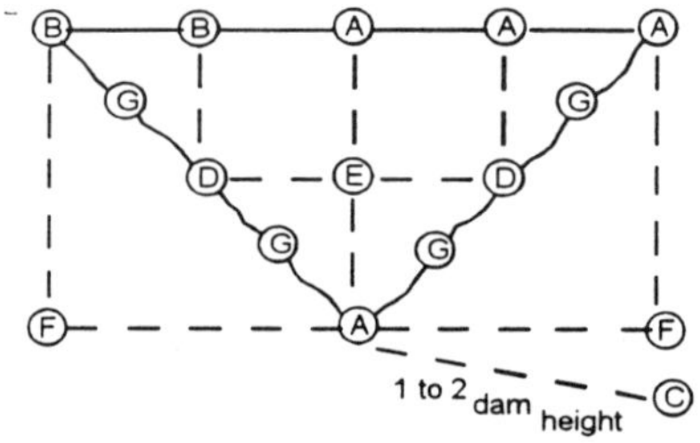

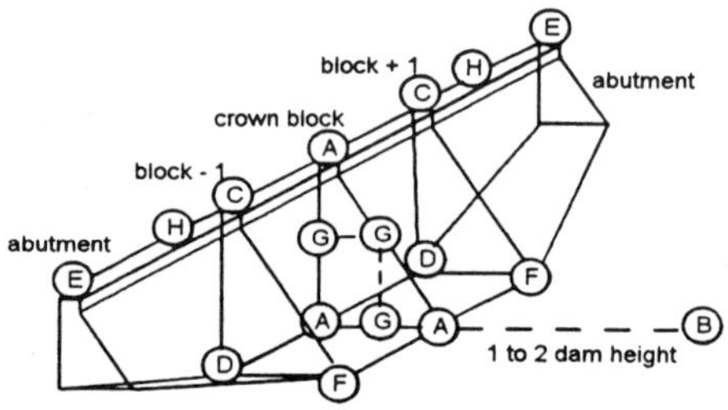

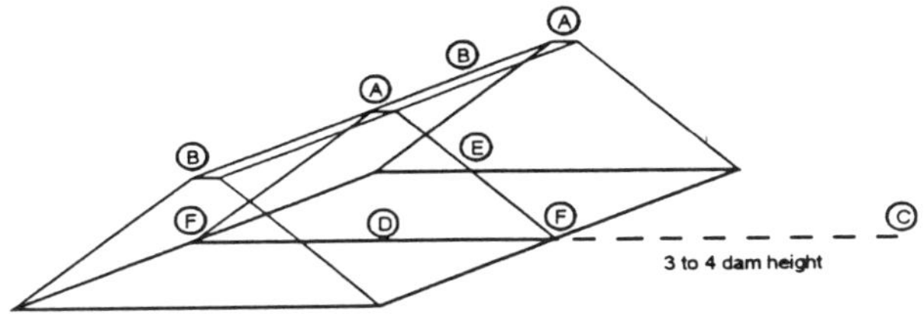

Figure 2. Accelerograph arrays:

a. Arch dams;

b. Gravity dams;

c. Embankment dams.

Embankment dams. Embankment dams are generally designed and analyzed assuming a bidimensional behavior, as in gravity dams. A similar instrumentation concept thus applies (instruments for free-field motion, effective base input motion and tridimensional response). The instrumentation schemes for embankment dams, Figure 2c, are adapted in part from [7]. An absolutely minimal instrumentation consists in placing instruments at one abutment and at the middle of the dam (crest level), thus obtaining information on the excitation and on the dam response (scheme A). The development of settlements can also be followed by the instrument at the crest. This instrumentation can be complemented by the one of scheme B to observe the variation of the excitation from one abutment to the other and the tridimensional response (instrument at a quarter point). A more detailed information on the excitation and on the response is obtained through schemes D and E, respectively, and on the effective base input motion through scheme F (middle section). Scheme C provides for the free-field motion, at a distance of 3 to 4 times the dam height (a distance equal to the period of the first mode of vibration of the dam multiplied by the shear wave velocity of the foundation is also suggested in [7]). This instrumentation can be complemented by bore-hole instruments in the dam, in priority at the middle section.

Remark. A more basic instrumentation can be performed when economical or other constraints prevent the installation of the arrays shown in Figure 2. It includes the observation of the free-field motions (one free-field instrument), of the total motions at the dam-foundation interface (one instrument at the base or at an abutment) and of the dam response (one instrument at the crest of the crown section) [8]. Further considerations on strong-motion instrumentation of dams are found in [9] and [10].

2.3 Specifications

The following main specifications were developed in the course of the instrumentation project considering the technology available in the early nineties. Most important is the requirement of a reliable functioning of the instruments under the prevailing environmental conditions. Instruments of simple design and construction that have been thoroughly tested under field conditions are thus recommended. The accelerographs should be digital with pre- and post-event recording and have a resolution of at least 0.001 g with 12 data bits. The lowest frequency recorded should be less than 10 per cent of the first resonance frequency of the dam, and less than the predominant frequencies of the earthquake motion (preferably 0 Hz). The highest frequency recorded should be larger than the frequency of the highest mode contributing to the dynamic dam response and also higher than the predominant frequencies of the earthquake motions (at least 30 Hz). The maximum recordable acceleration should be set according to a target observational return period (of the order of 100 years), accounting for the position of the accelerographs at or near the dam (amplification of the motions). All electronic and sensor components should have stable short-term characteristics, such as peak total recording noise of less than one least significant bit or, ideally, less than one half.

Appropriate measures should be taken against possible disturbances as a result of power surges and electrical ground loops, and the electromagnetic compatibility of the instruments should be adequate.

Common trigger and common sampling should be enforced within an array. If common sampling cannot be enforced, the records should have precise time marks. Local recording is preferred to central recording as this reduces the probability of malfunctioning.

3. Swiss National Dam Network

The Swiss national dam network was created in 1992 and 1993 for research purposes (installation of accelerographs in dams is not compulsory in Switzerland). It encompasses 4 dams and 29 accelerographs as summarized below. The instrumentation complies with the specifications given above (12-bit 3-component digital accelerographs, Δg = 0.00025 g, recording bandwidth 0 – 50 Hz, sampling rate of 200 per second). Particular attention was paid to the electromagnetic requirements as regular monitoring equipment installed in the dams was known to be prone to problems of this type.

A standard station is sketched in Figure 3. The recorder and the external accelerometer are mounted on the gallery walls (easy installation and instruments being "out of the way"). They are supported by electrically isolated plates. The power cable is connected to a power-surge protection and the cable segment between the protection and the accelerograph does not touch the wall nor cross other electric cables. All components are thus grounded to the same point what prevents ground loops from occurring. A space of a few millimeters is left between the mounting plates and the wall to allow for condensation water to run freely behind the instruments.

A protective roof is placed above the accelerometer to prevent the occurrence of saturated analog signals when water drops or small stones fall on the accelerometer (the sensors have a very high natural frequency and thus react to light shocks).

All the stations of an array are connected to a central unit by two optical fibers (this eliminates electric disturbances due to power surges in relation with lightning for example and due to differences in electric potentials). The central unit manages common trigger and common sampling of all the accelerographs and synchronizes their internal clock. It is connected to a public telephone line thus enabling remote communication with the central unit itself and, through it, with all the accelerographs.

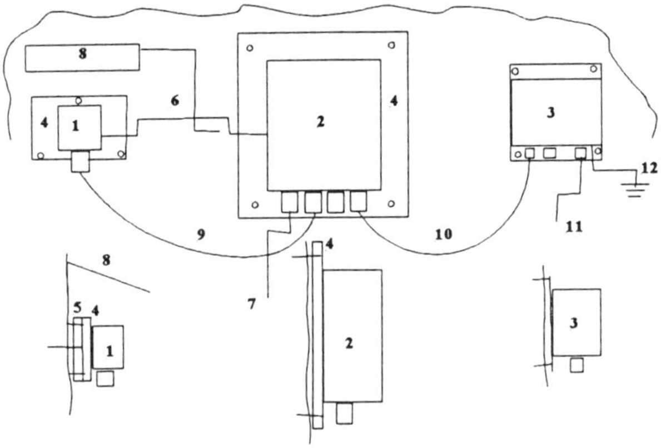

Figure 3. Standard accelerograph station (top: front view, bottom: cross sections)
1: accelerometer; 2: recorder; 3: power-surge protection; 4: electrically isolated supports; 5: metallic support; 6: grounding cables; 7: interconnection cable (optical fibers); 8: protecting roof; 9: accelerometer/recorder cable; 10: protected power supply; 11: unprotected power supply; 12: grounding.

3.1 Arch Dam Of Mauvoisin

The arch dam of Mauvoisin is the 4th highest concrete dam in the world. It has a height of 250.5 m, a crest length of 520 m, a base thickness of 54 m and a crest thickness of 12 m (Figure 4). It has been instrumented with 12 accelerographs. The main objective is the observation of the total effective input motions (including wave propagation) with 5 instruments installed at the abutments and two in the foundation rock (in exploratory galleries). Other objectives are the observation of the global dam responses (3 instruments in the upper gallery placed at location of maximum modal deflections and 1 in an intermediate gallery), and of the free-field motions with an accelerograph placed at the entry of a winter gallery 600 m downstream from the dam. The accelerographs are connected to the central unit located in a cavern at the right abutment (crest level) through a total of 6'494 m of fiber-optic cables.

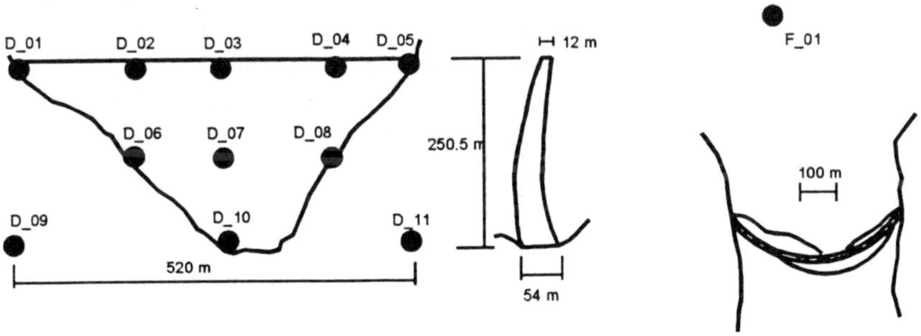

Figure 4. Sketch of the dam of Mauvoisin

3.2 Arch Dam Of Punt-Dal-Gall

An array of seven accelerographs (5 at crest level, 1 at the base and 1 in the free field) has been installed in the 130 m high arch dam of Punt dal Gall (540 m crest length, 10 to 12 m crest thickness and 24.5 m base thickness). Emphasis is put on the observation of the dam excitation and response. The free-field instrument is located 325 m away from the dam.

3.3 Dam Of Grande Dixence

The world's highest concrete dam of Grande Dixence (gravity dam of 285 m, base thickness of 198 m, crest length of 695 m) has been instrumented with an array of six accelerographs for minimal observation of dam excitation and response as well as of independent block behavior, Figure 6. 3 accelerographs are located at the crown section (1 at the crest and 2 on either side of the base) and 2 additional ones at the crest one block away from the central block (on either side). The free-field station is located in an access tunnel 290 meters downstream from the dam.

3.4 Dam Of Mattmark

The embankment dam of Mattmark is 120 meters high, 780 meters long, 9 meters wide at the crest and 373 meters wide at the base. An array of four accelerographs has been installed to obtain a minimum observation of dam excitation and response (1 at the right abutment and 1 in the middle of the crest, 1 in the middle of the base, and 1 in the free field). The free-field instrument, located 775 m downstream from the dam, is connected to the central unit by a fiber-optic cable of 2'170 m in length installed as a single segment (without signal amplification).**Dam Records**

3.5 Overview

The events listed in Table 1, none of which induced any damages to the dams, have been recorded.

The Domodossola quake occurred shortly after the accelerographs had been installed and before the network was commissioned. Instruments installed at Grande Dixence, Mauvoisin and Mattmark registered motions of up to 1.5 % g (Mattmark). The records were used to assess the functionality of the arrays. Shortcomings in the instrumentation were identified [11] and subsequently corrected.

Table 1. Events recorded at the dams up to early 1998.

Date	Name of Quake	Magnitude Ml	Focal Depth [km]	Dam affected	Epicentral Distance [km]
June 14, 1993	Domodossola	4.4	19	Grande Dixence Mauvoisin Mattmark	67 71 24
Nov. 1, 1994	Mauvoisin	Unknown	Unknown	Mauvoisin	~ 0
March 31, 1996	Valpelline	4.2	2	Grande Dixence Mauvoisin Mattmark	18 13 40
July 15, 1996	Meythet	5.2	5	Mauvoisin	100
Dec. 7, 1997	Mattmark	2.0	9	Mattmark	6

An event of unknown magnitude occurred in the close vicinity of Mauvoisin on November 1, 1994. 3 accelerographs were disconnected at that time for maintenance reasons. Thus, only 9 stations out of the 12 installed at the dam recorded the event.
The Valpelline quake triggered all stations at Grande Dixence and at Mauvoisin as well as the station at the crest of the crown section at Mattmark.
The Meythet quake triggered only the station at the crest of the crown section at Mauvoisin. The corresponding station at Mattmark was also the only one to register a motion due to an earthquake that occurred in the vicinity of that dam.

3.6 Identification Of Resonance Frequencies

Resonance frequencies have been identified from the power spectral density functions (*PSD's*) and from transfer functions (*TF's*). The resonance frequencies identified from the *PSD*'s are those of the dam-foundation-reservoir system. Those identified from *TF*'s are to some extent influenced by the resonance frequencies of the dam-reservoir system without foundation.

3.7 Dam Of Mauvoisin

The peak accelerations recorded during the local event of November 1, 1994, and during the Valpelline quake are reported in Figures 5a and 5b, respectively (x = stream direction, y = cross-stream direction and z = vertical direction). In the first case no amplification of motion in the dam is observed. Inspection of the records and of the *PSD*'s reveal that the motions are of high frequency content, larger than the resonance frequencies of the dam (it is in fact not certain that the event is an earthquake; although unlikely, an explosion or rock fall near the site is not excluded). On the other hand, an amplification clearly occurs during the Valpelline quake.

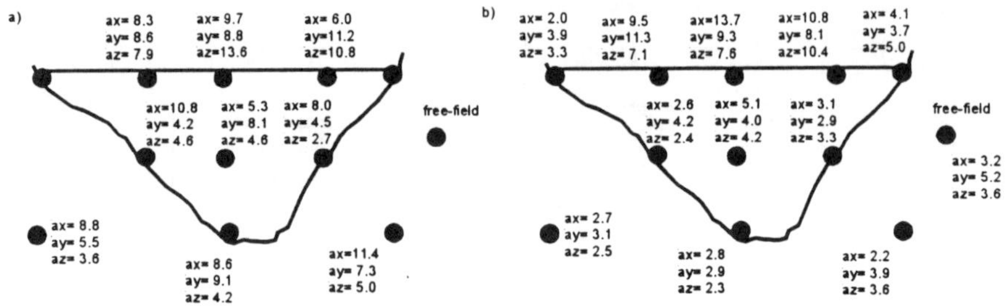

Figure 5. Peak accelerations at Mauvoisin in [cm/s**2]:
a) Local event of Nov. 1, 1994 b) Valpelline quake

The peak accelerations recorded at the crest (station D_03) during the Meythet earthquake are 5.3, 1.5 and 1.8 [cm/sec**2] in the stream (x-), cross-stream (y-) and vertical (z-) direction, respectively.
The water level was at 234 meters during the local event (reservoir almost full) and at 138 meters at the time of the Valpelline quake (slightly over mid-height). It was at 183 meters during the Meythet quake (more than two thirds full). The resonance frequencies listed in Table 2 are identified from the *PSD's*. They are in parenthesis whenever doubts on the correctness of the identification remain.
The differences in readings from one set of records to the next do not only stem from the interpretation made by the analyst (complicated by the occurrence of closed-space frequencies), but also from the differences in water level and in temperature conditions. An attempt has been made in the Table to associate the frequencies that correspond to one another. This was done by comparing the frequencies identified to those allocated during ambient vibration tests (see below).

Ambient vibration tests. A series of ambient vibration tests has been initiated at Mauvoisin to complement the results obtained by the array. Seven tests have been conducted at different water levels in 1995 and 1996 [12]. The resonance frequencies that have been identified are also indicated in Table 2 (T1 to T7). They initially increase with rising water level, and then decrease with a further rise. Because the principal effect of a rise in water level is to augment the mass of entrained water, one would have expected a reduction in resonance frequencies. That this is not the case at lower water levels is attributed to the vertical construction joints closing under the increasing hydrostatic pressure and thus to the dam becoming stiffer, what leads to an augmentation of the resonance frequencies. These two effects compete with one another, the former prevailing at higher water levels and the latter at lower ones.

Table 2. Mauvoisin: resonance frequencies [Hz].

T1	Valpelline	Meythet	T5	T2	T3	Local	T7	T6	T4
WL 124	WL 138	WL 183	WL 187	WL 199	WL 231	WL 234	WL 234	WL 237	WL 238
2.18	2.10	(2.24)	2.28	2.21	1.92		1.93	1.89	1.88
2.35	2.30	(2.39)	2.40	2.36	2.16	(2.13)	2.15	2.13	2.12
		(2.47)							
			2.76	2.68	(2.37)	2.36	2.37	2.36	(2.36)
		(2.98)	2.86	(2.80)	2.55	(2.59)		(2.52)	(2.49)
	3.28	(3.33)	3.27	(3.20)		(2.87)		2.71	(2.70)
3.36		3.53	3.55	3.53	3.28	3.20	3.21	3.20	3.18
	(3.77)	(3.94)							
	3.95					(3.64)			
4.22	(4.17)		4.30	4.20	(4.00)			(3.98)	
(4.39)	4.52	(4.52)	4.45	4.39	(4.14)	4.14		(4.09)	4.11
(4.62)			4.70	4.58	(4.22)		4.17	4.14	(4.19)
	(4.79)					4.47			
						(4.71)			
						(4.80)			
(5.03)	(5.06)		5.51	5.46					5.04
5.11		(5.90)	5.87	5.89		(5.12)		(5.09)	(5.08)
(5.22)			6.27	6.22					
	(5.76)					5.77			
	(6.09)	(6.57)				(6.10)			
6.23	(6.24)		6.77	6.71		6.32		(6.25)	(6.25)
	6.56					(6.72)			
	(6.78)					(7.29)			
	(7.35)		8.01	8.03		7.53			
7.29	(7.50)			(8.22)					(7.77)
	(7.79)								
8.43			9.48	9.56					9.44

3.8 Dam Of Grande Dixence

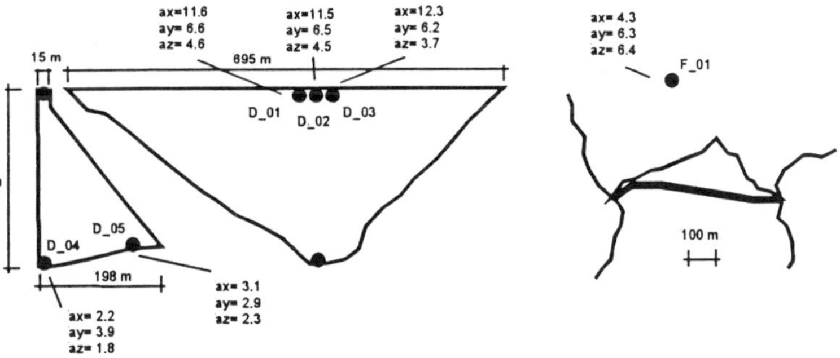

Figure 6. Grande Dixence: sketch and peak accelerations during the Valpelline quake [cm/sec**2]

The peak accelerations recorded during the Valpelline quake at Grande Dixence are reported in Figure 6. The water level reached 160 meters above the base at the time of the event, i.e. slightly more than mid-height. The resonance frequencies identified up to 8 Hz are reported in Table 3. The frequencies are again indicated in parenthesis whenever their identification is not sufficiently certain. Good correspondence exists between the two sets of identifications (*PSD's* vs. *TF's*), although a frequency may be identified with confidence from one function and not from the other (this was also the case for the Mauvoisin's records). It is noted that the resonance frequencies are also visible in the *PSD*'s of the free-field records, although the station is 290 meters away from the dam.

Table 3. Grande Dixence: resonance frequencies identified during the Valpelline quake [Hz].

From PSD's	2.05	2.23	(2.52)	3.02		(3.67)	(4.05)	(4.21)	(4.38)	(4.67)	(5.00)
From TF's	2.00	(2.30)	(2.55)	3.10	3.44	3.67		4.16	4.45	(4.79)	
From PSD's	(5.22)	5.56	(5.73)	(6.00)	(6.28)	6.70	(6.94)	(7.18)	(7.46)	7.83	
From TF's	5.20	5.61			(6.17)	(6.68)	(7.01)		(7.35)	(7.86)	

3.9 Dam Of Mattmark

The peak values recorded at the crest of the crown section during the Valpelline quake reached 6.5, 6.0 and 6.6 [cm/sec**2] and 4.7, 5.6 and 2.9 [cm/sec**2] during the local event of December 7, 1997 in the stream, cross-stream and vertical direction, respectively. The resonance frequencies of Table 4 have been identified. The water level was at 41 meters during the Valpelline quake (reservoir two thirds empty) and 95 meters during the local event (four fifths full).

Table 4. Mattmark: resonance frequencies [Hz].

Valpelline	(2.52)		3.11			(4.41)	4.96	(5.35)
Local event	(2.53)	2.83	(3.07)	3.46	(3.86)			5.27
Valpelline	5.77		(6.70)	(7.14)	7.46	(7.75)		
Local Event	(5.94)	(6.41)	6.90	7.16	(7.41)	(7.70)	(7.88)	

4. Final remarks

The arrays of the Swiss National Dam Network have the potential of providing most useful and sought-after information on the behavior of dams during earthquakes. While the network is running rather smoothly, the interpretation of the data collected so far has received too little attention. Potentially important information on total effective input motion and on dam response

during small earthquakes are available but not processed. This is the single most important issue at the moment.
The differences in resonance frequencies obtained from the *PSD*'s and from the *TF*'s as well as the large number of frequencies identified with uncertainty show that the simple identification techniques used are not suited to a comprehensive identification of the dynamic characteristics of dams. More advance techniques have to be used (see e.g. [13]), also addressing energy dissipation and mode shapes.
No attempt has been made to interpret the records in terms of effective input (uniformity of the abutment motion or lack thereof) nor of wave propagation. These aspects were also targeted when designing the arrays. There is some evidence that the non-uniform input might have adverse consequences that have not been sufficiently looked into in the past [14].
The dam arrays installed can also be used for warning purposes. A signal is generated as soon as oscillations of pre-set amplitudes occur at the dam. It is transmitted to the operating rooms, from where the appropriate safety checks are initiated.

5. Acknowledgments

Installation of the Swiss National Dam Network was funded by the owners of all Swiss dams. Additional support was provided by the Swiss Government and the owners of the instrumented dams. The network is being run by the Swiss Seismological Survey. The support of these institutions is sincerely appreciated.
The ambient vibration tests were performed as a joint project of the Concrete Structures Department of the Swiss Federal Laboratories for Materials Testing and Research and the Division of Safety of Dams of the Swiss Federal Office for Water Management.

6. References

1. National Research Council (1990) Earthquake engineering for concrete dams: design, performance, and research needs. National Academy Press, Washington D.C.
2. Darbre, G.R.(1995) Strong-motion instrumentation of dams, Earthquake Engineering and Structural Dynamics 24, 1101-1111.
3. Darbre, G.R. (1995) Strong-motion arrays at four large Swiss dams. Hydropower & Dams, Nov., 65-70.
4. Darbre, G.R. (1998) Instrumenting large dams for earthquake response, Proceedings of the Workshop on Geodynamical Hazards Associated with Large Dams (Luxembourg 1997), Cahiers du Centre Européen de Géodynamique et de Séismologie 16, 83-92.
5. Darbre, G.R. (1998) Strong-motion records at Swiss dams, 11th European Conference on Earthquake Engineering, Paris.
6. Fok, K.L. and Chopra, A.K. (1985) Earthquake analysis and response of concrete arch dams, Report No. UCB/EERC-85/07, Earthquake Engineering Research Center, University of California, Berkeley.
7. Fedock, J.J. (1982) Strong-motion instrumentation of earth dams, Open-File Report 82-469 (preliminary), U.S. Department of the Interior, Geological Survey.

8. Clough, R. W. and Ghanaat, Y. (1993) Concrete dams: evaluation for seismic loading, Proceedings International Workshop on Dam Safety Evaluation, Grindelwald, Vol. 4, 137-169.
9. Bolt, B.A. and Hudson, D.E. (1975) Seismic instrumentation of dams, Journal of Geotechnical Engineering Division, ASCE, 101, 1095-1104.
10. U.S. Committee on Large Dams (1989) Strong motion instruments at dams: guidelines for their selection, installation, operation and maintenance.
11. Darbre G.R. & Studer J. (1993) Verhalten der Talsperrenstarkbebenmessnetze beim Domodossola Erdbeben von 14. Juni 1993. *DACH-Mitteilungsblatt* 10.
12. De Smet C.A.M., Krämer C. and Darbre G.R. (1998) Ambient vibration tests at the dam of Mauvoisin, 16th International Modal Analysis Conference, Santa Barbara CA.
13. Loh C.-H. & Wu T.-S. (1996) Identification of Fei-Tsui arch dam from both ambient and seismic response data. Soil Dynamics & Earthquake Engineering 15, 465-483.
14. Mojtahedi S. & Fenves G.L. (1996) Response of a concrete arch dam in the 1994 Northridge, California earthquake. 11th World Conference on Earthquake Engineering, Mexico.

STRONG MOTION NETWORKS: A TOOL FOR THE ASSESSMENT OF EARTHQUAKE RESPONSE OF HISTORICAL MONUMENTS

E. DURUKAL, M.ERDIK, S.CIMILLI
Boğaziçi University, Kandilli Observatory and Earthquake Research Institute, Department of Earthquake Engineering, 81220, Çengelköy, Istanbul
e-mail: durukal@boun.edu.tr

Abstract

Strong motion networks installed in two historical monuments in Istanbul, Turkey, namely in Hagia Sophia Museum and in Süleymaniye Mosque, yielded valuable information regarding their time-domain and frequency-domain dynamic characteristics. The existing information was enhanced by recordings obtained during the recent August 17, 1999, Kocaeli (M_w=7.4) and November 12, 1999, Düzce, Turkey (M_w=7.2) earthquakes and during the following aftershock activity. Analysis of data yielded following results: The two buildings with similar structural support systems have different vibration characteristics. There is an earthquake specific variation of modal frequencies in the two structures which exists during low amplitude events as well. Local problems in terms of excessive vibrations exist in the southwest main pier of Hagia Sophia and possibly in the west main arch of Süleymaniye. Significant vertical vibrations at the crowns of the east and west main arches in Hagia Sophia are probably indicating locations where most of the damage are to be expected during a major earthquake close to Istanbul. The strong motion accelerometer networks are valuable tools for monitoring of historical structures and help to identify potential problems and/or to explain past/present structural problems. It is expected that such networks will become a standard tool for investigations towards the preservation of historical buildings in seismic areas in the future.

1. Introduction

Structural preservation of historical buildings in seismic areas has evolved to being one of the important and relatively new issues in earthquake engineering. The development of high dynamic range digital accelerometer systems paved the path to monitor and to study the dynamic response of historical monuments under actual earthquake ground motions. The results allow for the comparison of the dynamic behaviour of architecturally similar structures with different structural systems and/or different material properties and to make projections regarding their future earthquake

M. Erdik et al. (eds.), Strong Motion Instrumentation for Civil Engineering Structures, 209–228.

performance under scenario earthquakes. The prerequisite to a reliable and optimal strong motion instrumentation is to conduct a through study of the structural system of these monuments. This study should encompass the linear dynamic analysis, ambient vibration testing, soil and foundation investigations and the identification of zones of previous repair, weakness, cracking and other structural discontinuity. The number of sensors, their locations, orientations and common timing needs of the strong motion array can then be assessed in a reliable and optimal manner. The dynamic range and the frequency band of the accelerometers can be determined on the basis of the modal frequencies of vibration and the physical parameters of the scenario earthquake. The strong motion network directly measures the absolute accelerations at sensor locations. The total absolute velocities and displacements can be obtained with proper integration techniques. Under common timing the differential motions between the sensor locations can be easily computed. Further augmentation and redundancy of the earthquake response measurements can be realized with the installation of laser reflectometers, laser interferometers, displacement meters, short (for existing cracks) and long base-length strain gages, tilt- or inclinometers and triggered high-resolution time-lapse photography. The strong motion measurement system can also be extended to the foundation media through borehole accelerometers, tiltmeters and, if appropriate, piezometers to measure the pore water pressure.

Strong motion data obtained from the accelerometer networks in structures are used in structural system identification. Spectral system identification techniques are widely used by the engineering community, whereas parametric system identification techniques are relatively less preferred. The aim at the application of the technique is to estimate the natural vibrational characteristics of the building in question experimentally. The results can be used to calibrate numerical models, such as finite element models, which in turn yield valuable information about the static and dynamic behavior of the structure, hint on the reasons of past structural damages and deformations, if there are any, and can help to estimate the dynamic response during a seismic event or the feasibility and the reliability of structural interventions for restoration purposes. The spectral system identification makes use of Fourier amplitude, power, cross-power and phase spectra as well as transfer functions and transfer function phases for the identification of natural modes of vibration and natural modal frequencies. The parametric system identification, on the other hand, consists of choosing a parametric model, such as ARX or ARMAX models, calculating the model parameters using the input-output or output data, depending on the model chosen and data available, and validating the model by methods like zero-pole cancellations, comparing the real output with calculated output, calculating the residuals and their auto- and cross-correlations with the input. The on-line identification, which can be explained as determining the system parameters at every time step starting from an initial guess, can be a powerful tool for tracking the changes in the system parameters or in other words the non-linearity.

A strong motion recording system and the attendant investigations has been initiated by joint teams from Princeton (USA) and Boğaziçi (Turkey) universities (Mark *et al.*, 1992) for the Hagia Sophia Museum. Similar investigations were then extended to

Süleymaniye Mosque (Selahiye *et al.*, 1995). Both of the edifices are in Istanbul and constitute important monuments of the world's architectural heritage. In this paper results obtained from the strong motion arrays and the consequent analytical/numerical investigations will be covered, anomalous dynamic behaviors will be identified and a comparative study of the earthquake response of these two historical monuments will be presented.

2. Hagia Sophia And Süleymaniye

Historical monuments in Istanbul have been the living evidence of the damaging earthquakes in the city. Among these Hagia Sophia Museum and Süleymaniye Mosque have experienced several earthquakes and the assessment of their dynamic properties and earthquake worthiness is of both academic and practical concern

The location of the two structures in the Old City of Istanbul is shown in Figure 1. During their life span Hagia Sophia and Süleymaniye were effected by a series of earthquakes. As a result, western and eastern parts of the main dome and adjacent parts of the main arches and of the semidomes of Hagia Sophia collapsed three times. No structural collapses occurred in Süleymaniye due to earthquakes.

2.1. HAGIA SOPHIA

Hagia Sophia (Figure 2) was dedicated in 537 after five years of construction. It is one of the most important buildings of the world heritage and requires urgent concern due to the high seismic hazard in the western part of the North Anatolian Fault Zone. A network of strong motion accelerometers consisting of nine interconnected Kinemetrics SSA-2 units has been operational in the structure since August 1991 as part of the efforts for the determination of its earthquake worthiness. The configuration of the accelerometers is shown in Figure 2. The instruments cover three levels. One instrument is at the ground level; four of them are located at the cornice level next to the springing points of the four main arches; and four instruments are at the dome base level on the crowns of the main arches.

The main support system of Hagia Sophia consists of, from top to ground, main dome supported by the four main arches; two semidomes on east and west sides of the main dome supported by two secondary piers and two exedrae domes with their own colonnade system each; four main piers supporting the four main arches and four huge buttress piers on the north and south sides rising up to the dome base level (see Figuer 1). The plan dimensions of Hagia Sophia are 32 m by 80 m. Its main dome has a diameter of 31 m, which rises 56 m above the ground level.

2.2. SÜLEYMANİYE

The Süleymaniye mosque (Figure 3) was constructed between 1549-1557 and is celebrated as one of the biggest achievements of the Ottoman architecture. It survived several earthquakes without any significant structural damage since then. An accelerometer system consisting of eight individual Kinemetrics SSA-2 units operated in the Süleymaniye mosque between 1994 and 1999. The system was replaced by an interconnected array of eight GeoSys recorders in 1999. The locations of the accelerometer units in Süleymaniye can be seen in Figure 3. The positioning of the instruments is similar to the Hagia Sophia array except that there is no instrument at the ground level.

The vaulting system of Süleymaniye is similar to Hagia Sophia's with a main dome, four main arches, two semidomes and accompanying support system and four buttress piers. Its plan dimensions are 63 m by 73 m. It has a main dome of 27.5 m diameter that rises 48 m above ground.

2.3. COMPARISON OF STRUCTURAL SYSTEMS

The structural differences between the two buildings can be cited as follows; in Hagia Sophia the diameters of the main dome and the semidomes are the same whereas in Süleymaniye the semidomes are smaller in diameter relative to its main dome. The main arches are probably of cut stone in Süleymaniye, while they are of brick masonry in Hagia Sophia. There are no galleries on each side of Süleymaniye along its longitudinal axis, so that the main-and the buttress piers are connected only above the aisles. In Hagia Sophia the buttress piers are connected with main piers at the gallery level and at the first cornice where the main arches start to rise. In Süleymaniye the buttress piers rise in steps until a level close to the dome base, whereas in Hagia Sophia they continue until the dome base level as a single unit. The use of brick masonry with thick mortar layers starting from too low a level in Hagia Sophia has been replaced by the use of cut stones with closer joints in Süleymaniye.

3. Strong Motion Data Set

Between 1991-1995 the arrays have provided data pertaining to events with magnitudes varying between 3.9 to 4.8 M_b and epicentral distances between 22 to 140 km. The biggest event recorded by the Hagia Sophia array occurred on 22.12.1993 with M_b=4.8 at an epicentral distance of 59 km to the north of Istanbul. The biggest event recorded by the Süleymaniye array occurred on 8.2.1995 with M_b=4.4 at an epicentral distance of 140 km to the southwest of the city. Practically no records were obtained in the period of 1995-1999. In 1999 the Kocaeli and Düzce sequences were recorded by the two arrays with magnitudes changing between 4.3 and 7.4 M_d.

During the events predating the disastrous 1999 earthquakes maximum recorded accelerations were 13cm/s^2 in Hagia Sophia and 28 cm/s^2 in Süleymaniye. During the Kocaeli and Düzce events recorded peak accelerations reached 77cm/s^2 in Hagia Sophia and 227 cm/s^2 in Süleymaniye (see Table 1 and Table 2 for Hagia Sophia and Süleymaniye respectively). Data processing applied uniformly to all records involves baseline correction, high-cut filters at 10 Hz and low-cut filters at 1 Hz. As an example from the data set, the 12 November 1999 Düzce earthquakes as recorded by the Hagia Sophia and Süleymaniye arrays can be seen in Figures 4 and 5.

4. Earthquake Response of Hagia Sophia and Süleymaniye

4.1. TIME DOMAIN CHARACTERISTICS OF EARTHQUAKE RESPONSE

Based on pre-1995 earthquake recordings made in Hagia Sophia two features were found notable regarding its time domain behaviour; (1) different response characteristics of the SW pier relative to the other three piers (SW pier corresponds to station 4), (2) the high amplitude vibrations in the order of lateral vibrations at the crowns of east (station 9) and west (station 7) main arches in vertical direction (Durukal and Erdik, 1996).

Newly acquired data supported the observations mentioned above. Accelerations recorded at the top of the SW pier were associated with higher levels (see Figure 4 for the 12.11.1999 recording) and with different frequency characteristics when compared with data from stations 2, 3 and 5 at the same level. Displacements calculated at the same stations, not shown here, indicate that consistently higher displacements are sustained by the SW main pier. Higher response levels can also be seen from acceleration power spectra of 17.08.1999 records shown in Figure 6. The reasons could be soil conditions at the base of the SW pier, a foundation problem or localized structural or material problems. The bedrock topography beneath the main arches has been determined by Gürbüz *et al.* (1993). According to this study the bedrock is approximately 1 m below the ground level beneath the NW and NE piers. It dips towards south, such that the foundations of the SW and SE piers are at about 3 m below the ground level. This 2 m difference might have been an explanation if the SE pier had behaved similar to the SW pier as well. Emerson and Van Nice (1943), looking at the almost symmetrical structural deformation of the four main piers (an average of 0.15 m outwards in the E-W direction and an average of 0.45 m outwards in the N-S direction for a single pier), conclude that this can not suggest any differential foundation material behavior neither statically nor as a result of earthquakes. It remains only as one possible explanation, that this atypical behaviour is because of localised structural or material problems. It is almost impossible that materials different from other piers were used in the construction of the SW pier. Whether there is a structural deterioration or damage from the past earthquakes can not be studied at this stage since the body of the SW main pier is covered with marble panels. Hence, whatever the reason, this behaviour remains to be explained.

Table 1. Records obtained during the Kocaeli and Düzce earthquakes in Hagia Sophia

Event Date	Event Time	Event Location local	Event Coordinates North	East	Magnitude Md	Max.Acc-x* cm/s2	Max.Acc.-y* cm/s2	Max.Acc.-z* cm/s2	Triggered Stations S1	S2	S3	S4	S5	S6	S7	S8	S9
17.08.1999	3.02	Izmit	40.76	29.97	7.4	53,6 (4)	76,5 (2)	53,7 (2)	-	Y	-	Y	-	-	-	-	-
17.08.1999	3.17					5,4 (2)	6,1 (4)	2,4 (4)	-	Y	-	Y	-	-	-	-	-
17.08.1999	4.34					6,7 (2)	9,4 (4)	5,1 (2)	-	Y	-	Y	-	-	-	-	-
17.08.1999	4.49					5,4 (4)	3,5 (4)	3,1 (4)	-	Y	-	Y	-	-	-	-	-
17.08.1999	5.52					18,6 (4)	16,4 (4)	9,4 (4)	-	Y	-	Y	-	-	-	-	-
17.08.1999	6.15	Akyazı-Adapazarı	40.64	30.65	5.5	6,4 (4)	8,8 (4)	2,9 (4)	-	Y	-	Y	-	-	-	-	-
17.08.1999	7.15					5,6 (2)	6,2 (4)	5,3 (4)	-	Y	-	Y	-	-	-	-	-
17.08.1999	8.55	Adalar	40.78	29.05	4.3	7,0 (4)	7,1 (4)	2,3 (2)	-	Y	-	Y	-	-	-	-	-
13.09.1999	14.55	Izmit	40.77	30.1	5.8	67,8 (9)	57,5 (6)	74,2 (7)	Y	Y	Y	Y	Y	Y	Y	Y	Y
21.09.1999	0.28	Marmara Sea	40.69	27.58	5	12,9 (7)	12,3 (9)	14,7 (7)	Y	-	Y	Y	Y	Y	Y	Y	Y
29.09.1999	3.13	Yalova	40.7	29.34	4.8	31,6 (9)	14,2 (8)	14,4 (9)	Y	-	-	Y	Y	Y	Y	Y	Y
21.10.1999	2.08	Adalar	40.79	29	4.4	30,1 (9)	35,7 (8)	37,5 (7)	Y	Y	Y	Y	Y	Y	Y	Y	Y
07.11.1999	19.55	Hendek-Adapazari	40.71	30.7	5	2,7 (7)	4,9 (8)	4,6 (7)	Y	-	Y	-	-	-	Y	Y	-
11.11.1999	16.41	Sapanca-Adapazari	40.74	30.27	5.7	16,7 (7)	21,4 (8)	20,5 (7)	Y	Y	Y	Y	Y	Y	Y	Y	Y
11.11.1999	16.56	Sapanca-Adapazari	40.88	30.3	4.4	3,0 (7)	2,2 (6)	2,5 (7)	-	-	-	-	-	Y	Y	-	-
12.11.1999	18.57	Duzce-Bolu	40.74	31.21	7.2	22,6 (7)	35,1 (6)	26,4 (7)	Y	Y	Y	Y	Y	Y	Y	Y	Y
12.11.1999	19.07					2,8 (7)	3,5 (7)	2,7 (7)	-	-	-	-	-	Y	Y	-	-
12.11.1999	19.17	Duzce-Bolu	40.75	31.1	5.2	4,3 (9)	4,5 (8)	5,3 (7)	-	-	-	Y	-	Y	Y	Y	Y
12.11.1999	19.18	Duzce-Bolu	40.74	31.05	5.4	5,6 (7)	8,5 (6)	5,1 (9)	Y	Y	Y	Y	Y	Y	Y	Y	Y
12.11.1999	20.15	Kaynasli-Bolu	40.75	31.36	5	2,0 (7)	3,3 (7)	2,4 (7)	-	-	-	-	-	Y	Y	-	-

** Values in paranthesis indicate the stations where the relevant PGA is recorded,*
Note: Event information from Kandilli Observatory and Earthquake Research Institute

Table 2. Records obtained during the Kocaeli and Düzce earthquakes in Süleymaniye

Event Date	Event Time	Event Location	Event Coordinates		Magnitude	Max.Acc.-x*	Max.Acc.-y*	Max.Acc.-z*	Triggered Stations								
	local		North	East	Md	cm/s2	cm/s2	cm/s2	S0	S1	S2	S3	S4	S5	S6	S7	S8
17.08.1999	3.02	Izmit	40.76	29.97	7.4	227,2 (5)	161,7 (5)	165,1 (5)	-	Y	Y	Y	Y	Y	Y	Y	Y
17.08.1999	3.17					12,8 (5)	10,8 (6)	6,8 (5)	-	Y	Y	Y	Y	Y	Y	Y	Y
17.08.1999	4.34					20,2 (5)	20,9 (8)	13,3 (6)	-	Y	Y	Y	Y	Y	Y	Y	Y
17.08.1999	4.49					7,8 (5)	11,0 (6)	4,9 (5)	-	Y	Y	Y	Y	Y	Y	Y	Y
17.08.1999	5.52					29,6 (5)	18,8 (6)	10,3 (7)	-	Y	Y	Y	Y	Y	Y	Y	Y
17.08.1999	6.15	Akyazı-Adapazarı	40.64	30.65	5.5	10,9 (4)	8,5 (2)	4,0 (5)	-	Y	Y	Y	Y	Y	Y	Y	Y
17.08.1999	7.15					9,2 (5)	10,9 (6)	20,8 (5)	-	Y	Y	Y	Y	Y	Y	Y	Y
17.08.1999	8.55	Adalar	40.78	29.05	4.3	13,1 (5)	11,1 (8)	9,9 (5)	-	Y	Y	Y	Y	Y	Y	Y	Y
13.09.1999	14.55	Izmit	40.77	30.1	5.8	83,8 (5)	97,1 (6)	52,1 (8)	-	Y	Y	Y	Y	Y	Y	Y	Y
21.09.1999	0.28	Marmara Sea	40.69	27.58	5	14,6 (7)	15,7 (6)	5,8 (6)	-	Y	Y	Y	Y	Y	Y	Y	Y
29.09.1999	3.13	Yalova	40.7	29.34	4.8	14,5 (5)	12,8 (6)	9,3 (5)	-	Y	Y	Y	Y	Y	Y	Y	Y
21.10.1999	2.08	Adalar	40.79	29	4.4	38,0 (5)	43,8 (8)	28,3 (6)	-	Y	Y	Y	Y	Y	Y	Y	Y
07.11.1999	19.55	Hendek-Adapazari	40.71	30.7	5	7,8 (5)	4,7 (6)	2,2 (5)	-	Y	Y	Y	Y	Y	Y	Y	Y
11.11.1999	16.41	Sapanca-Adapazari	40.74	30.27	5.7	29,1 (5)	19,9 (6)	9,7 (6)	Y	Y	Y	Y	Y	Y	Y	Y	Y
11.11.1999	16.56	Sapanca-Adapazari	40.88	30.3	4.4	5,6 (4)	4,9 (6)	2,1 (5)	-	Y	Y	Y	Y	Y	Y	Y	Y
12.11.1999	18.57	Duzce-Bolu	40.74	31.21	7.2	19,1 (5)	20,4 (6)	7,1 (4)	Y	Y	Y	Y	Y	Y	Y	Y	Y
12.11.1999	19.17	Duzce-Bolu	40.75	31.1	5.2	4,9 (5)	4,3 (6)	2,1 (5)	-	Y	Y	Y	Y	Y	Y	Y	Y
12.11.1999	19.18	Duzce-Bolu	40.74	31.05	5.4	5,6 (5)	5,6 (8)	2,3 (5)	-	Y	Y	Y	Y	Y	Y	Y	Y

** Values in paranthesis indicate the stations where the relevant peak acceleration is recorded*

Note: Event information from Kandilli Observatory and Earthquake Research Institute

The structural history of Hagia Sophia hints the eastern and western arches and adjacent portions of the main- and semidomes as the most vulnerable locations. Indeed the eastern arch collapsed twice as a result of earthquakes in 557 and 1346 and the western arch collapsed during the 989 earthquake along with the portions of the main dome and the semidomes. However the main structure always remained intact. It was observed from the recordings from Hagia Sophia array that the vertical accelerations at the crowns of the west and east main arches (stations 7 and 9 respectively) are almost in the order of horizontal accelerations recorded at the same level and sometimes the peak acceleration is recorded in the vertical direction (see Table 1 and Figure 4). It is evident that especially the western main arch experiences high acceleration and displacement levels. These vertical vibrations of the arches could not be simulated by the finite element technique, although lateral dynamics of the building could be captured satisfactorily using this method.

Unlike in Hagia Sophia, in Süleymaniye the vertical motions at the crowns of the main arches are generally lower than their lateral counterparts (see Table 2 for the comparison of peak accelerations and Figure 5). Except maybe station 5 (see Figure 3), where higher out-of-plane accelerations were recorded in comparison to other stations, Süleymaniye in general seems to have a more uniform behaviour, which lessens the possibility of unexpected localised effects during stronger ground motions.

4.2. FREQUENCY DOMAIN CHARACTERISTICS OF EARTHQUAKE RESPONSE

For consistency the E-W axis of Hagia Sophia and the N-S axis of Süleymaniye will be called longitudinal axis hereafter. Similarly the N-S axis of Hagia Sophia and the E-W axis of Süleymaniye will be named as their transverse axis.

From the analysis of ambient vibration data 1.85 Hz and 2.10 Hz have been identified as the modal frequencies of Hagia Sophia by Durukal and Erdik (1994) along its longitudinal and transverse axes respectively. 2.35 Hz has been identified as the torsional modal frequency. It has been found that there is an earthquake specific variation of modal frequencies of Hagia Sophia. The highest values are obtained from the ambient vibration studies. They tend to drop with the increase in the acceleration level and with the increase of the duration of the strong motion part of the record. The first modal frequency of 1.85 Hz dropped as low as 1.56 Hz during the event on 12.12.1993 and the second frequency of 2.10 Hz dropped to 1.77 Hz. Analyzing the records of a single event by separating the data into three parts called pre-, co- and post strong motion parts, it could be shown that the variation can actually be tracked during a single specific event and that the modal frequencies recover their pre-strong motion values after the co-strong motion part dies out. The phenomenon is shown in Fig. 5 for the second modal frequency of Hagia Sophia based on records from station 6 and utilizing on-line parametric system identification technique. This had been an unexpected observation considering the low amplitude ground motions experienced by the structure. The reason was attributed to the micro-cracking throughout the structure (Durukal and Erdik, 1994) and to the existence of a jelly-like material in the mortar of

Hagia Sophia specific to the era in which Hagia Sophia was constructed which has the ability to emit the vibrations and indeed improves the earthquake performance of the building .

A similar study on the data from the Süleymaniye yielded however that the variation is not specific to Hagia Sophia, but exists in Süleymaniye with stone masonry arches as well. 3.35 Hz and 3.50 Hz have been determined as the first two natural frequencies along the longitudinal and transverse axes of the structure from the ambient vibration data using the procedure given by Gavin *et al.*, 1992. Analysis of the earthquake data showed that the first two modal frequencies drop as low as 3.17 Hz and 3.36 Hz respectively. The drop of the values in accordance with the increase in the ground motion amplitudes is also evident from the Süleymaniye data. It is clear that the explanations for the case of Hagia Sophia regarding the drop in these frequencies (Durukal and Erdik, 1994, Cakmak, 1995) are not sufficient for both cases now, since they were based on the assumption that the effect is due to the response of the brick masonry.

To assess the variation of modal frequencies during a single event, the time-localised modal frequencies were calculated using 50% overlapping, 5 second long segments. The change of modal frequencies are estimated for each station for a given event and then averaged to have a general picture for the whole structure. Using this approach, the variation of first and second modal frequencies of Hagia Sophia and Süleymaniye was calculated from recordings of events on 17.08.1999, 13.09.1999, 21.10.1999 and 12.11.1999. Figure 7 summarises the results for Hagia Sophia and Figure 8 gives the results for Süleymaniye. The trend followed by modal frequencies has been accentuated by curve fitting. It can be seen from the figures that variation can actually be tracked during a single specific event and that the modal frequencies almost recover their pre-strong motion values after the co-strong motion part dies out.

To study the effects of events of different magnitude, distance and acceleration levels, average of all transfer functions from nine stations in Hagia Sophia and average of power spectra from eight stations of Süleymaniye was taken for a given event. Estimated transfer functions are plotted in Figure 9 and Figure 10 for Hagia Sophia's EW and NS directions respectively for the events of 13.09.1999, 21.09.1999, 29.09.1999, 21.10.1999, 07.11.1999, 11.11.1999, 12.11.1999 (18:57) and 12.11.1999 (19:18). See Table 1 for the details of these earthquakes. The main shock of the 17 August 1999 Kocaeli earthquake was recorded by two stations only at the first cornice level of Hagia Sophia. Therefore no transfer functions were calculated for the main shock. The power spectra of the recordings at stations 2 and 4 are shown in Figure 6 for two directions. Estimated power spectra are presented in Figure 11 and Figure 12 for Süleymaniye's X and Y directions for the events of 17.08.1999 (3:02), 17.08.1999 (4:34), 17.08.1999 (05:52), 13.09.1999, 21.10.1999, 11.11.1999 and 12.11.1999 (18:57)). See Table 2 for the details of these earthquakes.

Identified modal frequencies from the newly acquired data for the first and second mode shapes are summarised in Table 3 for Hagia Sophia and in Table 4 for Süleymaniye.

There is a 25% drop of the first modal frequency, 27% drop of the second modal frequency in Hagia Sophia, in Süleymaniye 18% drop in the first modal frequency, 13% drop in the second modal frequency.

Presently the common reason causing the decrease of natural frequencies in both structures is unknown. Clearly the material non-linearity can not, by itself, be a rational explanation for the phenomenon, since it was observed during low amplitude events with very low strain levels as well.

Table 3. Modal frequencies, Hagia Sophia

Event	Hagia Sophia	
	1st modal frequency direction EW	2nd modal frequency direction NS
Ambient	1.85 Hz	2.10 Hz
17.08.1999	1.41 Hz	1.59 Hz
13.09.1999	1.43 Hz	1.61 Hz
21.09.1999	1.51 Hz	1.71 Hz
29.09.1999	1.80 Hz	1.74 Hz
21.10.1999	1.42 Hz	1.71 Hz
07.11.1999	1.61 Hz	1.79 Hz
11.11.1999	1.45 Hz	1.65 Hz
12.11.1999 (18:57)	1.38 Hz	1.53 Hz
12.11.1999 (19:18)	1.46 Hz	1.66 Hz

Table 4. Modal frequencies, Süleymaniye

Event	Süleymaniye	
	1st modal frequency direction X	2nd modal frequency direction Y
Ambient	3.35 Hz	3.50 Hz
17.08.1999 (03:02)	2.75 Hz	3.09 Hz
17.08.1999 (04:34)	3.16 Hz	3.15 Hz
17.08.1999 (05:52)	3.07 Hz	3.20 Hz
13.09.1999	2.84 Hz	3.15 Hz
21.09.1999	1.51 Hz	1.71 Hz
29.09.1999	1.80 Hz	1.74 Hz
21.10.1999	2.95 Hz	3.10 Hz
11.11.1999	2.98 Hz	3.15 Hz
12.11.1999 (18:57)	3.03 Hz	3.05 Hz

5. Conclusions

The time-domain and frequency-domain analysis of data summarized in this paper yielded following results: The earthquake specific variation of modal frequencies noted previously in the two structures, was displayed by the recent data as well. A drop up to 25% was observed in of the first modal frequency, a drop up to 27% was observed in the second modal frequency in Hagia Sophia as compared to the values determined based on pre-earthquake ambient vibration survey. In Süleymaniye a 18% drop in the first modal frequency, 13% drop in the second modal frequency was seen. Local problems were once more noted in Hagia Sophia in the south-west main pier and possibly in Süleymaniye in the west main arch. Significant vertical vibrations at the crowns of the east and west main arches in Hagia Sophia are probably indicating parts of the structure where most of the damage are to be expected during a major earthquake close to Istanbul. A post-earthquake, quick ambient vibration survey conducted in Hagia Sophia showed a frequency drop in the first two modal vibrational frequencies in the order of 4-5%, indicating the need for a comprehensive ambient vibration study.

In a major earthquake likely to occur in İstanbul, damaging stress levels can be reached in Hagia Sophia. Therefore a concerted interdisciplinary action is definitely needed to carry it over to the next generations in an intact form.

The strong motion accelerometer networks are valuable tools for monitoring of historical structures. They can help to show localized problems such as in the case in Hagia Sophia, on a quantitative basis they can hint on similarities and differences between two structural systems which otherwise seem similar to many; augmented with appropriate analytical/numerical tools, the data obtained from such strong motion arrays can indicate possible reasons for the past collapses of structures; and they can bring up new phenomena the explanation of which may not be possible with our present knowledge and need to be investigated in more detail. It is expected that such networks will become a standard tool for investigations towards the preservation of historical buildings in seismic areas in the future.

Acknowledgements

We would like to acknowledge the help provided by the following institutions: Hagia Sophia Museum Directorate, Directorate of Restoration and Monuments, Central Laboratory for Restoration and Conservation, Ministry of Culture. The work of maintenance of the arrays is due to the endless efforts of Mr. Aydın Mert and Mr. Muzaffer Gül of Kandilli Observatory and Earthquake Research Institute. At last but not the least we would like to thank to our mentor Prof.Ahmet Cakmak of Princeton University who has initiated our research on the earthquake protection of Hagia Sophia and who has dedicated his time and energy to the protection of cultural heritage in Istanbul.

References

1. Durukal, E. and M.Erdik (1994). Comparison of system identification techniques: a case study for Hagia Sophia. In: *Earthquake Resistant Construction and Design* (S.A. Savidis, Ed.), 2, pp. 993-1000.
2. Emerson, W. and R..L. Van Nice (1943). Hagia Sophia, Istanbul: Preliminary report of a recent examination of the structure. *American Journal of Archeology*, 47, 403-430.
3. Erdik, M. and E. Durukal (1996), Use of Strong Motion Data for the Assessment of the Earthquake Response of Historical Monuments, Proceedings of the 11th World Conference on Earthquake Engineering, Acapulco, Mexico, June 23-28, 1996.
4. Gavin, H., S. Yuan, J. Grossman, E. Pekelis and K.Jacob (1992). *Low-level dynamic characteristics of four tall flat-plate buildings in New York City*. NCEER-92-34.
5. Gürbüz, C., M. Deal, J. Bouwer, T. Bekler, A.S. Çakmak and M. Erdik (1993). *Technical Report on the Seismic Refraction Studies in Hagia Sophia* (in turkish). Bogaziçi University, Kandilli Observatory and Earthquake Research Institute.
6. Mark, R., A.S. Çakmak and M.Erdik (1992). Preliminary report on an integrated study of the structure of Hagia Sophia: past, present and future. In: *Hagia Sophia from the Age of Justinian to the Present* (R. Mark and A.S. Çakmak, Ed). Cambridge University Press. pp 120-131.
7. Selahiye, A., M.N. Aydinoglu and M.Erdik (1995). Determination of dynamic characteristics of Süleymaniye mosque by experimental and analytical methods (in turkish). In: *Proceedings of the Third National Conference on Earthquake Engineering*. pp. 284-293.

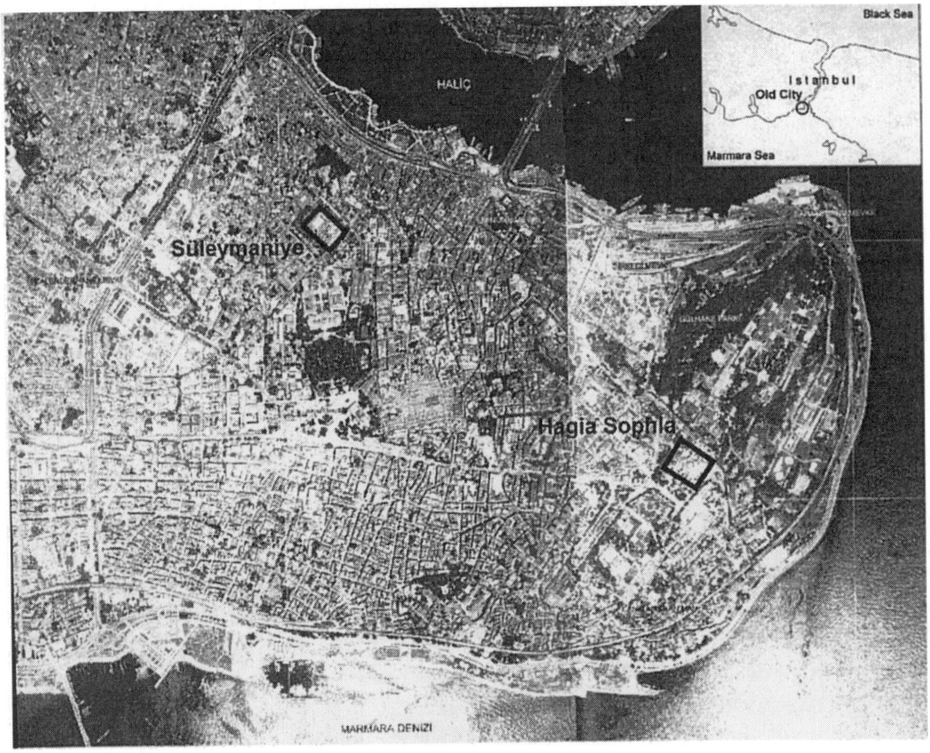

Figure 1. Locations of Hagia Sophia and Süleymaniye in the Old City of İstanbul

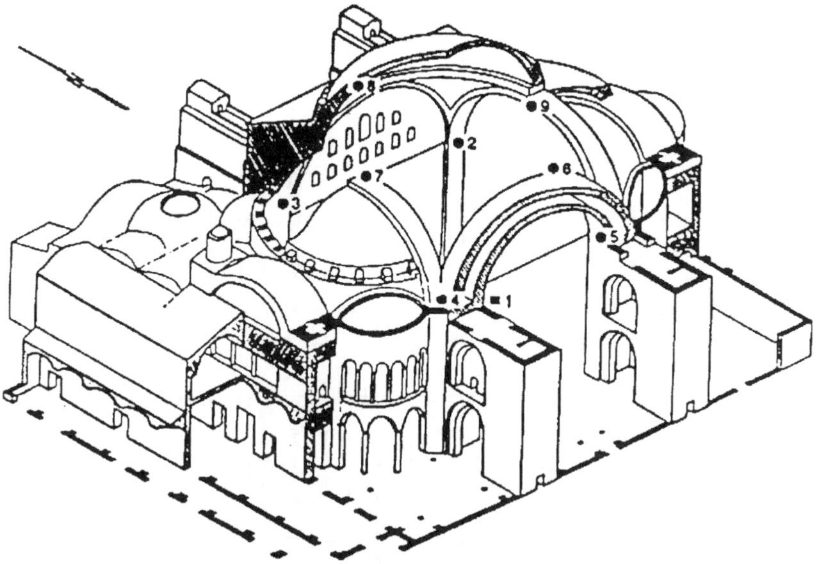

Figure 2. Main support system of Hagia Sophia and the strong motion array

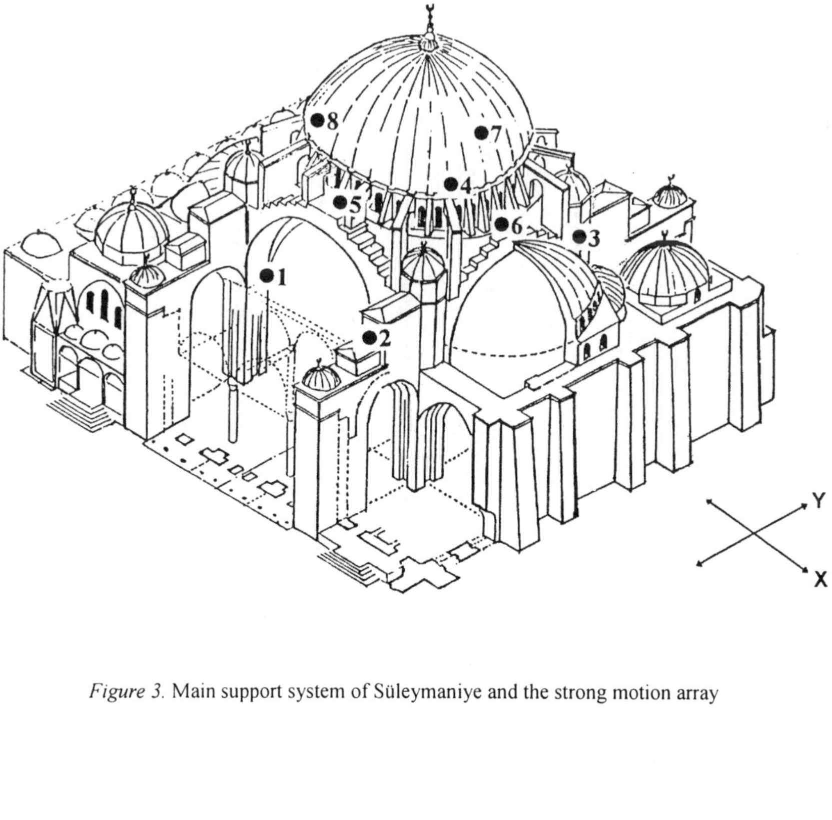

Figure 3. Main support system of Süleymaniye and the strong motion array

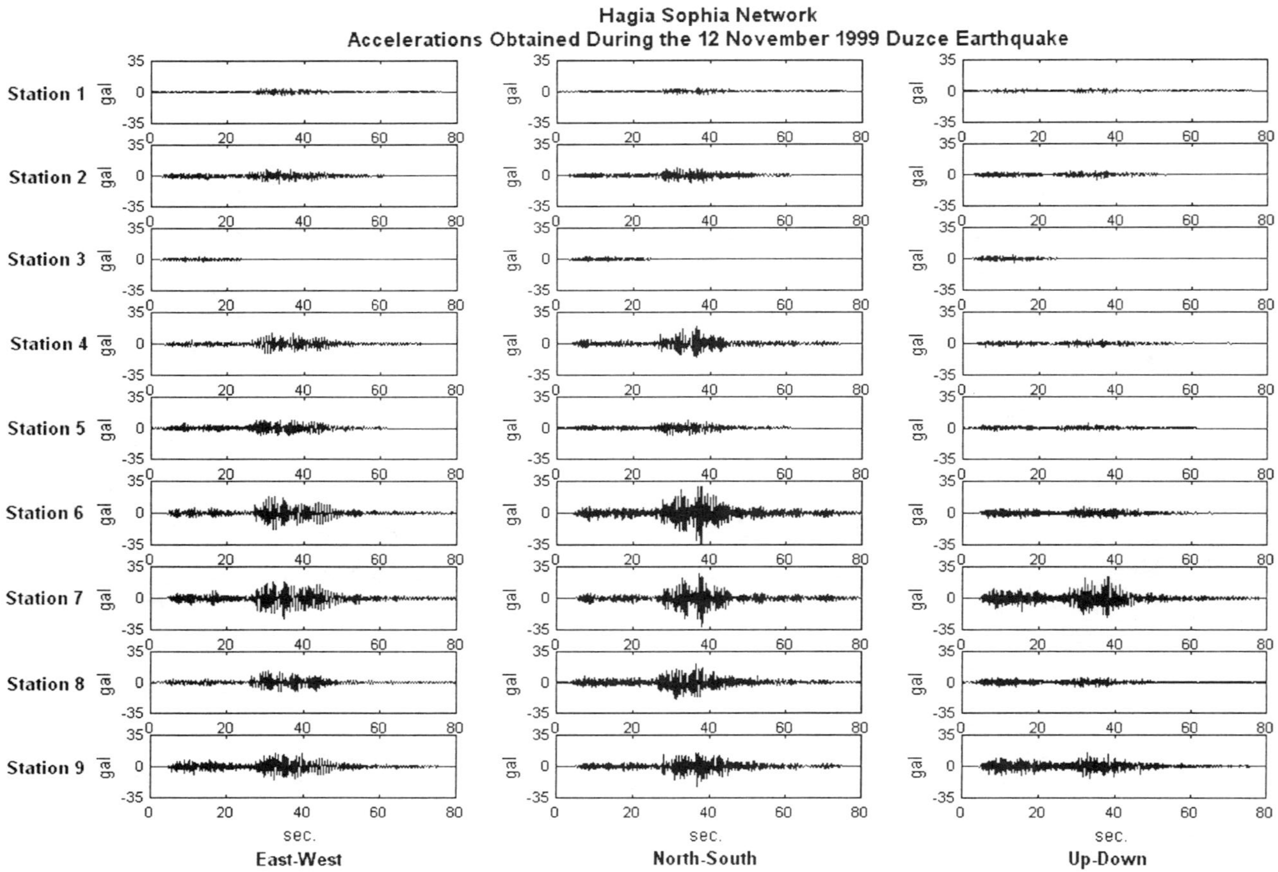

Figure 4. Accelerations recorded by the Hagia Sophia strong motion array during the 12 November 1999, Düzce earthquake

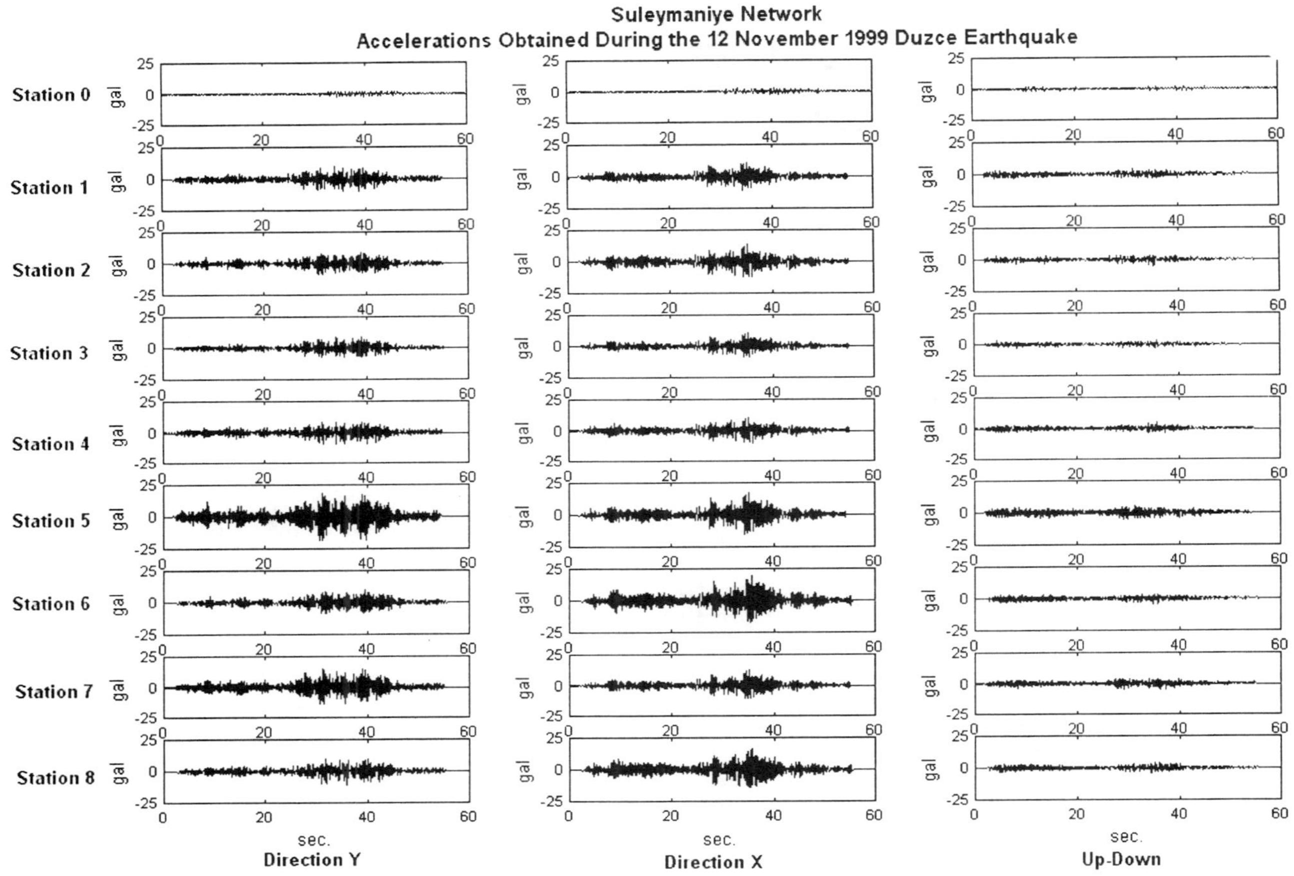

Figure 5. Accelerations recorded by the Süleymaniye strong motion array during the 12 November 1999, Düzce earthquake.

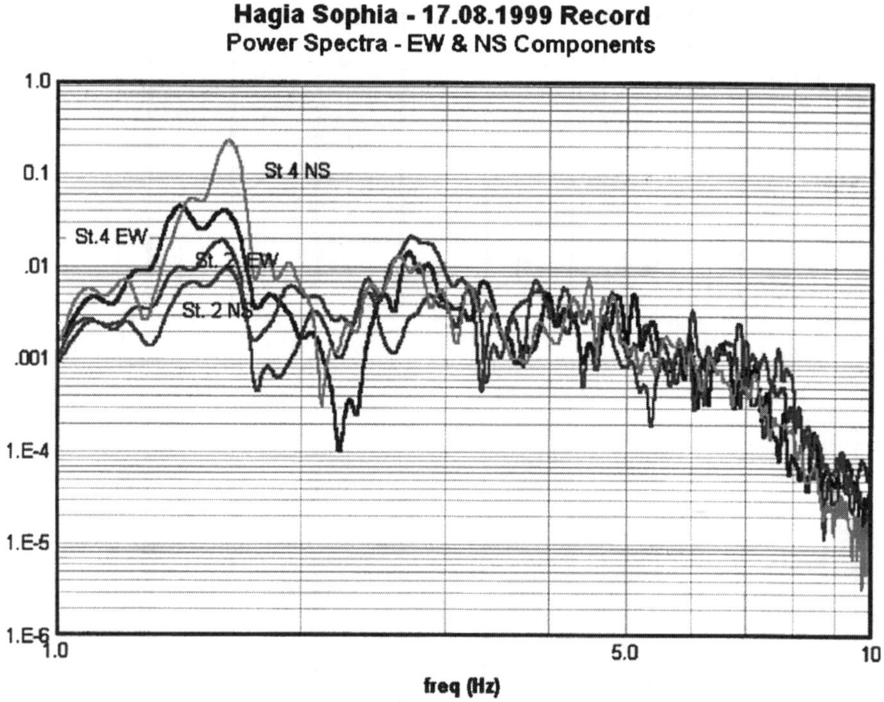

Figure 6. Power spectra estimations of the 17 August 1999, Kocaeli earthquake records obtained in Hagia Sophia

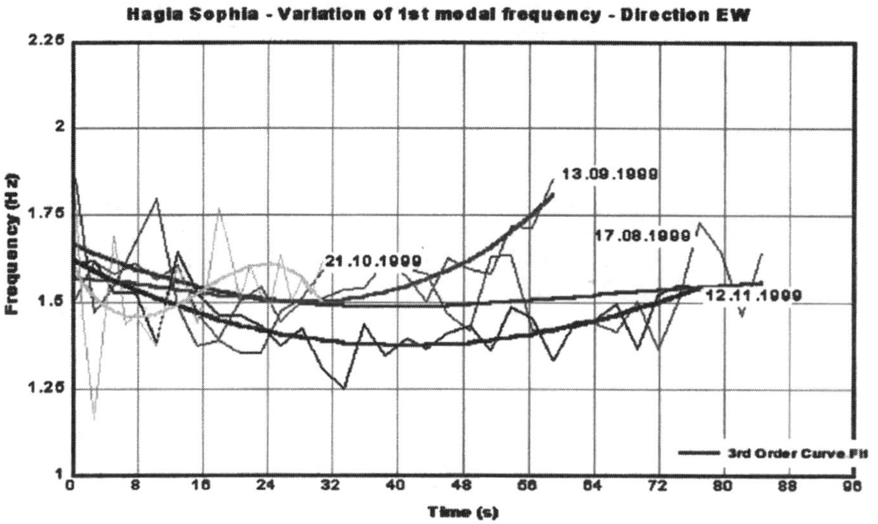

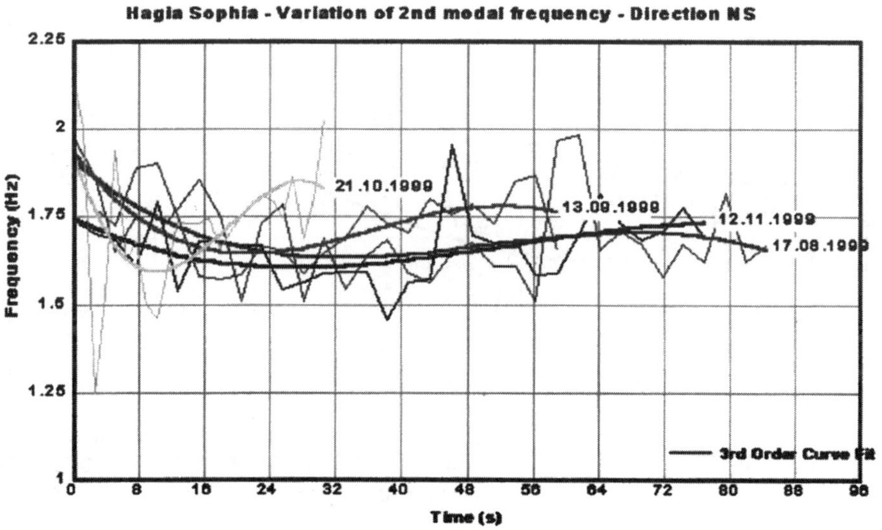

Figure 7. Variation of 1st and 2nd modal frequencies in Hagia Sophia

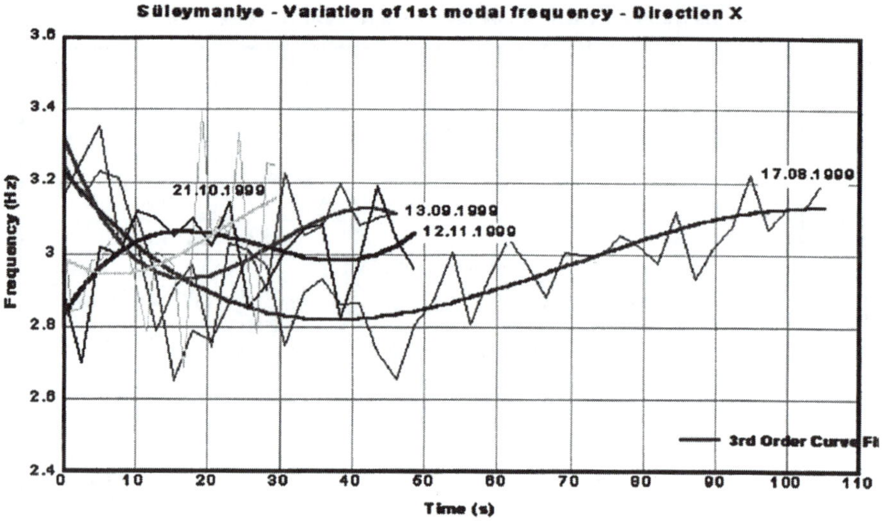

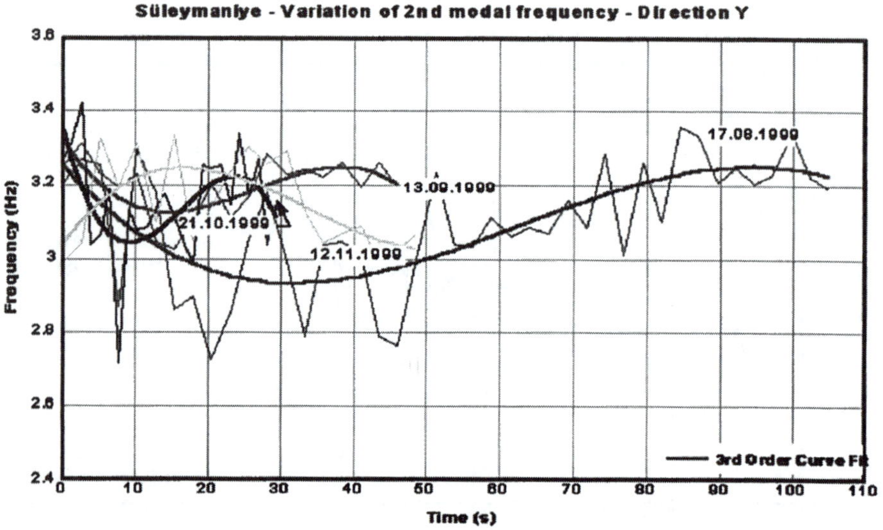

Figure 8. Variation of the 1st and 2nd modal frequencies in Süleymaniye

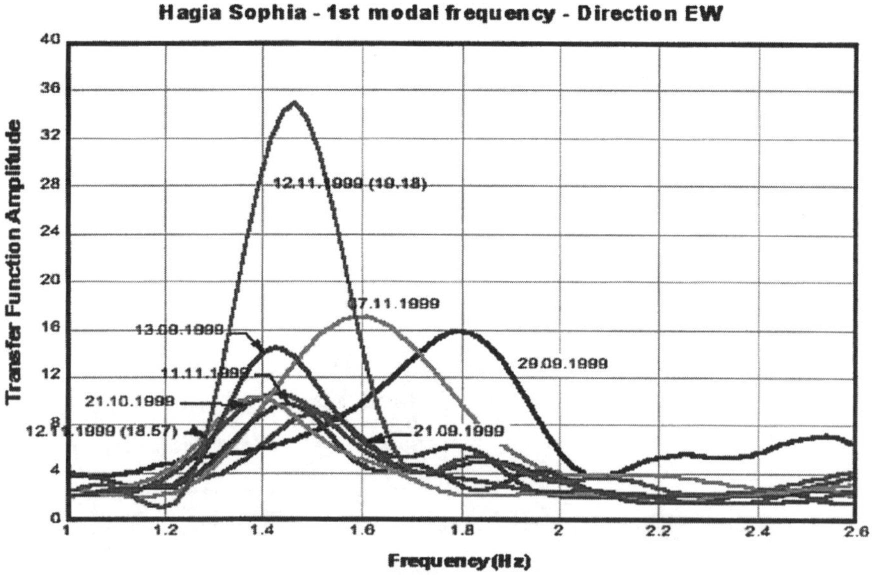

Figure 9. Estimated transfer functions for Hagia Sophia's EW direction for the events of 13.09.1999, 21.09.1999, 29.09.1999, 21.10.1999, 07.11.1999, 11.11.1999, 12.11.1999 (18:57) and 12.11.1999 (19:18).

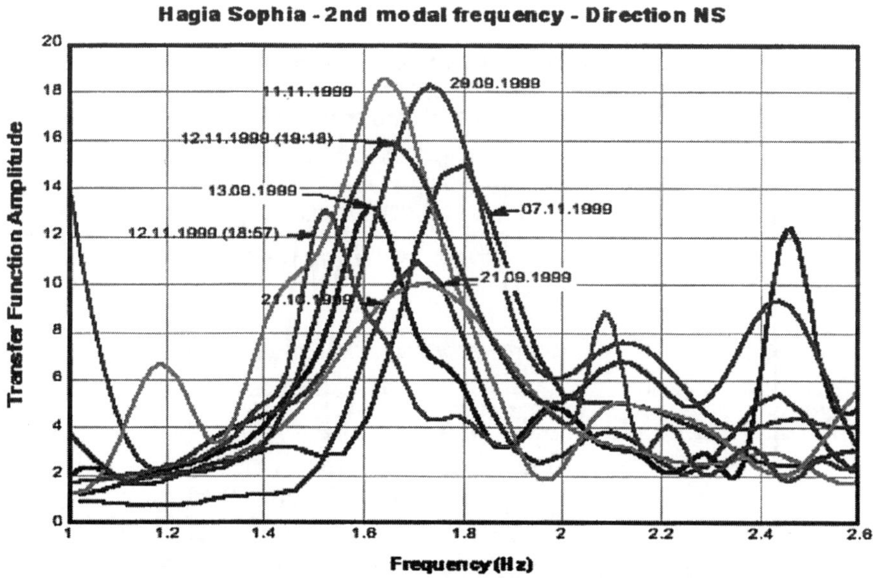

Figure 10. Estimated transfer functions for Hagia Sophia's NS direction for the events of 13.09.1999, 21.09.1999, 29.09.1999, 21.10.1999, 07.11.1999, 11.11.1999, 12.11.1999 (18:57) and 12.11.1999 (19:18).

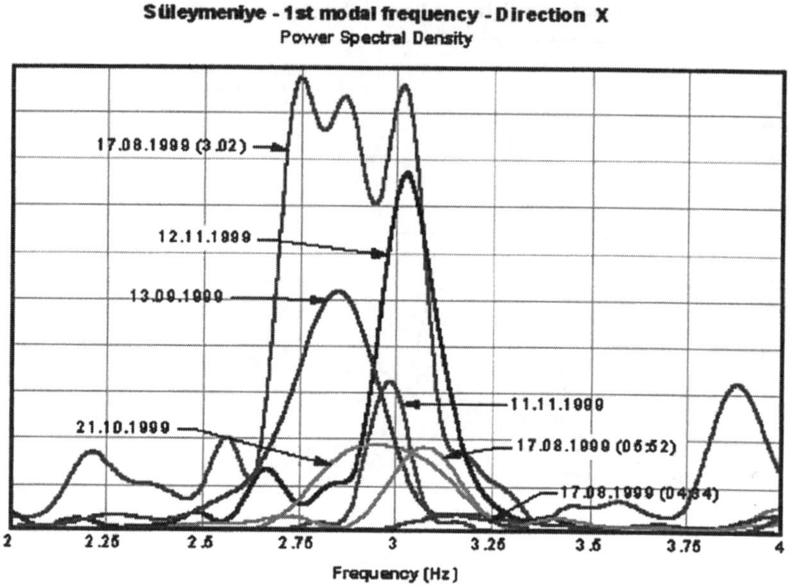

Figure 11. Power spectra estimations for Süleymaniye's direction X for the events of 17.08.1999 (3:02), 17.08.1999 (4:34), 17.08.1999 (05:52), 13.09.1999, 21.10.1999, 11.11.1999 and 12.11.1999 (18:57)).

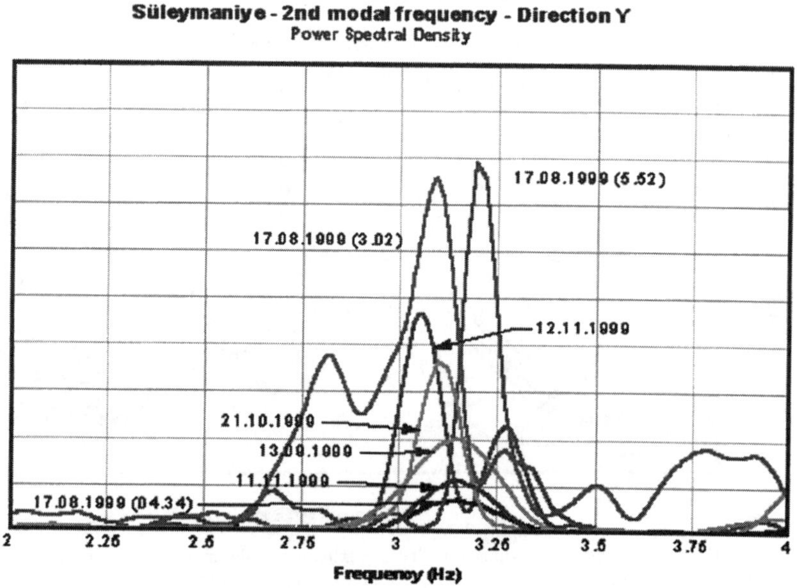

Figure 12. Power spectra estimations for Süleymaniye's direction Y for the events of 17.08.1999 (3:02), 17.08.1999 (4:34), 17.08.1999 (05:52), 13.09.1999, 21.10.1999, 11.11.1999 and 12.11.1999 (18:57)).

WIRELESS MONITORING AND LOW-COST ACCELEROMETERS FOR STRUCTURES AND URBAN SITES

JOHN R. EVANS
U.S. Geological Survey
345 Middlefield Rd, MS-977; Menlo Park CA 94025, USA;
Tel. +1-650-329-4753; FAX +1-650-329-5143 or -5163
e-mail: jrevans@usgs.gov.

1. Abstract

New technologies in wireless telemetry and low-cost micromachined silicon accelerometers enable lower cost and denser instrumentation of both structures and the urban environment. Applications include any site where robustness, capital or maintenance costs, spatial density, or ease of communication are controlling issues. This paper reviews these new implementing technologies and gives two practical examples.

Recently introduced techniques have vastly expanded the palette of practical telemetry options. These include unlicensed spread-spectrum, cell-phones in several variants, two-way paging, and a rapidly growing set of satellite services. Many products are available for each technique, so off-the-shelf solutions are routinely achievable. Relevant issues when selecting telemetry for a particular project include whether the service is offered in that region and how complete their coverage is, access from inside structures and urban canyons, capital and recurring costs, ease of installation, and viability after an earthquake. These issues vary greatly from one technique to another. However, an attraction shared by most is the elimination of most hard wiring, which can greatly ease installation and reduce the number of failure modes, both in routine maintenance and during an earthquake.

Low-cost micromachined accelerometers have reached price and performance levels of interest for structural and free-field monitoring, particularly to augment the spatial density of more expensive instruments in urban areas. For example, one can find dynamic ranges reaching from typical urban background noise levels to ±2 g for under US$100 per channel, or better the resolution of an SMA-1 for a few tens of dollars per channel. This paper includes performance details from USGS tests of the high-dynamic-range accelerometer. Most micromachined accelerometers exhibit small physical size, low cost, and great robustness and stability. In combination, these characteristics yield low capital and maintenance costs.

M. Erdik et al. (eds.), Strong Motion Instrumentation for Civil Engineering Structures, 229–242.

The new wireless technologies and micromachined accelerometers can contribute separately to structural monitoring, and they may be used separately in many situations. Additionally, the two technologies can be combined to sharply reduce the cost of arrays, and to permit arrays of unprecedented spatial density. The two examples presented here combine both technologies. In the first, spread-spectrum transceivers and micromachined accelerometers were used to monitor a New Mexico highway bridge for structural integrity. In the second, micromachined accelerometers and cell-phone Internet links (CDPD) are being used for a regional urban free-field array in Oakland, California.

2. New Telemetry Techniques

The gamut of telemetry options for structural and urban monitoring has exploded over the last decade to include unlicensed spread-spectrum, satellites, cell phones, and other methods. -This trend shows no sign of abating. In particular, many more satellite options are scheduled to come on-line over the next five years and may make this option more affordable than it now is.

2.1. SPREAD SPECTRUM

Unlicensed spread-spectrum is among the most interesting of the new telemetry options precisely because it obviates the difficulties of obtaining licenses, without the recurring costs of commercial telemetry providers. Spread spectrum includes two techniques: direct-sequence and frequency hopping. The latter is obvious, with the transceiver pair using familiar narrow-band methods but spending only moments on one frequency before hopping to the next in a prearranged or signaled pattern. Direct-sequence spread spectrum, such as that used to transmit GPS signals, uses a pseudorandom "chipping" sequence to dither the carrier across a large spectral range, typically on the order of one MHz. Receivers must know the pseudorandom sequence and correlate it with the incoming signal to recover the data stream. Both methods provide security and resist noise sources and, to some extent, multipathing. However, they add overhead and synchronization delays as well as lowering bit rates significantly. Nevertheless, the bit rates obtained are quite adequate for seismological data.

In the U.S., several "ISM" bands are now available for spread-spectrum transmissions at low power. They are centered at 915 MHz, 2.45 GHz, and several higher and lower bands. The two bands cited include nearly all products now available. Europe and Japan have similar bands. It is worth noting in passing that these frequencies are subject to significant attenuation by water, including foliage, and generally require a very clean line-of-sight between transmitter and receiver. However, some vendors claim to be able to use the multipathing typical of urban settings to transmit over bent paths. A brief survey of the Web and other sources finds dozens of manufacturers, consultants, IC makers, and related services for the ISM bands. These include, with little review and by no means limited to, those listed in *Table A1* of *Appendix A*.

The lower ISM band (915 MHz) is very similar in frequency and power levels to cell phones. Hence, the arrays of cell-phone towers in a typical city give an indication of the spacing, density, and elevations of receivers that would be required in a large, generalized urban deployment. These requirements are well beyond the resources of most researchers. Hence, the best scenarios for spread-spectrum telemetry probably are either a local deployment, with one or several nearby receivers, or a broader deployment with receivers on ridges and other high points using high-gain antennas. These antennas would illuminate patches of the urban area and should be able to hear faint signals transmitted from within non-metallic structures. Such a generalized deployment would be costly and could prove to be maintenance intensive. However, it could be done with relatively few well-placed high-quality receiver sites. Note that one could not legally use the high-gain antennas for transmission, so this is effectively a one-way telemetry solution, from sensors to lab but not the reverse.

2.2. CELL PHONES

Cell-phone networks offer at least three classes of telemetry solutions: circuit switched calls, CDPD (cellular digital packet data—a 19.2 kbps Internet node), and small-packet signaling options, such as Aeris™ (http://www.aeris.net/splash/index2.htm). Aeris' Microburst™ effectively allows one to transmit a 15-digit "telephone number" about once per minute, and is designed for vending-machine monitoring and similar low-bandwidth applications. Indeed, the bandwidth is far too low for waveform transmissions, but is sufficient for state-of-health messages and terse summaries of activity (e.g., PGAs). It is the least costly of the three options, and is primarily a one-way telemetry solution.

Circuit-switched connections are equivalent to landline modem connections, generally are charged by the minute, and bear no further discussion here. Robustness issues applicable to CDPD are relevant.

On the other hand, CDPD is a very interesting option offering both TCP (virtual connection) and UDP ("datagram") Internet signaling—a true, bidirectional Internet node. Some CDPD modems require the host computer (the seismograph) to carry the full TCP/IP stack, generally communicating *via* SLIP or PPP. Other modems contain this functionality and appear to the host as a simple AT modem. *Table A2* gives CDPD-modem vendor examples.

Issues that may affect CDPD performance include the use of landlines from many cell-phone base stations to their switches. These landlines typically include a partial or full T1 trunk line from the base station to the nearest telephone switching office, thence a "sonnet ring" to the cell-phone provider's main switch. From that point onward, service is generally over a redundant Internet backbone and should be reliable. Though the sonnet ring will function after a single-point failure, some earthquake-prone areas are likely to produce multiple failures, at which point reliability becomes an issue of rapid, automated transfers to other circuits, if they are still available. The trunk lines out to base stations probably cannot tolerate a single point of failure.

In my experience, CDPD modems sometimes have to be placed carefully, based on signal strength, but generally still work adequately from deep inside non-metallic structures or near the windows of metallic structures. It can be helpful, with wavelengths of around 35 cm, to shift the antenna by as little as 10 cm to improve signal strength.

2.3. TWO-WAY PAGING

In some locations, two-way paging may offer a solution. Paging systems generally have a relatively small number of high-powered base stationcommunicating with a central site over satellite links. The return leg, from the pager to the base station, is lower powered and more tenuous. So far, these return links have had limited uses, such as a set of standard responses ("yes", "no", ...) but they have greater potential versatility. One example is SkyTel™(http://www.skytel.com/SKYTEL.NSF/products/skywriter.html)andMotorola™Smart Pagers™ http://www.motorola.com/MIMS/MSPG/SmartPagers/).

2.4. SATELLITES

Lastly, some new satellite-based voice and data systems are already on line with many more planned for the next five years. The most famous probably is Iridium™, first of the "big LEOs" (that is, a large constellation of low-earth-orbit telecom satellites, in this case 66). At this writing, Iridium™ data services are not yet provided. They plan to begin these by the end of September, 1999, at ≥1200 bps (and more reasonable bit rates anticipated). Prices for data services are unavailable, but voice calls now range from US$2 to US$7 per minute, and within the US generally from US$2 to US$3 per minute. Iridium is below its planned customer base at this time, perhaps because of these prices and those of the telephones, which currently cost several thousand dollars.

Orbcomm™ (currently 24 microsatellites—41-kg each) may be more interesting to seismology and engineering, being entirely a small-packet data service. They use lower-frequency links than other satellite services, and may therefore be less sensitive to antenna placement. Orbcomm™ delivers data packets typically within two minutes, including over the Internet. Though relatively inexpensive, they still charge around one cent (US$0.01) per byte transmitted, which adds up quickly and for larger messages may exceed the cost of Iridium.

Teledesic™ (288 satellites planned) may eventually provide the most cost-effective solution, simply by virtue of its size and the intended market—every PC. It will provide a high-speed Internet connection. The target service date is 2003.

A number of other vendors are in various stages of development. Globalstar™ (48 satellites, about half now in orbit) probably is the farthest along. They plan to begin service by the end of September, 1999, although they have had some problems with failed launches. They are a direct competitor to Iridium™, but offering lower prices by simplifying the satellite portion of the system and connecting to the existing telephone system rather near the user.

They will sell services through partners, such as AirTouch™ in the US. ICO Global Communications™ is nearly as far along as Globalstar™, but is using only 10 satellites, so few because they will be in medium earth orbit. They anticipate beginning service in August, 2000. *Table A3* gives URLs for these organizations.

Prices in this sector remain quite high, but low initial user interest and increasing competition are likely to change this picture in the next few years. Most systems use rather high-frequency links to the user (mostly 1.6 to 2.4 GHz, but up to 29.1 GHz for Teledesic™), so very clear sight lines may be critical, and rain or other moisture may cause problems. The lowest-frequency links known to me are those of Orbcomm™ (137 to 400 MHz). Presumably, these will be the least sensitive to water and line-of-sight. Nevertheless, an outdoor antenna, typically mounted on a roof, may be required.

3. Silicon Accelerometers

Micromachining is a process similar to integrated-circuit manufacturing, but using anisotropic etchants and other techniques to create small mechanical structures, generally in silicon. Pressure sensors and accelerometers were the first broad commercial applications of micromachining technology, and many examples of each are now on the market. Accelerometers produced by micromachining generally have crystalline silicon cantilever springs, built-in over-range stops, and viscous gas damping. As a result, they exhibit great robustness against shock, vibration, and long-term spring sag. Shock resistance of several thousand g's is common, and spring sag in some decade-old test samples remains immeasurably small. Micromachined accelerometers are mostly open-loop devices, but at least one is an FBA. Most use capacitive sensing of the proof-mass offset, but a few use Wheatstone-bridge strainmeters. Most micromachined accelerometers are also low cost devices, so that both capital and maintenance costs tend to be well below those of traditional earthquake accelerometers.

Current examples of micromachined accelerometers range from rather noisy low-sensitivity devices (e.g., Analog Devices™ ADXL-05™; *Evans and Rogers* [1995]) to instrumentation-grade devices (e.g., ICSensors™ [*Evans*, 1998*a*], Kistler™), and many intermediate offerings. They range in price from around US$10 to several hundred dollars per component, with this price in most cases dropping sharply at larger purchased quantities.

Perhaps the quintessential application of micromachined accelerometers to seismic engineering, by far the largest I know of, is that of Tokyo Gas™. They are installing 3600 Sumitomo Precision Products™ triaxial micromachined accelerometers at about 1-km intervals throughout urban Tokyo (SIGNAL—Seismic Information Gathering and Network Alert System). The Sumitomo™ sensors are integrated into a complete instrument by Yamatake Corporation™. Tokyo Gas™ is using these instruments for automated shutoff of neighborhood gas supplies, based on a Spectral Intensity computation. The instruments also signal the shaking metrics to a central location to assist response optimization. Other commercial uses of micromachined accelerometers for earthquake response include: Kinemetrics™, in seismic triggers and low-resolution recorders; at least two small companies in southern California (SeisAlert™ of San Marino, and Geophysical Dynamics™ of Newberry Park); and the Geophysical Institute of Israel™, in Holon, which is producing *P*-wave alarms. Numerous other companies produce various earthquake triggers for shutting off gas or equipment, and some of these presumably are using silicon accelerometers. Lastly, Terra Technology™ of Redmond, Washington, and its partner GeoSys AG™ of Switzerland offer a low-cost single-board seismograph that is well matched to the capabilities of the ICS-3028™ and similar devices. They do not offer the corresponding sensors.

The most interesting micromachined accelerometers for structural engineering and seismology are those in the midrange. Those in the high-cost range offer little or no price advantage over traditional accelerometers (though they may be more robust), while those in the low-cost range often are too noisy to be of general interest. However, the lowest-cost sensors have been usefully applied by Kinemetrics™ as seismic triggers and are also offered by them as a 9- to 11-bit replacement for the SMA-1™. The most interesting mid-range example known to me is the ICS-3028™, which reaches from typical urban noise levels to ±2 g for around US$100 per component in parts cost. Assuming a factor of two markup to retail, this is still one third to one forth the cost of a state-of-the-art research accelerometer, such as the FBA-11™, and in large quantities may have a larger price advantage. The ICS-3028™ has a signal-to-noise ratio, conservatively, of about 16 bits (*Figures 1 and 2*, abstracted from *Evans* [1998*a*]). It begs a digitizer of at least 18-bit resolution to take full advantage of its capabilities.

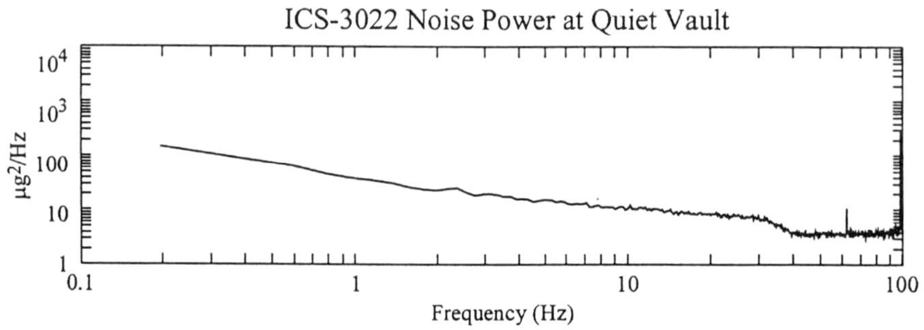

Figure 1. Sensor-noise power spectrum of the ICS-3028™ family of sensors (the ICS-3022™ is the same sensor in a slightly different package).

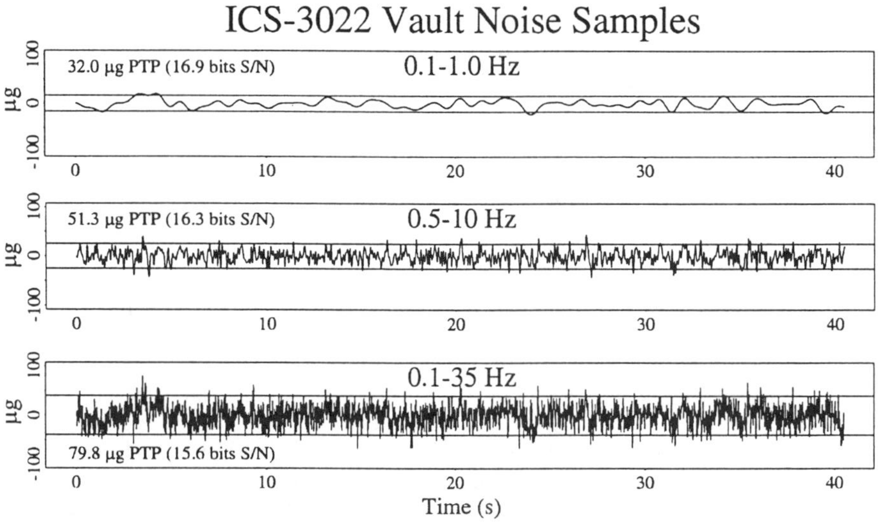

Figure 2. Sensor-noise examples in different frequency bands for the ICS-3028™ family of sensors.

Lower-performance examples include the ICS-3265™, Sumitomo's™ C3A-02™ (cf. Tokyo Gas™), and CSEM's™ MS-6100™, all of which are potentially less expensive per component than the ICS-3028™ though somewhat noisier, perhaps ~14-bits resolution. Even these lower-priced devices significantly exceed the performance of the venerable SMA-1™, upon which much of current earthquake engineering and code standards are still based. In other words, a great deal of interesting and useful work can be done with these sensors, particularly when used to increase the spatial density of existing arrays or otherwise augment more costly instruments. *Table A4* gives a summary of micromachined accelerometers known to this author.

For simplicity of reference, I often refer to state-of-the-art research instruments as "Class A", those comparable to the ICS-3028™ as "Class B", and those comparable to the C3A-02™ as "Class C" (*Table 1*). I envision an "ideal" mix of instruments as a graded, interleaved array. For example, a large, dense, urban array might have on the order of 100 "Class A", 1000 "Class B", and 10,000 "Class C" instruments, each group broadly distributed. This design yields a cost-effective mix of spatial resolution and amplitude resolution.

TABLE 1. A classification of strong-motion instruments.

Class	Bits Resolution*	Retail Cost†
A	≥20 bits	~$7000, quantity 100
B	16-19 bits	≤$2000, quantity 1000
C	12-15 bits	≤$1000, quantity 10,000

* For 4-σ peak-to-peak noise *versus* full-scale range.
† Complete from sensors through telemetry.

4. Example 1: Highway Bridge

Straser [1998] combined ICSensors™ and Kistler™ silicon accelerometers, embedded systems, and ISM spread-spectrum telemetry to produce a prototype structural-monitoring system designed to detect damage from both extreme events (earthquakes, hurricanes), and slow deterioration from corrosion, scour around bridge foundations, and so on. His prototype system includes multiple Sensor Units equipped with accelerometers and deployed on the structure, and a single PC-based Site Master controlling the telemetry exchanges, time synchronization, and routine monitoring, as well as providing the archival medium and much of the high-level analysis for the system. Sensor Units are capable of preliminary analysis, including FFTs, and are self-starting in earthquakes. The basic damage measure was modal analysis, though in related work Straser showed that modal variations due to temperature changes and other environmental conditions can exceed those caused by simulated earthquake damage by several times. He also applied time variation of normalized Arias Intensity to identifying unusual energy absorption in tall buildings, implying a damaged story absorbing the energy.

The largest single advantage identified by Straser in using a distributed wireless design was a factor of five decrease in installation time. Since some 80% of field time is used to install most conventional hard-wired systems, this speed implies a significant reduction in field costs. Elimination of long cables is also expected to mitigate rodent damage, cable failures during earthquakes, signal attenuation, and, in some cases, thermally induced noise in the cables (variable capacitance effects). For the off-the-shelf telemetry system he chose, Straser anticipated 300-m telemetry ranges for skeletal structures like bridges and 30-m ranges for enclosed structures like buildings. These ranges are typical of the ranges claimed for many ISM systems, though many reach 10-km and more with adequate line-of-sight.

5. Example 2: The TREMOR Array in Oakland, California

Both earthquake disaster preparations and emergency response benefit from clear knowledge of an earthquake's shaking strength and its effects. TREMOR is a project to produce real-time maps of free-field shaking strength down to a neighborhood level of detail by deploying a dense network of relatively inexpensive seismographs equipped with micromachined silicon accelerometers and rapid CDPD telemetry [*Evans*, 1998*b*]. These instruments will also capture and forward the waveforms for subsequent studies.

Earthquake shaking varies tremendously in strength over very short distances (*Figures 3a,b*), and seismologists are not yet able to predict all aspects of this complexity accurately. The number and reporting speed of strong-motion seismographs have been limited severely by the high cost of buying and operating them. Hence, it has been impossible to generate accurate, detailed maps of shaking strength at any time, least of all in the critical first minutes after a dangerous event.

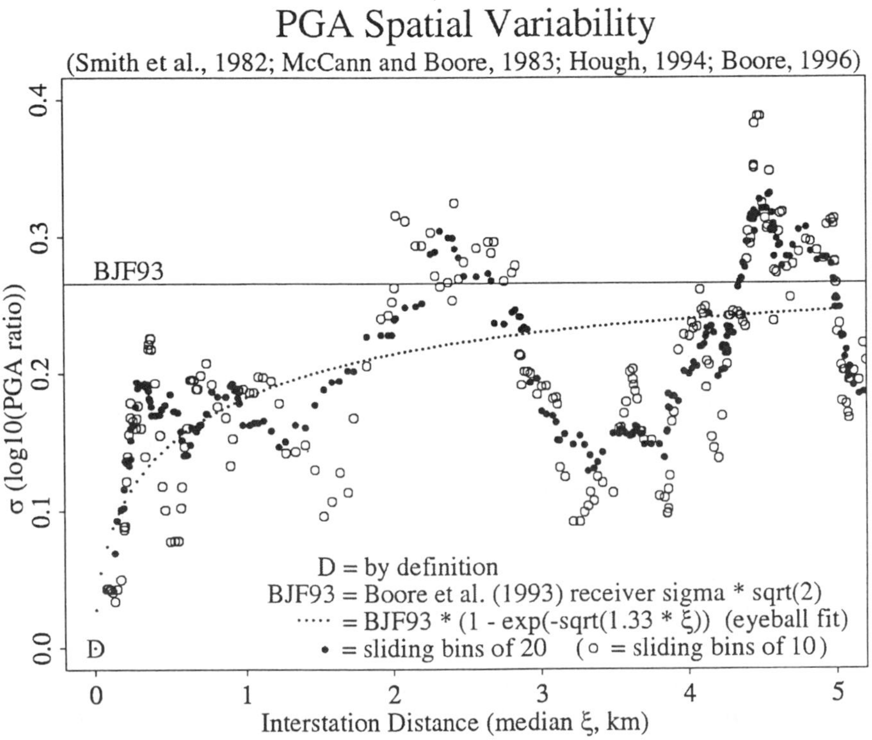

Figure 3a. Site-to-site PGA variance versus station separation for firm-soil sites.

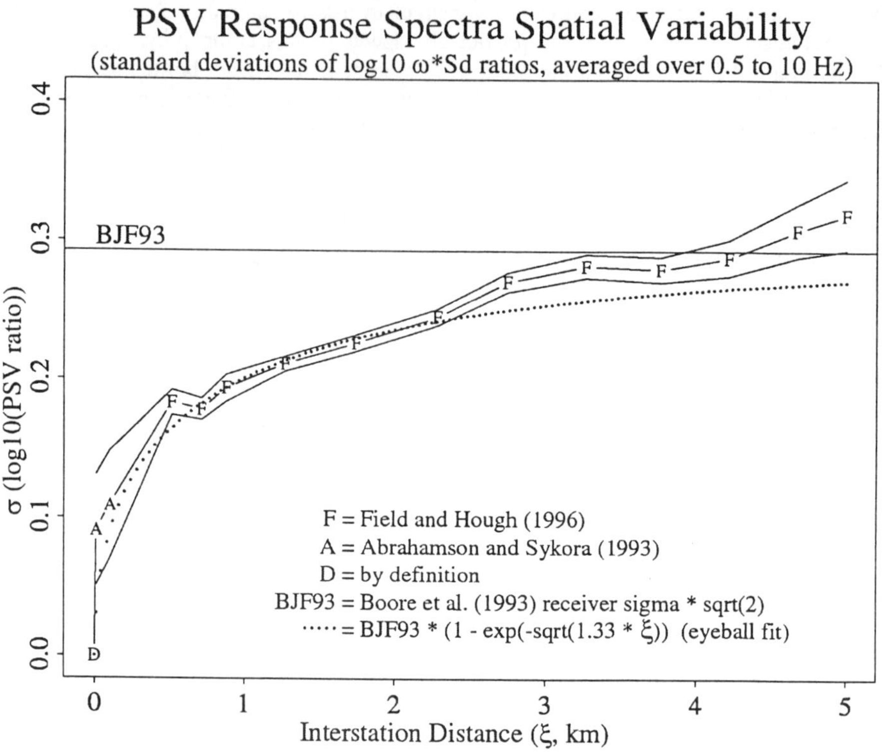

Figure 3b. Site-to-site PSV response-spectrum variance versus station separation for firm-soil sites. This spatial variability of shaking strength requires sub-km station spacing for accurate mapping of shaking strength.

TREMOR prototype instruments use a single-board JK Microsystems™ FlashLite™ PC (http://www.jkmicro.com/), an AirLink™ CDPD modem *via* GTE Wireless™, and a USGS-designed sensor block based on the ICSensors™ ICS-3028™ micromachined accelerometer. The result is a real-time strong-motion seismograph that captures the waveform and computes and forwards various measures of local shaking strength (currently Arias Intensity, PGA, and PSV response spectral values; eventually PGV). The real-time messages containing the shaking measures are transmitted about 75 s after the *P* wave arrives. They will be combined with other information to generate detailed, ground truthed maps of shaking strength. These maps will be available beginning within about 15 minutes after the earthquake and will improve in accuracy for about the next hour as additional information arrives and human data-review begins. With these maps, both detailed and regional damage estimates can be made and response efforts optimized.

The first seven prototype instruments were deployed in the third quarter of 1998 in the City of Oakland (*Figure 4*), with six more following in the Spring of 1999. We chose Oakland because of its high risk from the Hayward Fault [*Working Group*, 1990] and its high densities of population and critical infrastructure. Eventually, we hope to deploy hundreds to thousands of commercially produced TREMOR instruments with the support of local public and private organizations, including, for example, lifeline providers and banks.

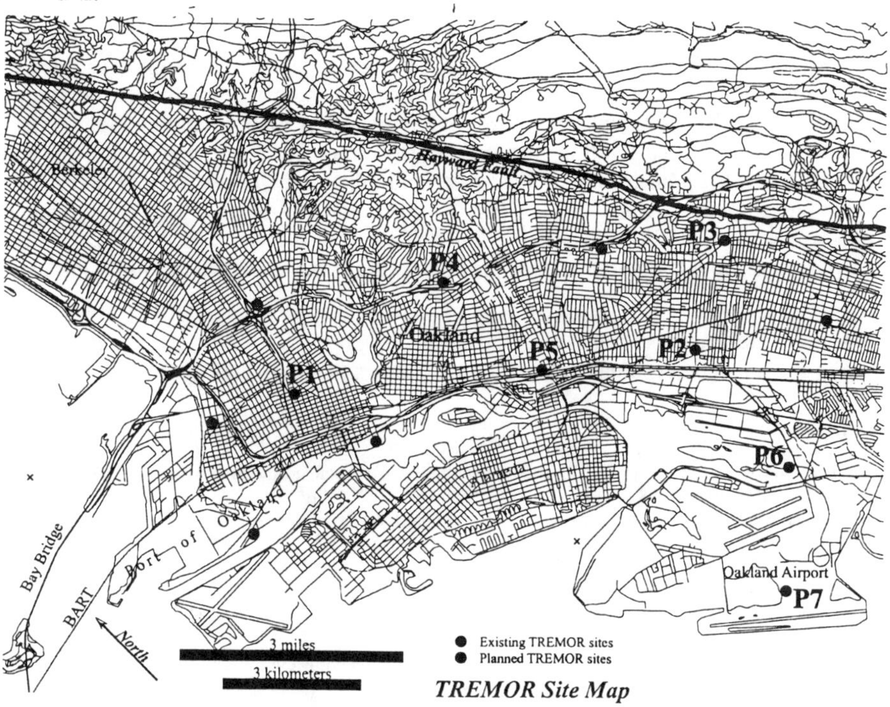

Figure 4. Map of TREMOR sites in Oakland, California. Light-gray sites are being deployed in the Spring of 1999. Black sites with names are already in service. With all 13 sites deployed, mean station spacing will be about 3 km. The eventual goal is station spacing of 1 km or less. For example, sites P6 and P7, though they are both on artificial fill, differed by a factor of two in shaking strength during an M4.1 event 25 km to the northwest.

6. Conclusions

Wireless telemetry can greatly reduce installation times for urban monitoring systems, reduce or eliminate rodent, vandal, and accidental damage to cables, reduce noise and signal attenuation caused by long cables, and reduce the number of failure modes in large earthquakes. Maintenance of wireless systems is also eased through state-of-health monitoring and even direct intervention from the laboratory.

Micromachined silicon accelerometers can lower initial capital costs and, by virtue of their robustness, also reduce maintenance costs in urban monitoring systems. They also can decrease the size of monitoring instruments, which should make them more acceptable to structure owners and should make the instruments easier to install. They may be adequate by themselves for some projects, or can be used to fill in spatial detail between more costly state-of-the-art "Class A" instruments in other situations. An ideal free-field urban network for a large metropolitan area might, for example, include ~100 "Class A" instruments, ~1000 "Class B" instruments based on a sensor like the ICS-3028, and ~10,000 "Class C" instruments. In combination, these instruments would provide sub-km spatial resolution and good to excellent amplitude resolution.

Used separately, each technology can reduce costs and improve reliability. Used together, these technologies can make practical much larger, denser arrays than was possible before.

7. References

1. Evans, J.R., and Rogers, J. A. (1995) Relative performance of several inexpensive accelerometers, *U.S. Geol. Surv. Open File Rep.* **95-555**, 38 pp.
2. Evans, J.R. (1998*a*) The design and performance of a low-cost strong-motion using the ICS-3028™ micromachined accelerometer, *U.S. Geol. Surv. Open File Rep.* **98-109**, 30 pp.
3. Evans, J.R. (1998*b*) The TREMOR Project: earthquake-shaking "radar" for the City of Oakland. A guide for users, sponsors, and the curious, *U.S. Geol. Surv. Open File Rep.* **98-586**, 6 pp.
4. Straser, E.G. (1998) A Modular Wireless Damage Monitoring System for Structures, Ph.D. thesis, Stanford University, 168 pp.
5. Working Group on California Earthquake Probabilities (1990) Probabilities of large earthquakes in the San Francisco Bay Region, California, *U.S. Geol. Surv. Circ.* **1053**, 57 pp.

8. Appendix A: Tables of manufacturers and corresponding Web sites.

Use of trade, product, or firm names is for descriptive purposes only and does not imply endorsement by the U.S. Government. The following are given only as examples to assist the reader who may wish to investigate these technologies.

TABLE A1. A sample of companies and web sites for spread-spectrum telemetry.

Company	URL
Acer	http://www.acerneweb.com/Html/oem/index.html
Airlinx	http://www.airlinx.com/
Aironet	http://www.aironet.com/
Axonn	http://www.axonn.com/index.html
Bentek Systems	http://www.bentek.ca/products.htm
Cascadia Systems	http://www.cascadiasys.com/dicam/info/wavelink.html
FreeWave	http://www.freewave.com/
GRE America	http://www.greamerica.com/metro.html
Lucent Technologies	http://www.waveaccess.com/
Metricom (UtiliNet)	http://www.metricom.com/organizations/utilinet/index.html
Micrilor	http://www.micrilor.com/index.htm
Root Inc.	http://www.root-hq.com/spread.html
Sirius	http://www.sirius.be/products.htm
Spread Spectrum RF Design Ctr.	http://www.spreadspectrumrf.com/index.html
Utilicom	http://www.utilicom.com/
Voyager Technologies	http://www.vgertech.com/index.html
Wi-LAN	http://www.wilan.com/products/
Wireless Scientific	http://www.natinst.com/daq/weather/wsi.htm
Xetron (Hummingbird)	http://www.xetron.com/900xcvr.html

TABLE A2. A sample of companies and web sites for CDPD modems.

Company	URL
AirLink	http://www.airlink.com/
INET	http://www.inet.com/spider/index.htm
Mitsubishi	http://www.mitsubishi.com/mea/index.html
Motorola	http://www.mot.com/MIMS/WDG/pdf_docs/35-.pdf
Novatel	http://www.novatelwireless.com/index2.html
Sierra Wireless	http://www.sierrawireless.com/
Uniden	http://www.uniden.com/

TABLE A3. A sample of companies and web sites for satellite voice and telemetry.

Company	URL
Globalstar	http://www.globalstar.com/
ICO Global Communications	http://www.ico.com/
Iridium	http://www.iridium.com/
Orbcomm	http://www.orbcomm.com/main.html
Teledesic	http://www.teledesic.com/

TABLE A4. A sample of companies and web sites for micromachined accelerometers.

Company	URL
Analog Devices	http://products.analog.com/products/info.asp?product=ADXL202
Crossbow	http://www.xbow.com/html/product.htm
CSEM	http://www.csem.ch/microsystems/mechanic/
EG&G ICSensors	http://www.egginc.com/egg/view_division.cgi/Divisions/Opto|112
Endevco	http://www.endevco.com/products.html
Forschungszentrum Karlsruhe	http://hbksun17.fzk.de:8080/liste.html
Kistler	http://www.kistler.com/acceleration.htm
Silicon Designs	http://www.silicondesigns.com/Product.html
Silicon Microstructures	http://www.udt.com/home.html
Sumitomo Precision Products	http://www.spp.co.jp/English/jigyou/jigyou-e.html
VTI Hamlin	http://www.vti.fi/Products/Sensors/sensors.html

THE NEED FOR DATA FROM INSTRUMENTED STRUCTURES FOR AN OPTIMAL CONTROL APPROACH TO THE SEISMIC INTERACTION BETWEEN ADJACENT BUILDINGS

ASTERIOS A. LIOLIOS
Democritus University of Thrace, Dept. Civil Engineering, Institute of StructuralMechanics and Earthquake Engineering, GR-67100 Xanthi, Greece, (tel.: +30 541 25687, fax: +30 541 26943, e-mail: liolios@demo.cc.duth.gr)

Keywords: Seismic interaction, Optimal control, Instrumentation data

Abstract

The paper deals with an optimal control numerical treatment of the dynamic inequality problem concerning the elastoplastic-fracturing unilateral contact between neighboring structures during earthquakes. The numerical procedure is based on an incremental problem formulation and on a double discretization, in space by the finite element method and in time by the Houbolt method. It is emphasized that the generally nonconvex constitutive contact laws can be simulated numerically by using data derived from response records of instrumented civil engineering structures. These interface laws are piece-wise linearized, and in each time-step a nonconvex linear complementarity problem is solved with a reduced number of unknowns. Finally, the method is applied to a civil engineering example of adjacent frames.

1. Introduction

Seismic interaction among adjacent buildings is often a m[illegible] cause of damages in seismically active regions, where, due to various socioeconomic reasons, the so-called continuous building system is allowed to be applied, see e.g. Newmark [11], Bertero [12] and Erdik, Çakmak and Durukal [29]. Thus the numerical estimation and the optimal control of the interaction effects to earthquake response of such buildings is significant for their earthquake resistant design, construction and repair.

Obviously the problem is very difficult from many aspects. Mathematically this problem of pounding of structures belongs to inequality problems of mechanics, where the governing conditions are equalities as well as inequalities, see e.g. Panagiotopoulos [1-4], Nitsiotas [5], Maier [6,7]. These so-called unilateral problems and their optimal control can be treated mathematically by the variational or hemivariational inequality concept, see Panagiotopoulos [1,2]. So, the seismic response of the interacting steel structures system investigated here is governed by a set of equations and inequalities, which is equivalent to

M. Erdik et al. (eds.), Strong Motion Instrumentation for Civil Engineering Structures, 243–252.

a dynamic hemivariational inequality in the way introduced by P.D. Panagiotopoulos [1]. As wellknown, the hemivariational inequality concept has been introduced into Mechanics and Applied Mathematics by P.D. Panagiotopoulos for first time in 1983, see [28], and constitutes now the basis of the so-called Non-Smooth Mechanics.

As regards the numerical treatment of such inequality problems in seismic mechanics, some numerical approaches have already been presented, see e.g. Wolf & Skrikerud [15], Liolios [9,10], Anagnostopoulos & Spiliopoulos [13], Papadrakakis et al. [14].

In the present paper, a special case of seismic building interaction is treated numerically. This case concerns the optimal control of the effects of the unilateral elastoplastic-softening contact between adjacent steel structures. So, the purpose here is to estimate numerically and to control actively in an optimum way the influence of the interaction effects on the seismic response of the adjacent steel structures. The latter can be obtained by suitably adjusting the gap between the buildings (if it is possible, e.g. for new constructions), and/or the contact material behaviour (hardening or softening) according to the optimal control theory in structural analysis, see e.g. Panagiotopoulos [3,4], Bisbos [19,20], Zacharenakis [21,22], Baniotopoulos [26,27]. The problem of the numerical simulation of the contact interface behaviour is the most complicated task. This can be treated effectively on the basis of data obtained by strong motion instrumentation for civil engineering structures.

The numerical procedure herein is based on an incremental problem formulation and a double discretization, in space by the finite element method and in time by the Houbolt method. The generally nonconvex constitutive contact laws are piece-wise linearized, as has been successfully done by Maier in structural elastoplasticity [6,7]. So, in each time-step a nonconvex linear complementarity problem is solved [8-10]. Finally, the method is applied to a civil engineering example of adjacent frames, and some conclusions useful for the praxis are discussed.

2. The Seismic Interaction Problem Without Control

For simplicity, a system of only two adjacent linearly elastic steel structures (A) and (B) is considered here. The extension to systems with more than two linear and/or nonlinear elastic buildings can be done in a straightforward way.

2.1 UNCOUPLED SYSTEM ANALYSIS

First the system of the two steel structures (A) and (B), considered as an uncoupled one, is discretized by the finite element method. So, assuming no interaction, the matrix equations of dynamic equilibrium are

$$\underline{M}_L \, \ddot{\underline{u}}_L + \underline{C}_L \, \dot{\underline{u}}_L + \underline{K}_L \, \underline{u}_L \text{-} \underline{M}_L \ddot{\underline{u}}_g \quad (L = A, B), \tag{1}$$

where $\underline{M}_L$, $\underline{C}_L$, $\underline{K}_L$ are the mass, damping and stiffness matrices, respectively; $\underline{u}(t)$ is the sought node displacement (relative to ground) vector corresponding to given ground earthquake excitation $\underline{u}_g(t)$ and appropriate initial conditions; and dots over symbols indicate time derivatives. Problem (1) can be solved by wellknown methods of Structural Dynamics. A response quantity of the uncoupled system is denoted by $Q^u(t)$, where superscript (u) means uncoupled system.

2.2 INTERACTION SIMULATION

Let j_A and j_B be two associated nodes on the interface (joint), where unilateral frictional contact can take place during an earthquake. These nodes are considered (see Liolios [10]) as connected by two fictive unilateral constraints, normal to interface the first and tangential the second one. The corresponding force-reactions and retirement relative displacements are denoted by r_{jN}, z_{jN} and r_{jT}, z_{jT}, respectively. They satisfy in general nonconvex and nonmonotone constitutive relations of the following type (2), expressing mathematically the unilateral elastoplastic softening contact with friction:

$$r_j(d_j) \in \partial R_j(d_j). \tag{2}$$

Here ∂ is the generalized gradient of Clarke, d the deformation and $Rj(.)$ is the superpotential function, see Panagiotopoulos [1,2]. By definition, rel. (2) is equivalent to the following hemivariational inequality:

$$R_j^{\uparrow}(d_j, e_j - d_j) \geq r_j(d_j) \,.\, (e_j - d_j), \tag{3}$$

where $R^{\uparrow}$ denotes subderivative and e_j virtual deformation.

By piecewise linearizing these relations as in [6-10] we obtain the following linear complementarity conditions:

$$r_{jN} = p_{jN}(z_{jN} - g_j + w_j) + c_j \, \dot{z}_{jN}, \tag{4}$$

$$w_j \geq 0, \qquad r_{jN} \leq 0, \qquad w_j . r_{jN} = 0, \tag{5a,b,c}$$

In (4a) c_j is the damping coefficient, p_{jN} the reaction function for the normal unilateral constraint, g_j the existing normal gap and w_j a non-negative multiplier. So, rels. (4) and (5) constitute linear complementarity conditions, which describe the unilateral contact behavior.

2.3 COUPLED SYSTEM CONDITIONS

Taking into account, now, the interaction, we write the incremental dynamic equilibrium conditions for the coupled system of the interacting buildings (A) and (B):

$$\underline{M}_A \Delta \underline{\ddot{u}}_A + \underline{C}_A \Delta \underline{\dot{u}}_A + \underline{K}_A \Delta \underline{u}_A = -\underline{M}_A \Delta \underline{\ddot{u}}_g + \underline{T}_A \Delta \underline{r}, \tag{6a}$$

$$\underline{M}_B \Delta \underline{\ddot{u}}_B + \underline{C}_B \Delta \underline{\dot{u}}_B + \underline{K}_B \Delta \underline{u}_B = -\underline{M}_B \Delta \underline{\ddot{u}}_g - \underline{T}_B \Delta r \tag{6b}$$

$$\underline{r} = \underline{r}_N + \underline{r}_T . \tag{7}$$

Here $\underline{T}_A$ and $\underline{T}_B$ are transformation matrices, and r is the coupling vector of the normal and tangential interaction forces, satisfying (4),(5). Geometric stiffness matrices, by which P-Delta effects are taken into account, could be also used in addition to normal stiffness matrices (see e.g. Maier[7], Anastasiadis [17], Chen & Lui [18]), To conditions (6) the usual initial conditions are adjoined. So the problem consists in finding the time-dependent vectors $\{\underline{u}_A, \underline{u}_B, \underline{g}, \underline{z}, \underline{r}, \underline{w}\}$ which satisfy the rels. (2)-(7) for the given earthquake excitation $\underline{u}_g(t)$.

2.4 TIME DISCRETIZATION AND PROBLEM SOLUTION

Further the problem of rels. (2)-(7) is discretized in time. Because this problem is nonlinear -due to inequalities- the mode superposition method cannot be applied. Thus, as suggested in Talaslidis & Panagiotopoulos [16], direct time-integration methods have to be used. Here the Houbolt method is preferred to other implicit schemes. In each time-step we assume that the unilateral constraints remain either active or inactive by adjusting suitably the time-step. To compute what is happening, the procedure of Liolios [10] is applied. So, a nonconvex linear complementarity problem of the following form is eventually solved:

$$\underline{v} \geq \underline{0}, \quad \underline{D}\,\underline{v} + \underline{d} \leq \underline{0}, \quad \underline{v}^T . (\underline{D}\,\underline{v} + \underline{d}) = 0 . \tag{8}$$

For most practical applications in structural mechanics, matrix $\underline{D}$ is a P-matrix and thus a unique solution of the problem (8) can be assured (Maier [6,7]). A response quantity for the coupled system is denoted by $Q^c(t)$, where superscript (c) means coupled system.

2.5 INFLUENCE COEFFICIENTS

Further, we introduce the influence coefficients

$$c = \frac{Q^c - Q^u}{Q^u} \tag{9}$$

where Q is the absolutely maximum value which takes a response quantity during the seismic excitation. Index (c) is for the coupled system and index (u) for the uncoupled one (i.e. without interaction).

By the influence coefficients c comparison is made between the uncoupled and the coupled cases. Thus, these coefficients show whether a structural element is overstressed or understressed due to interaction, and so for which elements an optimal control is desirable

3. The Optimal Control Approach

Next we consider the interface material behavior as a control for the interaction response of the system. Thus, the interface forces $\underline{r}$ of the unilateral contact constraints are the control vector, i.e. we have to adjust the interface behavior, see rel. (2), according to the optimal control principles [1-4, 23-25] in order to reduce the overstress effects. As an performance index we consider the following quadratic form of the influence coefficients:

$$I(r) = \int_{t_0}^{t_f} (c^T c) \, dt \,, \tag{10}$$

where usually $t_o = 0$ and $t_f = \infty$.

Further, we define

$$\underline{u}^T = [\underline{u}_A^{\,T}, \underline{u}_B^{\,T}], \quad \underline{M}^T = diag\,[\,\underline{M}_A^{\,T}, \underline{M}_B^{\,T}\,], \quad \underline{C}^T = diag\,[\underline{C}_A^{\,T}, \underline{C}_B^{\,T}\,], \tag{11a}$$

$$\underline{K}^T = diag\,[\,\underline{K}_A^{\,T}, \underline{K}_B^{\,T}\,], \quad \underline{B}^{\,T} = diag\,[\,+\underline{T}_A^{\,T}, \,-\underline{T}_B^{\,T}\,]\,. \tag{11b}$$

So, the system equations are in compact form:

$$\underline{M}\Delta\underline{\ddot{u}} + \underline{C}\,\Delta\underline{\dot{u}} + \underline{K}\,\Delta\underline{u} = -\underline{M}\Delta\underline{\ddot{u}}_g + \underline{B}\Delta\underline{r}. \tag{12}$$

Further, we define the state vector $\underline{x}^T = [\,\underline{u}^T, \,\underline{\dot{u}}^{\,T}\,]$, and so in the usual way [19,20,24,25] the state equations of the problem are:

$$\underline{\dot{x}} = \underline{A}\,\underline{x} + \underline{B}\,\underline{r} + \underline{d}\,, \tag{13}$$

where the suitable initial conditions have to be added.

So, after time discretization and following Panagiotopoulos [1,3,4], we arrive eventually to an optimal control problem of the following form:

Find $\underline{v}$ and $\underline{r}$ such that
$\underline{v} \geq \underline{0}, \quad \underline{D}\,\underline{v} + \underline{d} + \underline{B}\,\underline{r} \leq 0, \quad \underline{v}^T.(\underline{D}\,\underline{v} + \underline{d} + \underline{B}\,\underline{r}) = 0\,.$ (14)
and $I(r)$, given by (10), is minimum.

The mathematical and numerical treatment of the optimal control problem of rels. (12)-(13) follows Panagiotopoulos [1],[3],[4], as details will be given in a forthcoming paper.

4. Numerical Example

The steel system shown in Fig. 1a, consists of the two frames (A) and (B) in full contact (no gap), having three and eight stories, respectively. Their columns are IPBl 550 and IPBl 500 and their dimensions $L_A = 7$ m, $L_B = 10$ m, $h = 3.5$ m. The elastic modulus is $E = 21*10^7$ KN/m^2 and the damping ratio 3%. Both frames have rigid beams with total vertical loads $a_i\ m_0\ g$, where $m_0\ g = 98.1$ KN, ($g = 9.81$ m/sec^2), and a_i are coefficients shown in Fig. 1a. Unilateral frictionless contact can take place at the join-points i_A , i_B , where i = 1, 2, 3.

On the basis of experimental data obtained by instrumentation of civil engineering structures, the corresponding to rel.(2a) function p(.) is assumed to be as shown in Fig. 1b, where the branches AB and BCF are parabolas of 2nd and 3rd degree, respectively. The peak value p_p and the damping coefficient c for the i-th unilateral constraint are to be adjusted by the optimal control approach in order to be minimized the performance index given by the rel.(10). The remaining value is assumed to be 20% of the peak value p_p , i.e. $p_r = 0.20*\ p_p$. The above simulation of the unilateral contact is ceratinly a very complicated task and can be estimated on the basis of experimental results. The system is subjected to an horizontal earthquake ground displacement

$$x_g\ (t) = x_0\ e^{-2t} \sin(4\pi t), \quad x_0 = 10\text{mm}, \tag{15}$$

with corresponding diagram as in Fig. 1c.

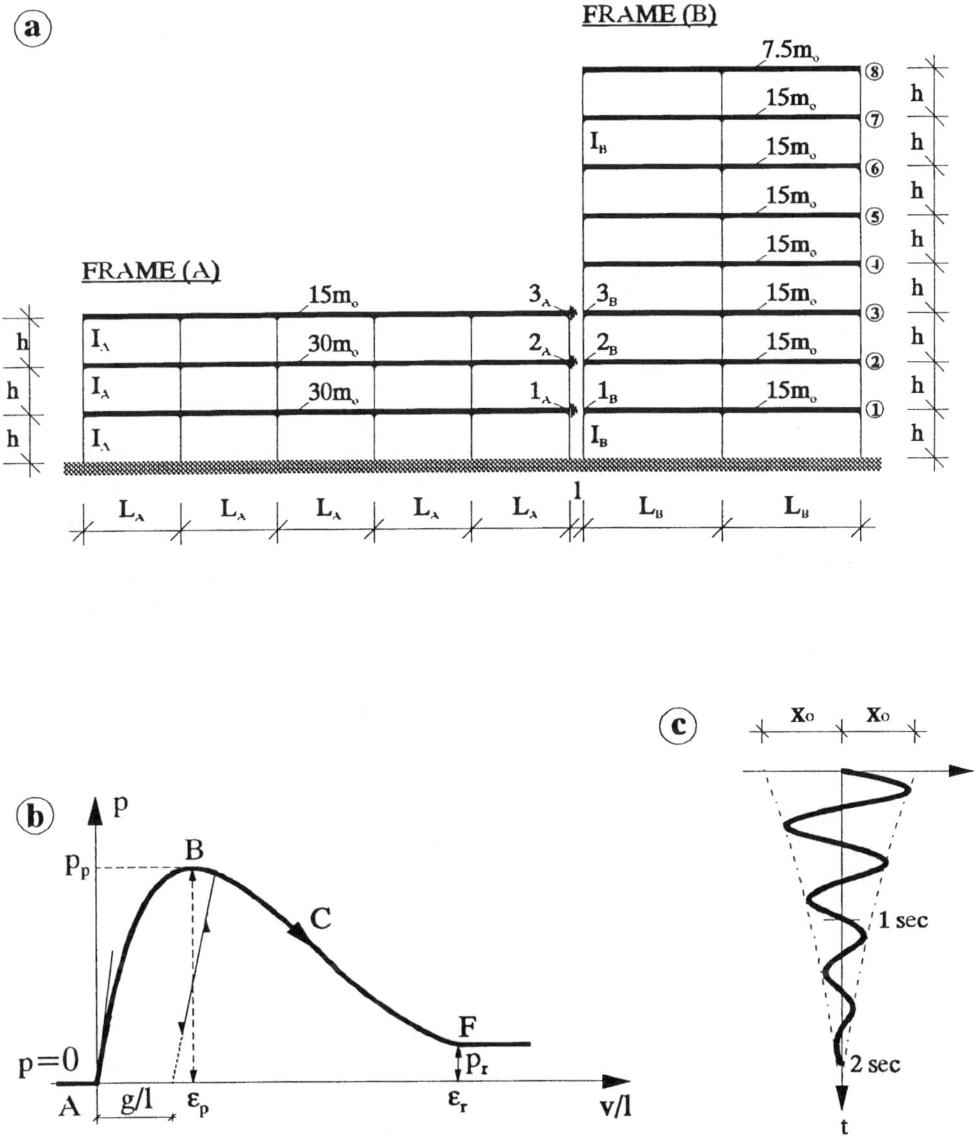

Fig. 1. Numerical example

We investigate the influence of the interaction effects first for the case of no control, where as peak value for the three unilateral constraints we take $p^P = 960$ KN, and secondly for the case where the peak values p^P will be adjusted by the optimal control approach.

Table 1. Influence coefficient c_i in (%) for the i-th floor shear force

FRAME	Floor		
	1	2	3
(A) : c_{nc}	-24.86	-17.05	-26.28
(A) : $c_{opt.}$	-16.43	-12.32	-18.29
(B) : c_{nc}	+68.51	+12.82	+26.93
(B) : $c_{opt.}$	+43.26	+7.38	+18.41

FRAME	Floor				
	4	5	6	7	8
(B) : c_{nc}	+119.45	+47.63	+46.97	+42.13	+45.82
(B) : $c_{opt.}$	+73.88	+31.64	+30.82	+25.97	+29.48

From the results obtained by applying the herein presented approach, they are shown indicatively in Table 1 only those concerning the influence coefficients for the floor-shear-forces. These coefficients are denoted as c_{nc} for the no-control case and as c_{opt} (the optimum values of the influence coefficients) for the case of optimal control. The uncoupled stress–state of the three-story frame (A) is reduced about 16%-26% due to interaction without control. On the contrary, the uncoupled stress–state of the eight-story frame (B) is increased about 12%-1186% due to interaction without control. The most remarkable increase is for the 4-th floor of (B), as was expected. So, if we design the 4-th floor of (B) without taking into account the seismic interaction effects, then the columns of this floor are overstressed about 120% more the designed capacity. Finally, as the values c_{opt} show, after application of the optimal control approach the overstress effects for steel frame (B) are reduced about 40%.

5. Concluding Remarks

As in the numerical example has been shown, seismic interaction, which is often not taken into account in the usual Civil Engineering design of adjacent buildings, can change significantly the earthquake response of steel structures in unilateral contact. A numerical estimation of the so caused seismic interaction effects can be obtained by the herein-presented approach. Further, using optimal control approaches can optimize these effects. Thus, the numerical procedure is realized by using available computer codes of the finite element method and the nonlinear mathematical programming (optimization algorithms).
Certainly the most complicated task in all the above cases is the realistic simulation of the dynamic unilateral contact behavior. To overcome this difficulty, experimental results and/or data obtained by strong motion instrumentation for civil engineering structures have to be used for the rational estimation of the parameters involved to simulate the

interface behaviour between adjacent structures. On the other hand, the herein presented numerical approach can be used effectively to estimate numerically and to control actively the influence of the interaction on the seismic response of adjacent steel structures. This can be obtained by using methods of the optimal control in order to adjust the gap between the buildings and/or the contact material behaviour (hardening or softening) of the structural interface elements.

References

1. Panagiotopoulos, P.D., 1993. *Hemivariational Inequalities. Applications in Mechanics and Engineering.* Springer-Verlag, Berlin, New York.
2. Panagiotopoulos⁼, P.D., 1985. *Inequality problems in mechanics and applications. Convex and nonconvex energy functions.* Birkhäuser Verlag, Boston-Basel-Stuttgart.
3. Panagiotopoulos⁼, P.D., 1980. Optimal control of unilateral structural analysis problems, In: Leipholz, H.H.E, (ed.), *Proc. IUTAM Symp. On Structural Control,* North-Holland, Amsterdam.
4. Panagiotopoulos⁼, P.D., 1984. Optimal control of structures with convex and nonconvex energy densities and variational and hemivariational inequalities, *Eng. Struct.*, **6**, 12-18.
5. Nitsiotas, G., 1971. Die Berechnung statisch unbestimmter Tragwerke mit einseitigen Bindungen. *Ingenieur-Archiv*, **41**, 46-60.
6. Maier, G., 1973. Mathematical programming methods in structural analysis. In: Brebbia, C. & H. Tottenham (eds.), *Variational methods in engineering, Proc. Int. Conf. Southampton University Press, Southampton,* Vol. **2**, 8/1-8/32.
7. Maier, G. 1972. Incremental Elastoplastic Analysis in the Presence of Large Displacements and Physical Instabilizing Effects, *Int. Jnl Solids and Structures*, Vol. 7, 345-372, 1971.
8. Klarbring, A., 1986. General contact boundary conditions and the analysis of frictional systems. *Int. J. Solids Structures*, **22**, 1377-1398.
9. Liolios, A.A. 1984. *A finite-element central-difference approach to the dynamic problem of nonconvex unilateral contact between structures.* In: B. Sendov, R. Lazarov and P. Vasilevski (eds.), *Numerical Methods and Applications*, pp. 394-401. Bulgarian Acad. Sciences, Sofia.
10. Liolios, A.A., 1989. A linear complementarity approach to the nonconvex dynamic problem of unilateral contact with friction between adjacent structures. *Z. Angew. Math. Mech.*, **69**, T 420-T422.
11. Newmark, N.M. & E. Rosenblueth, 1971. *Fundamentals of earthquake engineering.* Prentice-Hall, Inc, Englewood Cliffs, N.J..
12. Bertero, V.V. 1987. *Observations on structural pounding.* Proc. Intern. Conf. "The Mexico Earthquakes", ASCE, 264-278
13. Anagnostopoulos, S.A. and Spiliopoulos K.V. 1991. *Analysis of Building ounding due to Earthquakes.* In: Krätzig W.B. et al (eds.), *Structural Dynamics*, pp. 479-484, Balkema, Rotterdam.
14. Papadrakakis, M., Apostolopoulou, K., Bitzarakis, S. and Zacharopoulos, A. 1993. *A 3D model for the analysis of building pounding during earthquakes.* In: Moan, T. et al., (eds.), *Structural Dynamics-Eurodyn '93*, pp. 85-92, Balkema, Rotterdam.
15. Wolf, J.P. and Skrikerud, P.E. 1980. *Mutual pounding of adjacent structures during earthquakes.* Nuclear Engin. Design, vol. 57, 253-275.
16. Talaslidis, D. & P.D. Panagiotopoulos⁼, 1982. A linear finite element approach to the solution of the variational inequalities arising in contact problems of structural dynamics. *Int. J. Num. Meth. Enging* , **18**, 1505-1520.
17. Anastasiadis, K. 1986. *Méthode simplifiée de calcul du second ordre de bâtiments à étage*, Construction Metallique, No 4: 43-68.
18. Chen, W.F. and Lui, E.M. 1981. *Structural Stability*, Elsevier, New York.
19. Bisbos, C., 1982, *Optimal control of structures*, Doctoral Dissertation, Aristotle University, School of Technology, Thessaloniki, Greece.
20. Bisbos, C., 1985. Aktive Steuerung erdbebenerregter Hochhäuser,*ZAMM*, **65**, T297-9.
21. Zacharenakis, E.C.,1995. Input-Output decoupling and disturbance rejection problem in structural analysis, *Computers & Structures*, **55**, 441-451.
22. Zacharenakis, E.C.,1997. On the Disturbance Attenuation and H^{∞} -Optimization in structural analysis, ZAMM, Z. Angew. Math. Mech. **77**, no. 3, 189-195.

23. Kobori, T., 1994, *Current developments in active seismic response control of building structure*, paper presented in: Savidis, S. (ed. and chairman), ERCAD 94-2nd Int. Conf. Earthq. Constr. Design, June 1994, Berlin, (see also: Kobori, T., 1993, *Structural Response Control*, Kajima Public. Inc., in Japanese).
24. Paraskevopoulos, P.N., 1996, *Digital Control Systems*, Prentice-Hall Inc., London.
25. Boglou, A. K. and Papadopoulos, D. P., 1988, *Frequency-domain order reduction methods applied to a hydro power system*, Archiv für Elektrotechnik, vol. 71, 413-419.
26. Baniotopoulos, C.C., 1989, Optimal control of submarine cables in unilateral contact with the sea-bed profile, Comp. & Struct. **33**(2), 601-608.
27. Baniotopoulos, C.C., 1995, Optimal control of above-ground pipelines under dynamic excitations, Int. Jnl Pressure Vessel & Piping **63**, 211-222.
28. Panagiotopoulos⁼, P.D., 1983, Non-convex Energy Functions. Hemivariational Inequalities and Substationarity principles. *Acta Mechanica,* **48**, 111-130.
29. Erdik, M., Çakmak, A.S., E. Durukal, 1998. "*Assessment of the Earthquake Performance of Hagia Sophia*". In: M. Erdik, E. Durukal and A. Moropoulou (eds.), *Earthquake Performance of Historic Monuments and Protection Requirements*, Proceedings of the INCOMARECH 2nd Workshop, Istanbul, Boğaziçi University,2-3Oct.1998.

WIRELESS SENSORS FOR STRUCTURAL MONITORING

C. RICHARD LIU[1], LANLING ZHOU[1] , XUEMIN CHEN[1], S.T.MAU[2]
[1]Dept. of Electrical Engineering, University of Houston, 4800 Calhoun Road, Houston, Tx 77204-4793
[2]Newwark College of Engineering, GITC Building, STE 5704
New Jersey Institute of Technology, University Hights, Newwark, NJ07102
e-mail:cliu@uh.edu

Abstract

A wireless sensor system is developed for structural monitoring purposes. This experimental system includes 5 monitoring stations and a main data collection and system control station. Each monitoring station has an accelerometer of 3-axis. Total number of channels is 15. This system uses 900 MHz ISM band spread spectrum digital radio transceivers for data transmissions. At each monitoring station, the analog signal from sensors is digitized using a 16-bit A/D. A custom microcontroller (80C251 series) is designed and implemented to control the data collection and data transmission. A software for the sensor control, data collection, processing, coding, and transmission is inbedded into the microcontroller. A communication protocol is developed for the system. The central station is a PC with a radio module attached. Each monitoring station continuously collects data from the accelerometer, processes the collected data, and transmits processed data to main station when the monitoring station receives a data request command from the main station. The central station receives the data from each substation and stores them in a data file. The received data is analyzed by the main station. If one or more events exist in a period of time (e.g. 10 minutes), the data within this period will be stored. If no event is found, the data will be dropped. Received data can be analyzed using graphic display or computerized analysis. Initial lab tests show that this system is very efficient and reliable. The dynamic range of the wireless sensor is +/-4g, sensitivity is about 5 mg, and transmission distance is about 1000 ft indoor and 5 miles outdoor.

M. Erdik et al. (eds.), Strong Motion Instrumentation for Civil Engineering Structures, 253–266.

1. Introduction

Wireless sensors have been used in various areas of civil and environmental engineering. As the advancement of the telecommunication technology, wireless sensors quickly become practical. Due to the environment of the sensor applications, wireless telemetry system used to send dada from sensors has different requirements from the transceivers used in a regular data communication system. A structure health monitoring system using wireless sensors must be reliable, more noise resistant, and robust. In general, a structural damage monitoring system is composed of sensors, communication system, and a central master station that stores and processes data. The central master station collects data from each of the substations and estimates the damage of the structure being monitored. At each substation a less powerful microprocessor can be placed to perform analog to digital conversion, pre-processing the data, and control the communication system. The microprocessor at each substation can also store a short period of data measured by the sensors connected to the microprocessor. The substations can make simple decision on whether or not a transmission is necessary when there is no event detected.

The most critical component in the structural health monitoring system is the communication link. Traditionally, this link is a wired link via various kinds of wires. Since wires are vulnerable when damage occurs, a wireless system is preferred. Due to possible RF interference and power shortage during a disaster, the communications system must satisfy two basic requirements: (1) noise resistant and (2) automatic power switch from AC to battery. A RF device operating at a licensed frequency band will experience less interference. However, a licensed band radio is much more expensive than that operating in license-free band, e.g. industrial scientific and medical (ISM) band. Digital radios are available in several ISM bands: 49MHz band, 900MHz band, 2.4GHz band and 5.8GHz band. At higher frequency, less RF interference is expected. One the other hand, the higher the frequency we use, the shorter the transmission distance will be with the same transmitter power and receiver sensitivity. At the lower ISM band (49MHz), interference will be significant because too many devices use this band (walkie-talkie, cordless phone, etc.). Compared with 2.4GHz and 5.8GHz radios, the 900MHz radios are more mature and cost effective. To reduce interference with other devices working in the same frequency band, spread spectrum system is preferred.

Micro-machined electromechanical sensor (MEMS) accelerometers are now commercially available with reasonable cost. It gradually replaces traditional force balanced accelerometers due to its size, performance, and cost. Most of the MEMS accelerometers can survive 500g shock (0.5ms) during operation.

In this paper, we emphasize on the system integration and data communication protocol. A wireless structural health monitoring system is introduced in section 1. In section 2, detailed system information is given. Section 3 presents lab verification results and section 4 concludes this paper.

2. System Description

2.1. SYSTEM BLOCK DIAGRAM

A block diagram of the system is shown in Figure 1. There are 5 sub-stations and one master station. Each substation is conmposed of a 3-axis accelerometer, a microcontroller, a digital radio, and a power supply system. The accelerometers are ADXL05EM-3 MEMS sensors manufactured by Analog Devices. The size of the accelerometer is about 35mm x 25.4mm. Dynamic range is +/-4g with a bandwidth of 400Hz. Typical sensitivity is 500mV/g with a zero *g* drift of 60 mV at the temperature range of 0-70^{0}C. Notice that ADXL05 has a supply voltage of 5V single ended. Zero g output is 2.5V, which means that the measured voltage must be subtracted a 2.5V bias and then converted to acceleration. The spec of ADXL05 shows that the noise level is 5mg meaning that the minimum distinguishable acceleration is 5mg. However, due to the zero g drift and circuit noise, this data does not necessarily mean that this device is able to measure an acceleration of 5mg. To fully utilize the accuracy of the accelerometer, a minimum of 12bit

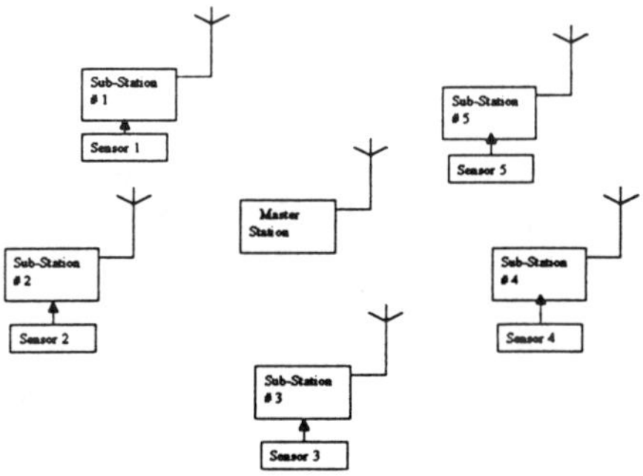

Figure 1. A concept view of the wireless sensor system

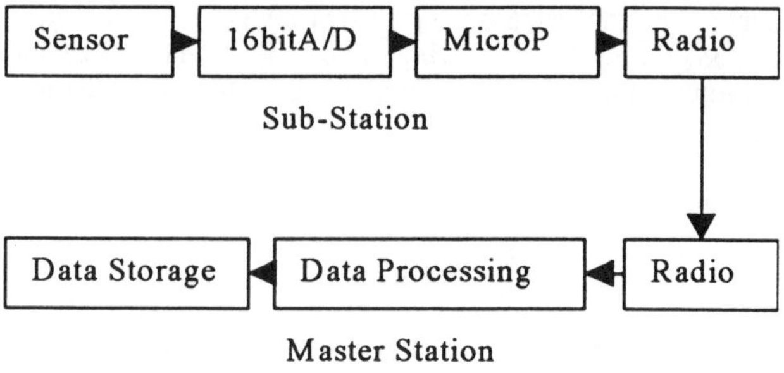

Figure 2. Block Diagram of the Master Station and Sub-Station

analog to digital (A/D) converter must be used. We use a 16bit A/D to convert the measured analog voltage to digital data. The block diagram of the main station and the sub-station is shown in Figure 2. The master station receives the digital data, process it and store it in the storage unit. The master station is a PC and the communication is through a serial port. The communication between the substations and the master station can be either sequential or query. Radios used in both master station and substations are half duplex which can be switched from transmit to receive mode back and forth. For a sequential protocol, the initial condition is that all the substations are at the listening mode. The master station transmits a start command to substation 1 requesting data. After the data is received and stored securely, the master station repeats above procedure for the sub-station 2. If data from sub-station X is not received correctly within *n* inquiries, an error message will be stored and displayed for that substation and the master station continues on the next substation. In the next iteration, substation X will be accessed only once. The value n is determined by the specific applications. In this system, *n* equal 2. If a correct data is received, the data is stored and processed. If no response is detected, an error message will be stored and displayed again. This procedure is called n-1 procedure. It is seen that using the n-1 procedure, a bad station will not cause the system to hold. On the other hand, the n-1 procedure gives a chance for the bad station to recover without deteriorating the overall sampling rate due to the bad station. This procedure is especially effective for very strong short burst type of noise, which may occur during a disaster. A flow diagram of the n-1 procedure is

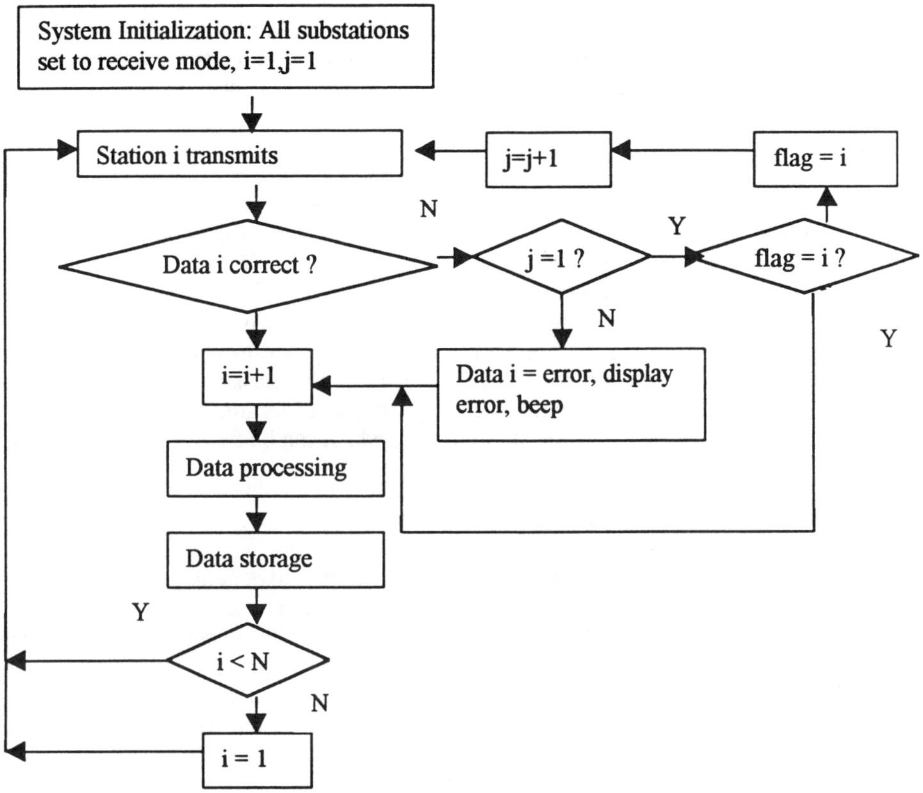

Figure 3. Flow chart of the control procedure using n-1 method. In this chart, n = 2.

shown in Figure 3. It is also possible to use a transmit-only radio at each substation and receive only at the master station. In that case, each transmitter must send data at all times. Several substations may transmit at the same time. There are several ways to avoid collision. The easiest method is to use different channels at each substation and the receiver at the master station switch channels to receive data from corresponding substation. The drawback of this method is that a synchronization period is needed for the receiver to be synchronized with the transmitter when the channel is switched. This synchronization time usually is not short (several tens of milliseconds). For slow data transmission, this method may be useful because the significant cost saving. The second method is to use a spread spectrum radio with long pseudo noise (PN) code to distinguish each substation as the case of CDMA. Since each station keeps transmit at all times, collisions are expected. The transmitted signal from station k present as noise if the receiver is receiving data from station i (i is not k). To implement the continuous transmission protocol, enough jamming margin must be designed to reduce data transmission error. The third way to avoid collision is to use a timed transmission sequence where each transmitter transmits at a precise time interval. To guarantee an accurate universal time base for each substation, a time standard signal must be used, which can be from a GPS receiver or an atomic time receiver.

2.2 RADIO LINK AND DATA FORMAT

To ensure a reliable data communication during a disaster, the radio system used in the structural monitoring must be noise resistant. Spread spectrum radio is a natural choice because of its noise resistant properties. There are two major types of spread spectrum radio available: direct sequence spread spectrum (DSS) and frequency hopping spread spectrum (FHS) radios [1]. The basic principle of the spread spectrum radio is to use a correlator and generate a processing gain. At the transmitter end, the narrow spectrum of the data is spread into wide band signal. At the receiver end, a correlation is performed which only despread the signal spectrum while the noise and interference spectrum is effectively spread. Thus, the energy of the desired data spectrum gets enhanced and the noise and interference energy is suppressed. The desired signal can be detected even with the presents of noise and interference. The ratio of the signal to noise ratio at the output of the receiver and the input of the receiver is defined as processing gain (PG). This process is illustrated in Figure 4. In this study, we use a DSS radio operating at 900 MHz ISM band (902 – 928 MHz) manufactured by GRE America, Inc. [2]. The maximum data rate is 57.6 kbps with half duplex. Modulation scheme of this radio is binary phase shift keying (BPSK). Most DSS radio uses BPSK because of its simplicity. DSS radio spread its spectrum by chopping the data bits into much faster pseudo noise (PN) sequences. It can be shown that the ratio of the PN code rate and the bit rate is the processing gain. Greater processing gain will increase the ability of the system to resist noise and interference. One the other hand, longer PN code will reduce the data rate because limited bandwidth resources. In the system presented in this study, an 8 bit PN sequence is used.

resulting a processing gain approximately 9dB. Since the radio system used in this project is half-duplex, to obtain data from each station the master station needs to switch from transmit mode receive mode, and each substation will switch from receive mode to transmit mode once. Each transition will need about 10 – 20 ms, which is very significant.

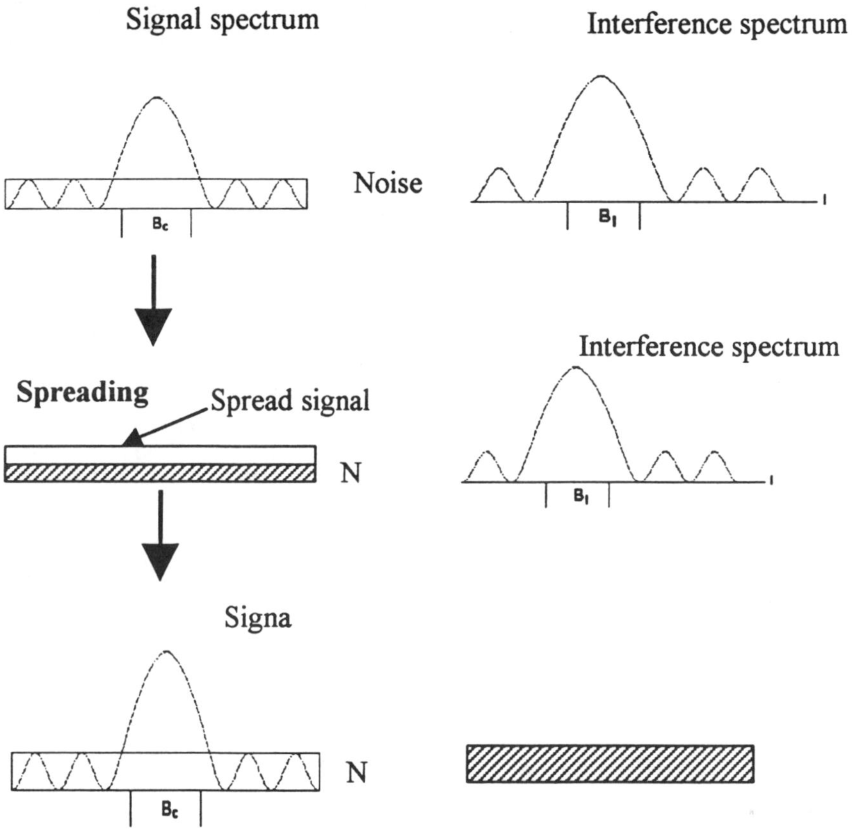

Figure 4. Illustration of noise resistant properties of a spread spectrum radio. B_c is the bandwidth of the signal and B_I is the bandwidth of the in-band interference. Horizontal axis for all the plots is the frequency.

To reduce the T/R switch times, the n-1 protocol is slightly modified and the modified protocol is like a quasi-token ring. At the beginning of the data transmission, the master station requests station 1 to transmit. Station one sends a data end signal to station 2 after the data is sent and station 2 starts transmission. Using this modification, the T/R switching time is reduced to half because the master station only at the listening mode after sending the start command. In case of a malfunctioning station appears, the master station will branch into the n-1 procedure as shown in Figure 3.

Data format sending to the radio is largely dependent on the user system requirement. It usually is a compromise among data rate, A/D accuracy, error correction ability, and protocol reliability. Higher data rate requires greater bandwidth. When the bandwidth is fixed higher A/D accuracy results in more bits in each sample and thus increases the total data length, which implies a slower overall sampling rate. In this project, we used a 16-bit A/D converter, which defines the data length of 2 bytes for each axis. For convenience, ASCII code can be used. However, the data bits becomes a lot longer using ASCII code because each ASCII code takes one byte and a 16-bit data will require 4 bytes of data to represent. Figure 5 shows the data frame if hexadecimal ASCII code is used. The total data frame will be 16 byte long. This kind of data frame is easy to implement because of the easy programming.

Start (1)	Station ID (1)	X data (4)	Y data (4)	Z data (4)	End (1)	Error Check (1)

Figure 5. Data frame using hexadecimal ASCII code. Numbers indicates bytes used.

More efficient ways of the frame design is to use binary and re-group into ASCII code. Figure 6 shows a frame using a binary regrouped ASCII code. Total length of the frame reduces to 8 bytes.

Start + Station ID (1)	X data (2)	Y data (2)	Z data (2)	End + Error Check (1)

Figure 6. Data frame using binary ASCII code. Numbers indicates bytes used.

2.3 MICROPROCESSOR

In each of the substations, a microprocessor is used to perform A/D conversion, system control, and data communication with the radio. The performance of the microprocessor will greatly affect the system performance of the wireless sensor system. The microprocessor used in this project is an 80C251 [3] based system. There are many microprocessors available off-shelf. However, to meet exact requirements for this project, we developed a microprocessor in house.

A 16 bit A/D converter is installed on board of this processor. It is found that by developing a microprocessor in house, the system becomes much more flexible in terms of A/D bits, I/O ports, special control and timing, RAM size and ROM size, board physical dimensions, and programming. The cost is also significantly reduced to about one third of a commercial processor board. A block diagram of the microprocessor and A/D board is shown in Figure 7. Detailed specifications of the board are as follows:

CPU	80C251 (8-bit processor)
Clock frequency	16 MHz
RAM size	256 kB
ROM size	256 kB
Serial I/O	2 (RS232)
Digital I/O	8 Channels
A/D channels	8 Channels
A/D resolution	16 bits
A/D conversion rate	166 kbps

The processor board is supplied with a +5V power supply and can be battery powered. The input of the A/D converter is connected to three output terminals (x, y, z) of the ADXL05 [4].

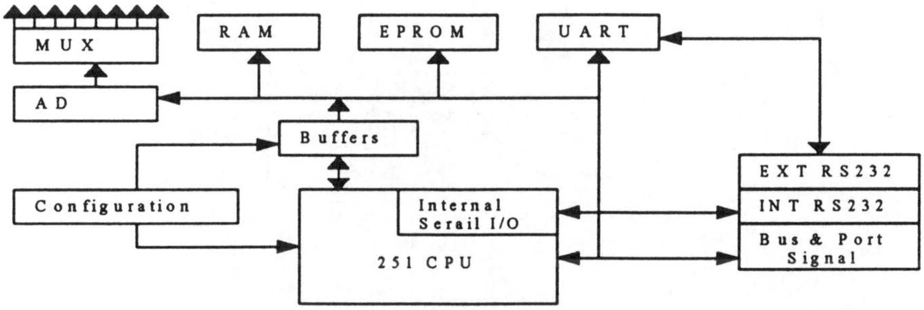

Figure 7. The block diagram of the microprocessor and A/D board.

3. System Performance and Initial Lab Tests

The wireless structural health monitoring system is implemented and initial tests were conducted. A software package implementing data communication protocols, perform data processing, storage and display was developed using C++ program language in the master station. At each substation, the microprocessor is programmed using C language. Software interface is implemented by using Visual C++ on a Pentium PC. A photograph of the master station and a substation is shown in Figure 8a, and 8b, respectively.

System verifications in terms of noise tolerance, transmission distance in door and out door, protocol effectiveness, bit error rate, and event data backup time were conducted. The results show that this system is very effective and reliable. However, due to lack of testing equipment related to shaking table and other structural related equipment, only very simple tests on the sensor performance were conducted. The purpose of the test is to demonstrate the functionality qualitatively. The system is tested by sitting the sensors on the lab bench near several computers to test acceleration noises on top of a lab bench. One minutes of data were taken and shown in Figures 9a, 9b, and 9c for the accelerations in x , y, and z axis respectively. It is clearly seen from Figure 9 that the system works relatively noise free when the environmental noise is excluded (between strokes of noise). A ten minutes test of acceleration detection is performed by shaking the sensors. The test results is shown in Figures 10a, 10b, and 10c for the data in x, y, and z, directions, respectively.

Figure 8a. A photograph of the master station

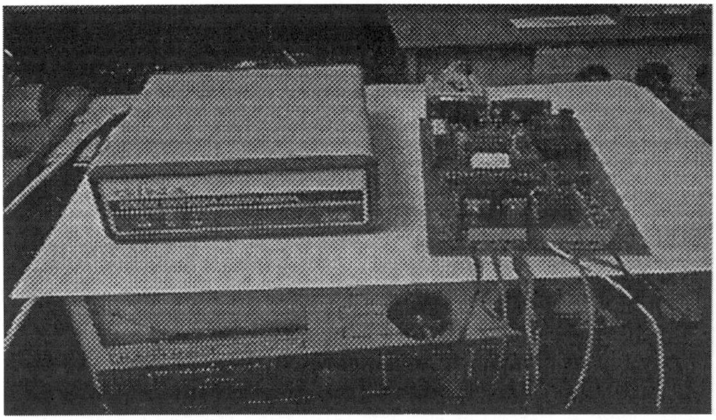

Figure 8b. A photograph of a substation

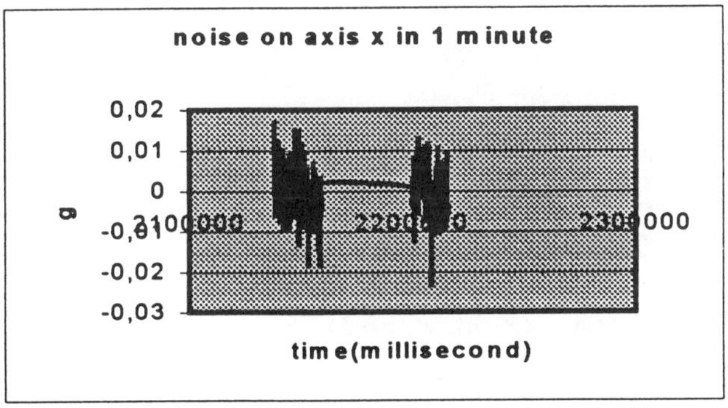

Figure 9a. Measured noise on top of a lab bench, x-axis

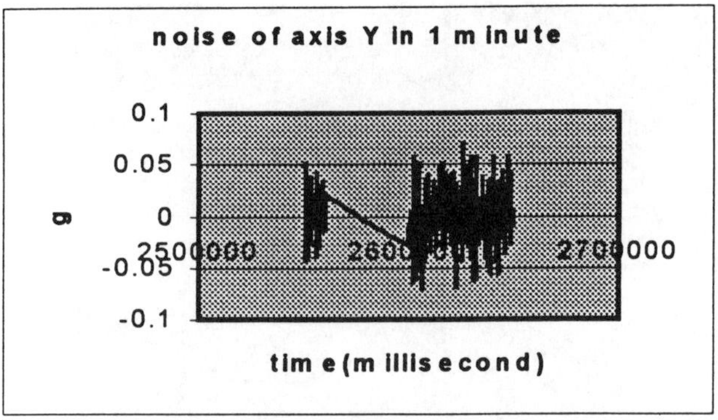

Figure 9a. Measured noise on top of a lab bench, y-axis

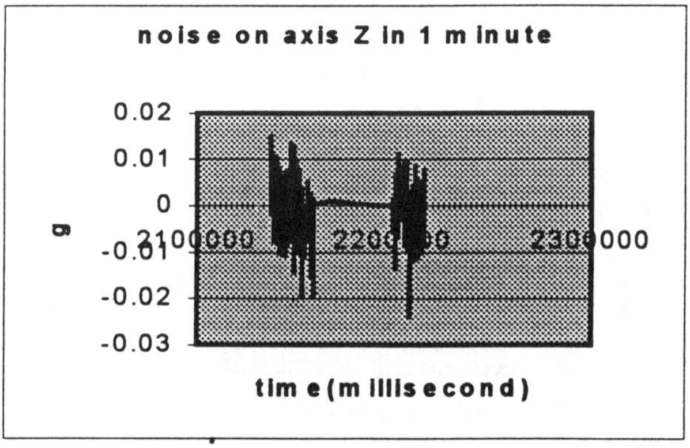

Figure 9a. Measured noise on top of a lab bench, z-axis

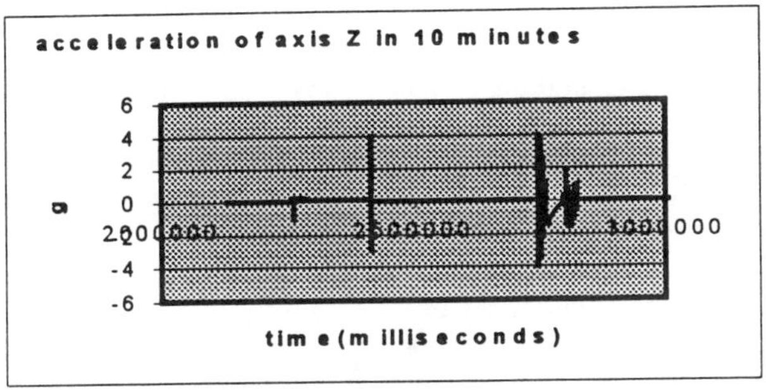

Figure 10a. Recorded 10 minutes data by shaking the sensor, x-axis

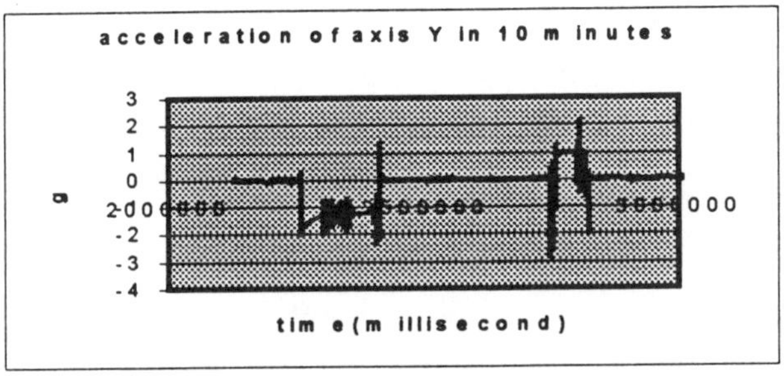

Figure 10a. Recorded 10 minutes data by shaking the sensor, y-axis

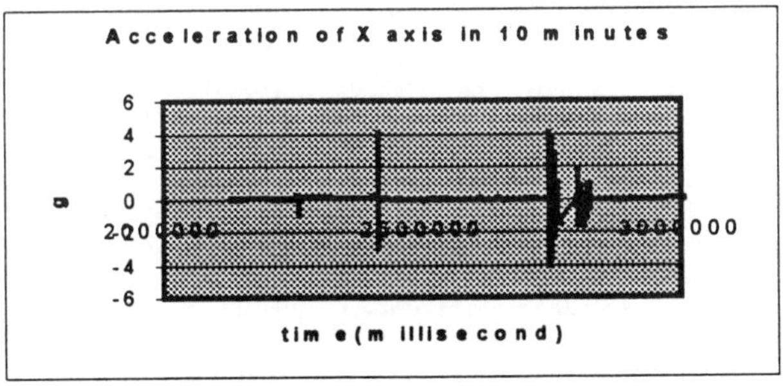

Figure 10a. Recorded 10 minutes data by shaking the sensor, z-axis

4. Conclusions

In this study, a wireless health monitoring system is described. This system uses a DSS radio system to transmit structural monitoring information from sensors to a master station. System principle and design, wireless protocols, microprocessors used in the system, and data acquisition components are discussed. The implemented system has 5 substations and 1 master station. Each substation measures 3-axis acceleration using an ADXL05 MEMS accelerometer. Tested results showed that this wireless structural monitoring system has the following specs:

Maximum transmission Distance: 1000 ft indoor
5 miles out door

Dynamic range: +/-4g

Noise level: 5 mg

Power supply: AC/UPS (up to 24 hours)

Resolution: 16-bit

Overall sampling rate: 25 Hz

Number of substations: 5

Communication protocol: modified n-1

References

1. Rappaport, T. S. (1996) Wireless Communications, Principles and Practice, Prentice Hall PTR, Upper Saddle River, New Jersey.
2. GRE America, Inc. (1998) Gina 500N/500NV User Manual, Belmont, CA.
3. Intel Co., (1996) 8xC251SA, 8xC251SB, 8xC251SP, 8xC251SQ Embedded Microcontroller Users Manual, Intel Co., Mt. Prospect, Il.
4. Analog Devices, Data Sheets (1998) ADXL50/ADXL05 Evaluation Modules, Norwood, MA.

PUERTO RICO STRONG MOTION NETWORK AND INSTRUMENTED BUILDINGS

JOSÉ A. MARTÍNEZ-CRUZADO, ESTEBAN L.LLOP-RAMİREZ
University of Puerto Rico at Mayagüez
P. O. Box 9041
Mayagüez, Puerto Rico, USA, 00681-9041
e-mail:cruzado@rmce02.upr.clu.edu

1. Introduction

Puerto Rico is the smallest of the four Greater Antilles. It is located between the Atlantic Ocean and the Caribbean Sea, approximately in the latitude 18.25N, longitude 66.50W. The population of the Island is about 3.52 millions (1990 census), while its area is of 8,875 km^2 (1990 census). The capital and the largest city is San Juan, located in the northern part of the Island. Puerto Rico is a self-governing Commonwealth in union with the USA.

The Island is on the edge of the Caribbean tectonic plate close to the contact zone with the North American Plate, making it a very seismically active region. The Caribbean Plate moves toward the Atlantic Ocean at a speed of 1.58 cm/year with respect to the North American Plate.

Figure 1 depicts the faults in the vicinity of the Island. The main fault at the interface of the two tectonic plates, called the Puerto Rico Trench, is mainly a left lateral strike slip with a dip angle of 45 degrees entering the North American Plate beneath the Caribbean Plate. The Mona Canyon is a strip of normal faults between Hispaniola and Puerto Rico, while the Anegada Trough is also a region of normal faults which separates Puerto Rico from St. Croix, Virgin Islands. South of the Island is the Muertos Trough, a subduction zone which dips under Puerto Rico at an angle of about 11 degrees. Finally, seismic zones of left lateral strike slip faults are located inland, among them, the most relevant ones are the Great Southern and Great Northern Puerto Rico Fault Zones. The seismic faults of the southwestern part of the Island are very active when compared to the seismic faults of the northeastern region. Therefore, Puerto Rico is surrounded by seismic fault zones forming the Puerto Rico Platelet, which moves to the right with a small counter clockwise rotational component.

M. Erdik et al. (eds.), Strong Motion Instrumentation for Civil Engineering Structures, 267–273.

The Island has a history of big earthquakes, the most relevant ones occurring in 1670, 1787 in the Puerto Rico Trench, 1867 in the Anegada Trough (estimated magnitude of 7.3), 1918 in the Mona Canyon, (Ms = 7.3), 1943 (Ms = 7.5) in the junction of the Mona Canyon and Puerto Rico Trench, and 1946 in the Puerto Rico Trench (Ms = 7.8).

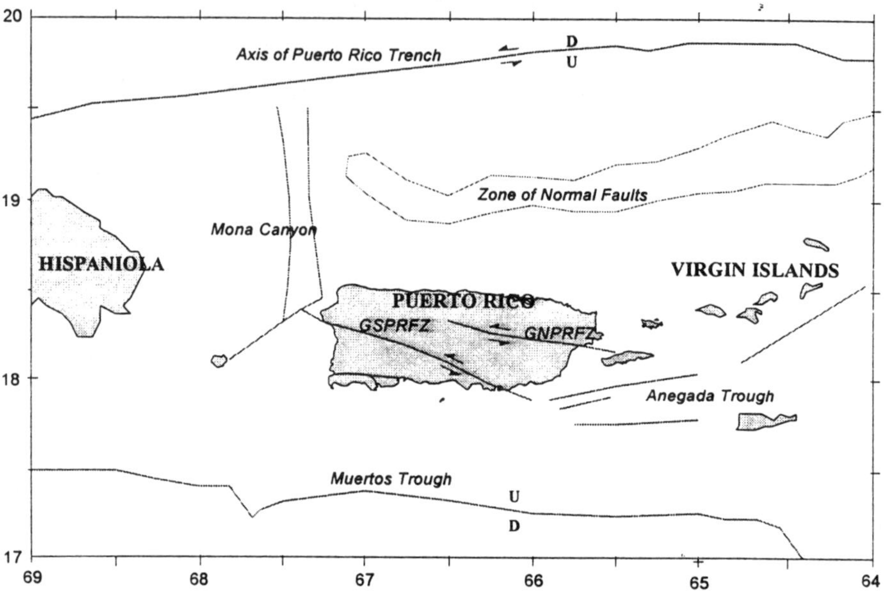

Figure 1. Main Seismic Faults Features in the Vicinity of Puerto Rico.

2. The Puerto Rico Strong Motion Network

The Puerto Rico Strong Motion Network was established during mid 1970's by the Association of Engineers and Surveyors of Puerto Rico with the collaboration of the United States Geological Survey (USGS). This initial network consisted of eight free field stations and an instrumented building. At this initial stage all the accelerographs were SMA-1.

At the end of the 1980's the Civil Engineering Department of the University of Puerto Rico at Mayagüez obtained a grant from the U. S. National Science Foundation to establish a more advanced strong motion network in the Island. The Strong Motion Network of the Association of Engineers and Surveyors of Puerto Rico was then transferred to the University of Puerto Rico at Mayagüez. It was not until January of 1994 that eight digital accelerographs SSA-2 were installed in the west side of the Island. The SMA-1 were relocated to the east side of the Island.

From 1995 to 1997 two more free field stations with SSA-2 were installed, an SMA-1 was moved to St. Croix at the Virgin Islands, and a building called La Inmaculada Plaza II was instrumented with the same digital accelerograph model. The Puerto Rico and Virgin Island Strong Motion Program (PRVISMP) was created under the Civil Engineering Department at UPRM. During 1997 the PRVISMP obtained a grant, this time from the U. S. Federal Emergency Management Agency, to support three projects. With this grant the free field stations with SMA-1 were replaced by digital accelerographs ETNA, and a local network of twelve free field stations were installed in

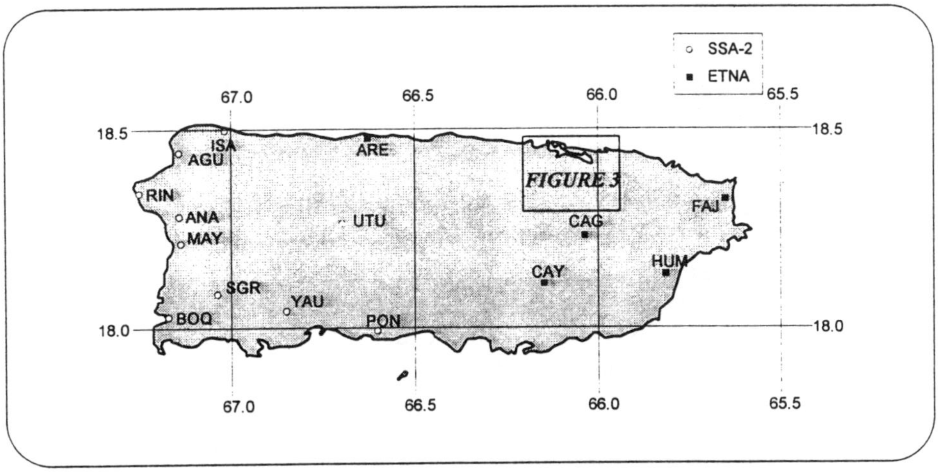

Figure 2. Current Strong Motion Stations in Puerto Rico.

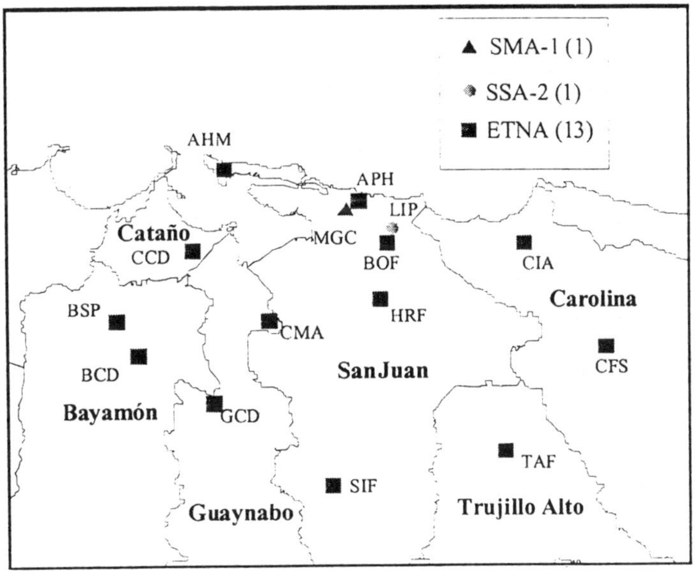

Figure 3. San Juan Metropolitan Area Strong Motion Network.

the San Juan Metropolitan Area, including six municipalities (population of 1.0 million). Figure 2 shows the current strong motion stations in the Island of Puerto Rico, while Figure 3 shows the details of the strong motion stations in the San Juan Metropolitan Area. It should be pointed out that the strong motion stations are working independently from each other.

3. Instrumented Buildings

As mentioned in the previous section the PRVISMP has two instrumented structures. By coincidence, each structure belongs to a different group of twin buildings. The first one is the Minillas Government Center (North Building), and the second one is La Inmaculada Plaza II (East Building).

3.1. MINILLAS GOVERNMENT CENTER, NORTH BUILDING

Figure 4 presents a general picture of the Minillas Government Center Buildings. These 18-story twin structures were built with reinforced concrete during the first half of the 1970's. The general dimensions of the North Building are 66.70 m tall, 65.30 m long, and 29.64 m width. The lateral as well as the vertical load resisting system consists mainly of a combination of a central core of reinforced concrete walls, where the elevator shafts and the stairs are located, and a perimetral reinforced concrete frame of 7 spans in one direction and 3 spans in the orthogonal direction. Each span is 9.07 m long.

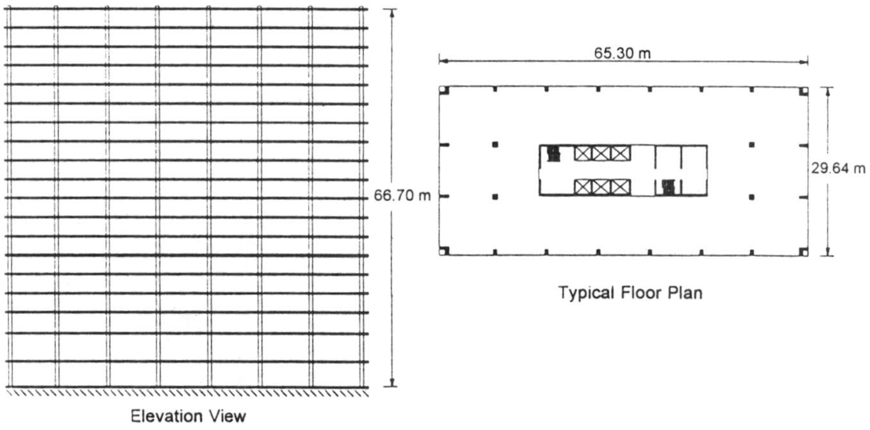

Figure 4. Elevation and Typical Floor Plan of Minillas Government Center, North Building.

The structure was instrumented around 1975 with two SMA-1. One triaxial sensor was installed on the basement, and another triaxial sensor was installed on the roof. The two sensors are interconnected, with the one on the basement being the master, and the one on the roof, the slave. After 24 years these instruments, which have a trigger level of 0.05g, have not triggered.

It is expected that this year the SMA-1 will be replaced by a 6-channel K2. The idea is to keep a triaxial sensor on the basement but to install three individual sensors on the roof, all three in the horizontal direction, one in the longitudinal and two in the transverse direction to be able to detect any torsional movement of the structure. The drawings of this building as well as the subsoil studies are available to interested investigators, solely for research purposes.

3.2. LA INMACULADA PLAZA II

La Inmaculada Plaza II is one of the two 26-story reinforced concrete wall twin buildings which were built in the early 1990's. Figure 5 shows a general layout of the building. The plan size is 55.06 m by 9.35 m while the structure is 68.88 m tall which means that the aspect ratio is 7.4. The ratio of wall area to plan area was calculated to be 4.75% in the transverse direction and 2.40% in the longitudinal direction. The structure was built over a total of 554 reinforced concrete piles. To resist the lateral loads, 80 piles were installed at an angle of 68 degrees with respect to the surface [1].

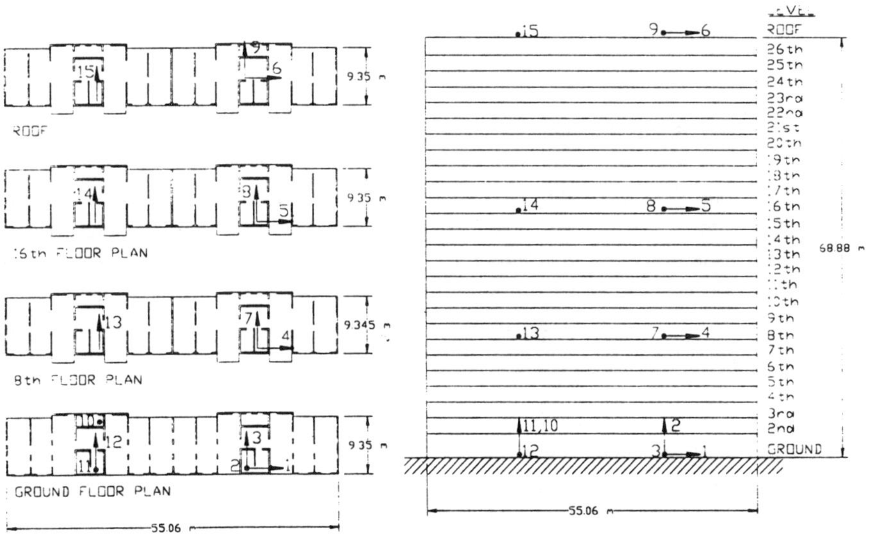

Figure 5. Plan, Elevation and Sensor Location for La Inmaculada Plaza II.

During the structure's construction several ties were omitted at the longitudinal reinforcement splice location of the wall boundary elements because of the tremendous congestion of reinforcement. Due to this fact, the government did not allowed the use of the structure. Consequently, an external active confinement was provided so that the requirements of the building codes regarding confinement were met.

The structure was instrumented during the summer of 1995. A total of fifteen sensors were installed. They were located on the ground floor, eight floor, sixteenth floor, and on the roof. With this instrumentation, features as the horizontal displacement in both orthogonal directions, the torsion and the rocking of the building can be recorded. Figure 5 also depicts the location and direction of the fifteen sensors.

On November of 1997, an ambient vibration test was performed for this building. The sensors were set in continuous recording, for about one minute, during a windy day with a wind speed of about 35 kph. The experimental period of vibration of the structure was estimated to be 1.51 sec in the transverse direction, and 1.27 sec in the longitudinal direction. The damping was estimated to be 1.32% in the transverse direction, and 1.13% in the longitudinal direction [1]. Even though the structure is quite symmetric, considerable torsion was observed. It was concluded that due to the fact that both twin structures are next to each other and separated by 12.47 m in the longitudinal direction a high speed wind travels between both structures causing higher horizontal displacement on the side of each building closer to the other. The structural drawings of this building and the ambient vibration test are available to the technical community only for research purposes.

4. Future Plans of the PRVISMP

During this year the PRVISMP is expecting to install six free field join stations in collaboration with the Puerto Rico Seismic Network, which is stationed at the Department of Geology of the University of Puerto Rico at Mayagüez. Each join station will consist of a broadband seismograph and a force balanced accelerograph (FBA-23). In this way, a wide range of frequencies will be recorded from the same site. The data will be used for the development of the wave attenuation law for the region.

The instrumentation of the Minillas Government Center, North Building, will be upgraded with digital accelerographs. Finally, with the SMA-1 Digital Retrofit Kit for the SMA-1, eight free field stations will be installed to set a Strong Motion Local Network for the city of Mayagüez, the third largest city of Puerto Rico (population of 100 thousand) located in the western region.

With respect to long term plans, the PRVISMP is looking for funding to communicate each accelerograph to the Civil Engineering Department via modem. In this way most of the maintenance as well as the data retrieving tasks could be done directly from the Department. Finally, the PRVISMP is interested in performing subsoil exploration at each strong motion station to determine features such as the S-wave velocity profile down to 30 m.

5. Acknowledgments

The PRVISMP is very grateful to the United States Geological Survey for their continuous support for about 25 years, installing and providing maintenance to the instruments as well as providing valuable guidance in every project. Also, the funding from U.S. National Science Foundation, Federal Emergency Management Agency, National Earthquake Hazards Reduction Program, the University of Puerto Rico, and Vetelba Inc. have been essential for the establishment of the PRVISMP. Kinemetrics Company, from who the equipment was bought, has always provided guidance with the instruments and their software. Thanks also to Prof. Luis E. Suárez who reviewed the manuscript of this short paper.

6. Reference

1. Cabrera-Iglesias, F. N. (1998) Evaluación del Rendimiento Sísmico del Edificio Plaza Inmaculada en San Juan, Puerto Rico, M.S. Thesis, Civil Engineering Department, University of Puerto Rico at Mayagüez, 112 pp.

STRONG MOTION INSTRUMENTATIONS OF DAMS IN MACEDONIA SOME EXPERIENCE AND RESULTS

VLADIMIR MIHAILOV, DRAGI DOJCINOVSKI
Institute of Earthquake Engineering and Engineering Seismology (IZIIS), University "St. Cyril and Methodius" - P.O.Box 101, 91000 Skopje, Republic of Macedonia
e-mail:mihailov@pluto.iziis.ukim.edu.mk

1. Introduction

A high dam is a particularly important element in seismic risk evaluation of the wider dam site area, i.e., in the definition of the seismic risk of the area in which a dam is planned to be constructed. A large number of high dams in the world and in our country are located within high seismicity zones or in zones close to areas which were affected by string earthquakes in the past.

The earthquake phenomenon involves almost always numerous problems which cannot be solved exactly due to the lack of instruments for recording earthquake intensities and response of structures. Without such a record, damage and behaviour of structures during strong earthquakes cannot be compared to the seismic design criteria nor proper decisions concerning rational repair and reconstruction could be made.

Data on the ground motion during earthquakes to which structures are exposed and behaviour of structures are fundamental for seismic hazard evaluation, definition of design parameters and criteria and for all other dynamic investigations in earthquake engineering. Without such data all investigations and analysis that follow would be based on assumptions. The irregularity in earthquake occurrence makes difficult the possibility to obtain immediately the most useful data.

One of the possible ways to solve these problems is to establish a network of a greater number of instruments for recording ground motion and response of structures during strong earthquakes. The main objective in seismic monitoring of structures (high-rise buildings, dams, power plants, bridges etc.) is to facilitate response studies that lead to improved understanding of the dynamic behavior and potential for damage to structures under seismic loading.

M. Erdik et al. (eds.), Strong Motion Instrumentation for Civil Engineering Structures, 275–291.

In this paper attention will be concentrated on code requirements for instrumentation of structures, particularly dams, and results obtained from some earthquake records.

2. General Background Information

The installation of networks for recording of strong earthquakes and the results which are obtained from them, has become an increasing need in the earthquake engineering and has considerable contribution to the overall activities for seismic risk reduction of existing urban media and for the minimizing of the damage to these structures under the effect of disastrous earthquakes. The application of the results is equally important both for the theoretical and fundamental investigations in the field of earthquake engineering and for application and practical investigations in the earthquake engineering.

The rapid development of networks for recording of strong earthquakes during the last twenty years has made a considerable contribution to the knowledge of the investigators that the theoretical investigations should have experimental proof - in the considered records - on the basis of which relevant decisions can be made.
Considering these reasons, in the beginning of the 70s, networks of strong earthquakes recording instruments were installed in several seismic regions in the world (USA, Japan, Italy, former Yugoslavia and others). This example was later followed by several other countries (Mexico, New Zealand, Iran, Turkey, Greece and others) thus, at present, there is a relatively high number of such networks, but still insufficient to cover all the seismically active regions in the world and to provide sufficient quantity of usable data. Therefore, a large number of countries in the engineering practice apply records obtained by other countries. However, having in mind that earthquakes are characterized by (1) the frequency and amplitude content, depending on the geological and the tectonic structure of the seismogene region; (2) the rate of the magnitude, i.e., the intensity of the earthquake; (3) the origin depth; and (4) the epicentral distance, it is obvious that they differ from those recorded in other areas, even in cases when earthquakes of the same intensity are considered. Therefore, it is necessary to use records from the actual seismogene region, or if used from another region, then one should be careful, and, if possible, use records from a region having similar seismo-tectonic characteristics.

A particular problem pose the records obtained by instruments placed at different types of structures. Apart from all the above state, in their case, care should be also taken about the type and the properties of the structure as well as the position of the instrument in it. Therefore, it is necessary, that the conclusions be brought very carefully and on the basis of more records.

3. Background Of The Macedonian Strong Motion Network

The strong motion instrument network installed on the territory of former Yugoslavia was one of the largest in Europe. It consisted of over 250 accelerographs type SMA-1 and about 130 seismoscopes type WM-1. The network was installed in the beginning of 1972 within the frames of the USA-Yugoslav scientific-research project entitled "Installation of Strong Motion Instrument Network on the territory of Yugoslavia". This project was realized by the Institute of Earthquake Engineering and Engineering Seismology (IZIIS) in Skopje in cooperation with the Californian Institute of Technology (CALTECH) in Pasadena until 1975 and then it was continued in cooperation with the United States Geological Survey (USGS) from Menlo Park until 1979. After this, the realization of the project was continued by IZIIS as a national project.

All the activities related to this instrument network were carried out by the scientific and professional collaborators of the Strong Motion Laboratory in IZIIS. Scientific-research in this field was also carried out simultaneously resulting in a number of bulletins and reports as well as numerous papers and presentations at scientific meetings.

Until 1991, the Macedonian strong motion instrument network was part of the instrument network of former Yugoslavia. It consisted of 110 accelerographs and 60 seismoscopes.
To provide a more detail insight, presented briefly further is the main concept and some of the main characteristics of this network.

4. Basic Concepts Of The Network

The basic concept of the Macedonian strong motion network enables obtaining of basic information required for predicting the dynamic response of various types of structures, improvement of codes for aseismic design, understanding of the ground amplification effects as well as for better investigation and perceiving of consequences caused by earthquakes.

In the design of major structures and facilities such as important buildings, dams, bridges and power plants, it is highly desirable to know the ground motion at a specific site that would result from a particular earthquake event. As the return period of major earthquakes associated with a given portion of a fault is generally quite long, it is impractical to wait for data from the particular event in question. Instead, it is necessary to extrapolate from data which have been obtained from other events which are thought to be in some sense similar to the particular event under consideration. This extrapolation process can only be reliable if there is an understanding of the individual factors which affect the character of strong ground motion such as: the nature of the source mechanism, the influence of the wave propagation path, and the effect of the local topographic and soil conditions. For this purpose, strong motion networks in Macedonia were developed with corresponding density in most active regions and with lower density in the regions with lower seismic activity, in order to study the following seismological and earthquake engineering aspects: earthquake source mechanism, wave propagation path, effect of local topography, free-field soil response at different soil conditions, site amplification factors, and structural response of different types of buildings and structures including soil-structure interaction. In areas of potentially unstable soils, strong motion records will help to determine the characteristics of the ground motions which might indicate landslides, subsidence, slumping and liquefaction.

The selection of detailed locations for establishment of these instruments makes it possible to obtain records on 1) bedrock, 2) on a surface of characteristic soils (alluvial and deluvial sediments, 3) on structures (multistory buildings, dams, etc.). the instrument distribution, of both accelerographs and seismoscopes was made following this basic concept. Table 1 shows this distribution.

Table 1. Distribution of instruments

Location	Instruments	
	Accelerographs	Seismoscopes
On bed rock	9	9
On characteristic soil	19	34
On structure	78	11
Total	106	54

Besides, the network also includes instruments installed by IZIIS for the requirements of other projects and financed by other investors. These instruments are mainly installed on characteristic structures and locations foreseen for structures of capital importance.

The strong motion program consists of five subactivities: (1) network design,(2) network operation,(3) data processing,(4) network management and (5) research as well as application. All these activities are under the responsibility of IZIIS - Skopje.

5. Seismic Monitoring Of Structures

One of the main purposes of the strong motion instrument network is providing of data on the dynamic behaviour of structures under the effect of earthquakes. To achieve this, instruments have been installed at previously defined points.

As can be seen from Table 1, included in the strong motion instrument network of Macedonia were also a total number of 78 SMA-1 accelerographs that were installed on different type of structures (high-rise building, dams, industrial facilities, etc.). Table 2 provides a review of structures on which these instruments were installed and the number of instruments.

Table 2. Instrumented structures

Type of Structures	Number of instrumented structures	No. of Instruments	
		SMA –1 Accelerographs	WM -1 Seismoscopes
High-rises building	10	32	4
Dams	11	34	7
Bridges	3	12	/
Total	24	78	11

Our past experience from 232 earthquake records taken on buildings in Banja Luka (Bosnia and Hercegovina), Skopje (Macedonia) and; Zagreb (Croatia) and 229 records on dams supports the experience of other countries such as USA, that more studious approach is necessary for determination of the number of instruments and their location in the course of instrumentation of buildings.

6. Seismic Monitoring On Dams

6.1. GENERAL NEEDS

The main purpose of seismic monitoring of high dams is the following: (1) precise definition of the seismic activity of the site, i.e., precise location of earthquake epicentres and their depth: (2) definition of main earthquake parameters: magnitude, frequency characteristics and some indications of focal mechanisms; (3) prediction of the mode of occurrence of future earthquakes; (4) provision of data on the dynamic behaviour of the dam body for the purpose of objective evaluation of its functioning immediately after the occurred earthquake, and (5) verification of design parameters by the actual behaviour of the dam body under an earthquake.

To provide these information that are of different nature, it is necessary to investigate and monitor the dam site by means of:

- a local network of seismological stations - seismographs,
- instruments for recording of strong earthquakes - accelerographs

The local seismological network distributed in a manner that it thoroughly covers the considered area provides an answer as to the spatial distribution of the earthquake epicentres and their energy characteristics, i.e., the local seismic activity. In these observations, importance is given to the period of filling of the reservoir due to the possibility of occurrence of "induced" local earthquakes as a result of the modification of the stress state of soil. This network should start being operational at least two years prior to the beginning of construction of the dam and should continue to the end of filling of the reservoir, i.e., for a minimum of three years after putting the dam into effect. It is desirable, particularly in areas of high seismic activity that this network works on a permanent basis.

The seismological stations distributed around the reservoir have to record the seismic activity in the region of the dam and the reservoir. There are several basic reasons justifying this seismic instrumentation among which is the investigation of the normal seismic regime by these observations with the purpose of contributing to the seismotectonic investigations for definition of the seismicity of the seismogene zone. Apart from this, this phase of investigations confirms or negates the existence of induced seismicity as a result of filling of the reservoir. If such a seismicity do exists, its relation with the normal seismic regime is defined. The results from these observations provide the possibility for making corrections of the main seismic degree. This type of investigations are performed by means of a network of seismological instruments distributed around the reservoir and telemetrically connected with the central recording station.

The strong motion instruments installed on the dam enable obtaining of basic data on its behaviour during an earthquake, i.e., making decisions about further exploitation or the need for repair of the dam immediately after the occurred earthquake.

This phase of seismic osculation of dams mainly refers to engineering aspects of the structure. It contains installation of instruments (in the ground and at the base of the dam) for recording of strong motions. The instruments are located at characteristic points of the base and the dam and the possibly obtained records are an invaluable parameter for verification of the mathematical model of the structure and its behaviour under the effect of a real earthquake.

An important element of these instruments are their output information. It is desirable that these be in such a form that they could provide an information on the intensity of an earthquake immediately after its occurrence. Based on this, a decision could be made regarding further exploitation of the dam. For instance, if the dam is designed for a = 0.15 g as a design parameter, and the maximum amplitude of recorded ground acceleration is less than this value, a decision can be made, with a great reliability, for further exploitation of the dam with no particular repair or strengthening However, when the recorded acceleration is greater than 0.15, it is desirable, in case when there are no visible signs of damage, to perform a special study and define the stresses and strains in the dam caused by the forces from the recorded earthquake.

6.2. TECHNICAL REGULATION

For the seismic monitoring of dams in former Yugoslavia, i.e., Macedonia, there is a Book of Regulations on Seismic Monitoring in which all the requirements related to instrumentation of dams are precizely defined. Some of them, along with the main parameters for preparation of a project on seismic monitoring of dams, are presented very briefly in the subsequent text because of the limited space.

According to technical regulations for seismic monitoring of dams of Macedonia, monitoring of dams was performed to provide data on (1) induced seismicity, and (2) dynamic behaviour of soil, the foundation and the dam body under the effect of earthquakes.

The induced seismicity is monitored by seismological stations (at least three) distributed around the dam reservoir. The dynamic behaviour of the dam discussed in this paper, was monitored by strong motion instruments distributed over the dam body.(Mihailov, Trnkoczy et al., 1990)

The seismic instrumentation of the dam body should provide exact information on the seismic input and the structural response during the earthquake. The distribution of the instruments is therefore of crucial importance. Their number varies depending on several parameters the most important of which are: the seismological and geological characteristics of the site, the foundation conditions, the type of the dam structure and the size of the reservoir.

The instrument locations are selected on the basis of dynamic analysis of the mathematical model of the dam and the experimentally defined dynamic characteristics of the structure by means of forced vibrations or measurement of ambient vibrations of the structure.

In principle, this method for definition of the instrument locations at the dam site is used for all dams in former Yugoslavia for which seismic monitoring is performed.
For each dam taken separately, projects on seismic monitoring were prepared in which particular attention is focused on the parameters stated below.

6.3. PROJECT FOR SEISMIC MONITORING

Seismic monitoring of dams and water reservoirs is planned, designed, carried out and organized for each structure taken separately. The parameters for elaboration of a seismic monitoring project can be classified into two groups:

- Parameters implicitly defined in the Book of Regulations on Seismic Monitoring involving general technical regulations;
- Parameters determined by the individual characteristics of each structure.

The parameters pertaining to the first group have already been described in (Chapter 6.3 in) "Book of Technical Regulations for Seismic Monitoring of High Dams"

The parameters of the second group are defined for each structure taken separately. By certain systematization , the parameters could be defined on the basis of:

- Seismic regime of the micro- and the macro-region;
- Dynamic and strength characteristics of the local soil;
- Mode of foundation;
- Type of dam and dynamic characteristics of the dam;
- Soil-foundation-structure interaction;
- Geometrical shape of the dam
- Dam lake capacity;
- Area of the dam lake.

All the above stated components have a partial and interactive effect upon the dynamic characteristics and the dynamic behaviour of the structures. However, in dynamic analyses of the mathematical model of the dam and experimental testing of full scale structures (full scale tests by forced vibrations and analysis by ambient vibrations), the structure and the local ambient are treated as an integrity so that the final concept of instrumentation for seismic monitoring is defined on the basis of results from dynamic analyses of the mathematical model of the dam and the experimentally defined values of dynamic characteristics of the dam, i.e.,

- natural frequencies
- damping capacity of the structure;

- vibration mode shapes of the structure.

Accordingly, it is necessary to elaborate a separate project for seismic monitoring of each structure in order to:

- establish an optimal system of seismic monitoring instruments;
- obtain practically usable and compatible data in case of an earthquake.

Data obtained by means of the seismic monitoring equipment may serve for multiple purposes, first of all, they are very useful for:

- verification of previous computations and analyses;
- analysis of the stress state and level of safety of the dam after the earthquake effect;
- analysis of the retroactive effect of the water reservoir on the seismic regime of the local territory;
- optimization of the process of design of future structures.

It is necessary to mention however that seismic monitoring of dams and reservoirs in many countries has been an obligatory part of osculation already for decades, particularly after the invention of seismic instruments with high quality electromechanical acceleration and velocity converters.

7. Data Processing

Data processing in central to the entire strong-motion program; it serves as a focal point for the functions of archiving records, processing of data, and dissemination both the data and information about the program to the user community. In the archival phase, all records are stored by station and cataloged both by event and station. In data processing, all significant records are digitized after which the raw digitized data are used to generate the following: uncorrected and corrected acceleration time histories; velocity and displacement time-histories; and various forms of frequency domain spectra (response spectra and Fourier spectra.

The ground motion characteristics during an earthquake represent the basis for solving of earthquake engineering and engineering seismology problems. These characteristics can be obtained from strong motion instrument recordings.

The three components, as recorded by the accelerographs the response of the instrument to the earthquake ground motion. Due to the limited capability of the instruments, the response gives a proper ground acceleration only for a very small frequency range. On the other hand, for calculation of many earthquake engineering and engineering seismological problems, the exact functions of acceleration, velocity and ground displacement at a wider frequency range, response spectra,

Fourier spectra and other ground motion information are required. The determination of all these ground motion characteristics necessitates detail processing of the recorded accelerograms.

The most serious problem for strong motion data processing is the determination of the exact ground acceleration function, the so called correct acceleration data at wider frequency range. In the past, these data used to be obtained by application of several methodologies. Recently, in order to obtain more correct information of the accelerograms many authors work on the development and unification of a standard strong motion processing procedure.

The IZIIS data processing system involves with in-house development, but much of it is patterned after the methodology developed by CALTECH-Pasadena, USA.

The application of the data processing method developed at IZIIS-Skopje, has been presented by the data processing results considering the Montenegro and Banja Luka earthquakes.

8. Obtained Results

Since 1973, several strong earthquakes have occurred on the territory of Yugoslavia, (the Montenegro coastal area - 1979, M= 7.0; Kopaonik - 1980, =6.3; Banja Luka-1981, M=5.4) and the neighbouring countries (Friuli, Italy-1976, M=6.5; Vranchea, Romania-1977, M=7.2, Thessaloniki, Greece-1978, M=6.3, etc.). Many earthquakes of moderate intensity ($M\geq 2.5$) have occurred, too.

In this period, 1206 accelerograms have been obtained out of which 823 by instruments installed on free field and 383 by those installed on different structures. (Mihailov, 1985)

Presented below are therefore only the results from the records obtained from instruments located on dams in Macedonia.

Significant set of data have been obtained from extensively instrumented structures during past earthquakes. Data sets obtained from well-instrumented structures, particularly dams, durin- the Negotino earthquake of Sep.28, 1985 (M=5.1). Bitola earthquake of Sep. 01, 1994

(M=5.2) and many other earthquakes of moderate intensity (M=3.5 - 5.5) will provide significant contribution to response studies.

The results from the records of these earthquakes shall be presented in the tables and the graphic presentations (records and response pectrum) that are given below. Apart from this, the locations of the instruments installed on the dam body shall be presented for the purpose of getting a thorough insight into the seismic monitoring of these dams and the results from such a monitoring.

More detailed data on these records and the obtained results can be obtained from the publications of the Institute of Earthquake Engineering and Engineering Seismology.

Table 3. Earthquake data obtained from instrumental data for Dam "Tikves" - Negotino

Earthquake Data								
Location	Date	Time (GMT)	Coordinate N	E	Io MSK-64	M Richter	Depth (km)	Note
Demir Kapija - Negotino	28-09-1985	$14^h50'$	41.510 ^{o}	22.266 ^{o}	VII	5.1	4	Obtained from microseismical data

Table 4. Maximum acceleration, velocity and displacement for Dam "Tikves" - Negotino

Data on The SMA-1 Instrument		Date on Record								
		Component N 150 S			Component N 60 S			Component Up.		
Location	No.	Acc. (cm/s/s)	Velocity (cm/s)	Displ. (cm)	Acc. (cm/s/s)	Velocity (cm/s)	Displ. (cm)	Acc. (cm/s/s)	Velocity (cm/s)	Displ. (cm)
Dam Crest	1158	-46.7	-2.299	0.300	40.4	1.453	-0.093	-33.3	-0.828	-0.033
Loc-2 Central Part of Dam	1156	-20.0	-0.637	0.049	17.0	-0.434	-0.024	14.2	0.409	0.015
Loc.-3 Lover Part of Dam	1155	-12.8	-0.253	0.007	12.4	-0.235	-0.008	6.2	-0.151	0.006
Loc.-4 Soil	2809	5.3	-0.131	-0.005	-3.6	0.109	-0.003	-4.0	0.067	0.002

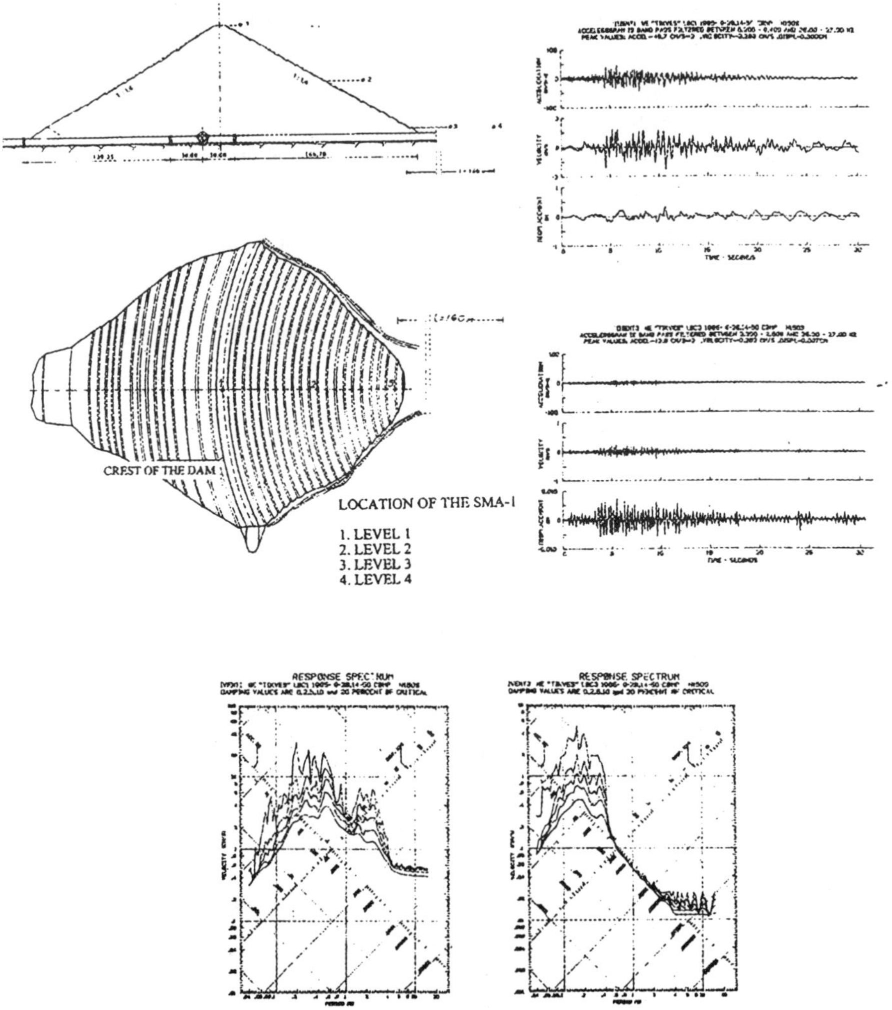

Fig.1. Seismic monitoring on the Tikves Dam. Location of Instruments, Accelerograms and Response Spectrum from Negotino Earthquake of Sept. 28, 1985.

Table 5. Earthquake data obtained from instrumental data for Dam "Strezevo" - Bitola

Earthquake Data								
Location	Date	Time (GMT)	Coordinate N	E	Io MSK-64	M Richter	Depth (km)	Note
Bitola-Resen	01-09-1994	$18^h12'$	41.13 °	21.24 °	VII	5.2	23	Obtained from microseismical data
Bitola-Resen	01-09-1994	$18^h12'$	41.10 °	21.18 °	VII^+	5.4	20	Obtained from microseismi-cal data

Table 6. Maximum acceleration, velocity and displacement for Dam "Strezevo" - Bitola

Data on The SMA-1 Instrument		Date on Record								
		Component N 40 E			Component N 50 E			Component Up.		
Location	No.	Acc. (cm/s/s)	Velocity (cm/s)	Displ. (cm)	Acc. (cm/s/s)	Velocity (cm/s)	Displ. (cm)	Acc. (cm/s/s)	Velocity (cm/s)	Displ. (cm)
Dam Crest	4777	-246.7	20.492	-1.992	166.3	-10.030	0.719	179.9	7.691	-0.449
Loc-2 Berma - I Central Part	5200	-244.3	-12.773	1.129	-149.4	-5.896	0.401	107.9	-2.407	0.091
Loc-3 Berma -II Left Side	4259	199.9	16.818	-1.478	103.2	-6.158	0.508	216.8	17.981	-1.576
Loc-4 Berma -II Right Side	5116	313.1	18.312	-1.865	-161.9	6.586	0.595	128.6	-3.740	-0.176

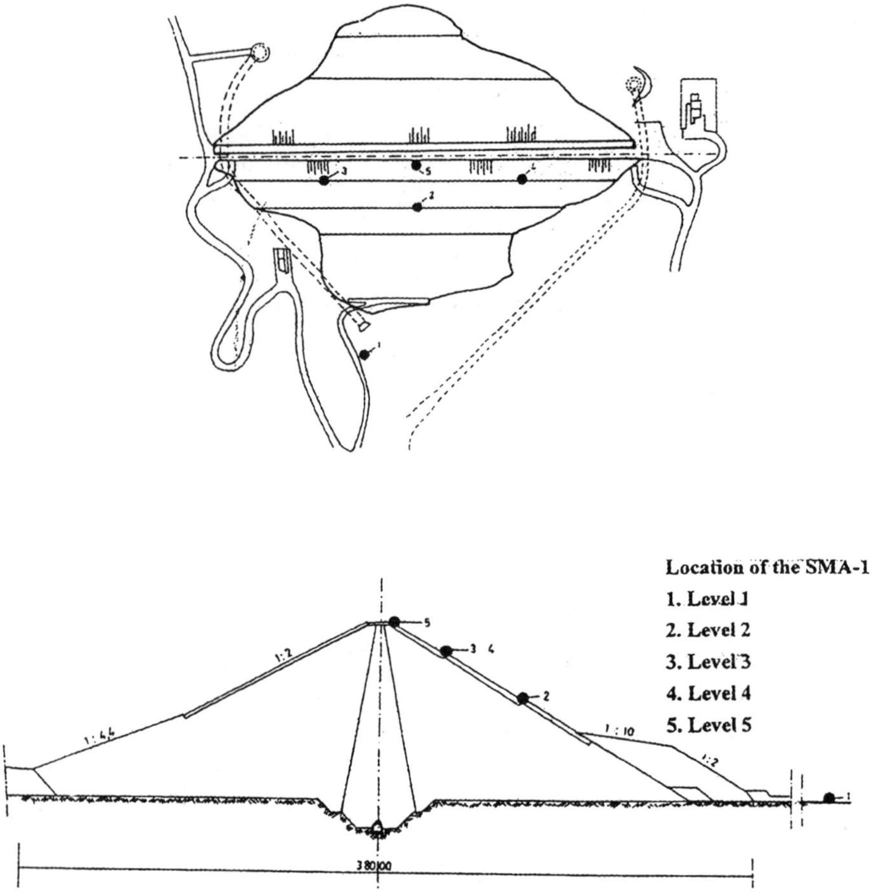

Fig. 2. Seismic monitoring on Strezevo Dam. Location of instruments

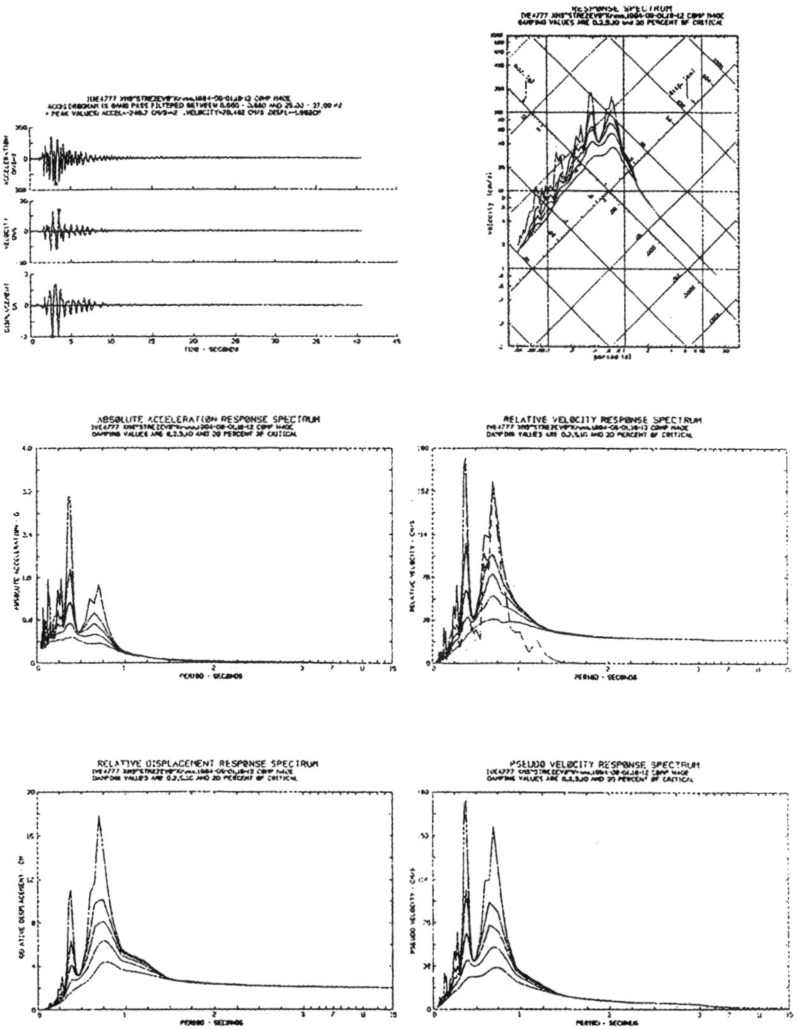

Fig. 3. Time histories and response spectra (Loc1), at the crest of the dam, N40E component of the Eathquake of September 1, 1994

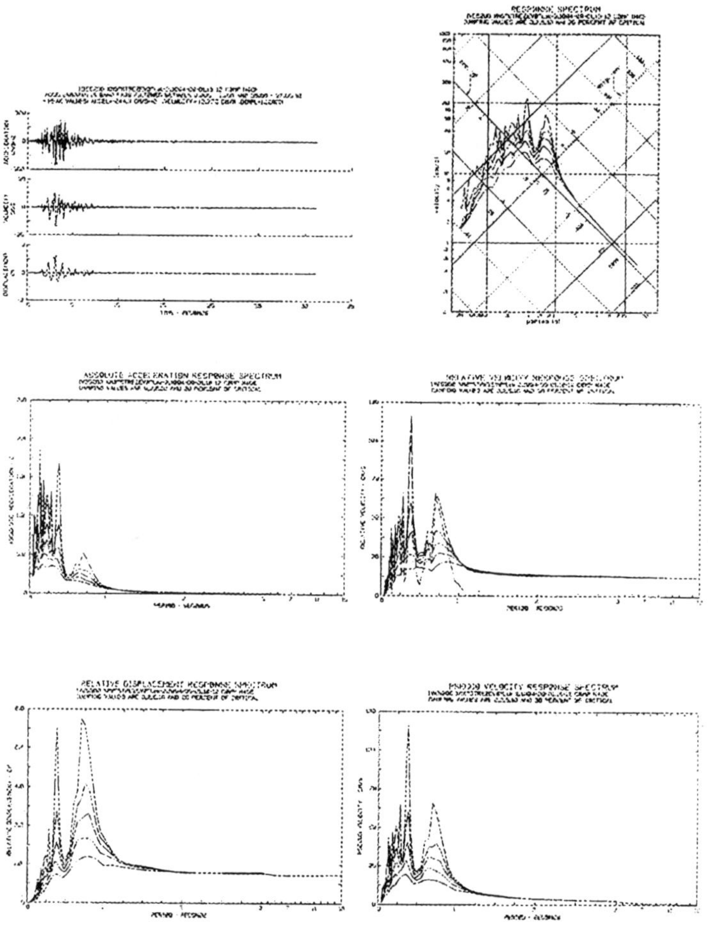

Fig. 4. Time histories and response spectra (Loc2), at the lover part of the dam, N40E component of the Eathquake of September 1, 1994

9. Conclusions

On the basis of the above presented results and the long term experience of seismic instrumentation of dams in former Yugoslavia and Macedonia, the following conclusions can be drawn:

The seismic monitoring of dams is still an actual problem in earthquake engineering. The results from processed earthquake records make considerable contribution to experimental and analytical studies of the dynamic behaviour of dams. All these contribute, directly, towards optimization of the process of design and construction of aseismic dams.

- The existing number of instrumented dams is relatively small, even on world scale. Its increase is necessary by instrumentation of dams, particularly those constructed by the application of newer technologies and methods of design. It is technically and economically justified, since the cost of the instruments, compared to the total investment value of the dams is symbolic.
- The existing instruments, discussed in this paper, are technologically relatively out of date, which very much affects the process of maintenance of the instruments and the data processing. Their frequency bend and dynamic range are relatively small, compared to new generations instruments.
- According to the above mentioned examples, the seismic monitoring of dams in future should be conducted by instruments of the latest generations, since, despite other, the approach for processing of the data obtained by them is completely automatic.
- Significant efforts are required to provide rational protection of dams against seismic effects. Seismic instrumentation is one of the most rational ways of protection.

References

1. Mihailov, V., D. Petrovski , & T. Kirijas , 1975. "Strong Motion Instrument Network in Yugoslavia " Bulletin No. 1, IZIIS Publication No. 47, Skopje.
2. Mihailov, V. 1977. "Strong Motion Instrumentation of Structures", Publication No. 51, IZIIS, Skopje.
3. Mihailov, V. 1985. "Yugoslav Strong Motion Network", *Physics of the Earth and Planetary Interiors* 3:110-122, Elsevier Science Publishers B.. Amsterdam.
4. Mihailov, V. 1990. "Yugoslav Strong Motion Network- Some Experience and Results", *Cahiers du Centre Europien de Geodinamique et de Seismologie.* Volume 1, pp. 41-51, Walferange, Luxemburg.
5. Mihailov, V., Trnkoczy A., 1990. "Instalation of Network of Instrument for Recording of induced Seismicity and Dynamics Behavior of Body Dams", *Cahiers du Centre Europien de Geodinamique et de Seismologie.* Volume 1, pp. 103-112, Walferange, Luxemburg.
6. Mihailov, V., Dojcinovski D & D. Mamucevski, 1995. "Sesismic monitoring on structures: Some experence and related problem", *10th European Conference*

AN EFFECTIVE EARTHQUAKE MONITORING PROCESS FOR EMERGENCY RESPONSE

YUTAKA NAKAMURA
System and Data Research Co. Ltd.
SDR Bldg. 3-15-3 Fujimi-dai
Kunitachi-shi, 186-0003, Tokyo, Japan
yutaka@sdr.co.jp

Abstract

There is no doubt that, investigation of durability of surface ground and dynamic characteristics of structures are important steps to estimate damage that will be caused by earthquakes. Together with this, it has been widely recognized that, systematical and continuous monitoring of earthquakes is indispensable factor for early warning systems.

Japan has 10% of global seismic energy and suffers severely from earthquakes. There have been many efforts including early warning system since 1960's. For the important infrastructures and automated systems, having an early alarm and earthquake information, even a few seconds before the destructive wave arrives, is still an important issue for many countries. Starting with conventional alarm seismometers and continuing with UrEDAS (Urgent Earthquake Detection and Alarm System), early warning system has been extensively used in Japan Railways for stopping high speed trains. This early warning system has been improved recently and this new effective real time monitoring approach is proposed in present study.

Basic concept of this system is to catch earthquake motion by P waves in and around the earthquake occurring area. For this purpose, new characteristic parameters, named as PI (P wave index) and DI(Damage Intensity) of seismic motion are proposed to be able to define destructiveness of the earthquake and effectively realize P wave alarm system. DI is defined as an inner product of acceleration and velocity vectors, at each time step. Multiplication of DI with mass received seismic motion indicate a power of motion acting to the object. With a P wave arrival, DI increases drastically and after S wave arrives it reaches to its maximum value. This value(maximum of DI) is named as DI value. PI is defined as maximum of DI calculated at the time step around P wave arrival and this value is suggested to be used for realizing P wave alarm. DI value is directly related to the seismic intensity and earthquake damage. With a continuous calculation of DI, P and S wave alarm can be issued (when PI exceeds the preset level P wave, and when ordinary monitored values or DI value exceed preset level S wave alarm can be issued). Detailed explanation of P wave detection methos was described in Nakamura(1996).

M. Erdik et al. (eds.), Strong Motion Instrumentation for Civil Engineering Structures, 293–302.

This system is very simple and it works with single station. Because of this, it is easy to apply it to all important lifeline structures like pipelines, power plants, railways and bridges. Combination of this alarm system with vulnerability information of the ground and structures estimated with other methods is promising for future early warnings, since it makes possible to understand the areas that will be vulnerable during the earthquakes.

1. Introduction

Damaged points due to earthquake motions are weak points revealed by earthquake. These weak points can be investigated in advance by understanding the seismic characteristics of surface ground and structures, to reinforce the structures before hit by an earthquake. In addition to this, if seismic characteristics of ground and structures are already known, seismic intensity distribution of an earthquake area can be estimated precisely.

To estimate the vulnerabilities of both surface ground and structures, K value method was proposed by Nakamura(1996, 1997). This methodology is briefly summarized in the present study. K value is simply calculated from the amplification and predominant frequency characteristics obtained from microtremor measurements. Then strain of the ground and structures are calculated by using this value together with an acceleration value.

To define the destructiveness of the earthquakes and to realize P wave alarm system, starting from the formulation of K value, new parameters called PI (P wave Index) and DI (Damage Intensity) are proposed in this study. As soon as P wave arrives to the observation station, PI is calculated from the P wave initial motion. And DI value (DI_v) is calculated after the arrival of S wave. Formulation for the calculation of DI function and examples for the data recorded by K-Net(National Research Institute for Earth Science and Disaster Prevention) are given in the following sections. Using PI and DI_v values together with K value early warning can be issued quickly and actual damage can be defined which will be helpful for deciding the quick response.

2. Estimation Of Seismic Characteristics And Vulnerability Indices Of Ground And Structures (K-Values)

For practical investigation of ground and structural characteristics spectral ratio of horizontal to vertical component (QTS: Quasi-Transfer Spectrum or H/V ratio) of microtremors measured on surface ground introduced by Nakamura(1989). And the formulation of vulnerability indices (K values) from Nakamura(1996) is briefly given in following part.

Calculation of K-values for ground: Vulnerability indices (K-values) derived from strains of ground and structures is formulated as floows. Simplifying the shear strain deformation of surface ground as shown in Figure 1, average shear strain (γ) of surface ground can be estimated with following formula,

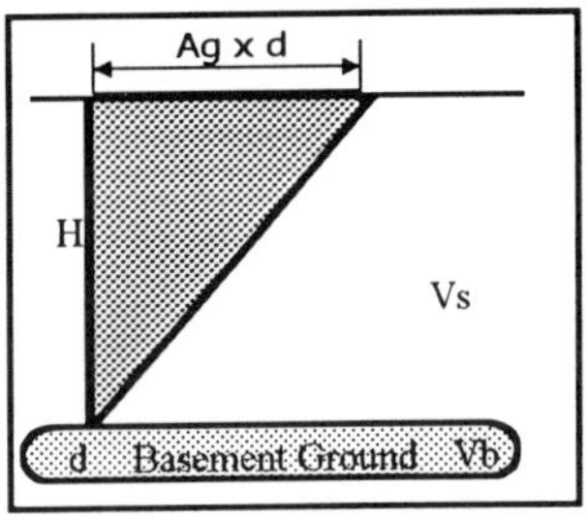

Figure 1. Deformation of surface ground

$$\gamma = A_g \cdot \frac{d}{H} \quad (1)$$

where, A_g is amplification factor, H is thickness of surface layer, and d is seismic displacement of the basement ground.

Replacing S-wave velocities of basement ground and surface ground as V_b and V_s , proper predominant frequency of surface ground (F_g) is approximately expressed as,

$$F_g = \frac{V_b}{4A_g \cdot H} \quad (2)$$

Acceleration of basement ground α_b is expressed as, $\alpha_b = (2\pi F_g)^2 \cdot d$

then γ is expressed by F_g, A_g and V_b as follows:

$$\gamma = \frac{A_g \cdot \alpha_b}{(2\pi F_g)^2} \cdot 4A_g \cdot \frac{F_g}{V_b} = \frac{A_g^2}{F_g} \cdot \frac{\alpha_b}{\pi^2 V_b} \quad (3)$$

If efficiency of applied dynamic force is assumed to be e % of static force, effective γ_e is,

$$\gamma_e = K_g(e) \cdot \alpha_b \quad (4)$$

and K_g is,

$$K_g(e) = \frac{e}{100} \cdot \frac{A_g^2}{F_g \cdot \pi^2 \cdot V_b} \quad (5)$$

The value of V_b is expected to be nearly constant in a broad area. Thus K_g which is expected to be useful to detect the weak points of the ground can be considered as an index to indicate easiness of deformation in measurement points.

As V_b is considered as 600 m/s, $1/(\pi^2 v_b) = 1.69\times10^{-6}$ (s/cm) can be obtained. If e is taken as, e = 60 %, then $K_g(e) \cong A_g^2/F_g$ and the effective strain can be estimated by multiplying $K_g(e)$ value with maximum acceleration of microtremor. Figure 2 indicates K_g-values obtained in San Francisco Bay Area after the 1989 Loma-Prieta Earthquake. The result along a line from sea coast to hillside for Marina district is shown. For highly damaged sites, K_g has a values bigger than 20. For non-damaged areas K_g is less then 5. Based on observations, maximum basement acceleration is estimated as 50 Gals around the area. From the difference in characteristics of the data in Figure 2, K_g value can be taken as K_g = 20, then $\gamma=1000\times10^{-6}$ according to the equation (4). This value separates the areas that are liquefied or not. This example shows that K values are directly related with damage.

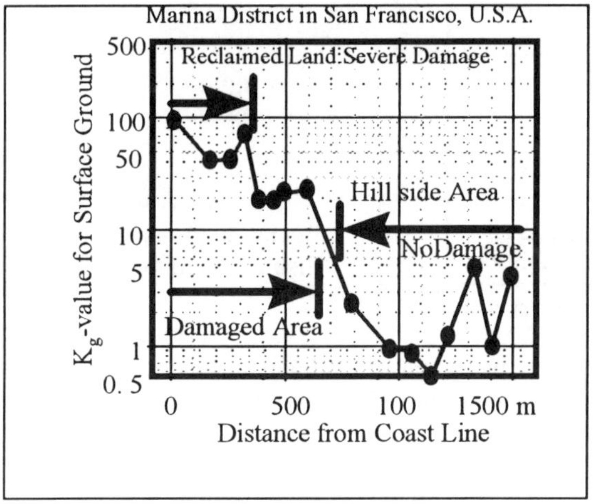

Figure 2. Kg-Values derived from QTS of basement ground in Gal (= cm/s^2).

After looking at the vulnerability of ground, same analysis was made for wooden houses. Figure 3 shows an example of comparison of calculated K values with damaged houses during the Kobe earthquake. Horizontal axis shows distance of damaged houses from JR-Kobe station and vertical left axis represents calculated K_g values for two different profiles and vertical right axis represents amount of damage. In this case also, it has been found that high K value is again in a close relation with the amount of damage, in other words K value is again high when damage is high. This example let author to search for the reason of the match between damage and K values which finalized with the relation of K values with a power of earthquake motion. This is briefly explained in following part.

3. Relation Between Kg And Power Of Strong Ground Motion, Definition Of Damage Intensity(Di) And P-Wave Index(Pi)

From the above examples we could see that there is a strong relation between K value and damage. And this relation is explained with the relation with the power of earthquake motion. It is important to understand the relation between Kg and power of strong ground motion before going to explain the new parameters of DI and PI. Figure 4 is simply drown

$$Power = Force \times Velocity$$

$$= Mass \times Acceleration \times Velocity \tag{6}$$

for explaining this relation. Power can be expressed as in equation (6),

$$= \frac{W}{g} \cdot A \cdot V = \frac{W}{g} \cdot A \cdot a \cdot A \cdot v$$

$$= \frac{W}{g} \cdot \left(\frac{a^2}{2\pi F} \right) \cdot A^2 = \frac{Wa^2}{4\pi^2 g} \cdot \frac{A^2}{F}$$

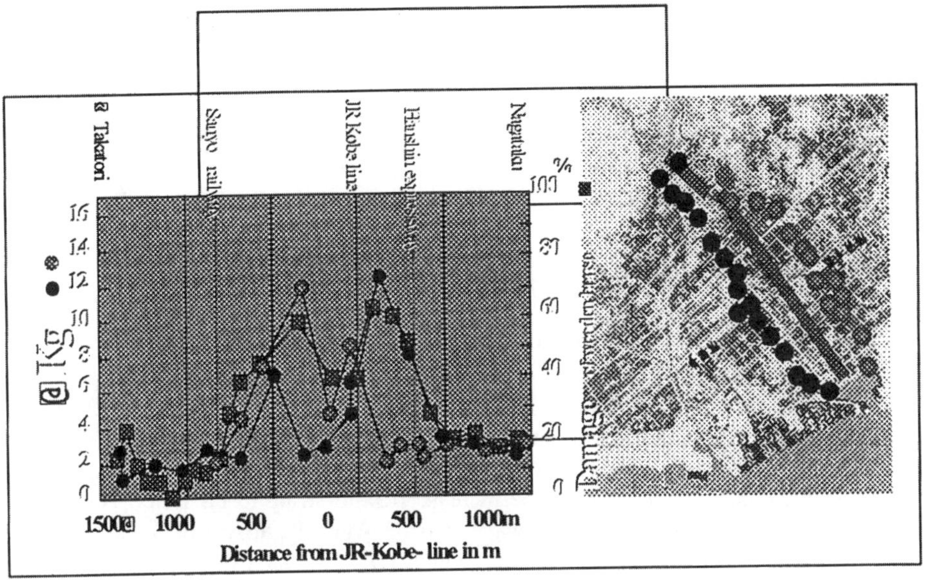

Figure 3. Relation between damage of wooden house and Kg-value

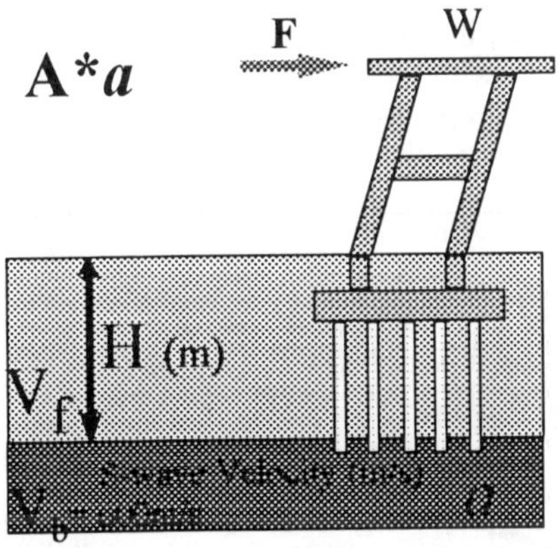

Figure 4 . Relation between vulnerability index and Power of strong ground motion

Starting from equation (5), DI function is also considered as an inner product of acceleration and velocity vectors and considered to be related with damage, same as K value. DI relation with acceleration and velocity can be simply written as,

$$DI = \log|\alpha \cdot v| \tag{7}$$

where α is an acceleration (in Gal = cm/sec^2) vector and v is a velocity vector(in mkine= 1/1000 cm/s). DI is practically calculated as a sum of products of acceleration and velocity for three components. Then the DI_v is defined as a maximum value of DI function after the S wave arrival.

A predominant circular frequency of earthquake motion ($\omega = 2\pi / T$) can be found from the ratio of α / v, and a period T can be expressed as $T = 2\pi \cdot v/\alpha$. Therefore, $v = \alpha T/2\pi$. Substituting this into equation (7), DI function can be expressed as,

$$DI = \log\left(\frac{\alpha^2 \cdot T}{2\pi}\right) = 2\log\alpha + \log T - \log(2\pi) \tag{8}$$

Figure 5 shows the change of DI as a function of time. When P wave arrives, DI increases drastically. PI value is defined as maximum DI within about t second after the P wave detection. This value is suggested to be used for P wave alarm. In this study, preset time t is taken as 1 second. Value of t can be increase maximum up to PS time. After this, DI continue to increase slowly until the S wave arrival. After the arrival of S wave, it reaches to its maximum level that is called as DI value (DI_v). This value can be related with damage and Instrumental Intensity scale of Japan Meteorological Agency (I_{JMA}).

I_{JMA} can be determined only after the earthquake terminated, on the other hand, DI_v has a very important practical advantage, since it can be calculated in real-time soon after the P wave arrives. This can be concluded as, with the continuous observation of DI, earthquake alarm can be issued efficiently and damage can be estimated precisely.

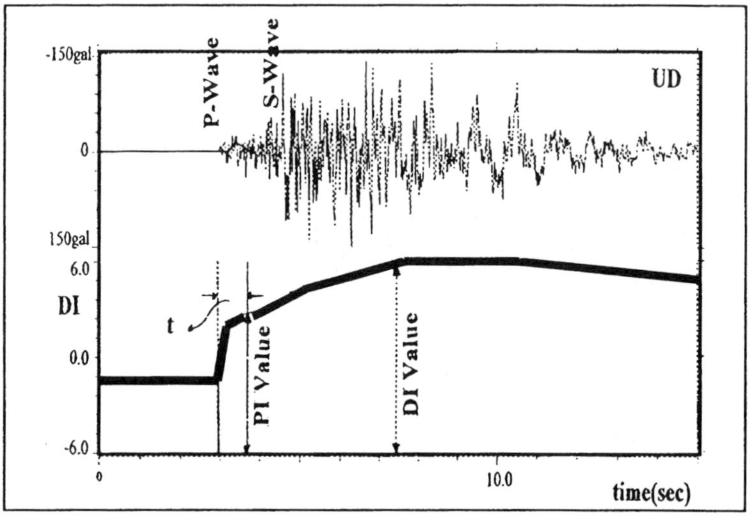

Figure 5. Change of DI as a function of time

Another important point of using PI, DI is that, seismic acceleration includes a high frequency vibration. This vibration masks the part of the record which is related with damage and make the relation between the maximum acceleration and damage unclear. Because of this problem, for example alarm seismometers of Japan Railways were set to cut the high frequency motion more than 5Hz. Even this condition is not enough to clarify the relation between maximum acceleration and damage. Knowing this problem, conventional alarm systems still issue the alarm using this relation and issue more than 90% unnecessary alarms. Instead of using maximum acceleration, the new PI and DI_v which have strong relation with damage are used for issuing alarm. A precise and quick damage estimation is possible with this approach as it was explained in previous section. Here maximum acceleration is given as a maximum value of vectoriel sum of two horizontal components filtered at 5Hz. The alarm can be issued soon after the P wave detection by using the value of PI. This is much earlier than issuing alarm with conventional approach. If P wave detection is not possible with PI, then the alarm can be issued after the S wave detection by using DI_v. This is same as issuing alarm by using maximum acceleration value. So in any case, issuing alarm is guarantied, if necessary.

Setting procedure for the P wave alarm is described in Figure 6. This figure shows the relation between PI, DI_v and maximum acceleration.

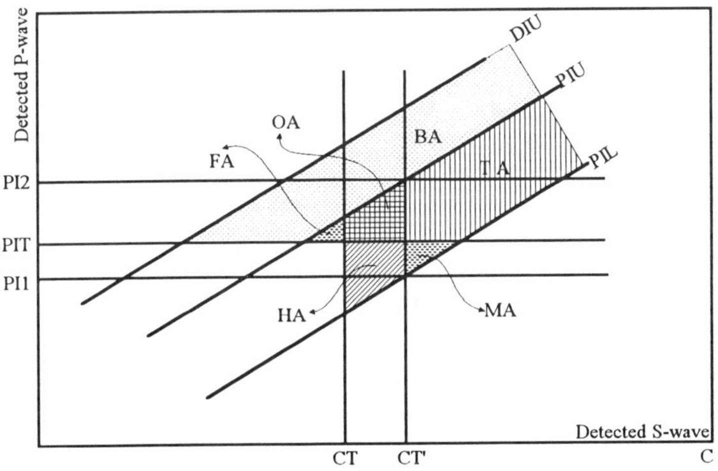

Figure 6. Relation between PI, DI and maximum Acceleration

Horizontal axis is related with the detected S wave. C represents the maximum acceleration or DI_v on this axis. Vertical axis represents the detected P wave and PI is the value calculated after the P wave detection as described in Figure 5, in previous section. According to the distribution of the data, the upper limit of DI_v is named as DIU and upper and lower limit of PI are named as PIU and PIL, respectively. PIT is the trigger level for the detected P waves. Depending on the wrong alarm tolerance, this level can be shifted to PI1 or PI2. For the high tolerance, value can be set as PI1 and for the low tolerance the setting value is PI2. *CT* is a trigger level for conventional alarm systems. *CT'* is selected as the minimum trigger level for detecting destructive earthquakes. *CT* level is smaller than *CT'* to be able to issue alarm earlier. As it is mentioned before, alarm is needed only for the earthquakes above the *CT'* level. Without considering *PIT*, setting the trigger level as *CT* (for issuing alarm earlier) the total area of *HA+OA* becomes over alarmed area, comparing with the *CT'* level. Adding the consideration of *PIT* to the same condition, the number of over alarms become *OA+FA* and depending on the *PIT* value, some missing alarm may appear in the area of *MA*. To reduce the number of the earthquakes in *MA* area, *PIT* level can be set to a smaller value. The area of *TA* represents the number of earthquakes that alarms will be issued for both cases. Setting trigger level as *CT'*, the issued alarm will be equal to *TA+MA*. Adding the consideration of PIT into the same condition, the alarm distribution will be TA, MA and OA+FA in which MA represents missing and OA+FA represents over alarm, respectively. As it is explained before, this is equal to issuing alarm with S wave detection but there will be a difference of ability of issuing alarm earlier for the latter case.

Detection of the P-wave is extremely important for making the system work properly. If the system fails for detecting P waves, then the number of the alarm may increase as the area of BA. The relation between the number of alarm issued and trigger levels can be summarized as in equation (9),

$$\left.\begin{aligned} &CT_{Alarm} = TA + OA + HA + MA \\ &PIT_{Alarm} = TA + OA + FA \\ &(PIT + CT')_{Alarm} = OA + MA + FA + TA \\ &CT_{Alarm} - PIT_{Alarm} = HA + MA - FA \\ &CT_{Alarm} - (PIT + CT')_{Alarm} = HA - FA \\ &CT'_{Alarm} = TA + MA \end{aligned}\right\} \tag{9}$$

If ($CT_{Alarm} - PIT_{Alarm}$) is bigger than 0 and also if ($HA - FA$) is bigger than 0, this means number of alarm is decreased comparing to the number of alarm which is issued by conventional alarm systems. *FA* and *HA* show the number of increased and decreased alarm, respectively. In the present paper, *CT* and *CT'* level for the alarm system are set to 40 and 80 Gals, respectively.

4. Application To Real Earthquake Records

Application of the method was made to the K-Net data recorded at three different stations from 1996 to 1997. These stations are selected as sample, since they include records with I_{JMA} up to VI. Distribution of the alarms for the case of setting maximum acceleration at 40 Gals and 80 Gals are shown in Figure 7. PIT vary between 2-2.5 in present study. As it can be followed from the figure, the number of alarms issued in the case of *CT* >40 Gals is 16. For *CT'* > 80 Gals together with PIT>2.5 the number of the alarm issued is 11, with the advantage of having alarm earlier and without having any missing alarm. It can be seen from this result that, usage of PI together with the maximum acceleration as *CT'*, the number of the alarms issued reduced about 30%.

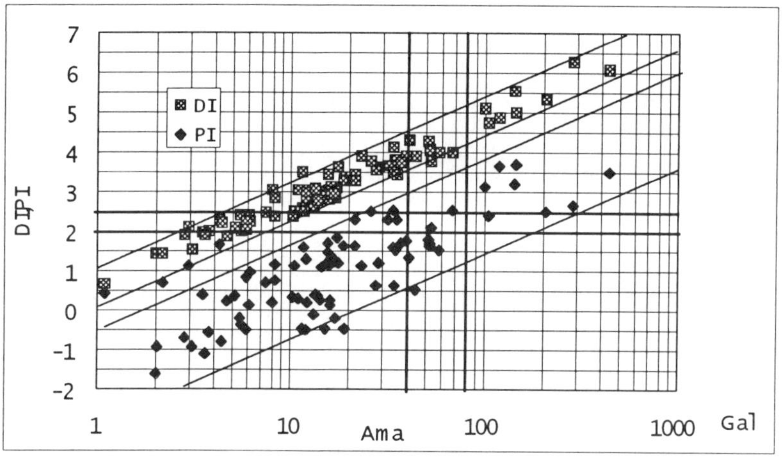

Figure 7. Application of the method to the K-Net data

In addition to this if grouping of the areas is made by using K value method, then it is also possible to decide the other alarm needed areas close to the observation point.

5. Concluding Remarks

Methodology which uses microtremor measurements of ground and structures, for estimating the vulnerability indices (K-values) is summarized. Then, starting from this approach, new characteristic parameters named as PI (P wave Index) and DI (Damage Intensity) are proposed together with its formulation, to define the destructiveness of the earthquake and be able to effectively realize the P wave alarm. These parameters can be calculated in real time and relating with damage, seismic alarm can be issued 1 second after P wave arrives. The formulation is also introduced. It has been clearly shown that by using P wave alarm (PI value) number of alarms decreased about 30%.

Combination PI, DI values together with K-value is promising for the future early warning systems, since it has been clarified with both methods that it is possible to estimate the vulnerabilities of all ground and structures concerned and it is possible to issue the alarm before the real damage occurs. This is extremely important since timely and effective response mobilization requires a well coordinated system of situation assessment and disaster knowledge. Because of this, for the effective emergency response and preparedness this intelligent system must be applied to the highly seismic regions.

References

1. Nakamura, Y., (1989) A Method for Dynamic Characteristics Estimation of Subsurface using Microtremor on the Ground Surface, Quarterly Rpt. of RTRI, Vol. 30, No.1, Railway Tech.Res. Inst.
2. Nakamura, Y., (1996) Real Time Information Systems for Seismic Hazards Mitigation UrEDAS, HERAS, PIC, Quarterly Report of RTRI, Vol. 37, No. 3.
3. Nakamura, Y., (1997) Seismic Vulnerability Indices For Ground and Structures Using Microtremor, World Congress on Railway Research in Florence, Italy, November 1997.

Acknowledgements

This research was supported by System and Data Research(SDR) Co., Ltd. I would like to especially express my sincere gratitude to Dr. Dilek E. Gurler for her discussion, cooperation and comments and to Mrs.Sawako Nakayama for her help during this study. I wish to extend my thanks also to National Res. Inst. for Earth Sci. and Disaster Prevention who maintain the K-Net service, for the data used in this study.

ESTIMATION OF DYNAMIC CHARACTERISTICS OF GROUND AND STRUCTURES WITH MICROTREMOR MEASUREMENTS- A SUPPORTIVE TOOL FOR STRONG GROUND MOTION INSTRUMENTATION

YUTAKA NAKAMURA , DILEK E.GURLER
System and Data Research Co. Ltd.
SDR Bldg. 3-15-3 Fujimi-dai
Kunitachi-shi, 186-0003, Tokyo, Japan
e-mail:gurler@sdr.co.jp

Abstract

Degree of damage during earthquakes strongly depends on dynamic characteristics of buildings as well as amplification of seismic waves. Determining these characteristics in advance and increasing durability of ground and structures beyond the presumed seismic force become a fundamental of earthquake disaster prevention.

One way of investigating dynamic characteristics is to install strong motion seismometers and monitor structure continuously. For the effectiveness of this installation, vibration mode characteristic of the structures should be clearly understand in advance. In this paper, usage of microtremor method as a supportive tool for strong ground motion instrumentation is presented. Since there is a growing interest to protect historical monuments, application of this method was made to the microtremor measurements performed at Suleymaniye Mosque located in Istanbul. In a short period of time present method provided several information including natural frequency, amplification factor and vibration mode characteristics of the mosque as well as its ground. Data showed that movement in longitudinal(NS) direction is larger than transversal(EW) direction. Vulnerability index, K values(Nakamura, 1997) is also calculated and several weak points of the mosque could be investigated. This approach help to understand the behaviors of old monuments as well as new buildings and can serve for retrofitting purposes.

1. Introduction

Since historical monuments are not suitable for sophisticated analysis, strategy for determining the characteristics of the structures should be selected carefully. One way of investigating dynamic characteristic is installing strong motion seismometers to the structures. Information coming from this method is valuable, but results are difficult to obtain in most cases, since decision of critical weak points and directions for the installation is extremely important. And also, getting results may take quite a long time,

M. Erdik et al. (eds.), Strong Motion Instrumentation for Civil Engineering Structures, 303–311.

because there may not be enough amount of earthquake record in the observation period or occurred earthquakes may not be strong enough to show the structural characteristics. An alternative method for analyzing dynamic characteristics of ground and structure using microtremor is introduced in this paper. Application is made at Suleymaniye mosque which was built by architect Sinan. Microtremor measurements were performed at several points on its ground and upper floor levels. With the analysis of this data, vibration characteristics of the structure are investigated. Formulation for the method for calculating vulnerability indexes, called K values (Nakamura; 1996, 1997) is also given in the present paper. K values basically derived from strains of ground and structures in the time of earthquakes, and can easily be calculated with natural frequency and amplification factor from ratio of transfer spectrums at higher and lower level of structures. This information allow one to determine weak points of structures as well as weak areas in the cities. Then important spots can be listed from most vulnerable one to the least and monitoring spots for the strong ground motion instruments can be decided.

2. Outline Of Measurement And Analysis

Measurement: An instrument named Portable Intelligent Collector (PIC) was used for microtremor measurements. PIC includes two sensors, connection cables, main body installed in a metal case that contains A/D converter, portable computer and amplifiers. Three components (two horizontal and one vertical) of microtremor are recorded at every measurement points. Sampling interval is 1/100 sec and the length of each record is 40.96 sec. Measurement was repeated three times at each observation point.

Analysis: After measurements, Fourier spectrum for each components are calculated and smoothed eighty times with Hanning spectral window. With this operation band width approximately become 0.5Hz. One frequency spectrum of one component was estimated by averaging the three Fourier spectra. Vibration mode characteristics are also investigated from spectral ratio of higher floors with ground floor.

Deformation of Structures During Earthquakes and Vulnerability Index (K values) for Structures:

It is important to know the present durability condition of the ground and structure correctly. Earthquake damage depends on strength, period and duration of seismic motions. And these parameters strongly reflect seismic response of surface ground and structure. Considering this, vulnerable weak points can be found by examining seismic motion characteristics. Deformation of structures are related with seismic motion of the basement, dynamic characteristics of surface layer and dynamic characteristics of the structures. Figure (1) simply shows the mode shape and deformation characteristics of nth floor structure.

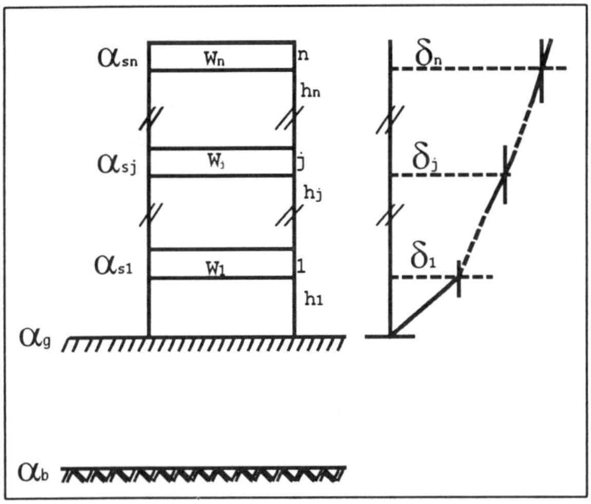

Figure 1. Derivation of mode shapes of nth floor structures. δ_j is the horizontal displacement and W is the weight of the jth floor and h_j is the height of jth column, α_b , α_g , α_{sj} are the horizontal acceleration of the basement, ground surface and j[th] floor of the structure, respectively.

Then, horizontal displacement δ_j can be written as in equation (1),

$$\delta_j = \frac{\alpha_{sj}}{(2\pi F_s)^2} \tag{1}$$

F_s is a predominant frequency of structure. Story drift angle for j[th] story γ_j is expressed as follows,

$$\gamma_j = \frac{\delta_j - \delta_{j-1}}{h_j} = \frac{\alpha_{sj} - \alpha_{sj-1}}{4\pi^2 F_s{}^2 h_j} \tag{2}$$

$$\begin{aligned} \alpha_{sj} &= A_{sj} \cdot \alpha_g \\ &= A_{sj} \cdot \mathrm{A_g} \cdot \alpha_b \\ &= A_{sgj} \cdot \alpha_b \end{aligned} \tag{3}$$

Here, A_g and A_{sj} are an amplification factor for ground and j[th] floor of the structure, respectively. A_{sj} is derived from S_{jh} and S_{gh} which are horizontal spectrum of jth floor and ground floor respectively. And A_{sgj} is derived from the ratio of S_j and vertical spectrum of ground floor and represents combined amplification factor of surface ground and structure. α_b and α_g are horizontal acceleration of basement and ground surface, respectively.

If unit of drift angle γ_j is 10^{-6}, h_j is meter and seismic acceleration is measured in unit Gal (cm/sec^2), than with the unit adjustment, equation (2) can be written in following form,

$$\gamma_j = 10^4 \cdot \frac{(A_{sgj} - A_{sgj-1})}{4\pi^2 F^2 \cdot h_j} \cdot \alpha_b \tag{4}$$

From equation (4) K_{Tgj} value (representing vulnerability index for ground and building) is defined as,

$$K_{Tgj} = 10^4 (A_{sgj} - A_{sgj-1}) / (4\pi^2 F_s^2 \cdot h_j) \tag{5}$$

unit of K values given above become 1/Gal. From this, vulnerability index for buildings can be written as,

$$K_{Tj} = 10^4 (A_{sj} - A_{sj-1}) / (4\pi^2 F_s^2 \cdot h_j) \tag{6}$$

Maximum allowable acceleration from jth column α_{baj} (in Gal) derived from equation (4) is,

$$\alpha_{saj} = 10^4 \cdot \frac{4\pi^2 F^2 \cdot h_j}{A_{sgj} - A_{sgj-1}} \cdot \gamma_{aj} \tag{7}$$

3. Information On Suleymaniye Mosque, Location Of Measurement Points And Conditions

Turkey has been the home of many civilizations since the beginning of history and with its rich past it stands as a resource for history. After the Ottoman took over Istanbul, contributed architect Mimar Sinan built Suleymaniye mosque between the year 1550-1557. This is one of the prime examples of the Islamic architecture and considered as the most beautiful of all imperial mosques in Istanbul. Figure 2, shows the bird eye view of the mosque. To ensure the absolute stability of the foundations, upon the bedrock of the site, three years of preparations were made, and three more years passed in the construction of these foundations. Figure 3, shows the location and side and front cross sections of the mosque, as well as position of the measurement points on the ground surface and inside the mosque. The mosque itself is almost square in plan, measuring 63x68 m. The height of dome is 53m. The central dome rests on four arches springing from four great piers; while a semi-dome is placed over the entrance portal. On either side of the main dome are five cupolas, supported on pillars set between the main piers, and resting on the arches from these piers. Mosque is illuminated by 138 windows which lighten the interior. The material used in the building of the mosque were brought around the site of Istanbul area, and even from the farthest part of the empire. Two of the piers originated from the city of Istanbul, one from Egypt and one from the ruins of Baalbeek.

Figure 2. Bird eye view of Suleymaniye mosque

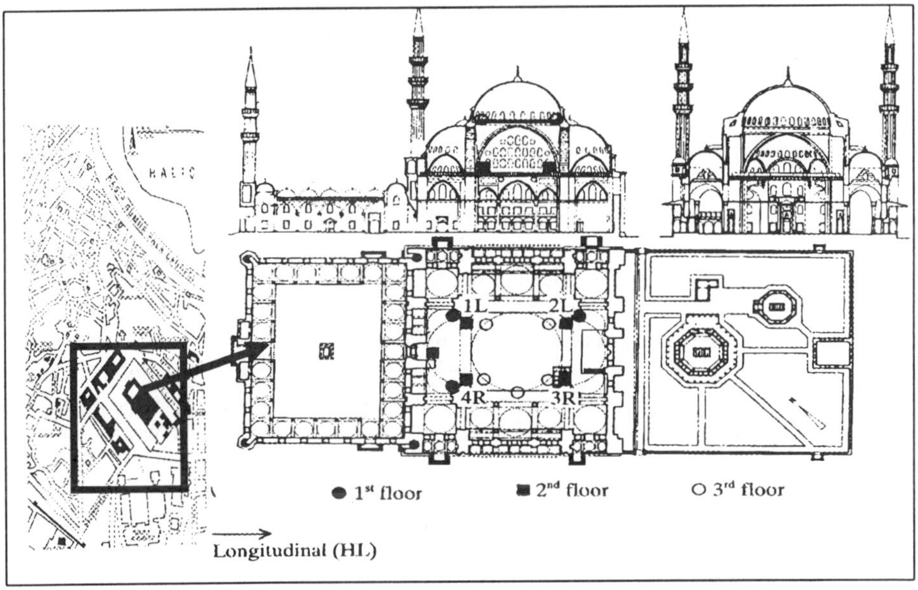

Figure3 Location and Cross section of the mosque in Longitudinal and Transversal directions.

●, ■, O shows measurements at ground, 1st, 2nd floors, respectively.

Measurements were performed at ground level outside the mosque close to the minarets, and also inside of the mosque, at the columns on semi-dome and big dome levels. Figure 4. shows some of the measurement conditions inside and outside the mosque.

Figure 4. Measurement conditions inside and outside the mosque

4. Results

Spectral ratios of second floor(dome level) and first floor levels with ground level are given in Figure 5. This ratio provided the predominant frequency and amplification factor of entire structure. There are peaks of different amplification levels showing different modes of structure in longitudinal and transversal components. Spectral ratio of second floor to ground floor shows three clear peaks at 0.8Hz, 3.5Hz and 4.0Hz, respectively. On the other hand in the ratio of first floor to ground floor amplification of peak at 0.8 and 4 Hz become very small but peak at 3.5Hz appears strongly. This can be interpreted as peaks at 0.8Hz and 4.0Hz is the effect of dome and 3.5Hz is the natural frequency of whole structure.

Figure 6. and Figure 7. also shows the vibration characteristics of structure at 0.8 and 3.5 Hz for both transversal and longitudinal directions. In these figures, it is easier to follow that vibration of first floor at 0.8Hz frequency is very small comparing with the second floor. On the other hand vibration at 3.5 Hz is high for both first and second floor levels, which shows that 3.5Hz is the natural vibration frequency of whole structure. In general amplification is higher in longitudinal direction than in transversal direction. Vulnerability index, K_{Tgj} values, are also directly calculated from the frequency and amplification information. Nakamura(1996, 1997) has already proved that K value is high where the damage risk is high. Calculated K_{Tgj} values are given in Table 1. Looking at these values, points where K_{Tgj} values are high such as H2L and H4R on the second floor(under the dome) can be interpreted as weak points of the structure. Maximum acceptable acceleration level for these points are also calculated from equation (7). And it is found that, maximum allowable accelerations at H2L point on dome level are 17Gal in longitudinal and 105Gal in transversal directions, for 0.8 Hz vibration frequency. Maximum acceleration limit of longitudinal direction is about six times smaller than the transversal direction. Again at dome level(second floor), another example can be given as point H4R, which is 18Gals in longitudinal and 23Gal in transversal directions which is very weak. Showing the weak points before a real earthquake damage occurs, present study showed that microtremor is a valuable tool to grasp the vulnerability of surface ground and structures.

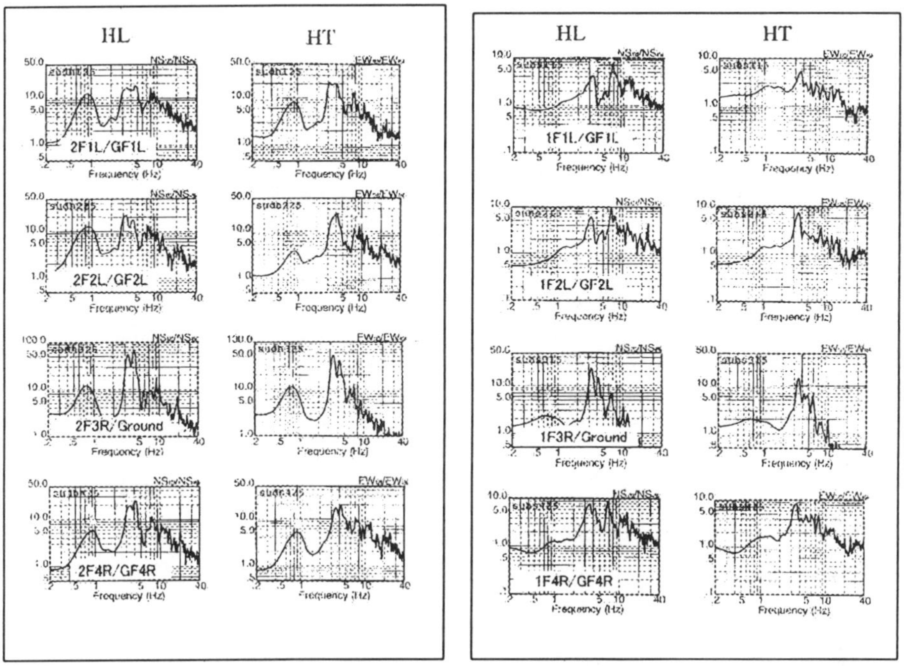

Figure 5 . Spectral ratios of second floor(left side) and first floor (right side) levels with ground level.

Table 1. Calculated K_{Tgj} values for 1st and 2nd floor levels

	K_{Tgj} value for 1st floor				K_{Tgj} value for 2nd floor			
	Fs=0.8Hz		Fs=3.5Hz		Fs=0.8Hz		Fs=3.5Hz	
	HL	HT	HL	HT	HL	HT	HL	HT
H1L	37	44	4.9	6	293	48	19	21
H2L	33	33	7.0	9	362	45	22	23
H3R	51	44	8.9	9	40	53	15	23
H4R	31	40	7.9	9	282	222	17	13

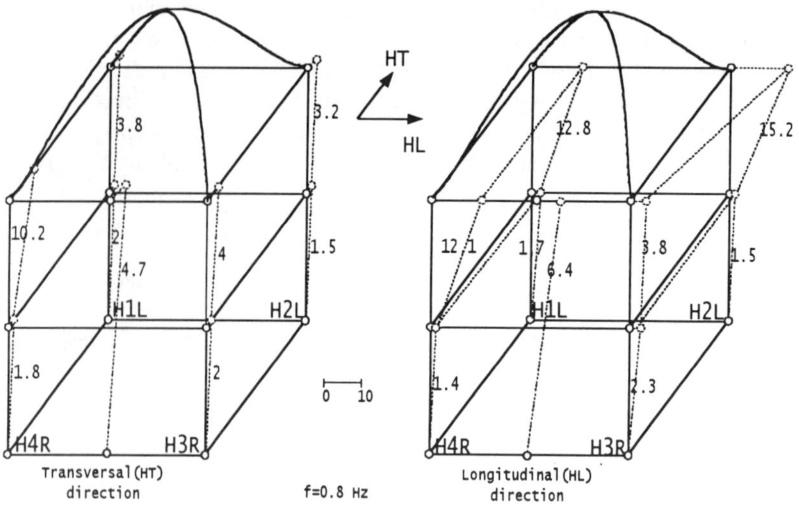

Figure 6. Vibration characteristics of structure at 0.8Hz.

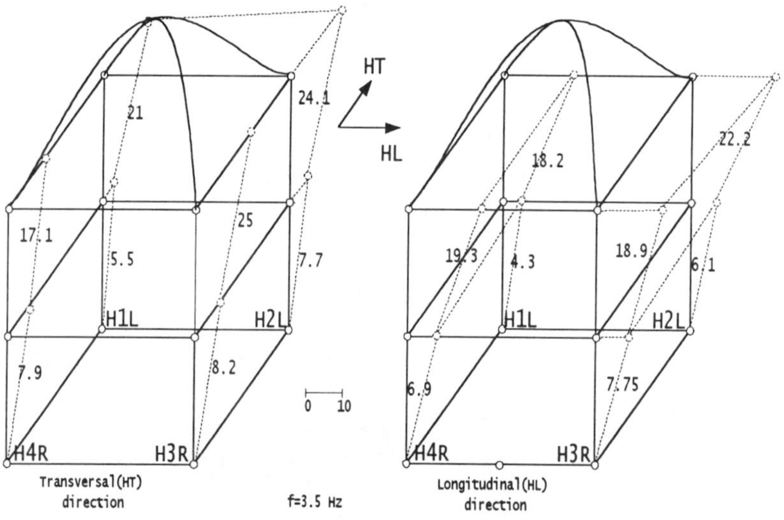

Figure 7. Vibration characteristics of structure at 3.5Hz.

5. Concluding Remarks

A method for investigating the vulnerability of structures is introduced. Using microtremor data, in a short period of time, present method is able to give valuable information about the natural frequency, amplification factor, vibration mode characteristics as well as week points of every type of ground and structures. With the present approach by defining weak points, it is possible to obtain real earthquake damage before earthquake occurs. This allow one to determine weak areas in the cities, and list important points from most vulnerable one to the least one. Then it is possible to decide the monitoring spots for the strong ground motion instruments. Making quick and precise damage estimation possible, giving early alarm for vulnerable areas if necessary and deciding monitoring points for setting strong motion, this type of preparatory study in highly seismic areas is very promising for the future disaster prevention activities.

6. References

1. Nakamura, Y., (1989) A Method for Dynamic Characteristics Estimation of Subsurface using Microtremor on the Ground Surface, Quarterly Report of RTRI, Railway Technical Research Institute(RTRI), Vol. 30, No.1.
2. Nakamura, Y., (1996) Real Time Information Systems for Seismic Hazards Mitigation UrEDAS, HERAS, PIC, Quarterly Report of RTRI, Vol. 37, No. 3.
3. Nakamura, Y., (1997) Seismic Vulnerability Indices For Ground and Structures Using Microtremor, World Congress on Railway Research in Florence, Italy, November 1997.

STRONG MOTION INSTRUMENTATION OF BUILDINGS

A Study on the Linear and Nonlinear Response of an Instrumented 52-Story Building

CARLOS E. VENTURA ,YUMING DING
Department of Civil Engineering
University of British Columbia,
2324 Main Mall, Vancouver, BC, Canada V6T 1Z4
e-mail: ventura@civil.ubc.ca

1. Abstract

A comparison of the recorded structural earthquake response of a building and its predicted response by dynamic analysis provides vital information to structural designers on the effectiveness of current methods of dynamic analysis. There have been a number of previous studies of this nature, but only a few have paid attention to investigating the three-dimensional nonlinear dynamic behaviour of instrumented tall buildings. The purpose of this paper is to study the seismic behaviour of a well-instrumented 52-storey steel frame building in Los Angeles, California. This building has been subjected to ground motions from several earthquakes among which the 1991 Sierra Madre earthquake and the 1994 Northridge earthquake were selected in this study. During both earthquakes the building responses appeared to be in the linear range. Frequency domain analyses of the recorded motions from these two earthquakes were conducted to determine the dynamic characteristics of the structure. Three-dimensional nonlinear dynamic computer analyses were then employed to evaluate the response of the structure induced by severe shaking. The results of this study show that by performing a linear three-dimensional analysis, the real response of a building during past earthquakes can be reproduced with confidence. By further performing a nonlinear three-dimensional analysis, the state and sequence of damage can also be predicted. The Nonlinear Static Procedure (pushover analysis) generally excludes higher mode participation, which can be important for high-rise buildings subjected to certain types of ground motions. Improvements to this procedure were explored.

2. Introduction

Comparison of recorded structural earthquake response and predicted response by dynamic analysis provides vital information to structural designers on the effectiveness of current

M. Erdik et al. (eds.), Strong Motion Instrumentation for Civil Engineering Structures, 313–330.

methods of dynamic analysis. This comparison could be either in the time domain or in the frequency domain, by utilizing system identification algorithms. It can also be done utilizing joint time-frequency analysis techniques, such as the Wavelet Transform or the Short Time Fourier Transform.

In the last decade, since the 1989 Loma Prieta earthquake in California, a significant amount of research on the response of instrumented buildings has been carried out in order to: 1) learn more about the seismic response of buildings by analyzing the recorded data and 2) improve dynamic analysis techniques to evaluate the performance of new and existing buildings. However, only some of this research has paid attention to the three-dimensional nonlinear dynamic behaviour. In particular, the three-dimensional nonlinear behavior of steel frame structures has not been studied as well as that of planar or two-dimensional behavior. The unavailability of computer programs capable of three-dimensional inelastic analyses has been cited as the justification for this simplification [8]. The study presented herein demonstrates that this justification is no longer valid.

The main objectives of this study were to use an analytical model, incorporated into an advanced computer program, to compute the 3-D linear response of a well-instrumented steel frame structure shaken by recent earthquakes; compare its computed response with its actual recorded response; and finally, to evaluate the nonlinear response of the structure during severe ground shaking. The emphasis of the study reported here is on the linear and non-linear global response and behavior of the structure, rather than on any of its elements. The recorded responses are for events in which the building behaves primarily in a linear elastic manner.

Figure 1. Overview of the FWT

3. Study Building

The study building, identified here as the FWT, is a 52-story steel frame office building in Los Angeles, California. This building was designed in 1988, constructed in 1988-90 and instrumented to measure seismic motions after construction. The FWT consists of a 52-story steel frame office tower and five levels of enlarged basement as underground parking. Due to architectural reasons, the floor plans of the tower are not perfectly square. On each floor, the tip of every corner is clipped and the middle third of each side is notched. Above the 36th story, at steps of about five stories, the corners of the floors were clipped further to provide a setback view to the exterior of the building. An overview of the building is shown in Fig. 1.

The FWT was instrumented by the California Strong Motion Instrumentation Program (CSMIP) in 1990. Twenty sensors in total were installed on different levels of the building to measure the global building response to earthquakes, including the translational, torsional, and vertical motions. The location of the sensors is shown in Fig. 2. To measure torsional motion of a floor, pairs of sensors were placed in the N-S direction rather than the E-W direction because the floor eccentricity is larger in the E-W direction. The selection of the floor locations for the sensors was based on the analytical mode shapes provided by the structural designer.

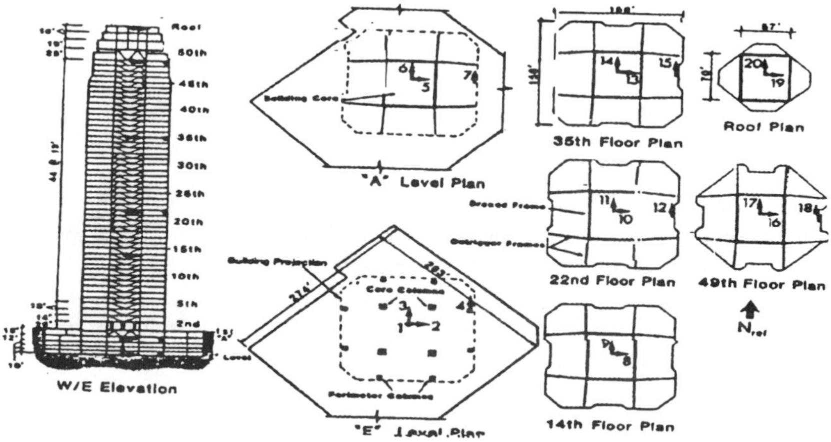

*Figure 2.*Sensor locations for the FWT (after Shakal, et al. (3)

4. Structural System

The structural system of the FWT has three main components: a braced core, twelve columns (8 on the perimeter and 4 in the core), and eight 914 mm (36 inch) deep outrigger beams at each floor connecting the inner and outer columns. The core, which is about 17.37 m (57 feet) by 21.34 m (70 feet), is concentrically braced between the "A" level (the level just below the ground level) and the 50th story. Moment resisting connections are used at the intersection of beams and columns.

The outrigger beams, about 12.19 m (40 feet) long, link the four core columns to the eight perimeter columns to form a ductile moment resisting frame. In each floor, the outrigger beams are laterally braced to prevent lateral torsional buckling and are effectively connected to the floor diaphragm by shear studs to transmit the horizontal shear force to the frame.

The interior core is concentrically braced. The sizes of mechanical ducts and door openings into the core dictated the configuration and sizes of braces (Fig. 2). The outrigger beams were designed to perform three functions [1]: 1) they have to support the design floor loads; 2) the outrigger beams along with core and perimeter columns have to act as a ductile moment resisting frame to carry a minimum of 25% of the design code level forces without the presence of interior core bracing; and 3) the stiffness of the beam should be such so as to create effective linkage between the interior core and the perimeter columns to provide effective overturning resistance to the seismic loads.

In high-rise buildings, it is always cost effective to minimize floor-to-floor height. The depth of outrigger beams was therefore dictated by a restriction on the ceiling cavity. To achieve these goals, the beams were offset into the floor and were notched at midspan to allow for the passage of the mechanical ducts.

Typical floor plan dimensions are 47.45 m square. Typical story heights are 3.96 m. The foundation consists of concrete spread footings (2.74 to 3.35 m thick) supporting the steel columns with a 127 mm thick concrete slab on grade.

The soil is very stiff shale or sandstone and has allowable bearing pressure of 718.2 kPa (15000 lbs/sq ft). All structural steel framing including columns use ASTM A-572 (grade 50).

5. Identification of Dynamic Characteristics

In this study, the natural periods, damping ratios and mode shapes were identified using computer program ME'scope [6]. The records were used to calculate the Frequency Response Function (FRF) and the coherence functions between pairs of records. From the frequency response functions the natural frequencies and the mode shapes were derived. The coherence functions were used to verify that the recorded motions were from the same excitation source or that nonlinearities of the system were not that significant as to affect results of the system identification analysis.

The two earthquakes considered in this study are the 1991 Sierra Madre earthquake and the 1994 Northridge earthquake. The building experienced its first strong ground motion during the Sierra Madre earthquake, which had its epicenter about 25 km north-east of the building. No building damage was reported after this event. The epicenter of the 1994 Northridge earthquake was located about 30 km north-west of the building. Again, no damage was reported after this earthquake. The acceleration response records obtained during the Sierra Madre earthquake had a peak ground motion of about 0.09g, and a corresponding peak structural response of 0.23g at the roof. The Northridge earthquake records are shown in Fig.3

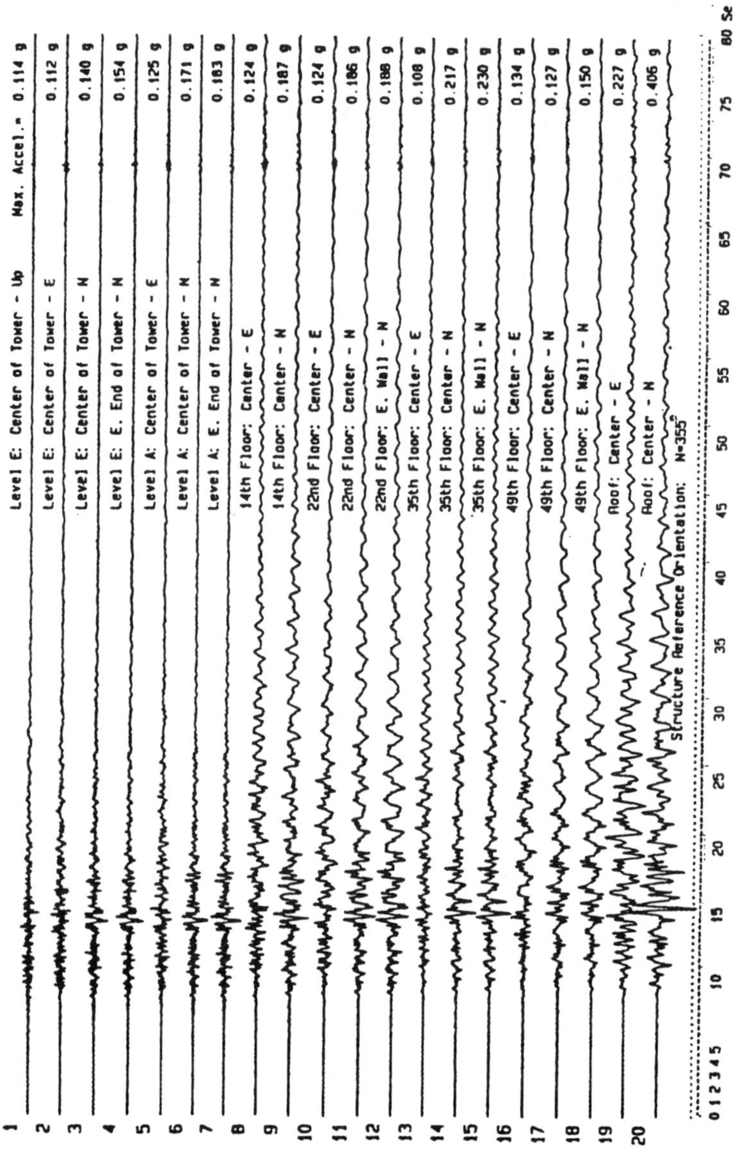

Figure 3. Recorded accelerations during the 1994 Northridge earthquake (CSMIP, 1994)

Recorded peak acceleration values are included at the right side of the record. The building periods derived using ME'scope program are shown in Table 1. For reference, the periods used for design of the building [2] are also included in the table. The last column of the table shows how the periods changed from one earthquake to the other earthquake. The largest change is less than 13%. Note that while the natural periods in the E-W direction in the two earthquakes remain almost constant, the natural periods in the N-S direction during the Northridge earthquake are longer than those during the Sierra Madre earthquake.

TABLE 1. Periods (sec.) of the first three modes in each direction

Direction	Design Value	Sierra Madre	Northridge	% of change
E-W(1)	6.73	5.47	6.06	+10.8
N-S(1)	6.59	5.85	5.75	-1.7
Torsional(1)	6.36	4.55	4.82	+5.9
E-W(2)	2.08	1.71	1.86	+8.8
N-S(2)	2.11	1.74	1.74	0
Torsional(2)	N.A.	1.62	1.82	+12.4
E-W(3)	1.13	0.92	0.96	+4.3
N-S(3)	1.18	0.91	0.91	0
Torsional(3)	N.A.	0.94	1.04	+10.6

6. Analytical Studies

Computer program CANNY-E [5] was used for the analytical studies. CANNY-E is a general purpose computer program for 3-dimensional linear and nonlinear static and dynamic analyses of building structures. The program is based on a lumped plasticity model and has the capacity for analyzing large structures. The primary objective of the analytical studies was to first perform a three-dimensional linear analysis in order to understand the behavior of the building during past earthquakes, then to calibrate the computer model until it represented the actual dynamic characteristics of the structure, and finally to perform nonlinear time history analysis to investigate the behavior of the structure during severe earthquakes. The building was modelled as a combination of braced frames and moment frames consisting of 14 main column lines. The computer model is shown in Fig. 4.

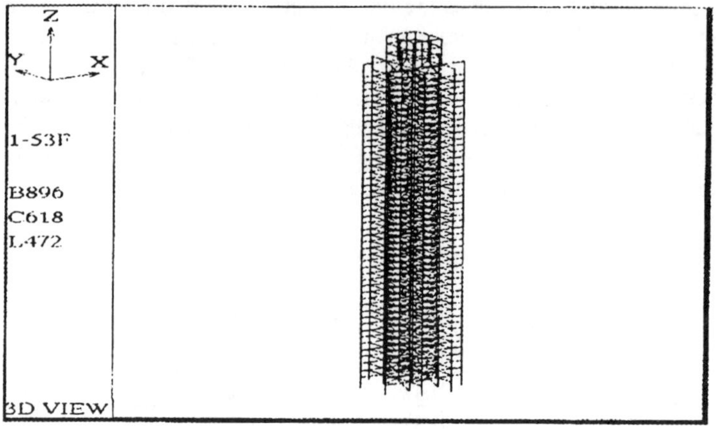

*Figure 4.*CANNY model of FWT

The floor diaphragms were assumed to be rigid. With this constraint, each diaphragm consists of three in-plane degrees of freedom, one rotational and two translational. The building was assumed to be fixed at the ground level. All the beam and column centre lines were considered to coincide with column line coordinates. The masses of the beam and column members associated with each floor were calculated from the specified dimensions of the elements and combined with the mass of the slab to determine the total story masses. The estimated mass and mass moment of inertia were lumped at the centre of mass at each floor.

6.1 DYNAMIC PROPERTIES AND CALIBRATION OF THE MODEL

The periods obtained from the CANNY model are compared with those from the recorded earthquake responses in Table 2. The periods inferred from the Northridge records were used to calibrate the model. From this table it is clear that the computer model has periods that are in good agreement (within 10%) with the building periods obtained from the Northridge earthquake records. The last column shows the difference between the computed periods and those obtained from the records. Mode shapes of the first six modes are shown in Fig. 5.

Damping ratios are an important component in the prediction of the analytical response of a building. After a detailed comparison of the analytical responses for different modal damping ratios, 5% of critical damping for mode 1 and 2% of critical damping for mode 9 were used to define a Rayleigh-type damping for the computer model. The analysis was carried out for 60 seconds of the Sierra Madre record and 180 seconds of the Northridge record. During both analyses two horizontal and one for torsional excitation were used as input ground motions.

TABLE 2. Periods (in sec.) of the first three modes in each direction

Direction	CANNY Model	Northridge	Percent difference
E-W(1)	6.00	6.06	+1.0
N-S(1)	5.61	5.75	+2.5
Torsional(1)	4.75	4.82	+1.5
E-W(2)	1.85	1.86	+0.5
N-S(2)	1.77	1.74	-1.7
Torsional(2)	1.71	1.82	+6.4
E-W(3)	0.97	0.96	-1.0
N-S(3)	0.97	0.91	-6.2
Torsional(3)	1.04	1.04	0

In order to evaluate the contribution of gravity load frames, an additional model was constructed including the perimeter gravity load columns. Comparison of periods, mode shapes and responses showed that including gravity load frames in this building has little effect on changing the dynamic characteristics of the computer model. Therefore, the computer model used for the analyses did not include the gravity load frames.

Fig. 6 shows the recorded and predicted absolute accelerations at the 49th floor during the Northridge earthquake. It can be seen that a good match between the two responses was obtained. Although not shown here, a good match was also obtained for the Sierra Madre record. Since the computer model fits well with the responses from two of the significant earthquakes that the building has experienced, the computer model was considered as a reliable representation of the structural system of the building during strong earthquake excitations. This sets the base for the nonlinear analysis described in the following sections.

6.2. NONLINEAR TIME HISTORY ANALYSES

The purpose of performing nonlinear analyses was to investigate the response of the building during severe shaking. One of the most important tasks of this effort was the selection of the variables defining the inelastic properties of the structural elements. An in-depth search of previous studies and design guidelines was carried out to define credible values for these variables.

All flexural members were modeled with a bilinear hysteresis model of the CANNY program, which is shown in Fig. 7. Yielding strengths were calculated based on the expected steel yielding strength. From the general notes on the site plans, all steel framing including columns are ASTM A-572 (grade 50).

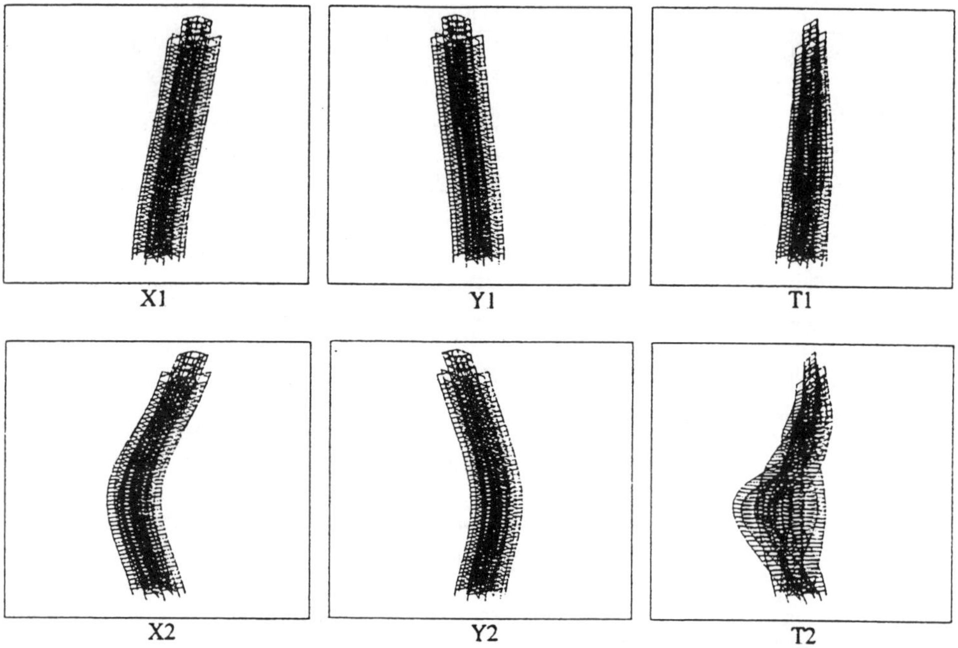

Figure 5. First six modes of vibration from CANNY model

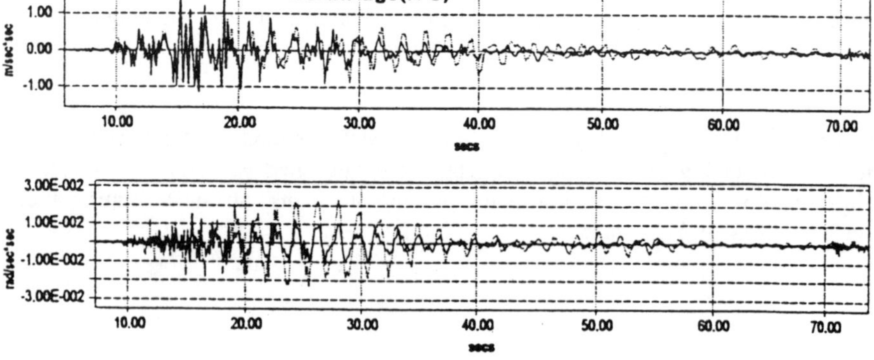

Figure 6. Accelerations of 49th floor during Northridge earthquake

Nominal yielding strength for A-572 (grade 50) is 345MPa (50 ksi). Expected steel yielding strength was assumed based on average values such as those cited in the NEHRP Guidelines (FEMA, 1997) in accordance with the 1995 SAC95-04 report: Fy = 380MPa (55ksi), and the post-yielding stiffness factor was chosen as 2%.
The unloading stiffness degradation factor was chosen as 0.3. The strength deterioration factor was chosen as 0.75. It was assumed that there would be 50% strength remaining at a ductility level of 3. For computation of the shear stiffness of beams and the axial stiffness of columns, these members were assumed to be linear elastic.

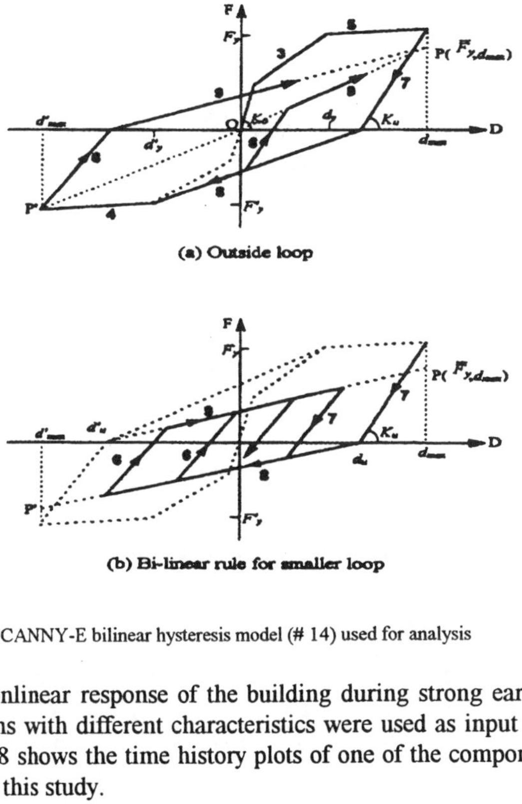

*Figure 7.*CANNY-E bilinear hysteresis model (# 14) used for analysis

In order to evaluate the nonlinear response of the building during strong earthquakes, three recorded earthquake motions with different characteristics were used as input to compute the building responses. Figure 8 shows the time history plots of one of the components of each of the ground motions used in this study.

First, a set of records from the SMAC-A station during the 1964 Niigata (Japan) earthquake was used as input excitation. The response spectrum for these records show that systems with periods in the 4.5 to 6 seconds range may be significantly excited. Thus, this record is expected to excite mostly the fundamental mode of the building. The second set of recorded motions used as input excitation was the Joshua Tree Fire station records during the 1992 Landers Earthquake in California. These records have a duration of at least 30 seconds of strong shaking. This set was selected because it may be of practical interest to observe the responses of high-rise buildings during this type of long duration shaking. Last, the Sylmar County Hospital station record obtained during the 1994 Northridge earthquake was used as input excitation. This is a short duration, strong impulse, near-source ground motion that may impose very high demands to tall buildings.

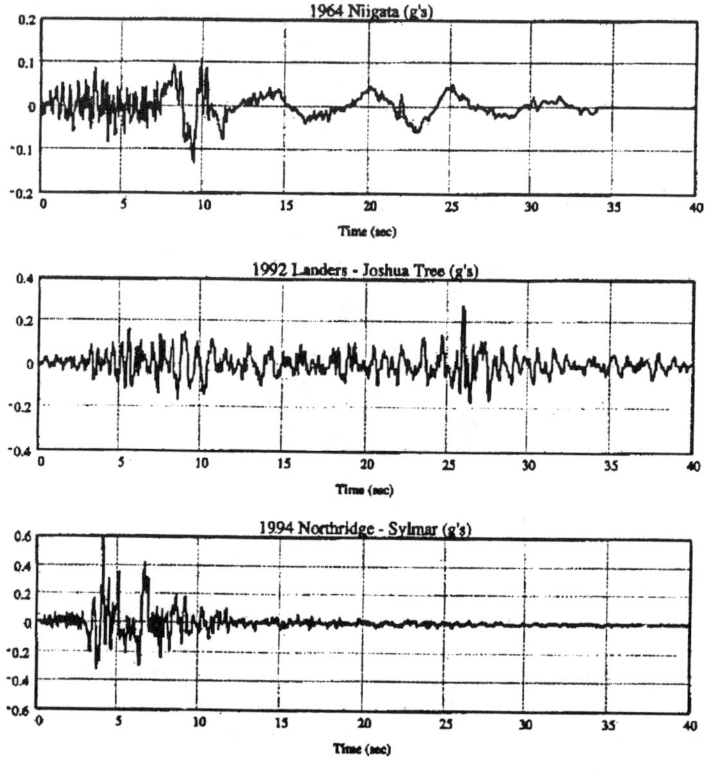

*Figure 8.*Acceleretions of ground motions selected hor nonlinear analysis

Results of the inelastic, time-history analyses are shown in Figs. 9, 10, 11 and 12 in terms of envelopes of lateral displacement, interstory drift, story shear and overturning moment. In these plots, the x-direction corresponds to the East-West direction of the building (see Fig. 2), while the y-direction corresponds to the North-South direction of the building. The plot of maximum lateral displacements shows the significant effect of the Sylmar motion on the response of this building. The maximum displacement plot also shows clearly the combination of at least two mode shapes in the responses of this building for the Joshua Tree and Sylmar motions. The input motions selected produce different envelopes of interstory drift in each direction of the building as shown in Fig. 10. The largest drifts occur for the Niigata motion in the North-South direction (y-direction). The maximum value for the interstory drift is 1.5% and it occurs at the 23rd story.

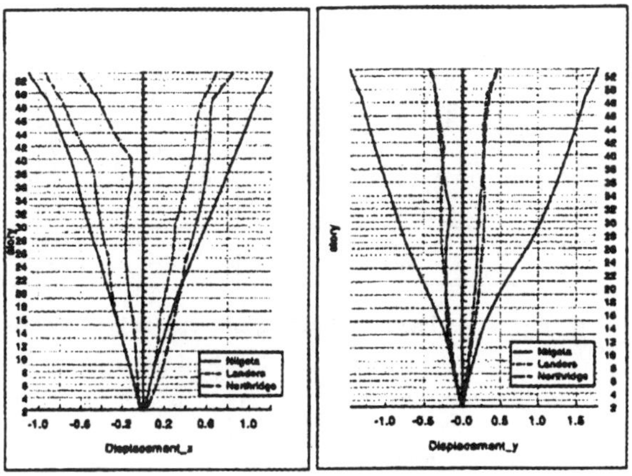

Figure 9. Lateral displacement envelope (in m)

It can be observed that, in contrast with the variation of story displacements, maximum story shear increases down from the roof to the 49th story (similar to the expected distribution for low-rise buildings). For the next 20 stories below there is no significant change of maximum story shear resulting from the Sylmar and Joshua Tree motions. This contradicts the code-defined inverted-triangular lateral load distribution along the height of the building. It is also an indication of higher modes effects. The maximum story shears in the lower part of the building show the usual increase towards the ground floor. However, one still could notice that in some cases the tip of the curve is growing outward instead of growing inward. This can only be explained by a load pattern which is nearly rectangular or trapezoidal.

The overturning moment plots for the Sylmar and Joshua Tree motions have peculiar onion-like shapes. The floors between the 20 to 25 story experienced the largest overturning moment during these two earthquakes. The overturning moment for the Niigata motion in the East-west direction show similar behavior, but in the North-South direction the moment distribution is as generally expected.

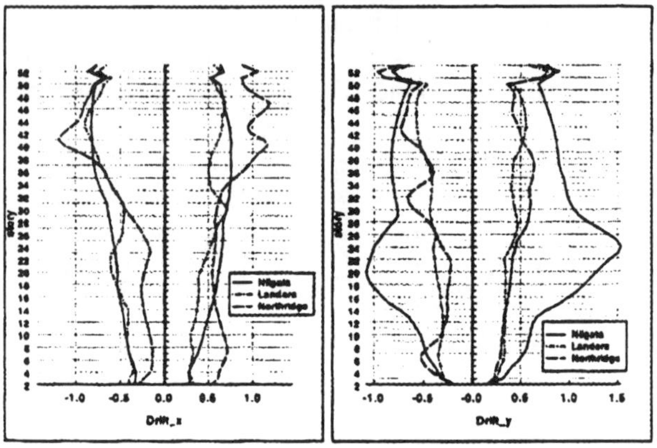

Figure 10. **Interstory drift ratio envelope (in % of story height)**

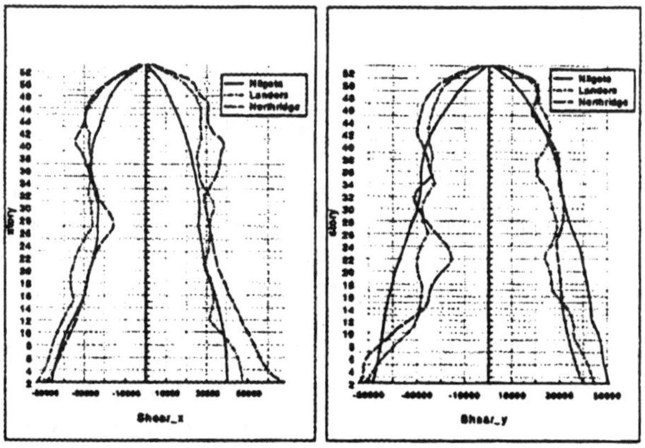

Figure 11. Story shear envelope (in kN)

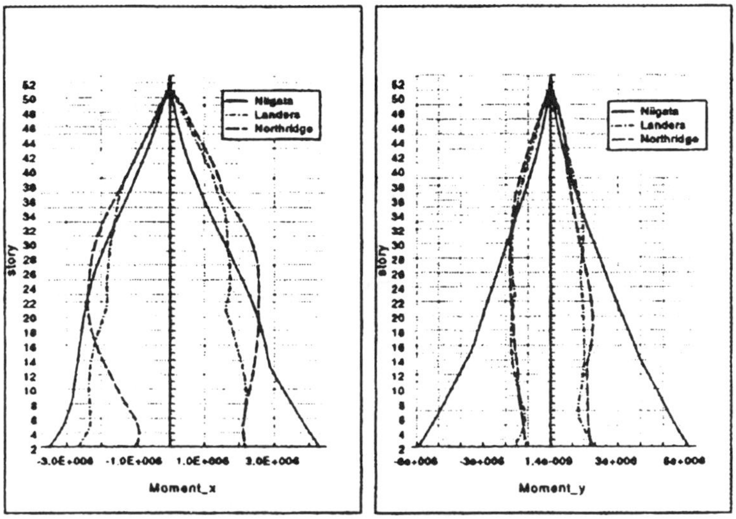

Figure 12. Overturning moment envelope (in kN-m)

6.3 PUSHOVER ANALYSES

In order to estimate the lateral resistance of the building at ultimate load, static nonlinear (pushover) analyses were conducted. The results of pushover analysis are sensitive to the choice of load pattern, particularly for this case of high-rise building. A predetermined triangular lateral force pattern was applied incrementally in a step-wise manner in each direction independently. This lateral force pattern from building design guidelines assumes the lateral force distribution as an inverted-triangular pattern which is in accordance with the first mode shape of multi-degree-of-freedom building structures.

The building was analyzed by applying a predefined lateral force pattern until a specified target roof displacement was reached. A target roof displacement of 4.5 m (3% of the height of building) and an increment of 0.1m to 0.001m (depending on the convergence requirements) was selected for this study. The results showed significant yielding in the structure at a roof deflection of approximately 2.5 m. First yielding members were the outrigger beams between the 23rd to 26th floors when the roof displacement reached 0.88 m. First group of yielding bracing members were those between the 22nd to 26th floors when the roof displacement reached 1.13 m. Maximum base shear was 47,108 kN. Maximum overturning moment is

6,717,308 kN-m. Fig. 13a shows the yielding propagation steps in the south perimeter frame in the east-west direction.

After completing the analysis for the code-defined lateral load pattern, a second lateral pattern defined in FEMA-273 [4] was adopted. This pattern, often termed the "uniform pattern," shall be based on lateral forces that are proportional to the total mass at each floor level. FEMA-273 is the first guideline to suggest the uniform load pattern for pushover analysis.

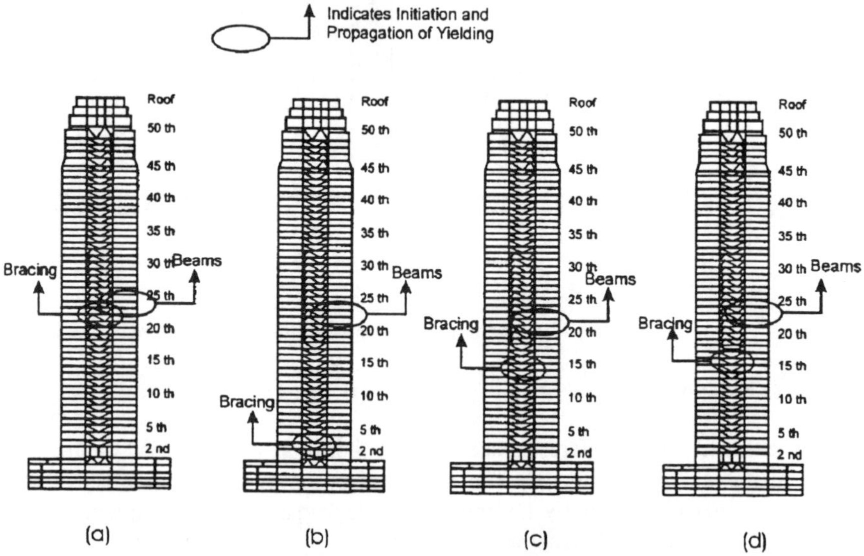

Figure 13 E/W elevation of south frame indicating yield propagation: a) time history - Sylmar record, b) pushover - triangular load, c) pushover - uniform load, and d) pushover -trapezoidal load.

The same target displacement as that used for code-defined lateral pattern was selected for the uniform load pattern. Figure 13b shows the yielding propagation steps in the south perimeter frame in the north-south direction. When compared with the time history responses of the building, it is clear that for this uniform lateral load pattern, pushover analysis reflects a realistic dynamic behavior of the building when subjected to near-field type motions such as the Sylmar record.

A third lateral pattern was investigated in order to make the response from the pushover analysis better match the response from the time history analysis. This proposed load pattern has a trapezoidal shape. The yielding propagation steps in Fig. 13c. clearly show that this is the best comparison of yielding steps between this load pattern and the time history response.

TABLE 3. Comparison of Shears and Overturning Moments

Event		Base Shear (*kN*)		Overturning Moment (*value x 10^6 kN-m*)	
		V	V/Vd	M	M/Md
Design		32,716	1.0	2.92	1.0
Time History Analysis	Northridge	14,182	0.4	0.53	0.2
	Niigata	50,961	1.6	6.43	2.2
	Joshua	39,623	1.2	2.08	0.7
	Sylmar	68,121	2.1	2.31	0.8
Pushover Analysis	Triangular	14,325	0.4	2.09	0.7
	Uniform	73,283	2.2	8.02	2.7
	Trapezoidal	56,043	1.7	6.80	2.3

The table 3 is a summary of the maximum base shear and overturning moments for the FWT based on design values, four time history analyses and three pushover analyses. It is clear in this table that during the Northridge event the structure stayed well within the design range. In contrast, the Sylmar record causes the largest base shear for the structure, and the Niigata excitation results in the highest overturning moment. For pushover analysis, a trapezoidal lateral load pattern is a compromise between the traditional inverted-triangular pattern and the FEMA defined uniform pattern.

7. Conclusions

The recorded motions of the FWT during the Sierra Madre earthquake and the Northridge earthquake were analyzed by utilizing a frequency domain system identification algorithm. Modal periods and mode shapes were derived and compared with those from the original design. This information was used later to calibrate a 3-D elastic computer model of the building. A three-dimensional nonlinear model of the FWT building was then developed and analyzed. The three-dimensional nonlinear dynamic analyses predicted the probable sequence of damage under certain types of ground motion. Traditional code-defined pushover analysis predicted fairly different ways of damage for near-field type input motions. An improved lateral load pattern was therefore derived and yielded better fit with the dynamic responses considered in this study. Pushover analysis could never be a perfect substitute of dynamic time history analysis. Pushover analysis serves as a simplified nonlinear design method that when properly implemented may adequately represent the dynamic responses of structure during severe ground motions.

8. Acknowledgements

The support for this project was provided by the Natural Sciences and Research Council of Canada. The authors would like to acknowledge the valuable information about the building provided by Dr. Banavalkar in 1993. The engineering staff as CSMIP is also acknowledged for providing additional information about the FWT building.

9. References

1. Banavalkar, P.V., (1991) Spine Structures Provide Stability in Seismic Areas, *Modern Steel Construction*, January, p.13.
2. Banavalkar, P.V. (1992) Personal communication with the first author, June.
3. Shakal, A., et al., 1994,) CSMIP Strong-Motion Records from the Northridge, California Earthquake of January 17, (1994) Calif. Dept. of Conservation, DMG, Office of Strong Motion Studies, Report OSMS 94-07, Sacramento, CA, 308 pages.
4. FEMA 273 (1997) The NEHRP Guidelines for Seismic Rehabilitation of Buildings, Federal Emergency Management Agency, Washington, DC.
5. Li, K. N. (1996) CANNY-E Users' Manual, Canny Consultants Pte Ltd., Singapore.
6. The ME'scope (1998) Operating Manual Ver. 4.0, Vibrant Technology, Jamestown, California.
7. SAC 95-04 (1995) Analytical and Field Investigations of Buildings Affected by the Northridge Earthquake of January 17, 1994".

CURRENT STATUS OF STRONG-MOTION MONITORING AND NOTIFICATION AT THE UNITED STATES BUREAU OF RECLAMATION

CHRIS WOOD, ANDY VIKSNE, JON AKE, DAVID COPELAND
U.S. Bureau ofReclamation
Geophysics, Paleohydrology and Sesimotectonics Group
Technical Service Center
P.O. Box 25007, MS D-8330 Deuver, CO USA 80225

Abstract

Acquisition of strong motion data at the Bureau of Reclamation provides near real time notification of strong shaking at our structures as well as recordings of earthquakes for use in the dynamic analyses of large, engineered structures. To facilitate rapid notification following strong ground shaking, we have developed a low-cost, automated system to retrieve event-triggered seismic waveform data from digital data loggers at remote sites. Only a modest infrastructure investment is required. The method relies on standard dial-up telephone circuits, a custom-designed serial interface device, and communications software that runs under a UNIX operating system. Event wave- forms can be retrieved in near real-time from strong-motion accelerographs as well as from general-purpose seismic data loggers. The system can perform automatic remote configuration and periodic state-of health monitoring of the field instrumentation. The system presently retrieves data from 70 digital strong-motion instruments at 49 dams and other lifeline facilities located throughout the western United States. It has also been successfully used in temporary site-response and aftershock studies to provide remote access to seismic data loggers with broadband seismometers. Deployment of instrumentation is also focused on acquiring the detailed recordings necessary for validating the increasingly sophisticated dynamic analyses of dams.

1. Introduction

Data obtained from strong-motion recording devices can address two basic needs, one short-term and one long-term. The short-term need is to provide rapid notification of strong-ground shaking, which enhances public safety by providing a rational basis for inspection and/or remedial actions at structures that have been subjected to earthquake loading. The long-term need is to expand the empirical data base of ground-motion recordings that can be used in the analysis and design of critical structures, both existing and proposed. This paper will focus on describing the current state of the strong-motion recording program at the U.S. Bureau of Reclamation within this framework.

Recording of strong-motion earthquake data at the Bureau of Reclamation (Reclamation) began with the installation of three strong-motion accelerographs at Hoover Dam in 1937. Between 1937 and 1971 strong-motion monitoring progressed slowly, with instruments

M. Erdik et al. (eds.), Strong Motion Instrumentation for Civil Engineering Structures, 331–341.

installed at only six additional sites. The 1971 **M** 6.6 San Fernando earthquake and the resulting failure of Van Norman Dam, lead to recognition of the need for enhanced understanding of dam response to strong earthquake loading. Clearly one of the first steps required in that process was the acquisition of strong-motion data. Thus, Reclamation initiated a program of upgrading existing instrumentation as well as adding new strong motion systems, focusing especially on critical structures located in seismically active areas.

At the present time the Reclamation Strong Motion Program operates 180 digital and analog systems at 69 critical water resources lifeline structures, including dams, power-plants, pumping plants and pipelines. Currently 49 of these sites are instrumented with digital accelerographs (Figure 1).

2. Strong Motion Notification System

Recent developments in earthquake instrumentation and automated data acquisition hardware and software have allowed a near real-time notification function to be added to Reclamation's Strong Motion Program, termed herein the Strong Motion Notification System. By providing near real-time evaluation and assessment of strong motions at critical structures that have been subjected to earthquake loading, the Strong Motion Notification System acts as an enhancement to public safety. At Reclamation this system is integrated within a broader dam safety program that allows for rapid evaluation of recorded ground motions. This evaluation is based on a set of specific ground motion thresholds and emergency response actions developed for each dam.

Why is this capability for rapid notification important for Reclamation? The history of dam failures and large flood events overwhelmingly suggests that the extent of downstream loss of life is directly related to warning time. In instances where significant warning time *has* been provided, life loss was generally significantly lower than cases where no appreciable warning was provided (Brown and Graham, 1988; DeKay and McClelland, 1993). With rapid notification of appropriate personnel of potentially adverse ground shaking levels, the potential for evacuation exists in most circumstances. Rapid notification of appropriate responsible parties of the ground motions experienced by nearby structures can also be effective in prioritizing post-earthquake inspections.

Currently, the Strong Motion Notification System is being developed in conjunction with the "Performance Parameters" process at Reclamation. This process seeks to evaluate high risk structure in the inventory, define appropriate specific thresholds for a number of different engineering parameters (pizeometric readings and ground motion levels, for example) and develop a data base containing these parameters. Then, when recorded ground motions are received and processed, the resulting values can be evaluated against these structure specific data base parameters and threshold specific notifications can be transmitted in near real time.

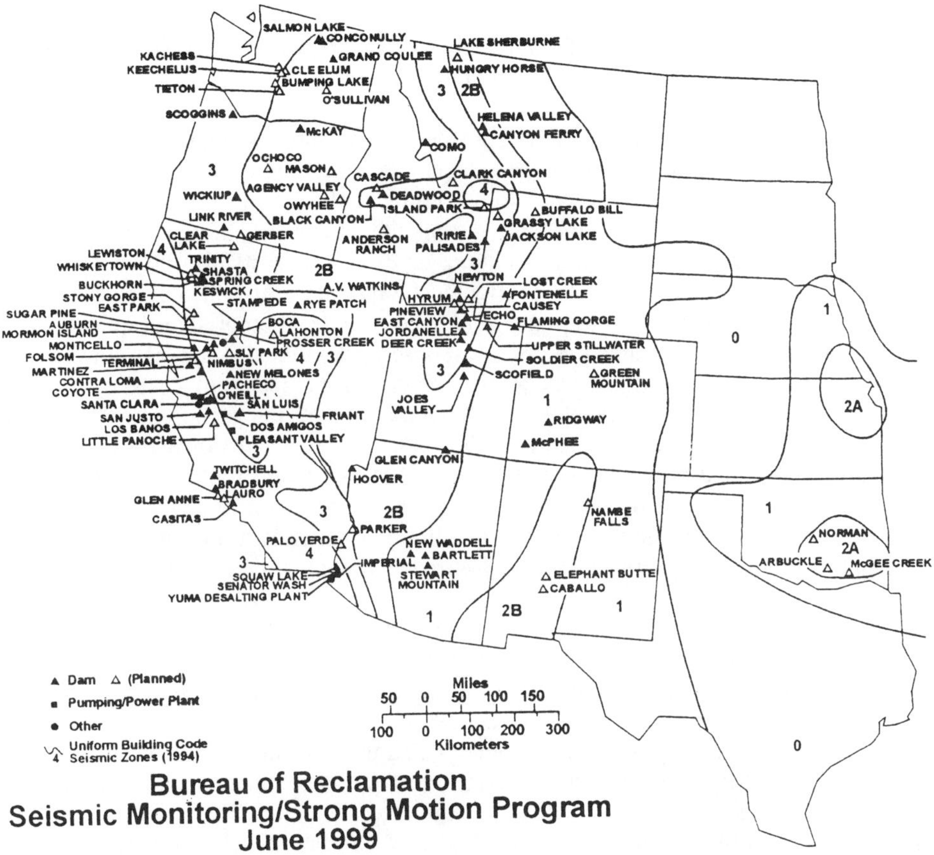

Figure 1. Reclamation lifeline structures with strong-motion instrumentation.

In developing a rapid notification system several studies questions were considered: How to :(1) retrieve the waveform data. (2) adaptively modify acquisition control parameters, and (3) verify that the instrumentation is functioning properly.

Several ways of retrieving the data were considered. The simplest approach for operating a network of remote data loggers is to service them solely through site visits. This technique was deemed inappropriate where truly rapid notification is required. Further, instrumentation failures are detected only during site visits, so data may go unrecorded if the interval between visits is large. In addition, repeated site visits may be very costly. An alternative retrieval technique is to configure the remote sites with continuous or iine-division multiplexed telemetry. Continuous telemetry can be appropriate for quickly issuing alert messages in early-warning and rapid-response applications. Poten- ial drawbacks are the equipment and installation costs of the infiastructure, recurrent costs of leasing or maintaining the telemetry channels, and the reliability of the technology. The system developed by Reclamation is one that provides many of the benefits of real-time telemetry, but at lower cost, and without the need for site visits, beyond routine naintenance. The system uses standard modems and dial-up telephone circuits. It retrieves waveform data from event-triggered seismic data loggers within a few minutes ıf a trigger. The telemetry and data logging hardware can be fully configured and moniored remotely.

Two basic innovations were needed to implement tha system: (1) a low-cost serial communications device that, when triggered, establishes a dial up connection with a remote host, and (2) host communications software to complete the retrieval process once the connection is established. Data telemetry hardware at each field site consists of a standard dial-up telephone line, surge protection, modem, and interface device. As many as five data loggers may be connected through the serial interface device to each modem and phone line. The data loggers and interface may be connected by cable directly to the modem or remotely by spread-spectrum radio links. The first configuration is used when the data loggers are co-located with the modem and phone line, and the second is used to handle small sub-nets of remote data loggers (up to five) that cannot be located at the phone line .

The interface device (called a triggered modem controller, or TMC) attemps to establish a dial-up modem connection with a remote host when one of the attached data logger detects an event. The TMC is triggered by the logical OR of the buffered trigger outputs.

Once connected, the TMC accepts commands from the host to switch the modem serial port to any of the serial ports of the attached data loggers.

We have developed a communications software package that is started up automatically on the host whenever a TMC logs in. It is also used to make outgoing calls needed to re-establish interrupted incoming calls, and for periodic polling of the TMCs. The software runs on UNIX platforms, and was written in ANSI-C. Drivers have been written for several models of commercially-available data loggers, including: (1) Kinemetrics model SSA-1, SSA-2 and K2, (2) Reftek model 72A, and Syscom/Sprengnether model MR2002. Incoming calls are handled by a modem pool using the UNIX login facility; calls from multiple remote sites are handled simultaneously. In addition to accepting incoming calls from triggered TMCs, the software can be run in a polling mode to perform automatic remote configuration and state-of-health monitoring..

If all of the equipment and the phone line are functioning properly, the time interval from event trigger to completion of data retrieval is usually five minutes or less for a single data logger, assuming typical event files of approximately 0.5 Mbyte and 28.8-kbaud connections. This includes the time required to make the connection, log into the host computer, interrogate the data logger, download the event data, and reset the data logger.

Each dam is typically instrumented with one to eight 3-component accelerometers at various positions on and away from the structure. The accelerometers (usually FBA- 23's) may be recorded either on several 3-channel data loggers, or one or more 6- or 12- channel data loggers. Data are usually sampled at a rate of 200 samples per second per channel. The data loggers are usually connected to a single TMC, modem, and dial-up telephone line (schematically shown in Figure 2). We have successfully used both direct wire connections between the TMC and modem, and spread-spectrum radio links with path lengths of 75 km or more. The system has also been successfully used for temporary site-response and aftershock studies to provide remote access to seismic data loggers with broadband seismometers.

The implementation of the Strong Motion Notification System began in 1990 with the deployment of digital accelerographs at two large embankment dams in California, Casitas Dam near Ventura and San Justo Dam near Fresno. Both systems were based on commercially available accelerographs that were linked via telephone lines to an early version of the data retrieval and analysis system described above. From 1991 through March, 1999, 47 additional sites were instrumented, for a total of 49 sites to date. Currently it is possible to transmit data to a number of different users (e.g., designers, emergency action personnel, inspectors) within minutes of strong shaking at a structure monitored by the Strong Motion Notification System. These data can consist of peak parameters such as peak acceleration or velocity, response spectra, derived values such as spectrum or Arias Intensity, or other critical information regarding the response of the structure. If predetermined site-specific thresholds (as defined by the Performance Parameters process) are exceeded, the Strong Motion Notification System automatically transmits a Seismic Alarm Message to the appropriate predetermined regional/area/district office via voice message, fax, and/or LAN providing all pertinent information. This rapid availability of data is in contrast to the several-day to several-week response that has been inherent in retrieving and processing records from existing analog strong motion instrumentation.

3. Acquisition of Strong Motion Recordings

In addition to the short term goal of providing rapid notification in the event of earthquake shaking, the long term goal of acquiring high-quality ground motion data continues to be an important requirement for the engineering community, as a whole, and Reclamation in particular. The Bureau of Reclamation operates or has dam safety responsibility for more than 350 dams in the western United States. Evaluation of the safety of those structures for

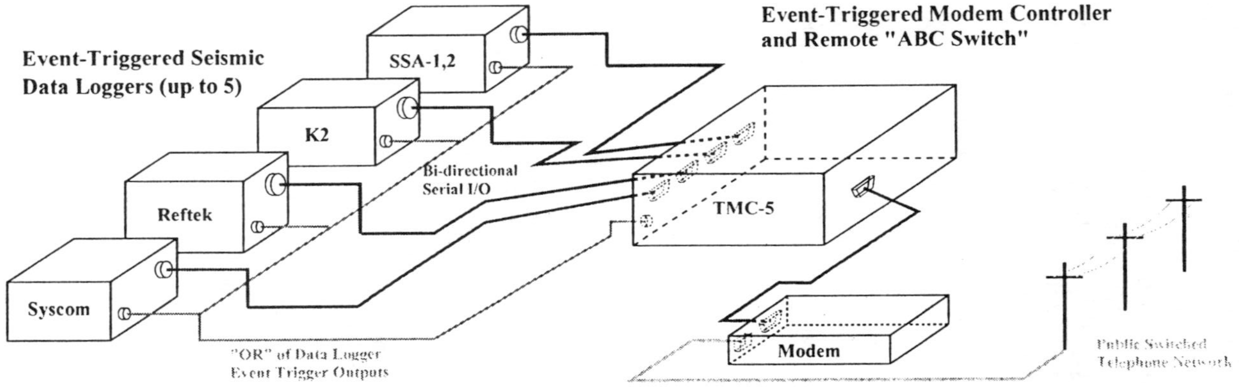

Figure 2. Typical field-site configuration using a Triggered Modem Controller (TMC), event-triggered seismic data loggers, modem and dial-up connection to the Public Switched Telephone Network. An event trigger from any of the data loggers causes the TMC to actively attempt to connect to a remote host computer by dialing up and logging in. If the connection attempt fails, the TMC retries the call after a delay, initially a few seconds, which is doubled following each failed attempt. Once the connection is established, the TMC becomes passive, accepting commands from the remote host software to switch between data loggers. The remote host interrogates each data logger, checks on its state-of-health, retrieves any event files, deletes successfully transferred events, and makes any scheduled parameter changes. Typical event files of less than 0.5 Mbyte can usually be retrieved within three to five minutes. Several models of commercially-available seismic data loggers are supported.

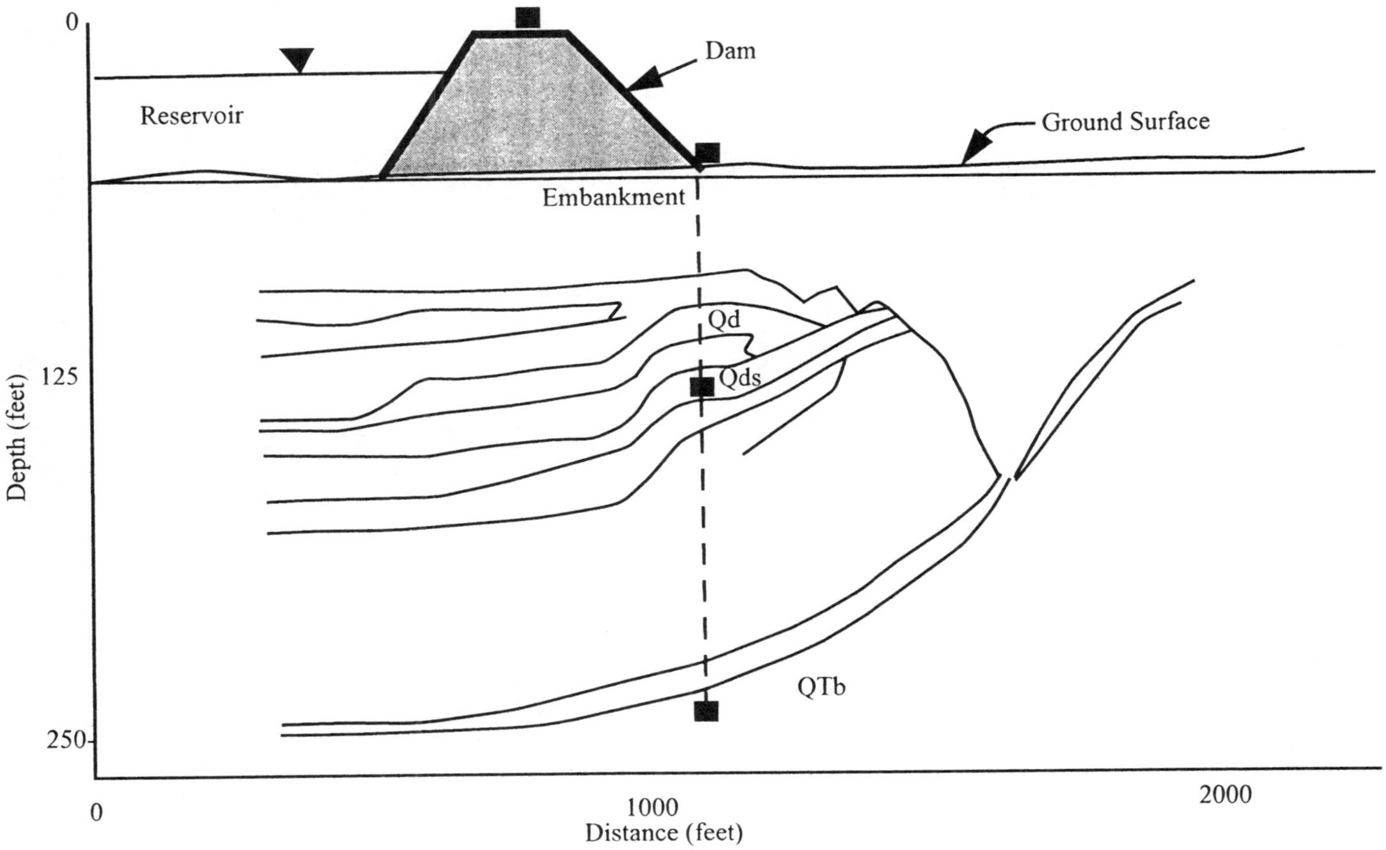

Figure 3. Schematic cross-section of strong-motion array, Wickiup Dam, Oregon. Three-component FBA-23 packages indicated by black squares. Units Qd and Qds are identified as potentially liquefiable, unit QTb is basaltic bedrock. Data logger located on dam crest.

earthquake loading is a high priority. To determine with confidence if a particular structure is capable of withstanding significant ground shaking requires knowing both geological/engineering properties of the site and specifics of ground shaking expected at that site. The details of what types of ground motion data are necessary to acquire for both site specific evaluation and future analyses depends in part on the type of dam under consideration. Elements of the system described above have provided remote access to seismic data loggers used with both accelerometers and broad-band seismometers for short-term, site-specific studies.

For embankment dams, liquefaction of foundation materials and/or the embankment continues to be a major safety issue. Evaluation of the likelihood of liquefaction occurrence at a given site depends on a number of factors. The impedance contrast between the soil and underlying basement rock, the potential for nonlinear behavior in the soil, frequency dependent scattering or exaggerated coherency of ground motions all influence the likelihood of liquefaction. Strong motion recording can help address all of these issues. Data acquired over a broad magnitude range can be used to determine weak/ strong motion amplification ratios (Su and others, 1998). This data can then be used to help assess the potential for nonlinear soil behavior at sites or depths of interest (Su and others, 1998; Hartzell, 1998). Instruments deployed in a horizontal array configuration can be used to help understand the spatial coherency of the ground motion wavefield (Abrahamson, 1988). Several recent investigations have focused on acquiring strong ground motion data in vertical arrays (Steidl, 1993; Steidel and others, 1996). At a site in Oregon, Reclamation is currently attempting to instrument specific soil layers that have been identified as potentially liquefiable in preliminary engineering evaluations using a vertical strong motion array (see Figure 3).

Presently, tew empirical data are available that provide a precise analog in terms of desired source-site geometry and site response for specific seismic safety evaluations. As a result, numerical simulations are often required for the evaluation of high-risk structures. Fault rupture simulations using one-dimensional Green functions are commonly used to synthesize near-field ground motions for dam safety evaluations. However, the premise that deterministic one-dimensional Green functions provide realistic amplitude and phase responses over an adequate bandwidth is not supported by recent research (O'Connell, 1999a). By obtaining broadband, on-scale recordings of smaller earthquakes at structures of interest, site-specific scattering/response functions can be developed. These functions can then be convolved with one-dimensional Green functions to reproduce observed broadband responses (O'Connell and Ake, 1995; O'Con- nell, 1999). Validation of this technique with observed ground motions from the M 6.7 1994 Northridge earthquake suggests neglecting site-specific scattering leads to unrealistic broadband enhanced direct S-wave coherency and predictions of physically implausible accelerations and velocities(O'Connell, 199b). For some scenario earthquakes this overprediction may be as much as 40% (Figure 4).

Factors that can influence the results of dynamic stuctural evaluations for concrete dams include spatial coherency of ground motions and effective impedance between dam and foundation. Evaluation of both of these factors requires an understanding of site response for the concrete structure in question. The dynamic interaction of foundation and structures

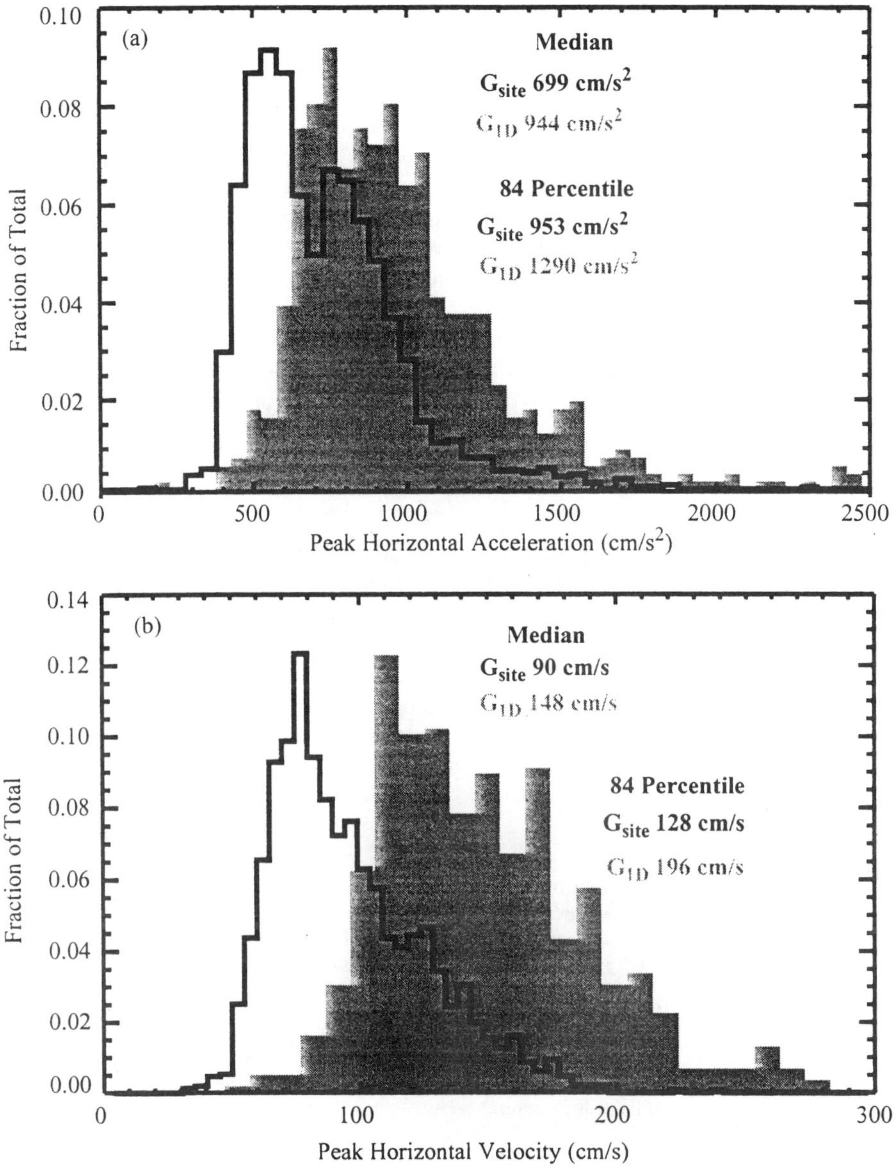

Figure 4. Comparison of site-specific Green function (G_{site}) (solid line histograms) and deterministic (G_{1D}) (shaded histograms) synthetic ground motion PHA (a) and PHV (b) distributions from 640 rupture simulations of a **M** 6.9 on the Red Mountain Fault at Casitas Dam, California. (From O'Connell, 1999b).

during earthquake shaking is critical to both the design of new structures and the analysis of existing structures (Erdik, 1987). In classical finite-element analysis, the equations of motion for a finite-element region are applied to the structure of interest and a portion of the foundation, the mass of the unbounded foundation is often omitted. During the time of simulation, energy must be removed from the structure as additional portions of the mesh are displaced, thus consuming energy (referred to as radiation damping). Neglecting radiation damping can lead to seriously erroneous results. Using material properties and ground motions recorded at several points on and near the structure, powerful finite-element methods (e.g. Wolf and Song, 1996) can be used to assess the degree of ground motion amplitude reduction as a fimction of frequency as a result of radiation damping (O'Connell, 1998). For two and three-dimensional finite-element models, the degree of spatial coherence assumed for input ground motions will have a profound impact on the results. The appropriateness of this assumption at a particular site can be evaluated using strong-motion recordings acquired at several locations on or near the structure of interest.

4. Conclusions

The Strong Motion Program of the Bureau of Reclamation is currently focused on providing near real-time notification of the occurrence of strong shaking at our structures. This process is integrated with the development of structure specific Performance Parameters and emergency action planning. In addition to the short term goal of providing rapid notification, the program is attempting to focus on the long term goal of instrumenting sites with the intention of acquiring data that can be added to the world wide data base of strong motion recordings, as well as being used for structare specific analyses.

5. References

1. Abrahamson, N. A., 1988, Statistical properties of peak ground acceleration recorded by the SMART 1 array, Bulletin of the Seismological Society of America, v. 78, pp. 26-41.
2. Brown, C. A., and W. J. Graham, 1988, Assessing the threat to life from dam failure, Water Resources Bulletin, American Water Resources Association, v. 24, No. 6, pg 1303-1309.
3. DeKay, M. L., and G. H. McClelland, 1993, Predicting life loss in cases of dam failure and flash flood, Risk Analysis, v. 13, pp. 193-205.
4. Erdik, M., 1987, Soil stnicture interaction effects on strong ground motion, *in* Strong Ground Motion Seismology, M. O. Erdik and M. N. Toksoz eds., D. Reidel Publishing Co., Dordrecht, Holland, pp. 559-580.
5. Hartzell, S., 1998, Variability in nonlinear sediment response during the 1994 Northridge, California earthquake, Bulletin of the Seismological Society of America, v. 88, No. 6, pp. 1426-1437.
6. O'Connell, D. R. H., and J. P. Ake, 1995, Ground motion analysis for Hoover Dam, Boulder Canyon Project, Arizona-Nevada, Seismotectonic Report No. 94-1, U. S. Bureau of Reclamation, Technical Service Center, Denver, CO,
7. O'Connell, D. R. H., 1998, Evaluation of radiation damping for Hoover Dam, Boulder Canyon Project, Arizona, Nevada, Technical Memorandum No. D8330-98-007, U. S. Bureau of Reclamation, Denver, CO, 15 pg.
8. O'Connell, D. R. H., 1999a, Replication of apparent nonlinear seismic response with linear wave propagation models, Science, v. 283, pp. 245-250.
9. O'Connell, D. R. H., 1999b, Ground motion evaluation for Casitas Dam, Ventura River Project, California, Seismotectonic Report No. 99-4, U.S. Bureau of Reclamation, Den- ver, CO, 117 pg.

10. Steidl, J. H., 1993, Variation of site response at the UCSB dense array of portable accel- erometers, Earthquake Spectra, v. 9, pp.289-302.
11. Steidl, J. H., A. G. Tumarikin, and R. J. Archuleta, 1996, What is a reference site?, Bul- letin of the Seismological Society of America, v 86, No. 6, pp. 1733-1748.
12. Wolf, J. P., and C. Song, 1996, Finite-element modeling of unbounded media, John Wiley and Sons, Chichester, England, 331 pg.

STRUCTURAL VIBRATION MONITORING SYSTEM FOR THE BOSPORUS SUSPENSION BRIDGES

NURDAN APAYDIN, MUSTAFA ERDİK
Bogazici University, Kandilli Observatory and Earthquake Research Institute, Department of Earthquake Engineering, Cengelkoy, 81220, Istanbul, Turkey
e-mail: napaydin@boun.edu.tr, erdik@boun.edu.tr

Abstract

The study will describe the structural health monitoring system installed on the Bosporus Suspension Bridge. The bridge, commissioned in 1973, has a main span of 1074 m. The bridge deck has a total width of 33.4m. The deck is supported by four steel towers of 165 m height. The ambient vibrations of the Bosporus Bridge has been investigated. For the structural health monitoring system, the following sensors are installed in the system: One wind speed transducer at the mid-span; Four vertical, three transverse and two longitudinal acceleration transducers at the mid-span, one-fourth of the span and at the top of Beylerbeyi tower; Four seismometers at the foundation of towers and cable anchorages; Two relative displacement transducers at the Ortaköy joint and; One optical displacement transducer at the mid-span. The response of these transducers, digitized at the source is transferred to central recording/monitoring system located at the control building. The triggering is based on the exceedance of the appropriately set threshold levels.

A new vibration monitoring system has been designed for Fatih Sultan Mehmet Suspension Bridge. Various techniques will be studied to optimally place 32 triaxial solid-state acceleration monitoring sensors. The data, digitized at the source will be monitored by two parallel PC systems operating under the "SEISLOG" data acquisition software, developed by University of Bergen. Damage to bridge will be inferred from the response of distributed sensors (vibrational data) and the analytical modeling of the structure through the use of system identification techniques. The structural models or properties (generally modal parameters) will be systematically determined by using the measured response and, if available, the corresponding excitation (ambient or transient) of the bridge through an integrated approach of experimental, analytical and computational techniques.

M. Erdik et al. (eds.), Strong Motion Instrumentation for Civil Engineering Structures, 343–367.

1- Introduction

Structural health monitoring can be defined as the diagnostic monitoring of the integrity or condition of a structure. The intent is to detect and locate damage or degradation in structural components and to provide this information quickly and in a form easily understood by the operators or occupants of the structure. The damage may result from fatigue, large earthquake, strong winds, explosion or vehicle impact. Early detection of damage or structural degradation prior to local failure can prevent "runaway" catastrophic failure of the system. In the event of an earthquake, the rapid detection of catastrophic failure of a strategic bridge can allow early and effective response by highway officials in rerouting traffic and defusing an otherwise difficult situation. The large physical sizes of the bridges necessitates extensive array of different sensors and appropriate technologies for data acquisition / reduction for rational health monitoring applications. The system should be able to automatically detect, locate and assess structural damage anywhere within the bridge system (health monitoring), and to communicate the status (alerting) to responsible authorities. At its simplest application, the threshold exceedance anywhere in the bridges is monitored and an alarm is provided if pre-set threshold levels have been exceeded. This alarm may be used to automatically divert the traffic if it indicates damage due to earthquake, explosion or impact.

1.1 PROPERTIES OF FIRST AND SECOND SUSPENSION BRIDGES:

The bridges link two parts of Turkey and also joins together the European and Asian parts of Istanbul

1.1.1 First Boğaziçi Suspension Bridge:

- The bridge was commissioned on October 29,1973.
- The main span is 1074m long.
- The bridge deck has a total width of 33m and carries six lanes of traffic
- The deck is supported by four steel towers of 165m height.

Figure1. The view of First Boğaziçi Suspension Bridge

The overall Structural parameters as follows;

Axial distribution of the dead load mains:

Deck	: 11000 kg/m
Cables(both):	3850 kg/m
Suspenders	: 120 kg/m
TOTAL	: 14970 kg/m

Area moment of inertia of the deck:

I_{xx}	: 1,3 m^4
I_{yy}	: 63,4 m^4

Torsional constant of the deck (k=3,4 m^4)
Nominal radius of the cables : r= 0.28 m
Modulus of elasticity (Steel) : E= $2.0x10^{11}$ N/m^2
Shear modulus of elasticity : G= $0.8x10^{11}$ N/m^2

1.1.2 Fatih Sultan Mehmet Suspension Bridge (Second Boğaziçi Suspension Bridge) :

- The bridge was commissioned in 1989
- The main span is 1090m long
- The bridge deck has a total width of 39,4m and carries eight lanes of traffic
- The deck is supported by four steel towers of 107.10m height

Figure2. The view of Fatih Sultan Mehmet Bridge

The overall Structural parameters as follows;

Axial distribution of the dead load mains:

Deck	: 12 074 kg/m
Cables(both)	: 6 194 kg/m
Suspenders	: 328 kg/m
TOTAL	: 18 569 kg/m

Area moment of inertia of the deck:

I_{xx}	: 1,73 m^4
I_{yy}	: 129 m^4

Torsional constant of the deck (k=4,754 m^4)
Nominal radius of the cables : r=0,34 m
Modulus of elasticity (Steel) : E= $2.0x10^{11}$ N/m^2
Shear modulus of elasticity : G= $0.8x10^{11}$ N/m^2

2-Ambient Vibration Tests On Suspension Bridges

Physical and analytical understanding of the dynamic response of a suspension bridge is very dependent on the knowledge of its natural frequencies, damping and normal mode shapes of vibration. Performing actual tests on full-scale structures is the only sure way of assessing the dynamic parameters. Dynamic parameters of a suspension bridge can be determined by means of Ambient Vibration Test (AVS). Bridge's motion is caused by traffic, wind and other environmental factors. Several Applications related to these tests have been reported.

Some of them can be listed as below:

- "Ambient Vibration of two suspension Bridges", Mclamore et. al. 1971

Test had been performed on the Newport and the William Preston Lane Memorial Bridge on August 1969
20 Modes of Vibration in the frequency range 0-60cpm
13 Modes of Vibration in the frequency range 0-60cpm

- "Ambient Vibration Tests of Suspension Bridge", Abdel-Ghaffar et. al. 1978

Test had been applied to the Vincent-Thomas Suspension bridge.
37 modes of vibration in the frequency range 0-3Hz.

- "Ambient Vibration Studies of Golden Gate:Suspended Structures", Abdel-Ghaffar et.al, 1984, Experiment had been conducted on the Golden Gate Bridge
91 modes shapes in the frequency range 0-1,5 Hz.

- "Ambient Vibration Studies of Golden Gate Bridge:II.Pier-Tower Structure", Abdel-Ghaffar et.al, 1984
46 modes shapes in the frequency range 0-5 Hz.

- "Ambient Vibration testing on the Humber Suspension Bridge", Browjohn et.al, 1986
41 modes of vibration in the frequency range 0-1,56Hz
67 modes of vibration in the frequency range 0-2Hz

- "Ambient Vibration Investigation of the Boğaziçi Suspension Bridge", Erdik et.al, 1989
5 experiment series, several measurements of the deck had been obtained at different locations and orientations at different traffic condition.Frequency range: 0,07-1,0Hz

- "Ambient Vibration Survey of the Bosphorus Suspension Bridge", Browjohn et.al, 1989
62 modes of vibration in the frequency range 0,1-1Hz

3- Ambient Vibration Surveys Of The Bosporus Suspension Bridge In Istanbul

3.1. AMBIENT VIBRATION SURVEY I

One of the studies was performed by Erdik and Uçkan in 1989. (Report No : 89-5, Department of Earthquake Engineering, Kandilli Observatory and Earthquake Research Institute, Boğaziçi University, Istanbul, Turkey, November, 1989) In this report, the results of the ambient vibration survey of the Bosporus Suspension Bridge in İstanbul were presented. The measurement campaign encompassed measurements along and across the main spans and both inside and outside of the deck and four series of measurements were analyzed These are respectively called Experiment Series 1, 2, 3, 4 (Fig.3,4,5,6) In each experiment series several measurements of the ambient vibration of the bridge deck were obtained: (1) at different locations and orientations, (2) at different wind and traffic conditions and (3) using accelerometers and seismometers. The results of these experiments have been analyses in terms of the peak frequencies. Some of these experimentally obtained mode shapes have been compared with the theoretical mode shapes obtained on the basis of the three-dimensional finite element analysis.

Theoretical analysis of the free vibration of suspension bridge indicate that the modes of structure can be divided into three groups with respects to the dominance of the respective displacements. In the first group the displacements of the deck is important (Cable Modes) and in the third group the displacements of the towers become important (Tower Modes). The deck modes can also be Classified as symmetric and asymmetric or as lateral, vertical and torsionally dominant. (Fig.7,8,9,10)

Theoretical 2D and 3D finite element analysis have produced results that correlate surprisingly well with the experimental data considering the crude assumptions made regarding the boundary conditions, equivalent shell elements used to model the deck and the equivalent frame elements used to model the cab. The experimental data were limited to the deck modes only.

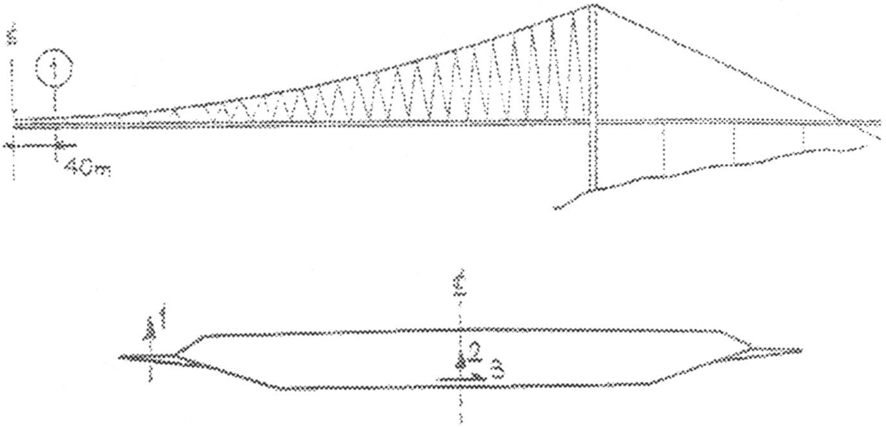

Traffic Condition : Very light
Section Location : At 40 m's from towards Beylerbeyi

Figure 3. Experiment Series 1

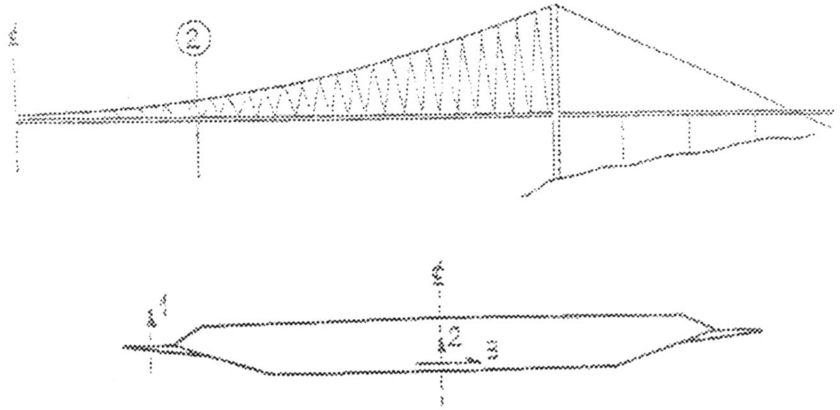

Traffic Condition : light
Section Location : At 179 m's from towards Beylerbeyi

Figure 4. Experiment Series 2

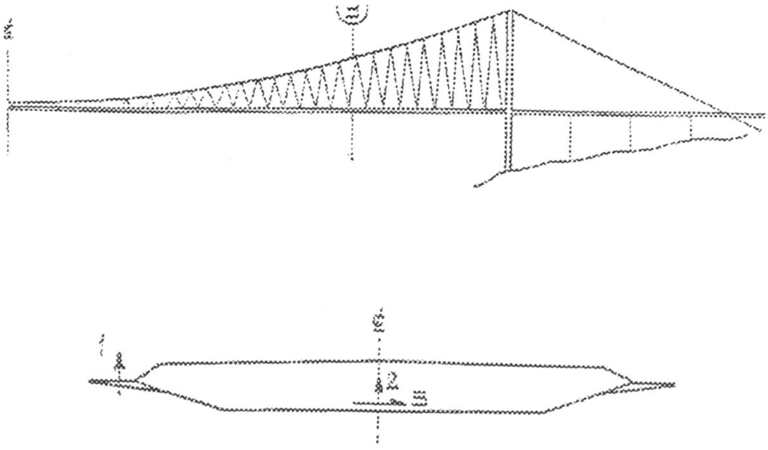

Traffic Condition : light
Section Location : At 1/3 of the main span from Beylerbeyi

Figure 5. Experiment Series 3

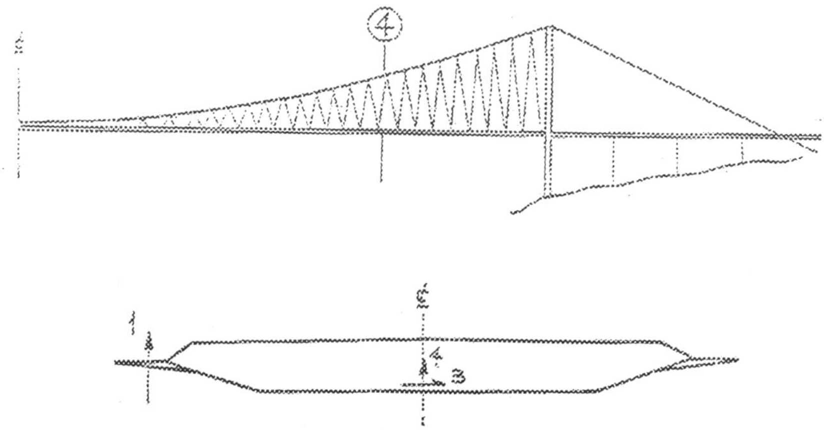

Traffic Condition : Normal
Section Location : At 1/3 of the main span from Beylerbeyi

Figure 6. Experiment Series 4

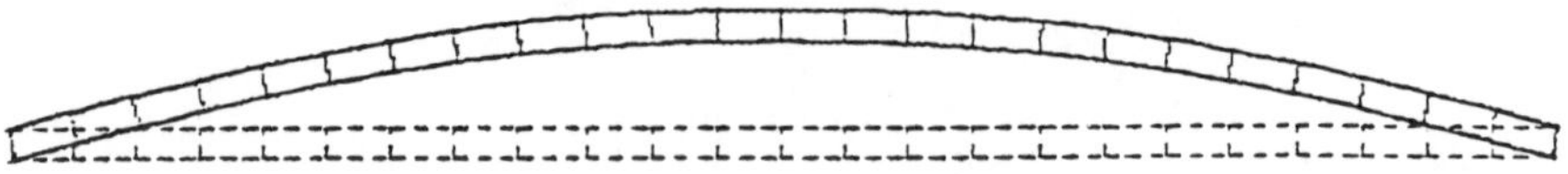

Theoretical

Experimental

Theoretical Frequency : 0.07 Hz.
Experimental Frequency : 0.07 Hz
Experimental Damping Ratio: 7 %

Figure 7. Comparison of the Theoretical and Experimental First Lateral Symmetric Mode Shape

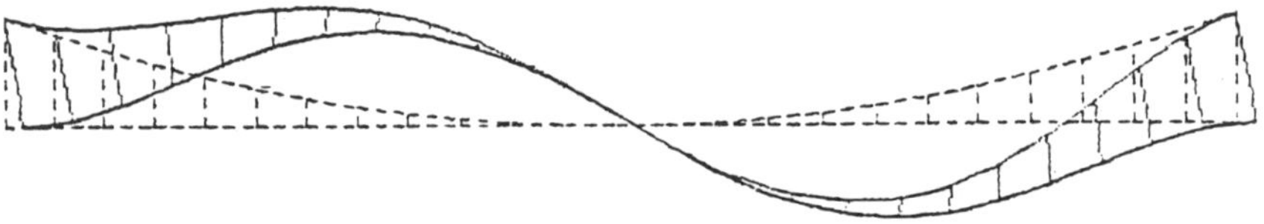

Theoretical

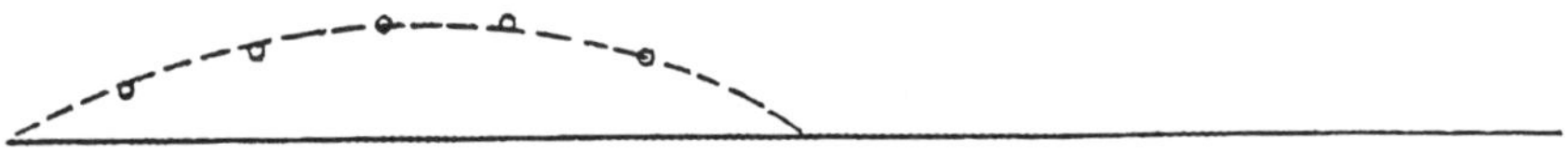

Experimental

Theoretical Frequency : 0.13 Hz.
Experimental Frequency : 0.13 Hz
Experimental Damping Ratio: 4 %

Figure 8. Comparison of the Theoretical and Experimental First Vertical Asymmetric Mode Shape

Theoretical

Experimental

Theoretical Frequency : 0.18 Hz.
Experimental Frequency : 0.17 Hz
Experimental Damping Ratio: 4 %

Figure 9. Comparison of the Theoretical and Experimental Second Vertical Symmetric Mode Shape

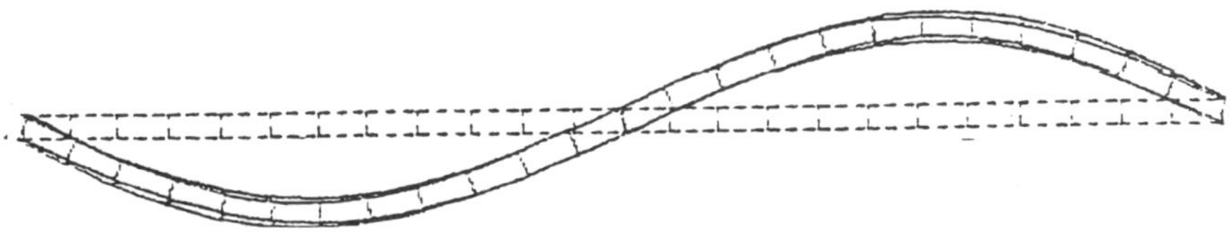

Theoretical

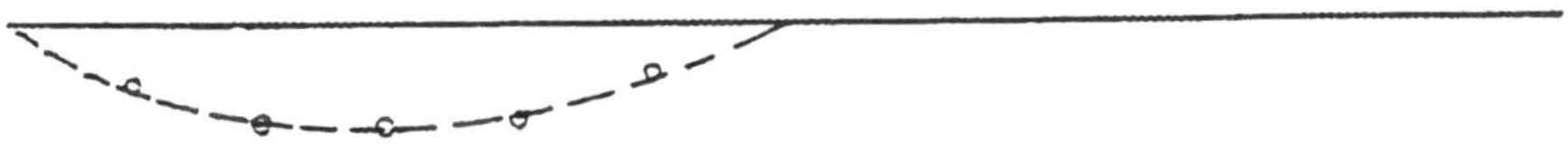

Experimental

Theoretical Frequency : 0.23 Hz.
Experimental Frequency : 0.22 Hz
Experimental Damping Ratio: 4 %

Figure 10. Comparison of the Theoretical and Experimental First lateral Asymmetric Mode Shape

3.2. AMBIENT VIBRATION SURVEY II

The other was performed by Brownjohn et.al (1988), (W. Brownjohn, A. A Dumanoglu, R T Severn, A. Blakeborough , Report No : UBCE-EE-88-1, Department of Civil Engineering, University of Bristol, January, 1988). In this study, ambient vibrations were measured on the main span and towers of the Boğaziçi Suspension Bridge and the experimentally obtained frequencies and mode shapes were compared with theoretical results. Vertical, torsional, lateral and longitudinal vibrations of the deck and one of the towers were measured. In general computed and measured natural frequencies and mode shapes are in close agreement. The measurements from the tower have shown that all the modes except weak mode occured at the same frequencies as main span vertical modes. No comparisons were attempted for the towers alone. Authors pointed out that for earthquake and aerodynamic response analyses the mathematical models can be used with confidence to represent the real structure.

3.3 COMPARISON OF RESULTS

Comparison between experimental and analytical studies of the Boğaziçi Suspension Bridge conducted by Erdik et. al. and by Brownjohn et al. are expressed below.

Table 1. Experimental and Analytical Modal Vibration Frequencies

	ERDİK ET. AL.		BROWNJOHN ET. AL.	
Mode Shape	Experimental Frequency	Analytical Frequency	Experimental Frequency	Analytical Frequency
1 st Lateral Symmetric	0.072 Hz.	0.0716 Hz.	0.070 Hz.	0.073 Hz.
1 st Vertical Asymmetric	0.150 Hz.	0.144 Hz.	0.160 Hz.	0.165 Hz.
1 st Lateral Asymmetric	0.215 Hz.	0.202 Hz.	0.209 Hz.	0.218 Hz.
2 st Vertical symmetrical	0.230 Hz.	0.225 Hz.	0.217 Hz.	0.225 Hz.
1 st Torsional Symmetric	0.320 Hz.	0.323 Hz.	0.324 Hz.	0.278 Hz.
2 nd vertical Asymmetric	0.330 Hz.	0.340 Hz.	0.324 Hz.	-
3 rd Vertical Symmetric	0.370 Hz.	0.365 Hz.	0.362 Hz.	0.372 Hz.
2 nd Lateral Symmetric	0.380 Hz.	0.385 Hz.	0.382 Hz.	0.437 Hz.

4- The Structural Health Monitoring System Of Boğaziçi Suspension Bridge

A structural health monitoring system was installed at Boğaziçi Suspension Bridge in October 1993.
The system consists of three subsystem;
The sensor subsystem,
The Data acquisition subsystem,
The Data recording/monitoring subsystem.
The system was provided by the government of Japan by the request of government of Turkey.
The purpose of the system for Boğaziçi Bridge is to continuously monitor structural motion and related environmental data and to alert the engineers if any parameters give the signals for the maintenance and construction.

The sensor subsystem consists of 28 channels of sensors. Wind speed and direction meter 1 set 2 signals, Accelerometer 10 set 10 signals, Seismometer 4 set 12 signals, Mechanical displacementmeter 2 set 2 signals, Optical displacementmeter1 set 2 signals. Total: 28 signals. One wind speed transducer at the mid-span; Four vertical, three transverse and two longitudinal acceleration transducers at the mid-span, one fourth of the span and at the top o Beylerbeyi tower; Four seismometers at the foundation of towers and cable anchorages; Two relative displacement transducers at the Ortaköy joint and; One optical displacement transducer at the mid-span (Fig.11) . The response of these transducers, digitized at the source is transferred to central recording/monitoring system located at the control building. The triggering is based on the exceedance of the appropriately set threshold levels.

Unit time for Recording :5 min.

Treshold Valve for Anomaly data saving:
Max wind velocity :20 m/sec
Max seismic acceleration : 4.9 m/sec^2
Analysis treshold value : 2.0 m/sec^2

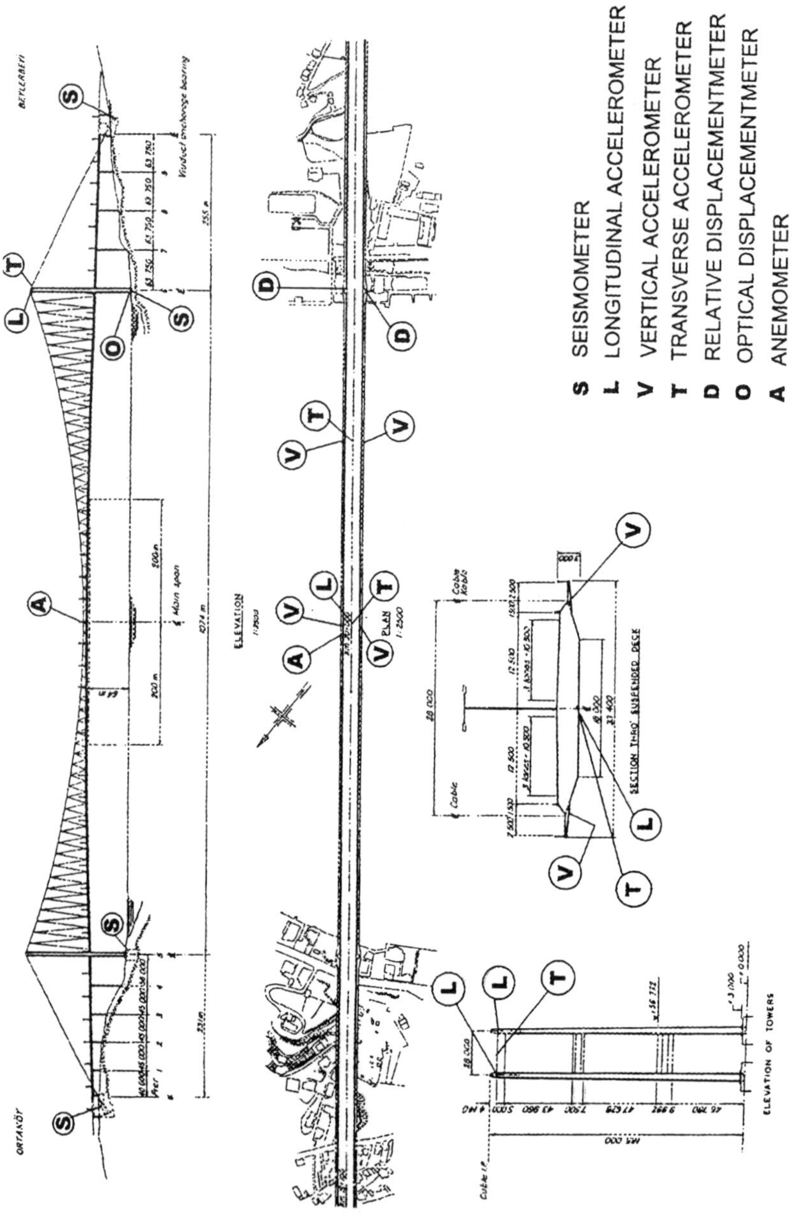

Figure 11 .Boğaziçi Bridge Sensors Position

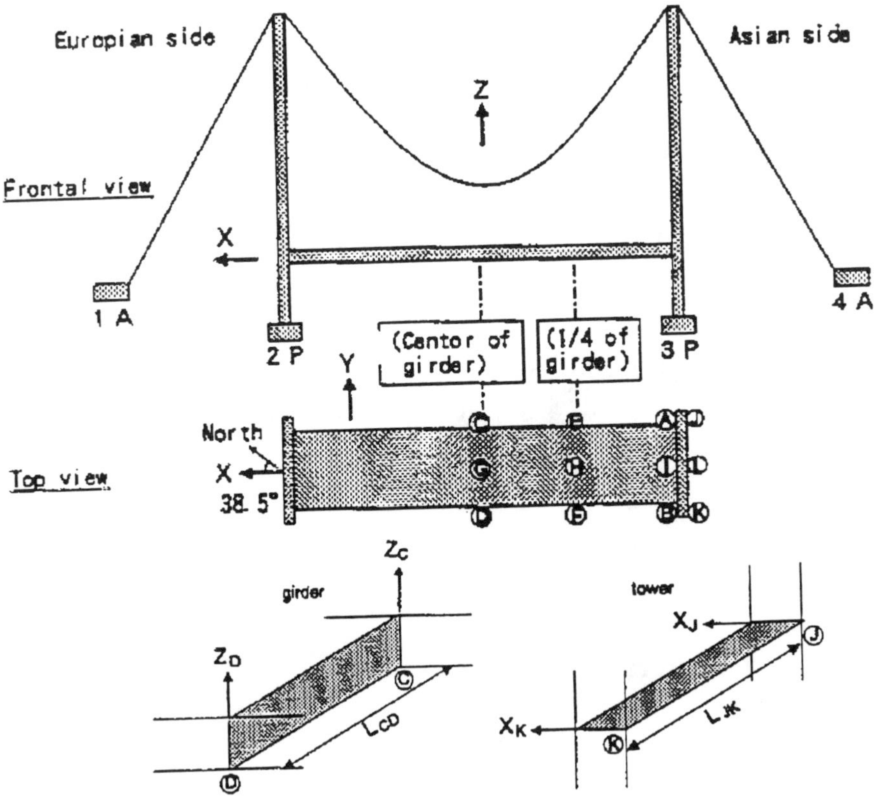

Bend of the girder

$$(Z_D + Z_C)/2$$

Twist of the girder

$$(Z_D - Z_C)/L_{CD}$$

Z_C : vertical displacement at point C

Z_D : vertical displacement at point D

Bend of the tower

$$(X_K + X_J)/2$$

Twist of the tower

$$(X_K - X_J)/L_{JK}$$

X_J : lateral displacement at point J

X_K : lateral displacement at point K

Figure 12 Definition of Measurement Points

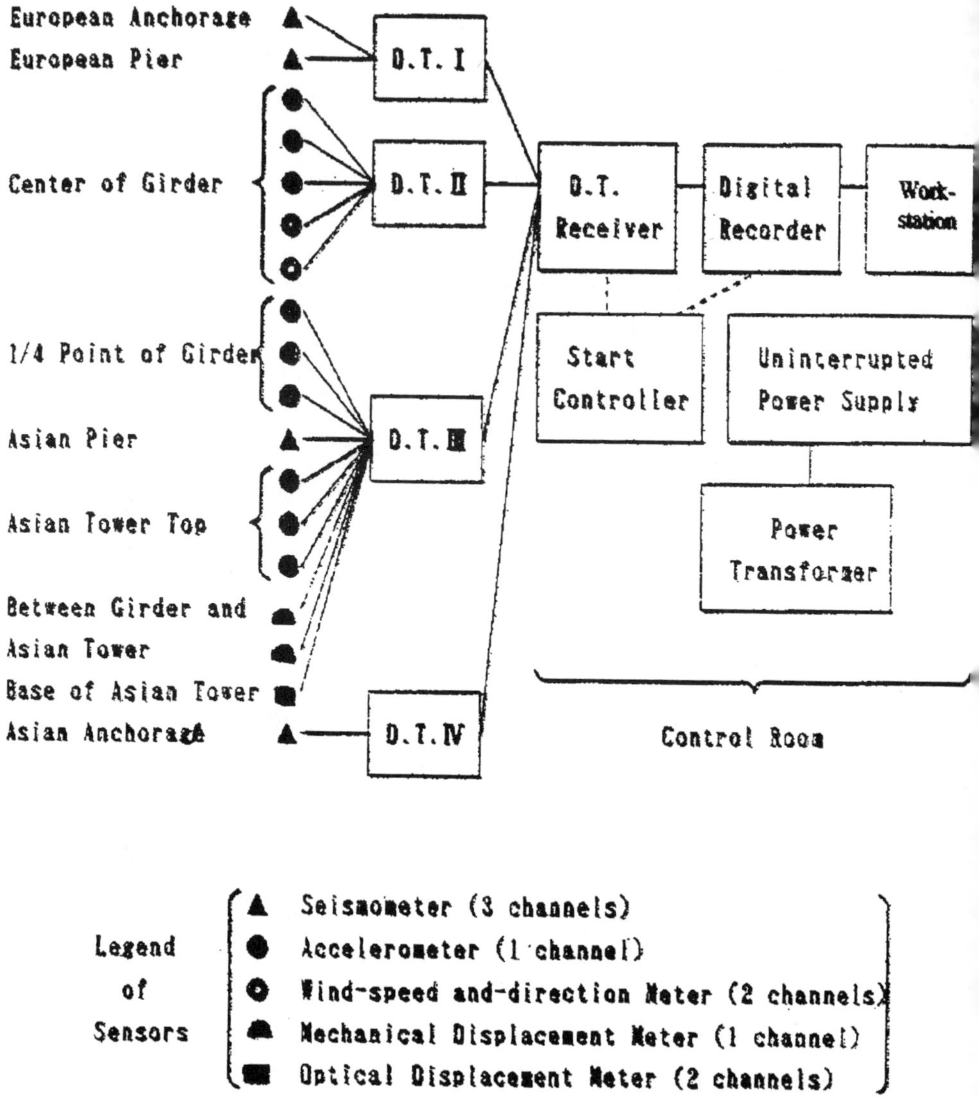

Figure 13 Composition of the Bridge Monitoring System

The Data acquisition subsystem installed on the bridge are capable of continuously digitizing and transmitting all signals to the central recording/monitoring system located at the control building all Sensor Signals. The Data acquisition subsystem consists of amplifiers and start controller. Each sensor has their own amplifier to amplify a signal to a appropriate certain level for transmitting.(Fig.13)

The Data recording/monitoring subsystem contains CRT display, workstation, external hard disk drive (HDD),external CD-ROM drive, external DAT drive, external MO drive, A/D converter SCSI terminator ,BNC connector panel, laser printer, serial printer, the power unit . Data acquisition software is to acquire and store measurement data continuously and to perform instantaneous analysis. Data acquisition process performs periodical (every hour) save, anomaly (hard wind or earthquake) save.

Figure 14 **Wind Speed and Directionmeter**

Figure 15 Seismometer

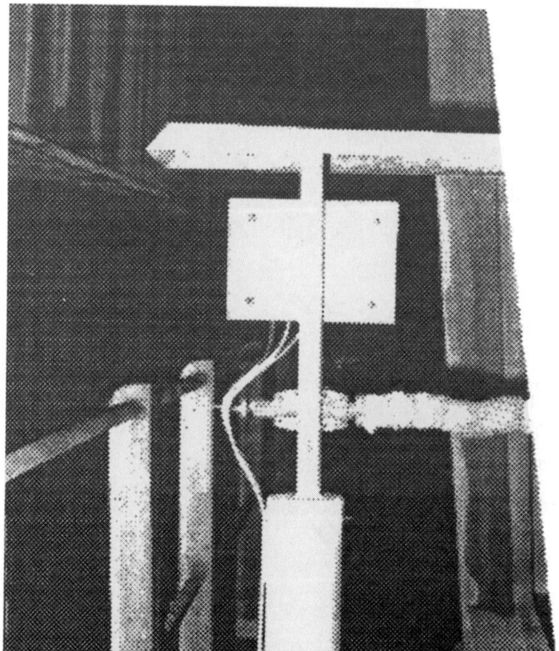

Figure 16 Mechanical Displacementmeter

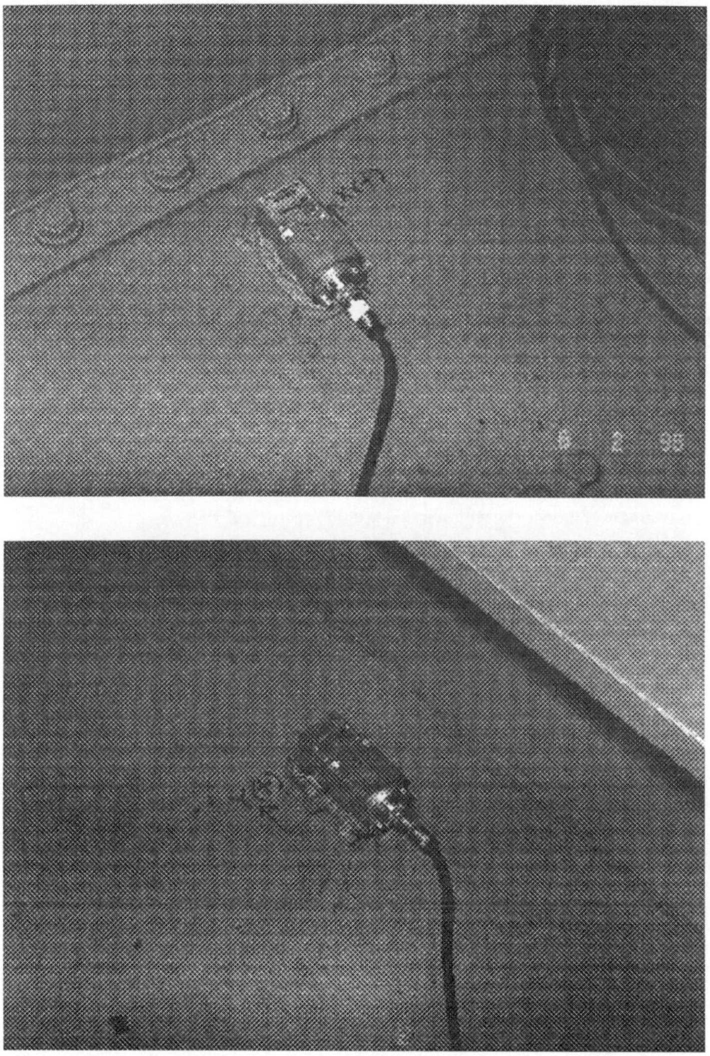

Figure 17 Accelerometers

Figure 18 Optical Displacementmeter

Figure 19 The view of The Bridge Optical Displacementmeter Point

Existing health monitoring system will be updated on the first Boğaziçi Bridge. It is planned to add the some new instruments to the system Additional Instruments are GPS (Global Positioning Systems) and acceleration transducers. GPS units will be used to monitor the bridge geometry. The central recording/monitoring system located at the control building is replaced by a PC based system.(Fig.20)

5. The Structural Health Monitoring System Of Fatih Sultan Mehmet Suspension Bridge

A new vibration monitoring system encompassing the latest technology is being designed for Fatih Sultan Mehmet Bridge.(Fig21)

12 acceleration transducers with x, y, z channels will be instrumented.The data, digitized at the source will be monitored by two parallel PC system operating under the "SEISLOG" data acquisition system, developed by University of Bergen.The system can record 1-64 channels with 12-24 bits resolution an is based entirely on off the shelf hardware. The system has two different analog to digital converter boards.A typical systems consists of a standard PC with MS-DOS and a real time operating System, QNX. MS-DOS with Windows may coexist with QNX on a separate disk partition.

Sensors:

Seismometers: Natural frequency: 0,5-1Hz.
Component: Lateral, vertical water proof
Signal Processor: Analog records will be modified by signal processor
Accelerometers: All accelerometers have a trigger criteria.
Range: +/- 0,59
Frequency range: 0-50Hz.
Natural Frequency: 600Hz.
Damping: 70%
Dynamic range: greater than 200sps
Input range: +/-2,5V
Filter: 50Hz Butterworth filter
Pre-event memory: 15 sec @ 200sps
Post-event memory: 10-90sec @200sps
Displacementmeter: between main span and mid span movement of expansion joint
Capacity: 50cm, Resolution: 0,01cm
GPS (Global Positioning System) : position updates: up to 10/sec
position latency: 30msec
Data acquisition/recording/monitoring system: Two parallel PC System
Analog/Digital converter
1-128 channels
12-24 bits resolution
200kHz sampling rate
Online-offline analysis
SEISLOG Data acquisition system

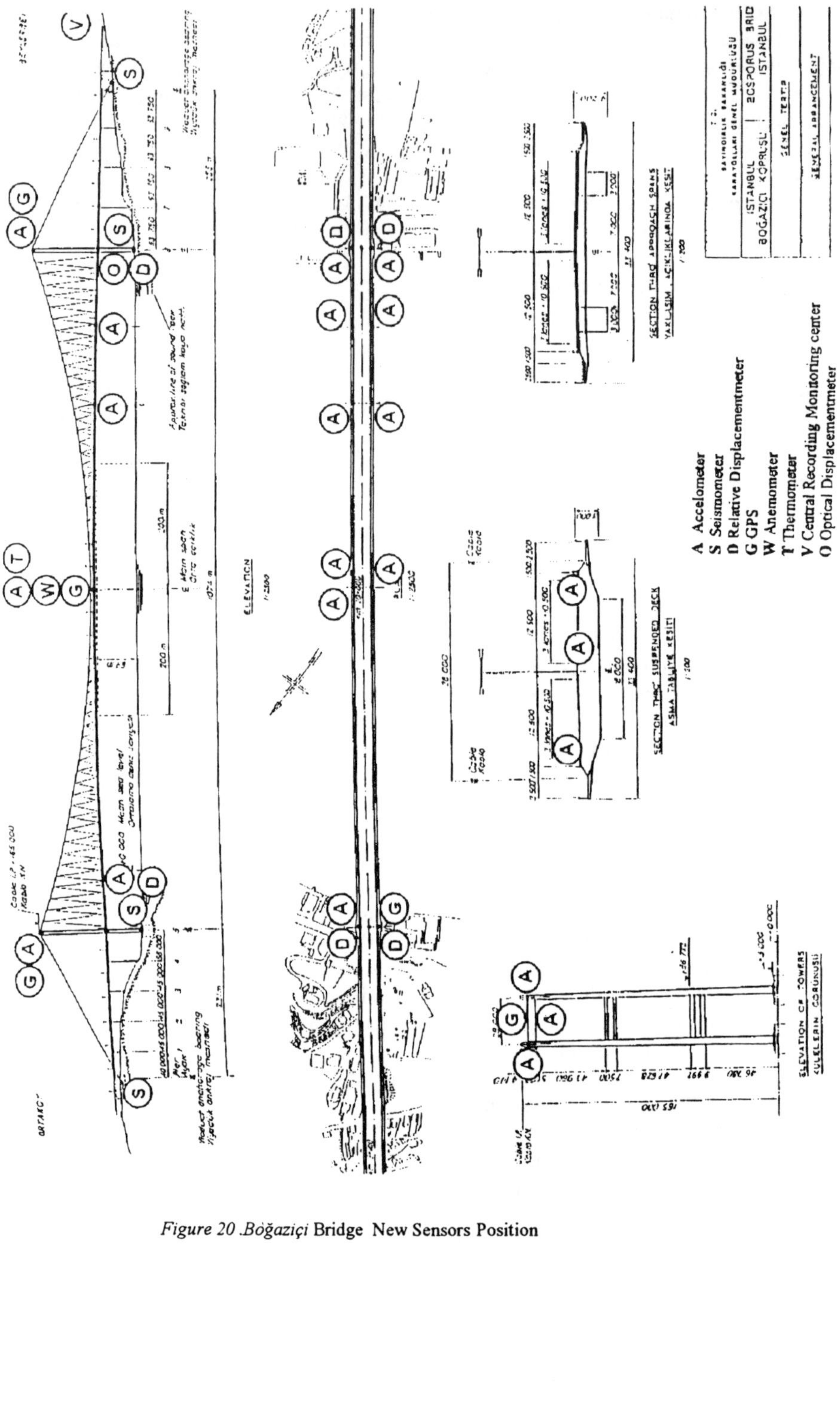

Figure 20 .Boğaziçi Bridge New Sensors Position

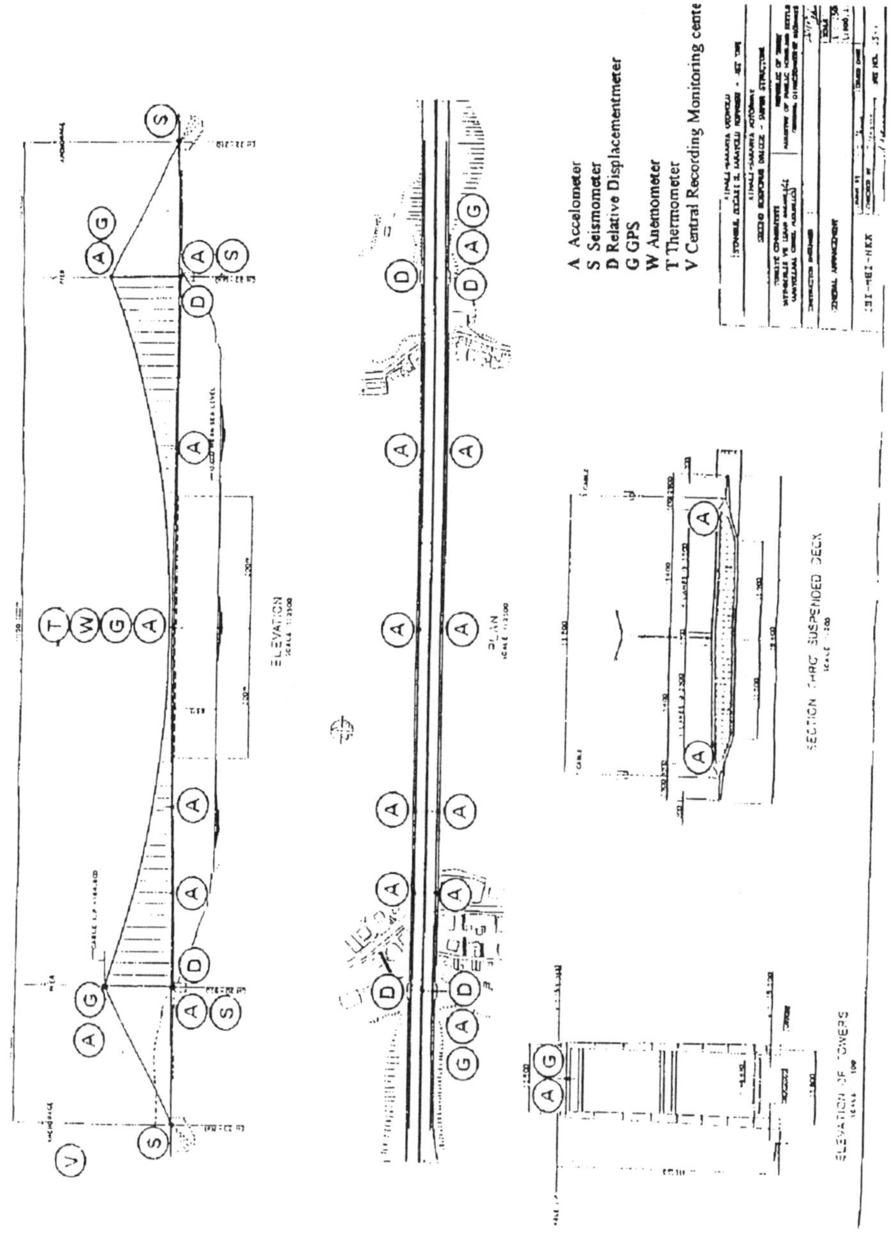

Figure 21 .Fatih Sultan Mehmet Bridge New Sensors Position

6. References

1. Abdel-Ghaffar A; R. H. Scanlan (1985),Ambient Vibration Studies of Golden Gate Bridge: I. Suspended Structure,*Journal of Engineering Mechanics,* Vol.111, No:4, pp: 19645
2. Abdel-Ghaffar A.M; S. F. Masri; R. L. Nigbor, (1996), Vincent Thomas Bridge Monitoring Tests , *Proceedings of the Fourth National Workshop on Bridge Research in Progress*, pp: 227-230, June-17-19, Buffalo, New York
3. Adanur S; A. A. Dumanoğlu ve A. Bayraktar (1997),Asma Köprülerin Lineer Olmayan Dinamik Davranışının İncelenmesi, *4. Ulusal Deprem Mühendisleri Konferansı*: 17-19 Eylül 1997- Ankara)
4. Aktan A. E.; A. J. Helmicki; D. Brown; V. Dalal; A. Levi; V. Hunt; M. Lenett; N. Catbas; A. Turer; T. Tankut, (1998) Structural ıdentification and health monitoring of an instrumented highway bridge, SEWC 98, *Structural Engineers World Confress*, July20
5. Aktan A. E.; K. Grimmelsman;A. J. Helmicki, (1998) Issues and opportunities in bridge health monitoring, SEWC 98, *Structural Engineers World Confress*, July20
6. Beck J.L; B. S. May and D. C. Polidori, (1994) Determination of Modal Parameter From Ambient Vibration Data For Structural Health Monitoring, *First World Conference on Structural Control*, 3-5 August, Los Angeles, California, USA
7. Beck J.L; M. W. Vanik and L. S. Katafygiotis, (1994) Determination of Stiffness Changes From Modal Parameter Changes For Structural Health Monitoring , *First World Conference on Structural Control*, 3-5 August, Los Angeles, California, USA
8. Beck J.L; M. W. Vanik; D. Polidori; S. May, (1998) Structural health monitoring using ambient vibrations, SEWC 98, *Structural Engineers World Confress*, July20
9. Beyen ; E. Uçkan & M. Erdik, (1994) Ambient vibration investigation of the Boğaziçi Suspension Bridge, Istanbul, Turkey, *Earthquake Resistant Cons. & Design*, Bakema, Rotherdam
10. Brownjohn J. M. W.; (1994) Observations on Non-Linear Dynamic Characteristics of Suspension Bridges, *Earthquake Engineering and Structural Dynamics*, Vol. 23, 1351-1367
11. Browjohn, J. M. W.; Severn, R. T.; Dumanoğlu, (1992),A. A. Full-scale dynamic testing of the second Bosporus suspension bridge, *Proceedings of the Tenth World Conference on Earthquake Engineering*, A. A. Balkema, Rotterdam, Vol.5 pages 2695-2700
12. Brownjohn J. M. W.; A. A. Dumanoğlu; R. T. Severn; (1989) A. Blakeborough,Ambient Vibration Survey of the Bosporus Suspension Bridge,*Earthquake Engineering and Structural Dynamics*, Vol. 18, 263-283
13. Brownjohn J. M. W; A. A. Dumanoğlu; R. T. Severn, (1992) Ambient vibration survey of the Fatih Sultan Mehmet (Second Bosporus) suspension bridge,*Earthquake engineering and structural dynamics*, 21,10, Oct. Pages 907-924
14. Çelebi M.; W. Prescott; R. Stein; K. Hudnut; S. Wilson (1998) GPS monitoring of structures in real-time:recent advances, International IDNDR-Conference on Early Warning Systems for the Reduction of Natural Disasters, Potsdam, Germany September 7-11
15. Dumanoğlu A. A.; J. M. W. Brownjohn (1992) ,Seismic analysis of the Fatih Sultan Mehmet (Second Bosporus) suspension bridge, *Earthquake engineering and structural dynamics*, 21,10 , pages:881-906
16. Erdik M.; N.Apaydın; (1998) Earthquake Early Warning and Structural Health Monitoring Systems for Bridges ,*Second Japan-Turkey Workshop on Earthquake Engineering* ,İstanbul
17. Erdik M; N. Apaydın, (1998) Early Warning System for the Bosporus Bridge , International IDNDR-*Conference on Early Warning Systems for the Reduction of Natural Disasters*, Potsdam, Germany September 7-11
18. Erdik M., E. Uçkan (1989) Ambient vibration survey of the Boğaziçi Suspension Bridges İstanbul-Turkey,Report No: 89-5 Department of Earthquake Engineering Kandilli Observatory and Earthquake Research Institute Boğaziçi University, November,
19. Higashihara H.; T. Moriya; J. Tajima , (1987), Ambient Vibration Test of an Anchorage of South Bisan-Seto Suspension Bridge, *Earthquake Engineering and Structural Dynamics*, Vol, 15,679-695
20. Jay-Chung Chen , (1995) Intelligent monitoring system for suspension bridges damage detections ,Bridges 21st Century, Hong-Kong, *proceedings the Hong-Kong Ins. of Engineers*
21. Heaton T.; R. Clayton; J. Davis; E. Hauksson; L. Jones; H. Kanamori; J. Mori; R. Porcella; T. Shakal, (1996) The trinet project, *Eleventh World Conference on Earthquake Engineering,* pp: 2136,
22. Lee G. C.; Z. Liang, Bridge Monitoring Systems: An Overview
23. McLamore V. M .; Gary C. Hart; A. M. ASCE, I. R. Stubbs, (1971) Ambient Vibration of Two Suspension Bridges, *Journal of the Structural Division, Proceedings* ASCE, 97, ST10,Oct., , Pages: 2567-2582

24. Nakamura Y: (1996) Real-Time Information Systems for Hazards Mitigation, *Eleventh World Conference on Earthquake Engineering*, pp: 2134
25. Nigbor, R. L., (1984) Full-scale ambient vibration measurements of the Golden Gate suspension bridge-instrumentation and data acquisition, *Proceeding of the Eighth World Conference on Earthquake Engineering*, Prentice-Hall, Inc., Englewood Cliffs, New Jersey, , pages 63-70, Vol. VI, Copyright with Earthquake Engineering Research Inst., El Cerrito, California.
26. Oasıs (1997) On-Line Alerting of Structural Integrıty and Safety Agbabian Associates,Inc. Pasedena California
27. Severn R.T. 81997) Dynamic studies on suspension and cable-stayed bridges, *Prof. Dr. A. Rıfat Varar Symposium*, Dec 10Istanbul Technical Univ., Istanbul
28. Tezcan S. S.; M. İpek; J. Petrovski; T. Paskalov, Forced Vibration Survey of Istanbul Boğaziçi Suspension Bridge, *Fifth European Conference on Earthquake Engineering*, Vol.II, pp: 152
29. Thomas Heaton, Robert Clayton , James Davis , Egill Hauksson, Lucille Jones , Hiroo Kanamori, James Mori , Ron Porcella, Tony Shakal, (1996)The TriNet Project , *Eleventh World Conference on Earthquake Engineering*, pp: 2136
30. Ventura; A. J. Felber; S. F. (1994) Stiemer,Dynamic Characteristic of Bridges By Experimental Investigations of Ambient Vibrations - Queensborough Bridge, *Conference on Earthquake engineering*, Chicago, July

SOIL-STRUCTURE-INTERACTION AND SEISMIC ISOLATION, AN INTER-DISCIPLINARY INVESTIGATION AT THE MULTIDISCIPLINARY SEISMIC TEST SITE INCERC, BUCHAREST, ROMANIA

M. BAUR [1], O. NOVAK [2], J. EIBL [1], D. LUNGU [3]

[1] *Institute for Concrete Structures and Building Materials, Concrete Structures Division, University of Karlsruhe, Am Fasanengarten, 76128 Karlsruhe, Germany*

[2] *Geophysical Institute, University of Karlsruhe, Hertzstraße 16, 76187 Karlsruhe, Germany*

[3] *Technical University of Civil Engineering - Bucharest, 124 Bd. Lacul Tei, 72302 Bucharest, Romania*

e-mail:gd25@ rz.uni-karlsruhe.de

Abstract

The Collaborative Research Center 461 (CRC 461) 'Strong earthquakes: A Challenge for Geosciences and Civil Engineering' (Germany) and the Romanian Group for 'Strong Vrancea Earthquakes' work on a multidisciplinary attempt towards the mitigation of the impact of the next strong Vrancea earthquake (Romania). The recently installed Multidisciplinary Seismic Test Site INCERC (MSTS) located in the eastern part of Bucharest serves as focus to study the entire sequence relevant in engineering seismology including structural dynamics, seismology, soil mechanics, and engineering geology. Within this co-operation, the complete chain for seismic hazard and seismic risk analysis and can be checked and verified.

For this purpose a base isolated test building was constructed, where acceleration will be measured at the base and at the top. Additionally, free field accelerometers in various depths were installed. Thus, free field accelerations and the response of the test building can be measured simultaneously.In this paper we want to focus on the test building. One mainly task, among others, is to investigate the dynamic behaviour of rubber bearings under seismic loading, including soil-structure-interaction.

Geotechnical parameters which will be available to a depth of 180 m ensure that the soil-structure-interaction incorporating the seismic behaviour of rubber bearings can be computationally studied with FE methods. These calculations include constitutive laws describing non-linear behaviour of rubber and soil material under cyclic loading.

M. Erdik et al. (eds.), Strong Motion Instrumentation for Civil Engineering Structures, 369–381.

1. Multidisciplinary Seismic Test Site Incerc (Msts)

The Romanian Vrancea area located in the southern Carpathians is the source region for strong earthquakes. Within the last 20 years Bucharest was affected by 3 events with moment magnitudes larger than 6.9.With a moment magnitude of 7.5, the 1977 earthquake caused more than 1500 casualties, most of them in Bucharest. The average recurrence rate make another strong event within the next 20 years highly probable. Under the umbrella of the Collaborative Research Center 461 (CRC 461) 'Strong earthquakes: A Challenge for Geosciences and Civil Engineering' and the Romanian Group for 'Strong Vrancea Earthquakes' German and Romanian scientists from various fields (geology, seismology, civil engineering, operation research) work on a multidisciplinary attempt towards the mitigation of the impact of the next strong Vrancea earthquake.

In this frame a Multidisciplinary Seismic Test Site (MSTS) is installed at the premises of the Building Research Institute INCERC which is located in Eastern Bucharest. The MSTS allows to study the only site in Romania where a complete set of strong motion records of previous Vrancea earthquakes is available: March 4, 1977, $M_w = 7.5$; August 30, 1986, $M_w = 7.2$, and May 30/31, 1990, $M_w = 6.9/6.4$ [11].

Three-component accelerometers are installed in the free field at the surface and will be installed at various depths (30 m , 100 m, 180 m). Geotechnical parameters of the subsurface soil layer are controlled to a depth of 180 m.

The key objectives of the research activities at the MSTS are:

- Research on the filtering of seismic signals through 180 m of soft soil layers
- Calibration of the geotechnical and seismological parameters for seismic microzonation
- Verification of seismic hazard and seismic risk analysis
- Performance of rubber bearings during expected moderate to strong earthquakes

The test building was designed as a rigid concrete block of 100 t supported by high damping rubber bearings on a rigid concrete footing of 72 t. The stiffness of the rubber bearings and the mass of the structure has an eigenperiod of 1.1 to 1.6 s. This period range represents maximum spectral accelerations during the 1977 and 1986 events (Fig. 1).

To record all 6 degrees of freedom (DOF) and the relative movement between footing and top of the concrete block, six 3-component accelerometers are installed as shown in Figure 1.

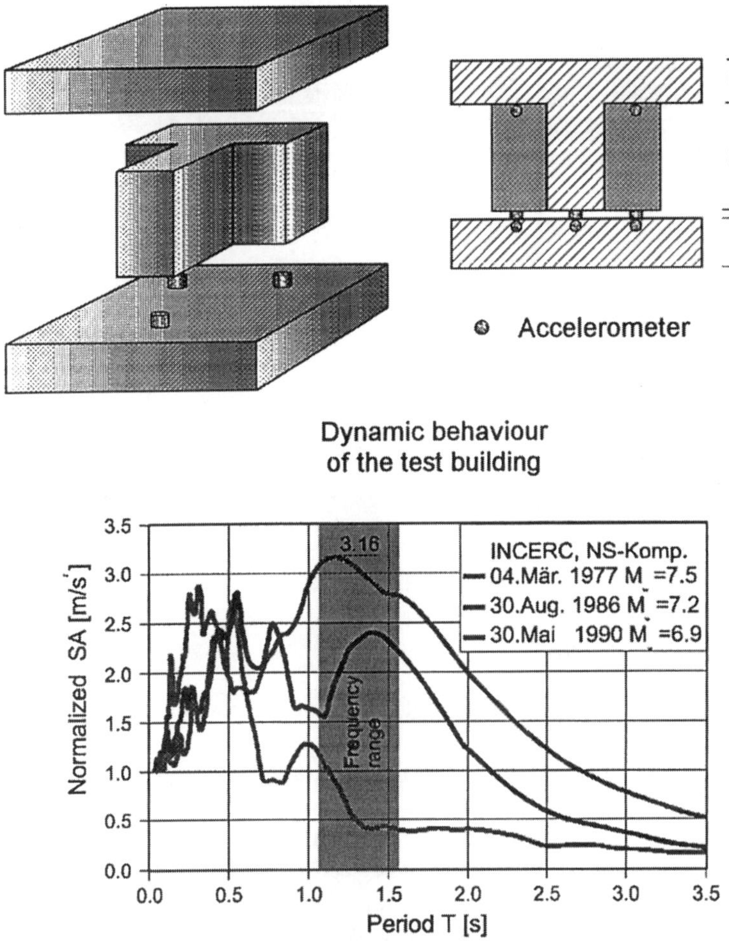

Figure 1: Concept of test building (top). Dynamic behaviour of test building (bottom).

Figure 2 shows the preparations for the drilling of the 180 m borehole at the MSTS (left) and illustrates the arrangement of test building and boreholes (right).

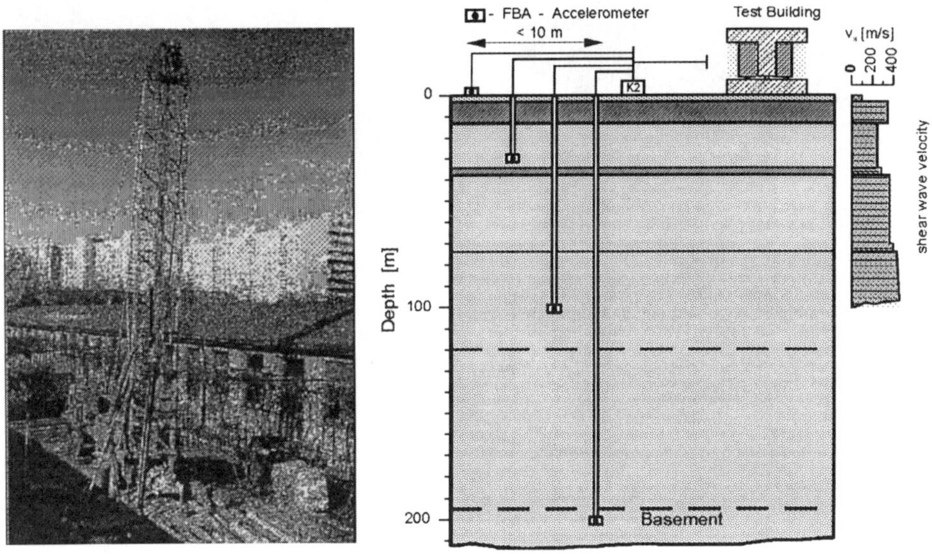

Figure. 2: Preparations for the 180 m borehole at INCERC (left). Concept of the MSTS (right).

2. Concepts Of Seismic Isolation

The conventional methods for earthquake-resistant structural design use high strength or high ductility concepts to mitigate damage from seismic impacts. In the first case, corresponding to shear wall structures, generally the design is problematic in that their fundamental frequency of vibration is in the range of frequencies where earthquake energy is strongest, resulting in a very high floor acceleration, which may cause damage to equipment or machinery. The second, the capacity design method, incorporates that a part of the energy transmitted into structure by an earthquake is dissipated by plastic deformations. The capacity method mostly used for flexible structures as frames, provided that plastic deformations occurred in structural elements, which are designed to undergo such large deformations. Therefore, the design of such yielding zones has to be planned carefully.

However, this concept may lead to a very high interstory drift, causing P-Δ effects and damage to non-structural elements. Thus, the costs for retrofitting or strengthening after a strong earthquake can be very high. An alternative approach consists in isolating the structure base from the ground by using rubber bearings.

It has to be pointed out that the dynamic system consisting of rubber bearings supporting a structure is predominantly governed by the non-linear behaviour of the bearings. Therefore we focus on the development of a constitutive law describing the behaviour of the rubber bearings under cyclic loading.

3. A Constitutive Law For Rubber-Like Materials

In the framework of EC Project 7010 on "Optimisation of design and performance of high damping rubber bearings for seismic and vibration isolation" a task at the Institute for reinforced concrete structures, University of Karlsruhe, was set-up to develop detailed numerical models for the analysis of such bearings [5], [6]. Following the most important formulations of a viscoelastic constitutive law for finite strains are illustrated, as they have been developed based on an approach by Simo (1987).

The law proposed makes the following assumptions [10], [15] :

- uncoupled volumetric and deviatoric response for finite strains
- viscoelastic properties represented by separable functions of time and strain
- no volumetric relaxation

The applicability of the first assumption to the high damping rubbers was confirmed in this work by tests examining the effect of hydrostatic pressure on the tensile behaviour.

The Simo law is here modified with a strain hardening function to cover the stiffening of rubber at higher strains, as observed experimentally, and a strain softening function to account for strain history effects.

Thus, the fully three-dimensional constitutive law takes the form:

$$\boldsymbol{\sigma}(\tau) = K \ln(J) \cdot \mathbf{I} + \int_{\tau=0}^{t} \Phi(t-\tau) \frac{\partial}{\partial \tau} \left[g(\varphi) \cdot H\left(\left\| \mathrm{dev}\, \overline{\mathbf{C}} \right\| \right) \cdot \mathrm{dev}\, \overline{\mathbf{C}} \right] d\tau$$

Here, the deformation gradient **F** is decomposed into a pure volumetric $J^{\frac{1}{3}}\mathbf{I}$ and isochoric deformation $\overline{\mathbf{F}}$, respectively [13]. Furthermore, the right Cauchy-Green-tensor $\overline{\mathbf{C}}$ for the isochoric deformation is given by:

$$\overline{\mathbf{C}} = J^{-\frac{2}{3}} \mathbf{C} = \overline{\mathbf{F}}^{T} \overline{\mathbf{F}}$$

The first term in the constitutive law covers the volumetric stress, while the deviatoric part described by a convolution integral containing:

- a relaxation function

$$\Phi(t) = G_\infty + (G_0 - G_\infty) \cdot e^{-\frac{t}{\nu}}$$

with a long and a short shear time modulus G_∞ and G_0 and a characteristic relaxation time ν.

- a strain softening damage function

$$g(\varphi) = \beta + (1-\beta) \cdot \frac{1 - e^{-\frac{\varphi}{\alpha}}}{\varphi / \alpha}$$

with the material parameters α and β and the norm of the Cauchy-Green-tensor

$$\varphi(\tau) = \max_{\tau \leq t} \left\| \text{dev}\overline{\mathbf{C}} \right\|.$$

- a strain hardening function

$$H(\left\| \text{dev}\overline{\mathbf{C}} \right\|) = \delta \cdot \left\| \text{dev}\overline{\mathbf{C}} \right\|^2 + 1$$

with a scalar parameter δ.

The material parameters: $K, G_0, G_\infty, \nu, \alpha, \beta$ and δ were determined for loading histories KOD and KOS, as shown in Figure 3.

Both loading histories are up to a rubber shear deformation of 200 % (tan □□= 2); where tan □□□= S/T. Here, S means the total rubber thickness and T the horizontal displacement.

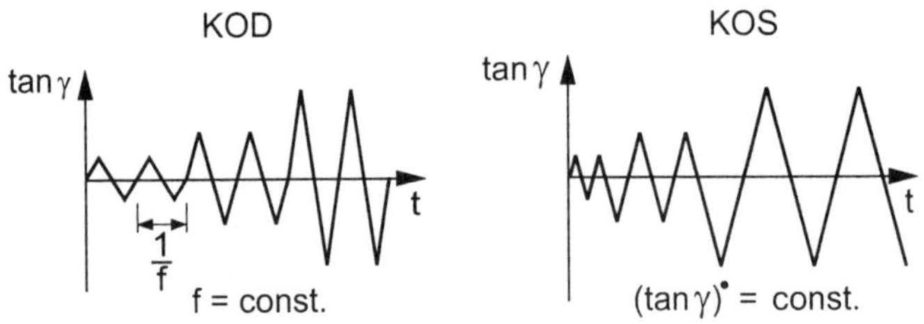

Figure. 3: Loading histories KOD (f = 0.1 Hz) and KOS (strain rate 7%/s)

The values of the material parameters appropriate to each test condition are listed in Table 1 [5].

Table 1: Material parameters for viscoelastic model

	KOS	KOD
K [MPa]	1000	1000
G_0 [MPa]	3.5	3.5
G_∞ [MPa]	1.3	1.3
ν [s]	5.0	0.5
α	0.1	0.1
β	0.2	0.2
δ	0.02	0.02

4. Fe-Simulation Of Rubber Bearings Compared With Test

An algorithm proposed by Kim et al (1991) was used to evaluate the convolution integral [10]. Minor changes to stiffness parameters in Table 1 were made to take into account the difference in strain rate experienced by the rubber in the bearing test. An example of a test result and the FE-analysis for a bolted bearing, D = 250mm and a total rubber thickness T = 60 mm, are shown in Figure 4 and Figure 5, respectively [5], [6], [7]. There is a good correspondence between the test results and the FE-analysis. The predicted loops are seen to reproduce the observed behaviour reasonably well.
The FE-model of the rubber bearing is generated with 20-node quadratic continuum elements for the rubber layers and the two anchor plates [1].

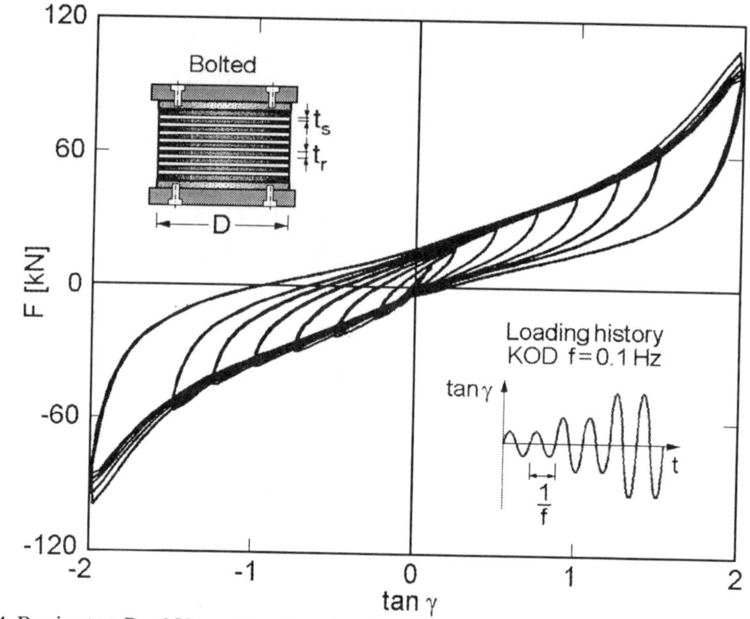

Figure. 4: Bearing test, D = 250mm, T = 60mm, Loading history KOD Rubber shear rate 1.8%/s

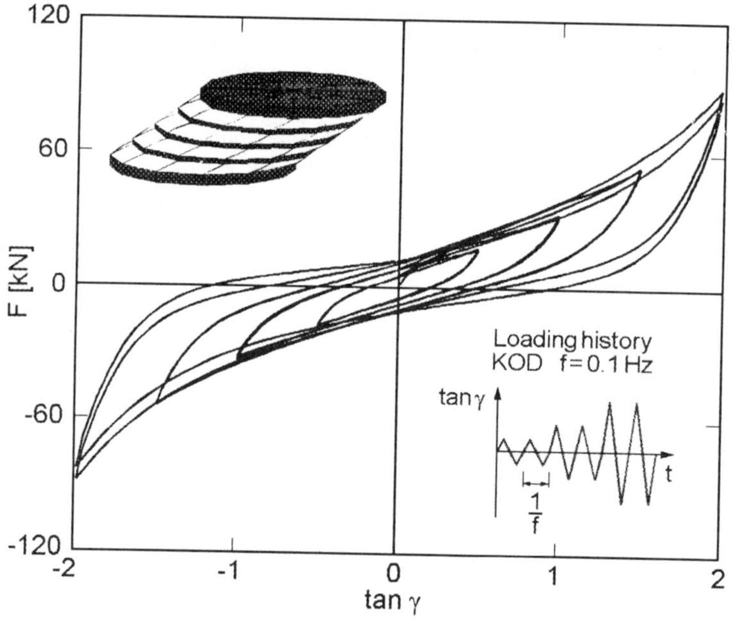

Figure. 5:FE-Simulation, D = 250mm, T = 60mm, Loading history KOD

Furthermore the response of time histories of two earthquakes with a different dominant frequency content, Friaul 1976 (3-5 Hz) and Vrancea 1986 (0.7-1 Hz) shown in Figure 6, has been calculated to test the calibrated KOD parameters. As loading mass, 33 tons were taken, which corresponds to a third of the mass of the test building (three bearings). Figure 6 illustrates that the 'low-frequency' Vrancea time history is amplified, which corresponds to the design of the test building, and the 'high-frequency' Friaul time history is attenuated.

For the Friaul earthquake it is shown that the seismic loading imparted to a structure can be efficiently reduced. The need to take into account the soil conditions before using rubber bearings as base isolation is obvious. In particular regions of soft soil which mostly reveal low frequency response to an earthquake are critical [3], [9].

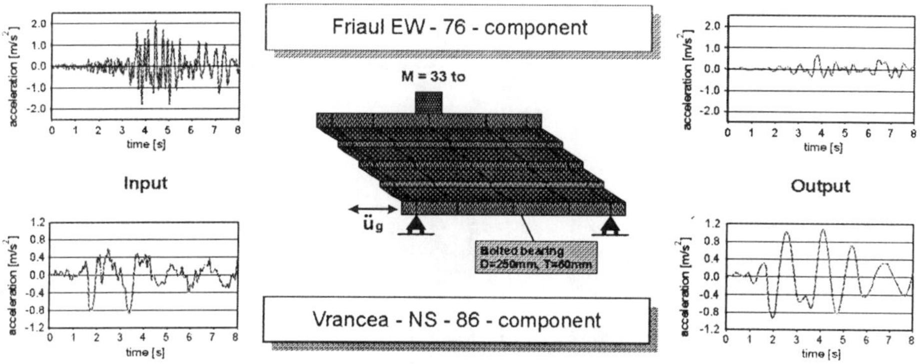

Figure. 6: Calculated response of Friaul 1976 and Vrancea 1986 earthquakes.

Figure 7 shows the spectral amplitudes and spectral ratio for the calculated response of the used bearing for the Vrancea 1986 earthquake.

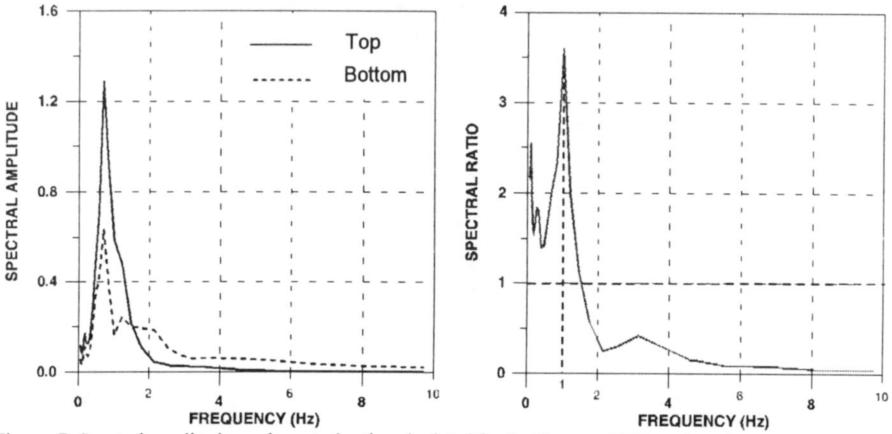

Figure. 7: Spectral amplitudes and spectral ratio calculated for the Vrancea 1986 earthquake.

The peak in the calculated spectral ratio for the Vrancea 1986 event in Figure 7 shows, that for moderate earthquakes the predominant frequency is about 1 Hz, while the calculated maximum shear deformation goes up to tan □□= 0.4. For stronger earthquakes the predominant period is expected in the range of 0.7 Hz and tan □□= 1.5, due to loss of stiffness of the rubber bearings for larger shear deformations.

5. Local Soil Conditions

The local soil conditions are governed by a predominately clayed soil profile, with a thick fine-coarse gravel - coarse sand - layer at the top, containing the water table level. The soil profile with in-situ measured shear and compression wave velocities up to 74 m depth is shown in Figure 8 [12].

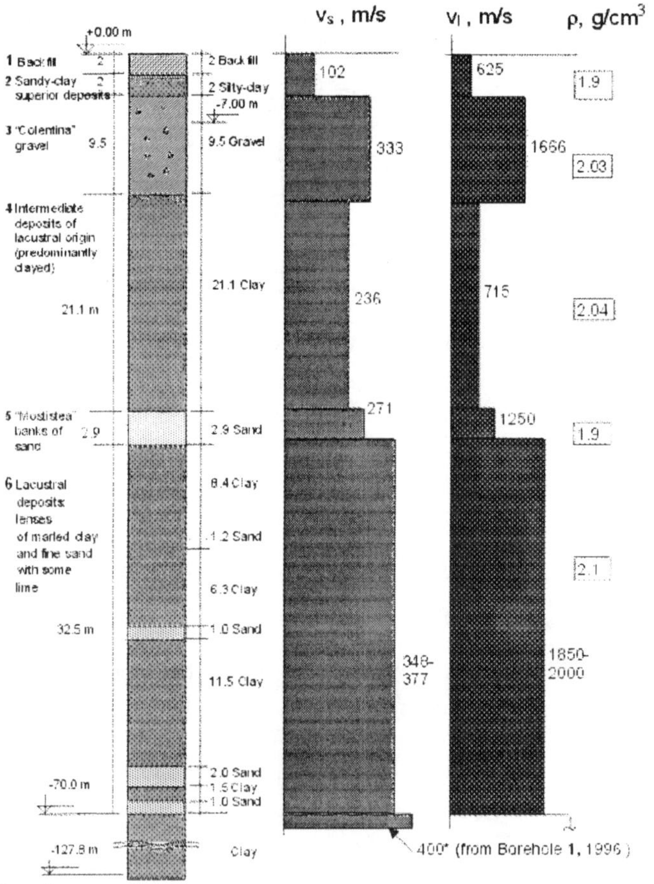

Figure. 8: Local soil profile

6. A Simplified Constitutive Law For Soil

To take into account the non-linear behaviour of soil material under cyclic loading, a simple pseudo-linear approach was chosen [2], [4]. In this approach, the non-linear behaviour of soil is modelled by using a constitutive relation, based on a variable moduli (VM) model.

It has to be point out, that these models can account for behaviour of a limited class of materials, loadings, and stress paths, they may not used as general models. Due to the circumstance, that it is rather difficult to get all parameters for more sophisticated constitutive relations in practice, these approach seems to be justified.

The basic assumptions of the incremental formulation are:

- uncoupled deviatoric and volumetric response

$$\dot{\sigma}_{ij} = \dot{p} + \dot{s}_{ij}$$

- tangential shear modulus G_t as a function of the second deviator stress invariant $\sqrt{J_{2D}}$ and the hydrostatic pressure p

$$G_t = G_0 + \beta_1 p - G_1 \sqrt{J_{2D}}$$

- tangential bulk modulus K_t as a function of the first strain invariant I_1

$$K_t = K_0 + K_1\ I_1 + K_2\ I_1^2$$

These leads to the following stress-strain-relation

$$\dot{\sigma}_{ij} = \dot{p} + \dot{s}_{ij} = K_t \dot{\varepsilon}_{kk} + 2G_t \dot{e}_{ij} = d\sigma_{ij} = K_t d\varepsilon_{kk} \delta_{ij} + 2G_t \left(d\varepsilon_{ij} - \frac{1}{3} d\varepsilon_{kk} \delta_{ij} \right)$$

The stress-strain-relation and calibration of parameters for different soil layers will be derived from in-situ measurements and dynamic laboratory tests, which will actually be done.

First results of a simulation for a simple shear test for different loading histories are presented in Figure 9. The non-linear behaviour of calculated $G-\gamma$ curve fit empirical derived $G-\gamma$ curves, given in the literature, reasonable well [8], [14], [16].

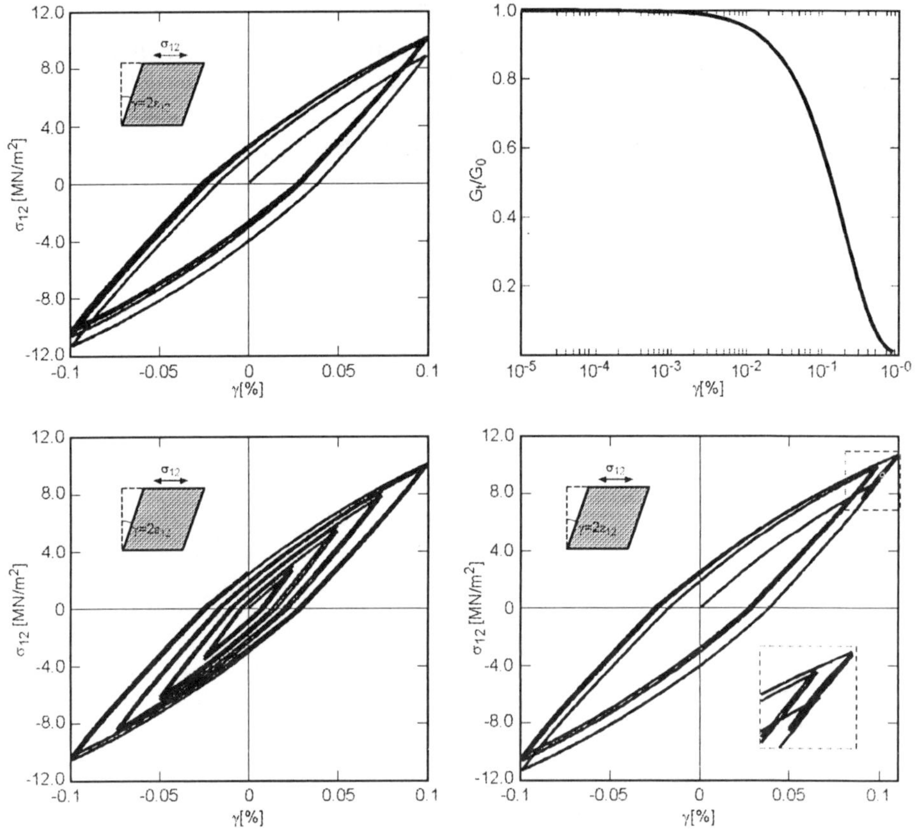

Figure. 9: Stress-strain-relation for different loading histories with the VM-Model

7. Conclusion And Future Directions

The results of the behaviour of rubber bearings under cyclic loading up to serviceability of tan □ = 2 presented herein have shown that non-linear effects as well as the hysteretic behaviour can be simulated in good accordance to the test results with the proposed constitutive law. What remains is to test such bearings under real conditions.

Therefore one of the tasks within the framework of the CRC at the University of Karlsruhe, Germany, is an experiment in downtown Bucharest, Romania, at the MSTS. For this purpose among others a rigid concrete bloc founded on rubber bearings and a rigid footing was erected. Free-field accelerations in various depths and the relative accelerations between the concrete bloc and the footing will be measured simultaneously.

Additionally, soil-structure-interaction will be taken into account to control the complete dynamic system.

8. Acknowledgements

The CRC 461 at the University Karlsruhe, Germany, is mainly supported by the grant of the German Research Council "Deutsche Forschungsgemeinschaft".

References

1. Abaqus, 1997 Abaqus Version 5.7, User Manual Hibbit, Karlsson & Sorensen, Inc., USA
2. Chen, W.-F; Saleeb, A.F., 1982 Constitutive Equations for Engineering MaterialsJohn Wiley & Sons, New York
3. Chopra, A.K., 1995 Dynamics of Structures Prentice-Hall, Englewood Cliffs, N.J.
4. Desai, C.S.; Siriwardane, H.J., 1984 Constitutive Laws for Engineering Materials – with Emphasis on Geologic Materials Prentice Hall Inc., Englewood Cliffs, NewYork
5. Eibl, J.; Hehn, K.-H.; Baur, M.; Böhm, M.; Schmidt-Hurtienne, B., 1996Task 8: Detailed Numerical Models of Bearings Technical report No.6, Brite-Euram II, Project BE 7010Institut für Massivbau und Baustofftechnologie, Universität Karlsruhe (TH)
6. Eibl, J.; Hehn, K.-H.; Zeller, W., 1996 Task 5: Tests on Individual Elastomeric Bearings -Diameter 250, 500, 800 mm-Technical report No.5, Brite-Euram II, Project BE 7010 Institut für Massivbau und Baustofftechnologie, Universität Karlsruhe (TH)
7. Fuller, K.N.G.; Forni, M.; Bettinali, F.; Mazzieri, C.; Eibl, J.,1997 Proc. of the International Post-SMiRT Conference Seminar Taormina, Sicily, Italy, August 25 to 27, 1997 Published by GLIS in cooperation with International Atomic Energy Agency (IAEA), pp 271-290
8. Idriss, I.M., 1990 Response of Soft Soil during Earthquakes Proc. a Memorial Symposium to Prof. Seed, Univ. of California, Berkley, pp 273-289
9. Kelly, J.M., 1993 Earthquake Resistant Design with Rubber Springer-Verlag, London
10. Kim, M.J.; Gupta, A.; Marchatas, A.H, 1991 Utilisation of Simo-Taylor Constitutive Model for Simulation of Isolation bearings Trans. 11th Conf. SMiRT, (ed.) Shibata, H., Vol. K2 Published by Atomic Energy Society of Japan, pp 169-174
11. Lungu, D., Cornea, T., Nedelcu, C., 1998 Probalistic Hazard Assessment and Site-Dependent Response for Vrancea Earthquakes in Bucharest, Romania, in: „Vrancea Earthquakes„ Wenzel, F., Lungu, D., Novak, O., (eds.), Kluwer Academic Publishers, Dordrecht, Netherlands
12. Lungu, D.; Aldea. A.; Moldeveanu, T.; Ciugudean, V.; Stefanica, M. , 1998 Surface geology and dynamical properties of soil layers in Bucharest in Bucharest, Romania, in: „Vrancea Earthquakes„ Wenzel, F., Lungu, D., Novak, O., (eds.), Kluwer Academic Publishers, Dordrecht, Netherlands [13] Malvern, L.E., 1969
13. Introduction to the Mechanics of a Continous Medium Prentice-Hall, Englewood cliffs, N.J.
14. Seed, H.B; Idriss, I.M., 1970 Soil Modulus and Damping Factors for Dynamic Response Analysis Report No. EERC 70-10, University of California, Berkley, California
15. Simo, J.C., 1987 On a Fully Three Dimensional Finite-Strain Viscoelastic Damage Model: Formulation and Computational Aspects Comp. Meth. Appl. Mech. Eng. 60, pp 153-173
16. Sun J.L.; Golesorkhi, R.; Seed, H.B., 1988 Dynamic Moduli and Ratios for Cohesive Soils Report No. EERC 88-15, University of California, Berkley, California

GPS IN DYNAMIC MONITORING OF LONG-PERIOD STRUCTURES

MEHMET ÇELEBI
U.S. Geological Survey,
345 Middlefield Rd. (MS977),
Menlo Park, CA 94025
e-mail:celebi@samoa.wr.usgs.gov

Abstract

Global Positioning System (GPS) technology with high sampling rates (~10 sps) allows scientifically justified and economically feasible dynamic measurements of relative displacements of long-period structures --- otherwise difficult to measure directly by other means, such as the most commonly used accelerometers that require post-processing including double integration. We describe an experiment whereby the displacement responses of a simulated tall building are measured clearly and accurately in real-time. Such measurements can be used to assess average drift ratios and changes in dynamic characteristics, and therefore can be used by engineers and building owners or managers to assess the building performance during extreme motions caused by earthquakes and strong winds. By establishing threshold displacements or drift ratios and identifying changing dynamic characteristics, procedures can be developed to use such information to secure public safety and/or take steps to improve the performance of the building.

KEYWORDS: monitoring, GPS, strong-motion, building, structural response, frequency, displacement, acceleration

1. Introduction

Seismic monitoring of structural systems constitutes an integral part of National Earthquake Hazard Reduction Program of the United States. In general, until recently, monitoring the response of structural systems for the purpose of assessing and mitigating effects of earthquakes (and also severe winds) has relied on measuring the shaking response by deploying accelerometers throughout a particular structure of interest to the scientific and engineering communities. In contrast, there are no efficient or feasible methods to measure displacements during an earthquake or severe wind. Recordings of the acceleration response of structures have served us well. Studies conducted on such records have been useful in assessing design/analysis procedures, improving code provisions, and correlating the response with damage.

M. Erdik et al. (eds.), Strong Motion Instrumentation for Civil Engineering Structures, 383–396.

Since the M_s=6.7 Northridge (17 January 1994) and M_s=6.8 Kobe (17 January 1995) earthquakes, drift studies and assessment of susceptibility to damage of tall buildings have become important issues, particularly because so many steel-framed buildings were damaged, some severely and some lightly. In the Los Angeles area, for example, following the Northridge event, several hundred steel-framed buildings had to be examined, assessed, and repaired or retrofitted. Only three of these buildings were instrumented prior to the event, providing some limited acceleration response data to be used for interpretation of the widespread damage. Additional data, if available in real-time or near-real-time, could have been very useful for studies and for design of repair and retrofitting projects that followed. Therefore, there is a great need for better and more extensive monitoring of tall buildings.

Relative displacements, which are key to assessing drift and stress conditions of structures, are difficult to measure directly. On the other hand, measuring acceleration response requires a double integration process to arrive at displacements. The integration process is not readily automated because of the nature of signal processing, which requires (a) selection of filters and baseline correction (the constants of integration), and (b) often substantial judgment when anomalies exist in the records. Consequently, this process can lead to errors in the calculation of velocities and displacements. This problem is more acute for permanent displacements. Accelerometer measurements cannot be used to recover the permanent displacements at the centimeter level; and even if they could, it is questionable if it can be done in real-time. That is, the level of accuracy of displacements calculated from accelerations has not been widely verified by observations (e.g., some shake table tests performed to compare the performances of accelerometers and accelerographs have not been directed at checking displacements).

An alternative method to measure relative displacements while monitoring structural systems can be accomplished by using GPS technology, now advanced to record at 10 sps with an accuracy of ± 1 cm horizontally and ± 2cm vertically. This provides a great opportunity to monitor long-period structures reliably (e.g., tall buildings that are 20-40 stories or more). The majority of the tall buildings are flexible steel-framed structures. The fundamental period of such a flexible-framed building can be estimated with the empirical formula[2] , T = 0.1 N, where N is the number of stories of the building. This means that at least 20-40 data points will be

---

[2] For most flexible buildings, the fundamental period (T) is approximated by 0.1N, where N is the number of floors of a building (even though the fundamental period can vary between [0.05-0.15]N depending on the flexibility of the building). Therefore, to simulate a 40-story building, we set the period (frequency) equal to 4 s (0.25 Hz) and proportioned the length, width and thickness of the cantilever.

recorded for one cycle of motion of a 20-40 story building vibrating at the fundamental period. This provides sufficient accuracy to assess the average drift ratio of a building. Such information can be very useful in assessing the damage to a building. In addition, the main value in using GPS technology to assess the condition of a building is that the displacement measurements can be made directly in real-time and with sufficient precision. As discussed later, we have made preliminary tests to prove the technical feasibility of the application of GPS to monitoring structures.

There is great potential for the application of GPS technology to monitor long-period structures during earthquakes. The application can also be extended to monitoring wind-induced deformation of tall buildings, long-span suspension and cable-stayed bridges, and tall chimneys. Furthermore, with future advances in GPS technology and improvements in sampling capability (e.g., higher than 10 sps), it will be possible to monitor short-period structures as well. Additionally, direct measurements of displacements will enable us to reliably detect structural movement caused by failure of the ground under a structure (e.g., liquefaction).

2. Technical Justification

In the last few years, there have been numerous studies related to the technical feasibility of using GPS to measure displacements of civil structures. Most of the initial work has been accomplished by aerospace atmospheric researchers. Studies related to the application of GPS for static or dynamic measurements of displacements of structural systems include but not limited to those by Hyzak and others (1997), Teague and others (1995), Guo and Ge (1997), Kondo and Cannon (1995), Lovse and others (1995), Hudnut and Behr (1998), Behr, Hudnut and King (1998) and Stein and others (1997). In our study, we direct the efforts to actual permanent deployment of GPS units alongside accelerometers, and also to use the displacement measurements with GPS as a health monitoring tool.

3. Model Tests Simulating A Tall Building

To investigate the feasibility of using GPS technology to monitor tall buildings (and other long-period structures), we conducted two experiments. Figure 1 depicts a photo and the overall set-up for a simple and inexpensive experiment designed by selecting a standard stock steel bar to simulate a 30-40 story flexible building. We selected the length, thickness, and width of the two bar specimens to yield a fundamental period of approximately four seconds in the weak direction. To make things simple, we purposefully selected the width and thickness of each of two bars with an extremely weaker axis in one direction. The width was varied to show the sensitivity of measurements during vibration and at 10 Hz sampling rate. Each bar was fixed at the base and the GPS unit was attached at its tip. By providing an initial displacement (simply by pulling the top of the bar and releasing), each bar was set into free vibration and its motion was recorded. Results are summarized in Table 1. Figure 2 shows the particle motion and time-history of one of the tests performed. The axes of the bar were at an angle to the NS (and EW) direction. Therefore, the NS and EW components of displacements are identical in phase and proportional in amplitude. Also, since the GPS unit is not symmetrically and concentrically mounted in the weak direction (photo in Figure 1), the amplitudes of positive and negative displacements measured are not the same. The detection of the effect of the eccentric mass adds to the assurance that the measurements are accurate and sensitive. These simple tests and results were and can be duplicated easily and readily.

Figure 3 is a plot of NS components of measured relative displacements and corresponding amplitude spectra of Bars A and B. The figure shows the accuracy and sensitivity of the GPS monitoring technology at ten samples/second. The measurements differentiate between the frequency of the free-vibration response of the two bars with different dynamic characteristics. From the data, the fundamental frequency (period) of the two bars are identified to be 0.245 Hz (4.08 s) and 0.296 Hz (3.38 s) respectively. Also, a damping percentage of approximately 2% is extracted. This simple test shows that sampling at 10 Hz with GPS units provides a clear and accurate displacement response history from which drift ratios and dynamic characteristics of the specimen can be derived (Çelebi et al., 1997, 1999 and Çelebi, 1998). The implications of this go beyond just the measurements. It can be shown that identification of variation of dynamic characteristics can be used to identify not only different structural systems, but also possible nonlinearities that occur during vibration (e.g., due to damage and plastic behavior of the structural members, components and/or joints, or to soil-structure interaction under varying amplitudes of input motions).

4. GPS Ambient Test of a 44-story Building and Strong-motion Acceleration Records

In a second test, we measured ambient vibration (due to winds and traffic noise) of a 44-story building with a GPS unit temporarily deployed on its roof. A reference GPS unit was located within 500 m of the building. The signals were very noisy and amplitudes very small; therefore, most common methods to identify structural characteristics did not work. Only the cross-spectrum of the two orthogonal, horizontal, low-amplitude, ambient displacement recordings (when the signals are coherent and approximately 180° degree out of phase) were used to identify the fundamental frequency of the building at 0.23 Hz (Figure 4), with another frequency at ~ 0.3 Hz. Despite the very small signal, these frequencies appear to be reliable when compared with the 0.23 Hz frequency calculated from accelerations recorded with a triaxial accelerograph on the 38th floor (accepted here as the roof response) during a small earthquake (Figure 5). A comparison of these frequencies is provided in Figure 6.

Despite the small displacements (<1 cm) and low signal to noise ratio of this experiment, the fact that dynamic frequencies could be identified indicates that during larger displacements, better identification of the dynamic characteristics as well as drift ratios can be made with higher confidence.

5. Looking to the Future: Real-Time Monitoring

5.1. PERCEIVED REAL-TIME MONITORING

Initially, we are planning to deploy the GPS units only on tall buildings that are already instrumented with accelerometers. This will facilitate comparison of absolute and relative displacements measured by GPS and calculated by double integration of accelerations. The GPS units will be configured to provide data to indicate the real-time average drift ratios and changes in the dynamic characteristics of the buildings. This information can be made available to building managers (or interested parties) in real-time or whenever a predetermined displacement threshold is reached. The building managers can assess the response of the buildings according to (a) different threshold displacements (e.g., A, B and C as shown in Figure 7), (b) drift ratios, or (c) changing dynamic characteristics. If a situation is serious, the management can make decisions to evacuate the building for additional inspection. Therefore, one by-product of the effort would be to secure the safety of the occupants and significant contents of the building. Thus, a real-time structural health monitoring environment will be created. At least three GPS units per building are required to monitor a tall building: two of the units should be deployed on the roof to detect translational and torsional response of the building, and the third unit will serve as a reference ground station to evaluate relative displacement. The ground site also needs excellent sky visibility.

Similar deployments are being planned for other types of long-period structures. One project in development at this time is for deploying GPS units on one of the long-period suspension bridges such as the Golden Gate Bridge and Bay Bridge (San Francisco) or Vincent-Thomas Bridge (Los Angeles). As in the case of tall buildings, changes in dynamic characteristics after the displacements at critical locations of a bridge have exceeded predetermined thresholds, can be calculated in near-real-time (Figure 8). When warranted, the management can make decisions to inspect the bridge (e.g., decisions can be made to stop the traffic, thus securing the bridge safety, which is one of the objectives of lifeline earthquake engineering). With the GPS technology, we can furnish time-dependent displacements for the relative movements of critical locations of structures. For example, for the bridges, GPS units placed at pre-selected locations of bridge elements can indicate, in real-time, the amplitude of the displacements of the decks and towers, as well as movements of key bridge elements relative to local bedrock reference points. Thus, movements of the piers relative to the abutments, the top of a tower with respect to its base, or the span with respect to the ground, can be made at a centimeter-level of precision, in real-time. We will recover both the dynamic motions that accompany the earthquake, as well as the static or permanent displacements experienced by the bridge once the shaking has stopped. Such permanent displacements affect the state of stress of a bridge, and provide evidence for distortion or failure of bridge elements or subsidence of piers due to ground compaction induced by earthquake shaking.

5.2. CURRENT DEPLOYMENTS ON TALL BUILDINGS

As this paper is being written, deployment of GPS units on the roof of two buildings in Los Angeles has been completed, and we are in the process of deploying GPS units on the roof of a building in San Francisco. Figure 9 shows the actual deployment on the roof of one of the buildings in Los Angeles. As it happens with field deployments, physical obstacles and constraints necessitate alternate approaches. In this case, the window cleaner machinery on the roof of the 44-story building travels on rails near the parapet and uses the parapet continuously. Therefore, any deployment had to be planned away from the parapet wall. To solve this problem, a stiff auxiliary support frame (Figure 9) was erected. The figure also shows the GPS antenna and the radio antenna to communicate with the reference station so that differential displacements are obtained.

5.3. REQUISITE SOFTWARE

Requisite software is being developed to assess and mitigate the two natural hazards (earthquake and severe wind) affecting the structures by using the displacements measured by the GPS units. Figure 10 shows a screen image of software being developed to assess displacements observed with GPS and provide alarm (warning). Such collected information on the response of the structure during strong motion events (or strong winds) can be used to make decisions for further evaluation of the susceptibility to damage of the structure, and future repair/retrofit schemes may be developed.

5.4. BENEFITS AND OTHER APPLICATIONS

- The collected information on the response of the structure during strong-motion events (or strong winds) can be used to make decisions for further evaluation of the susceptibility to damage of the structure, and future repair/retrofit schemes may be developed.

- The recorded data can be used to analyze the performance of the structure and such results can be used to improve future analyses/design procedures.

- The data collected can also be used to assess long-term displacements of critical locations of structural systems (e.g., permanent displacements, settlement of foundations, long-term deformations due to change of temperature and the plate tectonic deformation spanning the San Francisco Bay and parts of the Los Angeles Basin) and to develop methodologies on how the findings can be incorporated into useful practical design procedures.

6. Conclusions

It is shown in this paper that recent advances in sampling rates of GPS technology allow real-time monitoring of long-period structures such as tall buildings and long-span bridges. The advantage over conventional monitoring using accelerometers is that relative displacements can be measured reliably in real-time and with sufficient accuracy to assess potential damage to the structures. The technical feasibility is illustrated through two tests conducted on two vertically cantilevered bars that simulate tall buildings, and an ambient test of a 44-story building. Both approaches show that GPS monitoring of long-period structures provide sufficiently accurate measurements of relative displacements such that dynamic characteristics of the vibrating systems can be accurately identified. This capability can be used for structural health monitoring purposes. Procedures and software are being developed to permanently deploy GPS units on tall buildings and suspension bridges.

7. References

1. Behr, J. A., Hudnut, K. and King, N. (1998) Monitoring Structural deformation at Pacoima Dam, California using continuous GPS, Proc.11th International Technical Meeting of the Satellite Division of the Institute of Navigation [ION GPS-98; Nashville, TN], pp. 59-68.
2. Çelebi, M., Prescott, W., Stein, R., Hudnut, K., Behr, J., and Wilson, S. (1999) GPS monitoring of dynamic behavior of long-period structures : ***Earthquake Spectra*** (Journal of EERI), vol. 15, no. 1, pp. 55-66.
3. Çelebi, M., Presscott, W., Stein, R., Hudnut, K., and Wilson, S. (1997)Application of GPS in monitoring tall buildings in seismic areas, Abstract, AGU Meeting, San Francisco, CA, Dec. 1997.
4. Çelebi, M. (1998) GPS and/or strong and weak motion structural response measurements – Case studies, Structural Engineers World Congress-CD-ROM Proceedings, San Francisco, CA.
5. Guo, J. and Ge, S. (1997) Research on displacement and frequency of tall Building under wind loading using GPS, *ION Conference*, Kansas City, MO, (September).
6. Hudnut, K. W. and Behr, J. A. (1998) Continuous GPS monitoring of structural deformation at Pacoima Dam, California, Seismol. Res. Lett., Vol. 69, No. 4, pp. 299-308.

7. Hyzak, M., Leach, M. and Duff, K. (1997) Practical application of GPS toBridge deformation monitoring, Permanent Committee Meeting and Symp., Int'l Federation of Surveyors (FIG), (May).
8. Kondo, H. and Cannon, M. E. (1995) Real-time landslide detection system using precise carrier phase *GPS, ION GPS95 Conference*, The Institute of Navigation, Palm Springs, CA, (September).
9. Lovse, J.W., Teskey, W. F., Lachapelle, G., and Cannon, M. E. (1995) Dynamic deformation monitoring of a tall structure using GPS technology, *Journal of Surveying Engineering*, ASCE, **121**, 1, 35-40.
10. Stein, R. S., Hudnut, K. W., Satalich, J., and Hodgkinson, K. M. (1997) Monitoring seismic damage to bridges and highways with GPS: Insights from the 1994 Northridge earthquake, in *Natl. Seismic Confr. on Bridges and Highways*, Sacramento, Fed. Highway Admin. and Calif. Dept. Of Transportation, 347-360.
11. Teague, E. H., How, J. P., Lawson, L. G., and Parkinson, B. W. (1995)GPS as a structural deformation sensor, *Proceedings of the AIAA Guidance, Navigation and Control Conference,* Baltimore, MD, (Aug.).

Table 1. Results of Tests with GPS Units

[H=Height, B=Width, t=thichness, F=Frequency,T=Period]

BAR	L (m)	B (cm)	T(cm)	Measured		Damping
				F(Hz)	T(s)	[ξ](%)
A	1.82	3.8	0.32	0.245	4.08	~ 2.0
B	1.82	5.0	0.32	0.296	3.38	~ 2.0

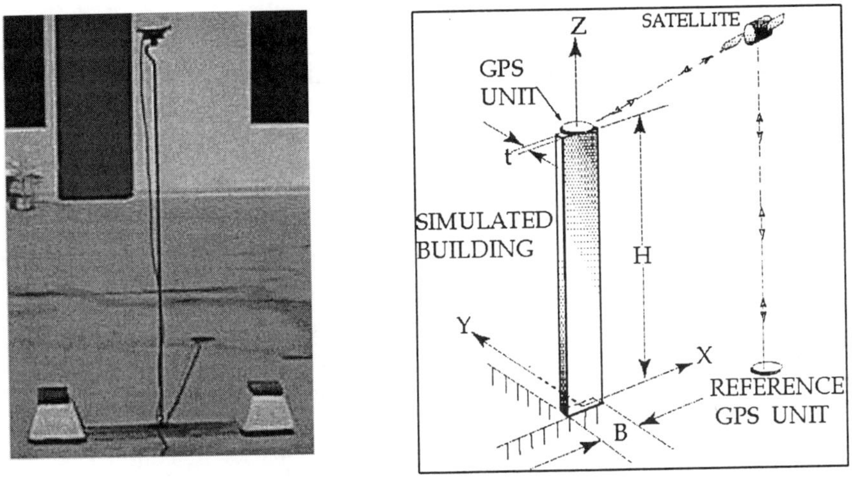

Figure 1. Photo and schematic of test set-up for using GPS for dynamic monitoring of tall buildings.

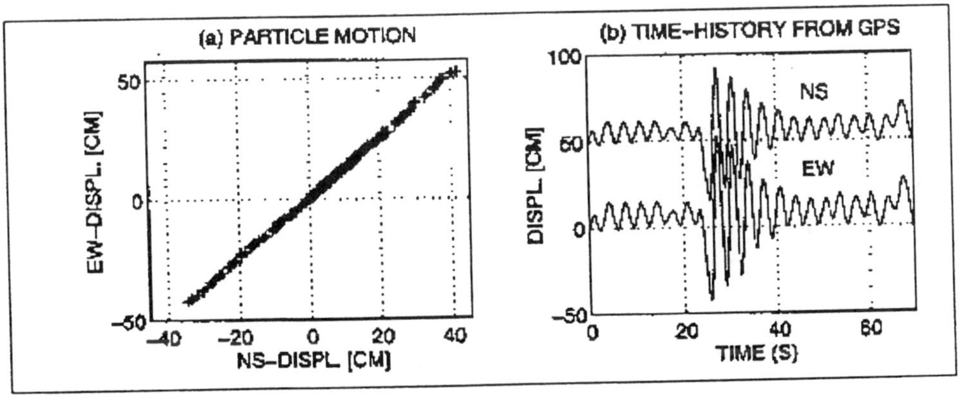

Figure 2. Particle motion and time-history of relative displacements (NS and EW components) of simulated test specimen.

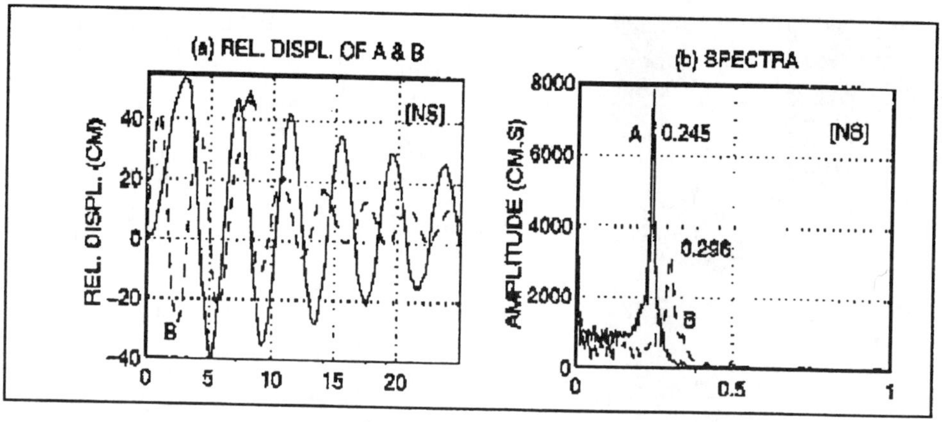

Figure 3. Relative displacements of two test specimens (NS components only) in free-vibration and corresponding amplitude spectra identifying the fundamental frequencies of the test specimens.

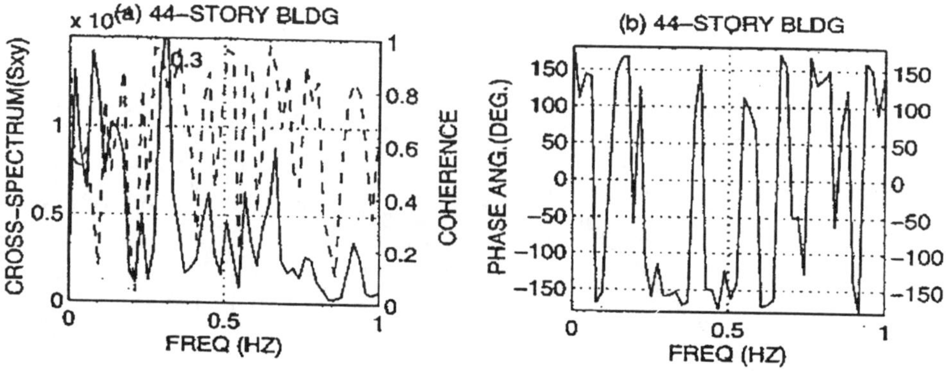

Figure 4. Ambient test of a 44-story building with GPS technology: Cross-spectrum (left-solid), coherence (left-dashed) and phase-angle (right) of two orthogonal horizontal motions.

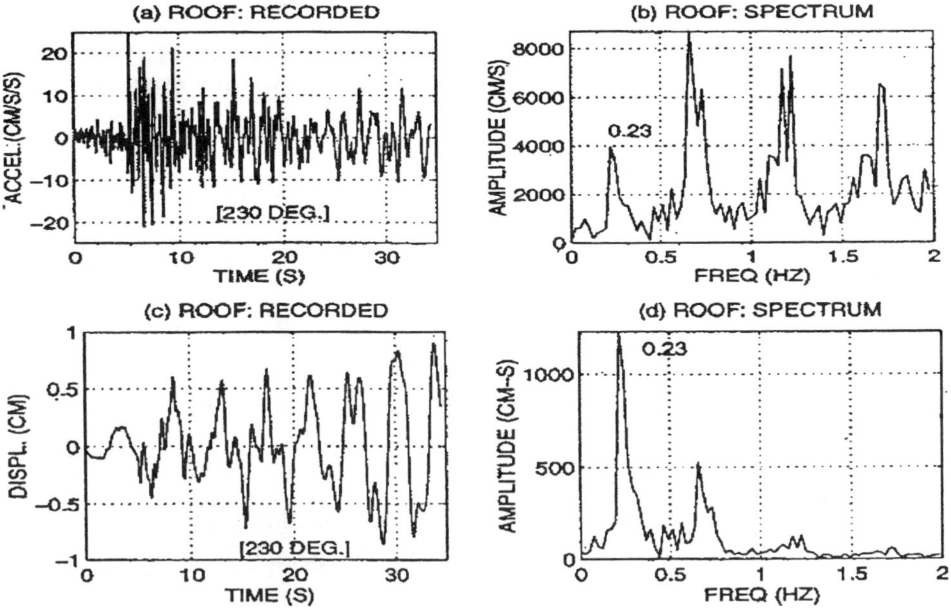

Figure 5. Recorded roof (38th floor accepted in lieu of roof) accelerations of a 44-story building and displacements (derived by double integration) and amplitude spectrum.

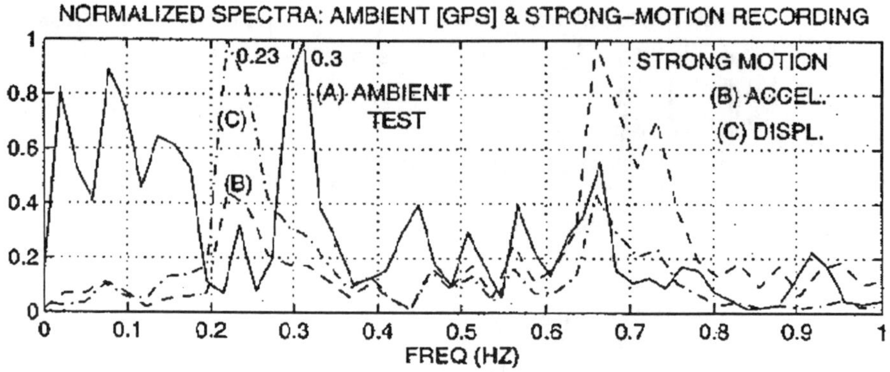

Figure 6. Comparison of normalized frequencies for ambient GPS recording (from cross-spectrum) and strong-motion recording (from amplitude spectrum).

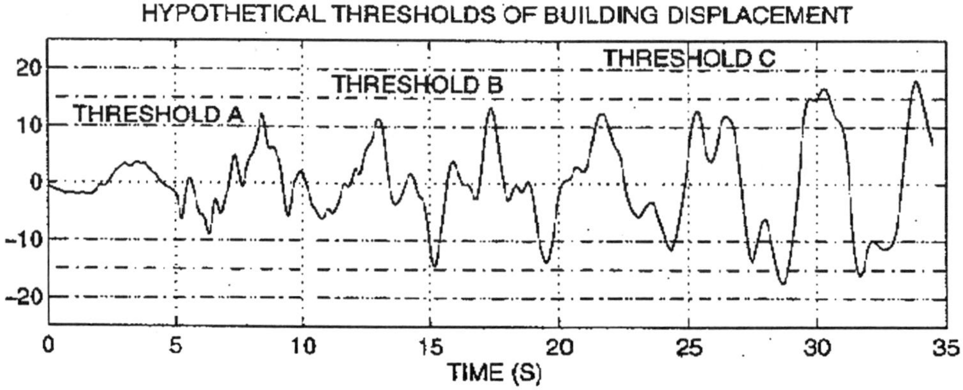

Figure 7. Hypothetical thresholds of displacements. The time-history of displacements shown is actually integrated from accelerations recorded at the 38th floor (accepted in lieu of roof) of a 44-story building. The actual record (in Figure 5) is amplified by 20 times for illustration purposes.

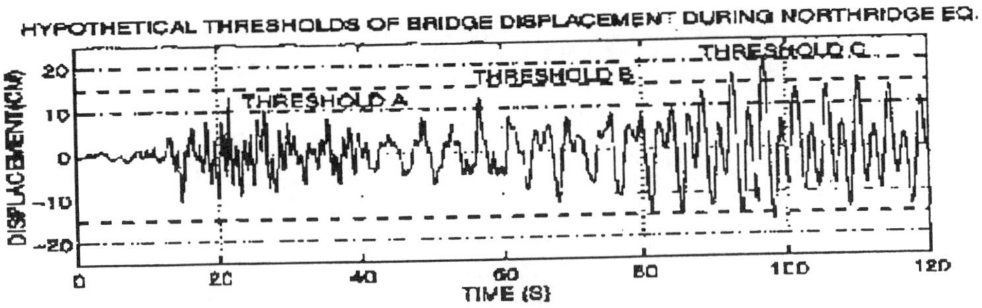

Figure 8. Hypothetical thresholds for displacement (from double-integration of recorded acceleration) of channel 21 (vertical at mid side-span) of Vincent Thomas Bridge (1994 [M=6.7] Northridge earthquake).

Figure 9. Deployed GPS Unit and Necessary Frame System

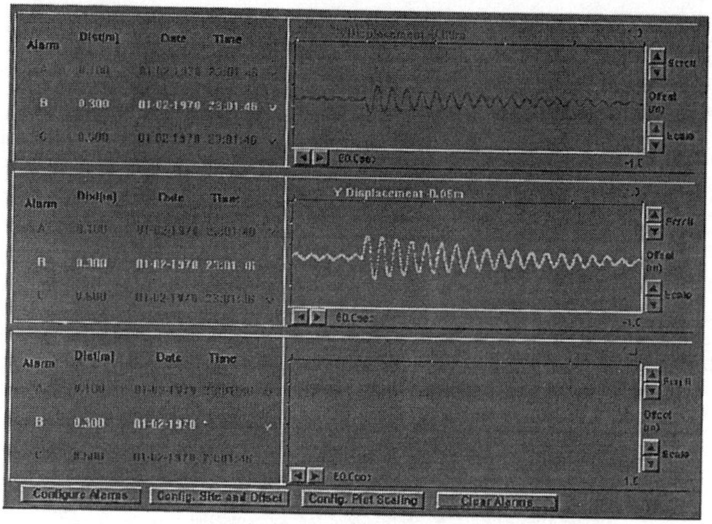

Figure 10. Software developed to provide alarms at three thresholds of displacement of an instrumented structures with GPS.

REAL TIME LARGE STRUCTURE MONITORING USING THE INCLINATION SENSOR

R. N. CELIK
ITU Civil Engineering Faculty, Department of Geodesy and Photogrammetry,Istanbul,Turkey
e-mail:celik@sariyer.cc.itu.edu.tr

Abstract
This paper gives the brief information about the precise inclination sensor called NIVEL20. Thereafter fields that the sensor might be used to determine flatness, torsion, deformation and etc are mentioned. A test carried out to determine the movements of a building and a pillar built on this building are examined using continuously collected data by two NIVEL20s. Therefore movements of the building and the pillar are simultaneously be monitored and effects of building movements onto pillar are investigated in micron level. Results obtained are expressed and interpreted.

1. Introduction

Turkey is a developing country that new large constructions are continuously built in every part of the country such as high rise buildings, large highway structures, dams and so on. These structures are built due to several reasons like renewing, replacements etc. However, all new things always come along with additional problems. The most important problem coming with these large structures are deformations occurred on such structures. If these deformations are not monitored with respect to time, some problems that can not be compensated can cause serious defects, and hence it could be too late to tell 'STOP'. Therefore, precaution must be taken beforehand.

Fortunately, now, the developing technology aids to find a way to sense the size and character of the movements even in real time. This new technology comes up with an instrument called NIVEL20. The new instrument is capable to be used as a very precise real time monitoring sensor for large structures. Internal data collection frequency of this instrument is less than a second which means the motion of large structures under a loading effect can also be observed and investigated in real time.

In the flowing sections, reader can obtain an explanation about working principle of NIVEL20. Thereafter, as an example, test, which is carried out in Civil Engineering Faculty building of Technical University of İstanbul for structure monitoring is explained in detail. Additionally, the results obtained from real time monitoring test is investigated, analysed and then interpreted. Moreover, the way of using NIVEL20s for monitoring purposes on highway bridges (viaducts) is briefly touched on.

M. Erdik et al. (eds.), Strong Motion Instrumentation for Civil Engineering Structures, 397–405.

2. Working Principle of NIVEL20

NIVEL20 is a precise inclination sensor. It is capable to measure inclination in two different independent directions. These directions (X and Y) are placed as perpendicular and intersect each other at an origin. A coordinate system is defined regarding to the origin. Therefore, inclinations measured based on X-axis and Y-axis can take minus and plus value, see Figure 1. As a result, the direction of inclination can easily be identified. This axis definition is displayed on the sensor for user information. The advantage of having inclination measurements in two directions is to observe inclination changes of a surface in every direction and also the interpretation of height information regarding inclination changes.

NIVEL20 inclination sensor works with an optoelectronic principle that has no moving part. The reference plane is provided by a fluid horizontal that is always perpendicular to the true vertical. The inclination is measured with the difference of an angle between the true horizon and the surface of an object on where the sensor placed [1]. In order to start measuring, no adjustment is required unless the sensor placed as out of its working range (± 1.5 mrad). Therefore, the surface on where the inclination sensor is placed must roughly be

Figure 1: Working Principle of NIVEL20

levelled. However precisely levelled surface can be much better because in some case a measurement must be able to be made close to edge of the working range limitation or small movements might put the sensor out of the working range due to roughly levelled surface.

NIVEL20 is a sensor that measures also the temperature of the environment to eliminate the errors occurred due to temperature changes. The sensor is capable to be used in continuous mode that provides continuous monitoring observation as it was used in the test described in this paper.

3. Where & How can The Inclination Sensor be Used?

In effect, the inclination sensor can be used anywhere where very precise inclination measurements are required. One of the main fields that the inclination sensor can successfully be used is industry for levelling large machines, controlling surface flatness and straightness, controlling deflection and torsion of the large machines and surfaces and etc.

All applications mentioned above are entirely related with industry. However, when large building construction and their deformation are considered, the inclination sensor is also a very useful tool to tackle with the problems come along with large constructions. Dam, viaduct, bridge and high rise building deformation can easily and successfully be monitored in real-time using NIVEL20 because the number of sensors (max 128 sensors) can be connected each other with cables. This capability provides that large areas can be covered with sufficient number of sensors and hence movements of large areas can be monitored. For example if a viaduct or a bridge deformation is considered, temporary and permanent movements of the construction can continuously be monitored. Temporary movements might occur due to temporary heavy loading like traffic jam or long vehicle passing or etc. Permanent movements might occur due to ground properties or natural effects as earthquake, land sliding etc. Both might be the deformation source, although permanent movements are more effective movements for deformation. However the temporary movements might also be observed to compere the act of construction with the modelled one against expected effects. All these movements can be identified using the inclination sensor in real-time continuous mode.

Also during destruction work, buildings and other constructions exists around the destruction area might seriously be damaged due to vibration and loss of static support taken by destroyed building. Therefore the inclination sensor can also be used during destruction work, to be able to monitor the other constructions within effective range of destruction side effects. There might be so many different fields than mentioned above, which the inclination sensor can aid to solve for many problems.

4. Definition of The Test

In order to see the capability of the sensors for structure monitoring a test has been carried out in ITU. The test area was the roof of the civil engineering faculty building. At the roof there is a pillar that is for GPS continuous reference station and is also dedicated as IGS (International Geodetic Service) site. There was some hesitation on location of the pillar since it is on a building that might not be stable; self-movements of the building might affect the position of the pillar. Therefore, self-movements of the building and the pillar should have been monitored with respect to time.

As a first step two NIVEL20 sensors have been placed: one on the pillar and one on the building. The pillar has been built on one of the piers of the building. The sensor on the building was placed very close to the pier. The sensor on the pillar was protected against direct sunlight and rain covering with a plastic cover, and this sensor was called as Sensor-1. The other one was in the building, and this sensor was called as Sensor-2. Both sensors' axis were oriented parallel to each other. Namely, X-axis of the Sensor-1 was parallel to X-axis of Sensor-2 and also Y-axis of the sensors. X-axis was oriented approximately East-West direction and naturally Y-axis was in North-South direction since X and Y-axis are perpendicular, see Figure 2.

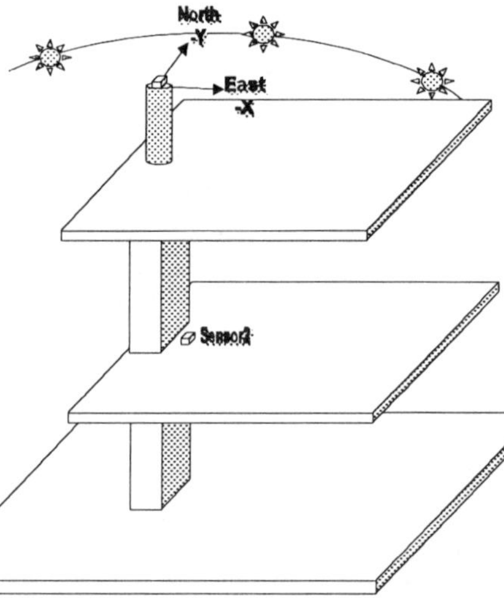

Figure 2: Location of the sensors

Both sensors have been connected each other via cable in RS485 standard and then to a notebook computer using RS232 standard. SOPOM software [2] was used to communicate with the sensors and to collect and evaluate the raw inclination data. Raw data have been collected on-line and stored every minute. Data collection was completed within approximately a week.

5. Results of The Test and Interpretation of The Results

The inclination measurement was carried out eight days. Table-1 shows the dates and the weather conditions during test. As is seen from the table, measurements were started in 19th March.1999 and completed 26th .March.1999. Within these dates system continuously collect data except the time of electricity was off. In whole period, electricity was cut 4 times and total number of minutes is 29, which is negligible.

During the test, first 2 days, the weather was cloudy and rainy, thereafter following 3.5 days was sunny. Afternoon of 24th of March the weather was cloudy and also rainy again. Following 2 days was sunny again. Temperature from night to day gradually changed. The lowest temperature was about 3°C in early morning in 24th and the highest was about 31°C in afternoon in 24th for Sensor-1. The lowest temperature was about 14°C in morning in 21st and 22°C in afternoon in 25th for Sensor-2, see Figure 3. Figure-3 shows the temperature changes, time (day basis) against temperature in Celsius degree.

Table-1: Test dates and weather conditions

Day	Date	Time	Weather Condition
1- Friday	19-3-1999	17:10	Cloudy & Heavy Rain
2- Saturday	20-3-1999		Cloudy & Rainy
3- Sunday	21-3-1999		Sunny
4- Monday	22-3-1999		Sunny
5- Tuesday	23-3-1999		Sunny
6- Wednesday	24-3-1999		Sunny - Cloudy & Rainy
7- Thursday	25-3-1999		Sunny
8- Friday	26-3-1999	18:25	Sunny

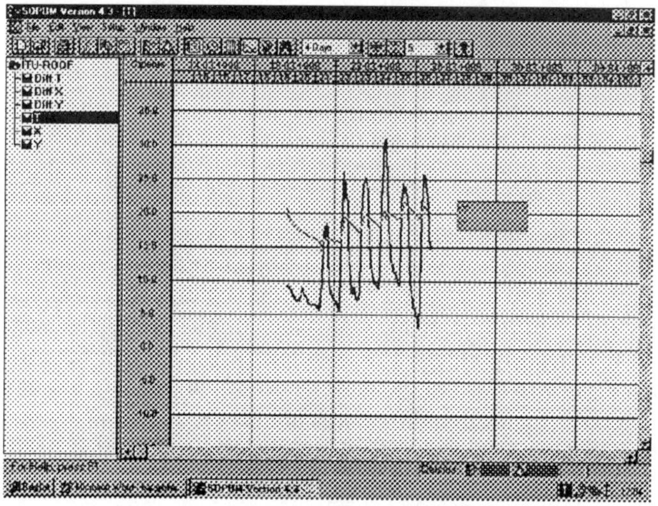

Figure-3: Temperature measurements of the sensors

The blue line, which has also high amplitude, represents Sensor-1 measurements and the green line, which has low amplitude, represents Sensor-2 measurements. Therefore the maximum temperature differences are 29°C for Sensor-1 and 8°C for Sensor-2. This can be seen also in Figure-4. Figure 4 shows the temperature differences regarding zero measurements of each individual sensor during test period, time (day basis) against

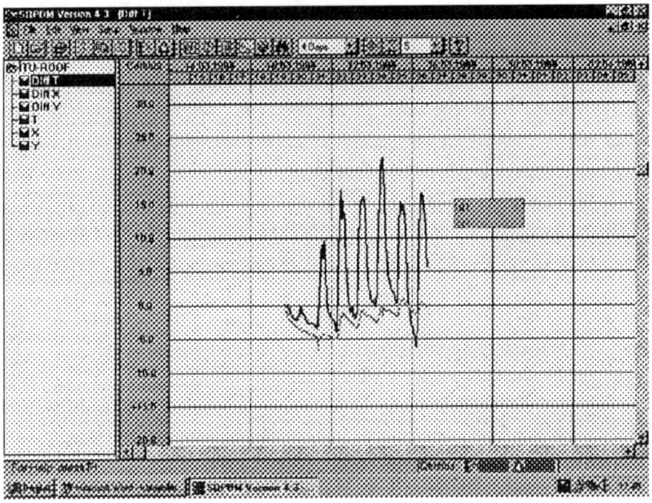

Figure-4: Temperature differences from the zero temperature measurement

temperature in Celsius degree. Blue line, which has also high amplitude, represents Sensor-1 measurement differences and Green line, which has low amplitude, represents Sensor-2 measurement differences.

Having examined temperature differences, inclination measurements can be investigated. Inclination measurements are also given as difference of inclination from zero measurements of each sensor. Figure-5 shows the inclination measurement differences with respect to X-axis, and Figure-6 shows the inclination measurements regarding Y-axis. Both graphs in Figure 5 and 6 shows the inclination differences time (day basis) against inclination differences in mrad. In both graphs the blue line, which has also high amplitude, represents Sensor-1 measurement differences and the green line, which has low amplitude, represents Sensor-2 measurement differences.

Before examining the graphs one should be remember that X-axis was in the same direction of East-West and the Y-axis was in North-South. This information is very important for interpreting the results, and also this should be known that minus X-axis directing East and minus Y-axis directing North. Having remembered all these, the graphs can be investigated. When the graph in Figure-5 is examined it is seen that Sensor-2 (the green line) which has been located to observe the movements of the building has not show much movements in X direction. Inclination differences are between -10 and +40 μrad, which is negligible. However, Sensor-1 (the blue line) which has been located on the pillar has shown large movements. It is interesting that, movements draw approximately same figure almost every day; but movements in first 2.5 days and 24th March are different. Why those days were different? When the weather conditions are investigated given with Table-1, it can be seen that in those days there was no direct sunlight on the pillar. First 2 days were totally cloudy and also rainy, however in 24th of March first half of the day was sunny and then it was cloudy

and rainy again. Therefore, the pillar began its periodic movements in first half of the day; but did not complete due to loss off direct sunlight effect and hence amplitude of the movements was less than expected one. Moreover the highest temperature differences were also recorded in 24th of March; but that did not aid the pillar to complete its periodic movements. This also shows that the movements of the pillar do not dependent on temperature differences, it mostly depends on direct sunlight effect. The inclination measurements on Y-axis also prove this result. In X-axis direction, pillar makes its maximum movements during daylight. The pillar begins moving about 6:30 in the morning

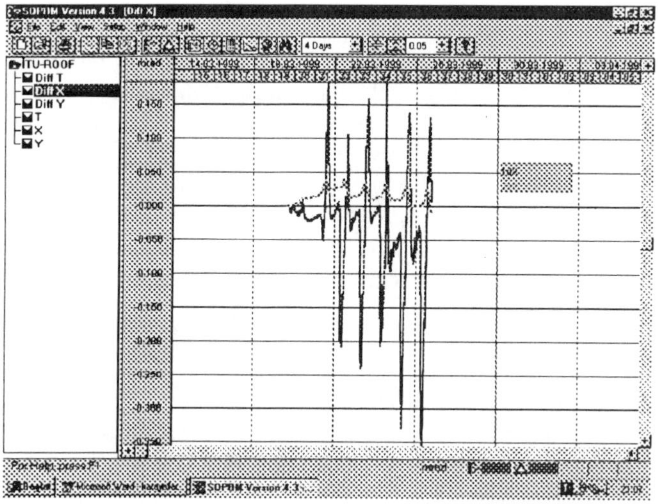

Figure 5: Inclination differences from the zero inclination measurement on X-axis

towards to positive X-axis direction and reach maximum movement between 9:00 and

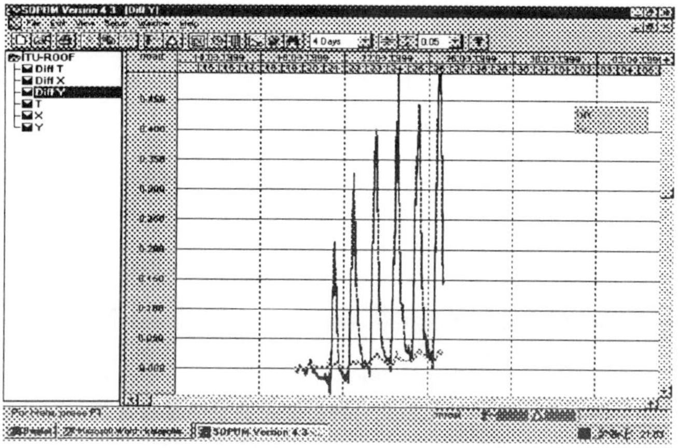

Figure-6: Inclination differences from the zero inclination measurement on Y-axis

10:00 hours. Thereafter it goes back to its morning position and finds this position between 13:00 and 14:00 hours. Then it begins moving towards negative X-direction and reach maximum movement between 16:00 and 17:00 hours and turning back its morning position again. This shows that the side of the pillar, which takes the direct sun light, expands due to heating up, and hence inclines towards to the side under shadow.

The graph in Figure-6 shows that Sensor-2 did not incline much in Y-axis direction. Inclination differences are between 0 and -35 μrad, which is negligible. However, Sensor-1 did record some interesting inclination differences with respect to time here again; but these movements were one directional. Here, pillar begin inclining towards to negative Y-axis direction about 6:30 in the morning and reach the maximum movement between 11:00 and 14:00 hours. Thereafter it turns back and completes its movement between 19:00 and 20:00 hours. These movements are also related with direct sunlight effects on pillar.

6. Conclusion

As is seen from the test results, the building, where the pillar built on, has no movement influence on the pillar. The building is stable, its maximum movements is less that 50 μrad that is negligible. However, the pillar has self-movements. When these movements are compared with the building ones, they are too large; but even so the size of the pillar movements are not big enough to be taken into account when GPS continuous station is consider. It moves between +180 and -355 μrad in X-axis direction and between +540 and -40 μrad in Y-axis direction. The total size of movement is about 535 μrad in X-axis direction and 580 μrad in Y-axis direction, In other words, it is about 0.8 mm/m. Therefore, these will not effect the position of the GPS site.

However, it is obvious that the pillar moves under direct sunlight effect. Its movements are directly related with sun movements. When the results are reviewed, it can be realised that the pillar follows the sun as a moonflower follows the sun but in opposite direction. In other words, the pillar inclines towards to opposite direction that points out orbital location of the sun, since when pillar take direct sunlight, only one side of the pillar heats up. Hence, pillar loses its homogeneity and it expands. Consequently, the pillar moves towards to the other side

This test shows that NIVEL20 is the right sensor to precisely monitor large structures. It is capable to identify even a very small movement and also capable of separating two different movements occur on a same structural body.

Acknowledgements

I would like to acknowledge Sistem Bilgisayar ve Teknik Hizmetler Sanayi AŞ and Leica that provide the sensors and technical support to carry out this test at ITU.

References

1. Anonym, 1998, Nivel20 Instruction Manual, 308 592 EN-IX.98, Leica Geosystems AG, Switzerland
2. Anonym, 1998, SOPOM Instruction Manual, U2-267-0EN-II.98, Leica Geosystems AG, Switzerland

MONITORING DEFORMATION ON KARASU VIADUCT USING GPS & PRECISE LEVELING TECHNIQUES

R. N. CELIK[1], T. AYAN[1], H. DENLI[1], T. OZLUDEMIR[1], S. EROL[1], B. OZONER[1], N. APAYDIN[2], M. ERINCER[2], S. LEINEN[3], E. GROTEN[3]
[1]*ITU Civil Engineering Faculty, Department of Geodesy and Photogrammetry*
[2]*General Directorate of Highways 17th Division of Directorate Istanbul*
[3]*Institude of Physical Geodesy, Darmstadt University of Technology*
email:celik@sariyer.cc.itu.edu.tr

Abstract
A project has been carried out to investigate the performance of GPS for determining deformations of large strong structures. As a test structure, a viaduct has been chosen 40 km away from West of Istanbul and built in a marsh area. Four epochs of GPS and levelling measurements have been done. Results obtained from epoch measurements are investigated and expressed in this paper. Also the needs of a special force centring equipment to achieve the accurate results are emphasised.

1. Introduction

The development of technologies affects all engineering fields. Now using technological developments large size of structures can easily be constructed and then these are opened and served to the use of people in a very safe condition. However, this service must also cover monitoring of structure against natural effects, since natural effects might cause serious defects when long period of time of use is considered.

Karasu viaduct, which is 2160 metres long, has been structured on a marsh area, and is on the Transport European Motorway (TEM). It is an important highway construction and is one of the longest viaducts in Turkey. Although geological and geotechnical ground analysis has been done before it was structured, the viaduct must be monitored by using geodetic techniques against possible deformations. Therefore a project has been agreed to determine the deformations of the viaduct.

In order to determine the possible deformations, 24 survey marks on the viaduct as deformation points and 6 pillars out of the viaduct as reference points have been established. Four epochs of GPS observations have been done on these points. First epoch was carried out in July 1996. Second one was in March 1997. Thereafter the third one was in October 1997 and finally the last one was in April 1998. All epochs have been individually evaluated and the combinations of the results obtained are investigated.

M. Erdik et al. (eds.), Strong Motion Instrumentation for Civil Engineering Structures, 407–415.

The following sections of this paper provide information about the preparation and planing process of the project, and the fieldwork including survey equipment specially designed for this project, and finally the determinations and results obtained are expressed.

2. Definition of The Project

Today, using developing technologies, specially the GPS techniques, large structures can successfully be monitored against any deformation effect. This project has been set up to monitor possible movements of the Karasu Viaduct, since the viaduct has been structured on a marsh area that increase the deformation risk on the structure by time. The Karasu Viaduct is a large, 2160 metres long, highway bridge about 40-km away from West of Istanbul in Turkey. It is a part of Transport European Motorway (TEM) and passing over the lake of Büyükçekmece dam. First 1000 metres of the viaduct have been built upon the lake. The viaduct has 2 x 2 lanes, and each lane is 13 metres wide, and whole structure has been built as curve based on approximately 100 piers [1], see Figure 1.

This project has been agreed as a joint project between the Department of Geodesy, Istanbul Technical University, and Institute of Physical Geodesy, Darmstadt University of Technology, under the sponsorship of the Volkwagen-Stiftung, Germany, and with the support of 17. Section of General Directorate of Highways Organisation, Turkey. The project aim is to monitor deformation on the viaduct using GPS and precise levelling techniques to understand especially the suitability of GPS for this purpose.

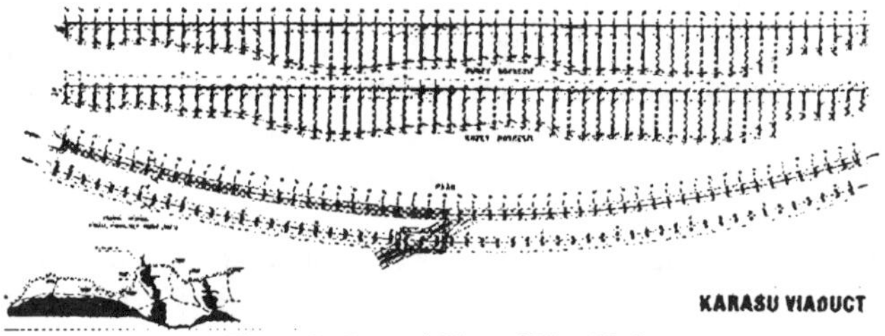

Figure 1: General Plan of The Viaduct

3. Surveying Campaigns & Equipment Used

As is very well known that the deck of all bridges always moves with in a certain limitation, that does not mean that these movements are deformations. Naturally, deformations occur on the body of the bridges. In order to be able to detect the deformation of the bridge using geodetic techniques, measurements must be carried out directly on the piers of the viaduct. Therefore, using a machine 24 holes have been opened on the deck to reach the piers, see Figure 2. Thereafter, totally 12 survey marks have been placed on

Figure 2: Viev of Survey Marks and Equipment used for Force Centring of GPS Antenna

every fifth piers in each side of the viaduct, northern and southern sides respectively. Addition to these 24 survey marks 6 more survey marks as pillars have been set up out of the viaduct as reference points for deformation measurements, see Figure 3.

In order to achieve required accuracy, forced centring is essential to set the GPS antennae on the survey marks, and also special equipment with a known constant from the antenna

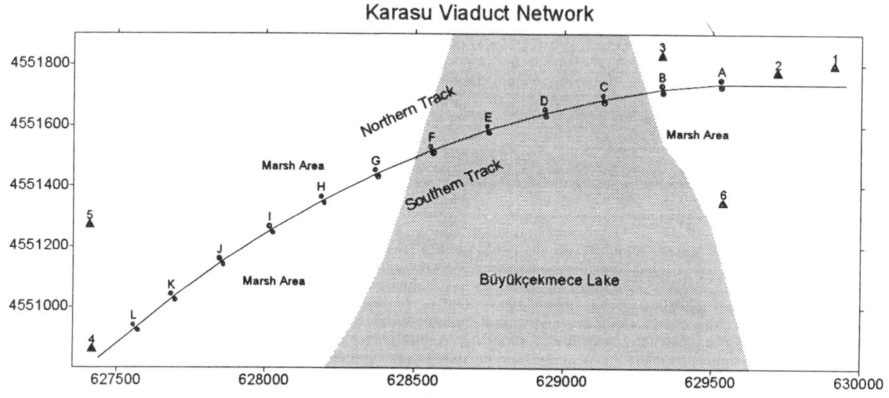

Figure 3: Location of the Survey Marks in Karasu Viaduct

phase centre should be used to determine the antenna heights as accurate as possible. To realise this, special equipment has been considered and developed as a fixed known length rod which can easily be placed into levelled tribrach and on which the antenna can be screwed, see Figure 2.

Four survey campaigns, that were timely independent each other have been carried out. In other words, four epochs of measurements have been done, since the expected deformation rate has been assumed to be few millimetres level per year. First campaign was in July 1996, second one was in March 1997, third one was in October 1997 and finally the last one was in April 1998. As is seen from the dates of measurement the intervals between the campaigns are about half a year, for exact intervals, see Figure 4.

Figure 4: Exact intervals betwen campaigns

Two separate survey groups carry out GPS campaigns and precise levelling measurements were simultaneously. For GPS survey, Leica SR399 and Trimble SSi double frequency GPS receivers were used. The detailed list of GPS receivers used and the survey points measured for each campaign can be obtained in Table 1.

TABLE 1: Summary of The GPS Campaigns and Levelling Measurements

No	Date of Epoch	Type of Measrm.	Available Measurement Data							Sessions	Receiver Type
			Bridge Points	Ref1	Ref2	Ref3	Ref4	Ref5	Ref6		
I	Jul'96	GPS	✓	✓	✓	✓	✓	✓	o	2x4 hours	2 Leica SR299
		Levelling	✓	✓	✓	✓	✓	o	o	Rf.Pnt 8h	4 Trimble SSi
II	Mar'97	GPS	✓	✓	✓	✓	✓	✓	✓	2x3 hours	6 Leica SR299
		Levelling	✓	✓	✓	✓	✓	o	o	Rf.Pnt 6h	4 Trimble SSi
III	Oct'97	GPS	✓	✓	✓	✓	✓	✓	✓	2x3 hours	6 Leica SR299
		Levelling	✓	o	✓	✓	✓	o	o	Rf.Pnt 8h	4 Trimble SSi
IV	Apr'97	GPS	✓	✓	✓	✓	✓	✓	✓	2x3 hours	6 Leica SR299
		Levelling	✓	✓	✓	o	✓	o	o	Rf.Pnt 8h	4 Trimble SSi

Two timely independent sessions were carried out on all survey marks in every epoch. In first campaign GPS observation collection were carried out 4 hours for every session at points on the bridge, and reference points were observed 8 hours. In the other campaigns session have been reduced to 3 hours since after the first campaign evaluation, it was seen that same base line accuracy can also be achieved less than 4 hours. The reference points have been observed always 8 hours except the second campaign because of the security reason of the survey team. Days are short in March, hence working on or around the highway and marsh area are very risky in dark.

4. Determination and Results of Data Collected

Evaluation of the GPS observation was carried out using Bernese software version 4.0, [2]. Observation of each campaign is individually processed. However that process was done in two levels: Firstly the single epoch solution was realised and then the solutions of the epochs produced are compared and investigated each other to identify the deformation between epochs at the second level. In effect, the results obtained from these two levels are interacting forward and backward regarding stochastic structure, network datum, fixed points and so on, [3].

The network established is a small network especially considering use of GPS technique, it is about 2.6 km long and 1 km wide. Therefore, no problem can occur during ambiguity fixing for baseline solution, when the amount of data collected in a session is considered, and also the ionosphere effect is neglected. As a result, it is sufficient to use standard processing parameters of the software to solve for the ambiguities and hence to determine the baselines.

GPS measurements were also supported with levelling measurements for two reasons. As is very well known that the weakest component of a coordinate obtained by GPS is the height component, this is mainly because of the weakness of geometric structure of GPS. Therefore the support from levelling survey will definitely benefit the GPS. Another benefit of levelling is to check antenna heights measured that directly affects the height obtained by GPS. Precise levelling measurements have been also evaluated and then the levelling network has been adjusted. Both results obtained from GPS and levelling are also compared and investigated, Figure 5 and 6 gives the final results of height components for both surveys, GPS and levelling, differencing the epoch solutions analysed. Horizontal figures are not given since significant horizontal deformation was not detected.

First of all GPS results are investigated given as northern and southern graph in Figure 5. These graphs include the results of reference points, that surprisingly the deformation is only detected on two of them, although they were expected to be stable points. As is mentioned, the area is a marsh area that this characteristic might widen also underneath of these two reference points, point 2 and 4. Location of these points is close to east-end and west-end of the viaduct respectively, see Figure 3. Generally, all points as reference and deformation are showing similar characteristic of movements from one epoch to other that they are going upwards and downwards. However, maximum movements obtained at reference point 2 and 4 can be interpreted as deformation. The other movements on the viaducts obtained cannot be interpreted as deformation since they are not significant movements regarding the accuracy achieved. This interpretation is also supported by the results obtained from leveling. Similar directional and similar amount of movements at each point were detected from both surveys, see and compare Figure 5 and 6.

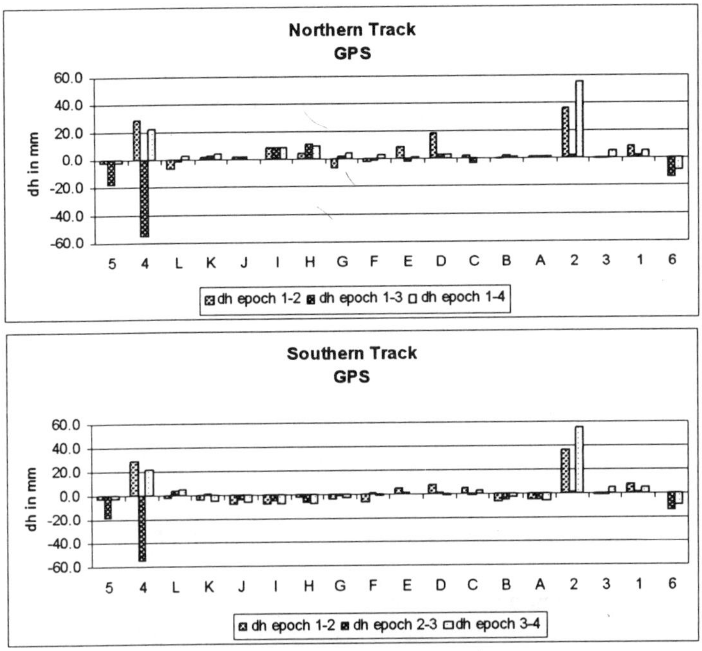

Figure 5: GPS height differences in Northern & Southern Track

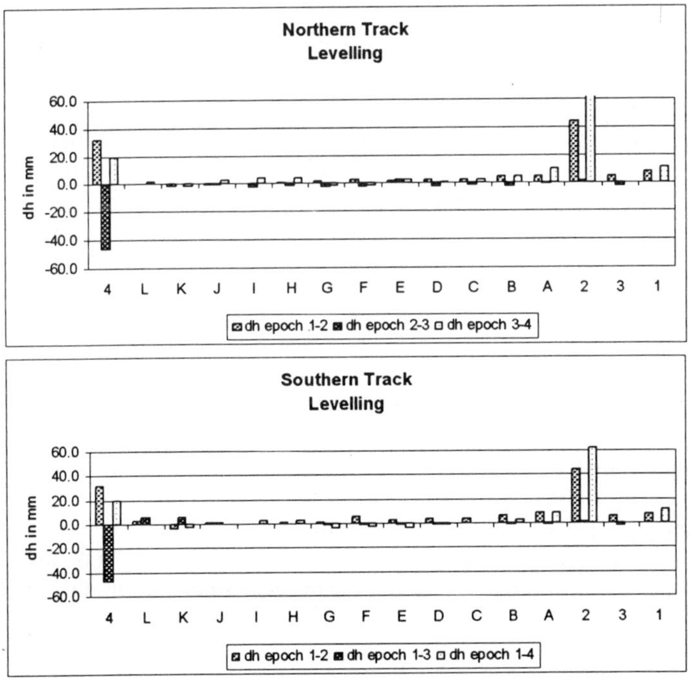

Figure 6:Levelling height differences in Northern & Southern Track

TABLE 2: Accuracy measures achieved in all epochs

Accuracy Measures	Epoc h	Nort h	Eas t	Up	Nort h	Eas t	Up
Group 1: Standard errors from GPS adjustment	1	0.1	0.1	0.3	0.2	0.1	0.4
	2	0.1	0.1	0.2	0.1	0.1	0.2
	3	0.1	0.1	0.2	0.1	0.1	0.2
	4	0.1	0.1	0.2	0.1	0.1	0.2
Group 2: GPS repeatability: rms from differences session solution-epoc solution	1	5.1	5.9	4.6	19.7	29.0	23.1
	2	1.6	1.4	5.8	6.2	6.6	18.5
	3	1.6	2.2	3.1	4.2	5.2	6.1
	4	1.9	1.8	2.4	9.5	12.9	5.7
Group 3: Height accuracy from levelling adjustment	1			0.49			0.66
	2			0.86			1.15
	3			0.78			1.29
	4			2.19			2.96

In order to interpret the meaning of differences, the accuracy achieved from surveys should be examined. Table 2 gives the information about accuracy achieved. Group 1 in Table 2 shows the standard errors of GPS results directly taken from Bernese software. As is seen from the figures, they are too optimistic since they only represent the position error excluding centring, antenna height measurement, and multipath errors. Therefore, they point out position error of antenna phase centre. In other words they represent user independent achievable error using the GPS system. However, Group 2 in Table 2 represents the rms obtained from differences between session solution and epoch solution. Thus, these errors are much more close to the reality and accepting these as the accuracy of the GPS results is more realistic. First epoch of Group 2 has higher errors than the other epochs; this is because the equipment used for centring and antenna height measurement in first epoch. For the following epochs, these problems were eliminated developing additional hardware tools for centring and antenna height measurements. Benefit of these developments is seen from the improved positional error on following epochs in Table 2 - Group 2.

5. Conclusion

GPS technique is sufficient to be used for determining or monitoring deformation on large strong structures. In order to achieve the best results the side equipment must be well designed for force centring and antenna height measurements. The experience with Karasu viaduct obviously draws this conclusion. Levelling also aids to improve height accuracy of GPS; firstly providing a possibility of checking the antenna height measurement, and secondly the combination of both GPS and levelling measurements improves the 3D positional accuracy. However this is not mentioned in detail in this paper since the levelling measurements were carried out to compare the height components of GPS position which is the weakest component in accuracy mean.

As is explained, four epochs of measurement collected were carried out. In every epoch 24 deformation points and six reference points were measured. All data collected were processed and then analysed. Thereafter, it was seen that two points moved, point 2 and 4; but these points are not the deformation points as expected to move, on the viaduct, they are the reference points. Their movements were in two directions, upwards and downwards. This result shows that marsh area widens underneath of these reference pillars and these pillars have not been founded as deep as piers of the viaduct. Consequently, they move but the piers not.

Acknowlegments

Volkwagen-Stiftung, Germany, that is sponsorship of the project is gratefully acknowledged. İzmir Grup AŞ that provided three more GPS receivers and 17th Division of General Directorate of Highway Organisation who provided the security of highway during measurement campaigns are gratefully acknowledged.

References

1. Anonym, Karasu Viaduct, Internal publication of General Directorate of Highways 17th Division of Directorate, Istanbul
2. Rothacher, M. and Mervat, L., (1996), Bernese GPS Software 4.0, Astronomical Institute, University of Berne, Switzerland
3. Leinen, S., Groten, E. And Ayan, T. (1998), The Combined Use of GPS and Levelling For Deformation Monitoring- Experiences From a Large Highway Viaduct, Symposium-Recent Crustal Movements, 1998, Cairo

SEISMIC MONITORING OF NUCLEAR POWER PLANTS; AN APPROACH TO OPTIMAL AND MORE ACCURATE SEISMIC DATA PROCESSING AND INTERPRETATION PROCEDURE

D.M.DOJCINOVSKI , D.J. MAMUCEVSKI AND V.P.MIHAILOV
Institute of Earthquake Engineering and Engineering Seismology at the University St. "Cyril and Methodius'; 91000 Skopje, Republic of Macedonia; 73, Salvador Aljende, Str.; P.O.Box 101; 91000 Skopje Republic of Macedonia; Tel: +389-91-111-344 (112-154); Fax: +389-91-112-163; e-mail: dragi@pluto.iziis.ukim.edu.mk

Abstract

The seismic monitoring systems on NPP-s have two main functions: (1) to provide recorded accurate data on seismic input and dynamic behaviour of structures and the vital hardlines and processing control and protection systems, and (2) to enable automatic alarm, and, in some cases, automatic stoppage of the processes that go on in NPP-s in case of earthquakes with intensities higher than the previously defined values.
The exact records of the seismic input and dynamic response are important from two aspects: (1) they should enable fast inspection of NPP-s after earthquakes and (2) they should enable checking of the previous design and analytical models for definition of the seismic safety of the NPP-s.

The new regulations for seismic design of NPP-s anticipates a complex concept for definition of intensity of seismic motion. Apart from amplitudes (peak acceleration) and frequency characteristics (dominant frequencies), the new regulations consider also the energy characteristics (cumulative absolute velocity).

Presented in the paper is a concept for processing of accelerogrammes based on comparative analysis of two software packages: Kinemetrics SWS (Seismic Work Station) and the procedure developed at the Institute of Earthquake Engineering and Engineering Seismology. Both concepts are based on the same theoretical basis and represents derivatives of the main procedure developed at CALTECH Institute from Pasadena, USA.

Elaborated is a concept for definition of the input parameters; band-pass filter corner frequencies and slopes, elimination of DC components in the recorded accelerograms and calculation and selection of 100 most relevant values of frequencies for calculation of response spectra. Used as seismic input were two types of data: recorded accelerogrammes generated by biaxial shaking table (the applied acceleration and displacement time histories were used as reference and comparative signals) and "in situ" record of blast experiments.

Key words: processing of accelerogram, time history, response spectra, band-pass filter

M. Erdik et al. (eds.), Strong Motion Instrumentation for Civil Engineering Structures, 417–431.

1. Introduction

The seismic monitoring systems in nuclear power plants (NPP) represent part of the safety instrumentation installed for the purpose of:

- Alarming and (or) turning off individual systems of the NPPs in case of an earthquake inducing motion of ground, engineering structures and equipment that exceed the previously defined amplitudes or spectral amplitudes of motion induced by strong earthquakes, and,
- Providing of certain and accurate records of earthquake ground motion, motion of the structures and the equipment for the purpose of verification, evaluation and possible correction of the results obtained from previously performed field, analytical and experimental investigations.

The seismic monitoring of NPPs is performed continuously, within a long time period, in several phases of construction and exploitation of the NPPs.

- In the phase of investigation of the site and design of the NPPs, monitoring of the site and its surrounding is performed by use of seismographs for the purpose of obtaining recorded data on the seismic activity of the immediate surrounding of the site of the future NPP. These investigations are part of the experimental-analytical investigations for definition of the seismic parameters of the site.
- In the phase of construction of the NPP, monitoring by use of seismographs continues, but local accelerographs are also installed for recording of strong ground motion for the purpose of possible verification of the design seismic parameters.
- In the phase of operation of the NPP, integral seismic monitoring systems are installed. These have at least two functions:

Recording of:

- input earthquake motion (on the free field of the NPP site or at the foundation),
- dynamic behaviour (response) of the engineering structures and the equipment (if seismic instruments or sensors are installed on these) in the course of the seismic effect.

Protection of:

- structures, equipment and personnel in case of exceedence of allowable acceleration levels (previously defined design values), through alarming or eventual automatic shut down of certain systems of the NPP.

In this study are presented briefly the theoretical bases of the procedure for processing of accelerogrammes and interpretation of processed data. Comparative results from application of two procedures for processing of seismic data of the type of accelerogrammes are also presented. Presented in the form of conclusions and recommendations is a global concept for using the processed data in the process of exploitation, maintenance, modifications and modernization of NPPs. Finally, some main recommendations for correct application of the programme package SWS for the needs of NPP "Kozloduy" are given.

2. Processing Of Digitized Accelerogrammes

The digitized accelerogrammes are processed by application of sophisticated and scientifically verified methodologies that enable obtaining of corrected and verified data on the amplitude-frequency content of the recorded earthquake motions.

The two principal tasks of this research programme are the following:

- interpretation of the mathematical bases and the theory of the procedure for processing of accelerogrammes, and,
- comparison of two methodologies for processing of accelerogrammes via processing of identical digitized accelerogrammes.

The subject of investigation are two programme packages:

- The programme package SWS (Seismic Work Station), block diagram (Fig. 1) of "Kinemetrics", Pasadena, USA which is a commercial product of the firm, and,
- The programme package for processing of accelerogrammes that is applied in the Institute of Earthquake Engineering and Engineering Seismology (IZIIS) Skopje, Republic of Macedonia, block diagram (Fig. 2), which is applied in scientific and applicative investigations realized in IZIIS.

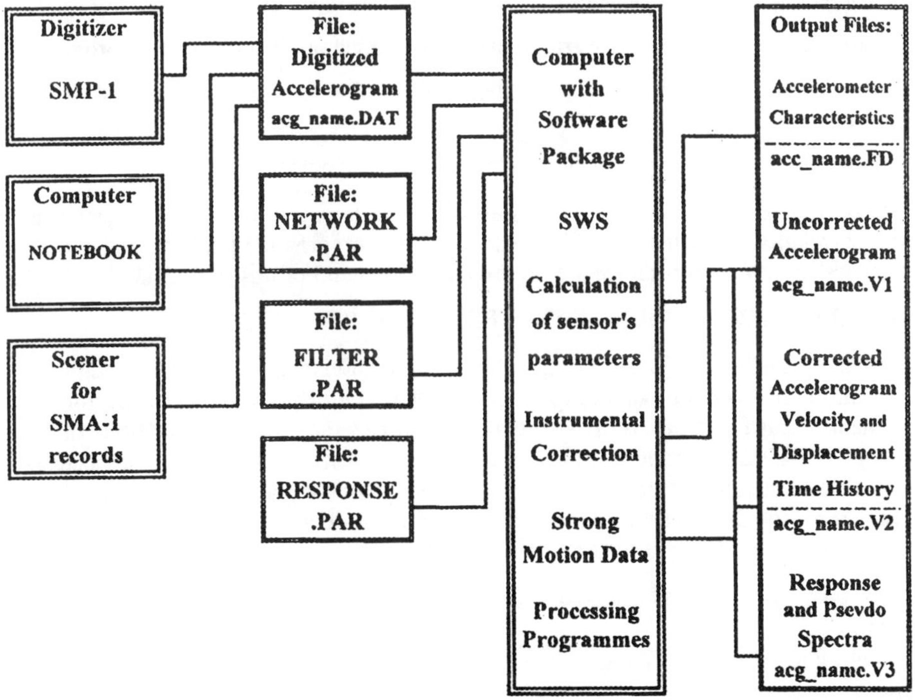

Figure 1. Functional block-diagram of the programme package "Kinemetrics" SWS for instrumental correction and processing of accelerogrammes

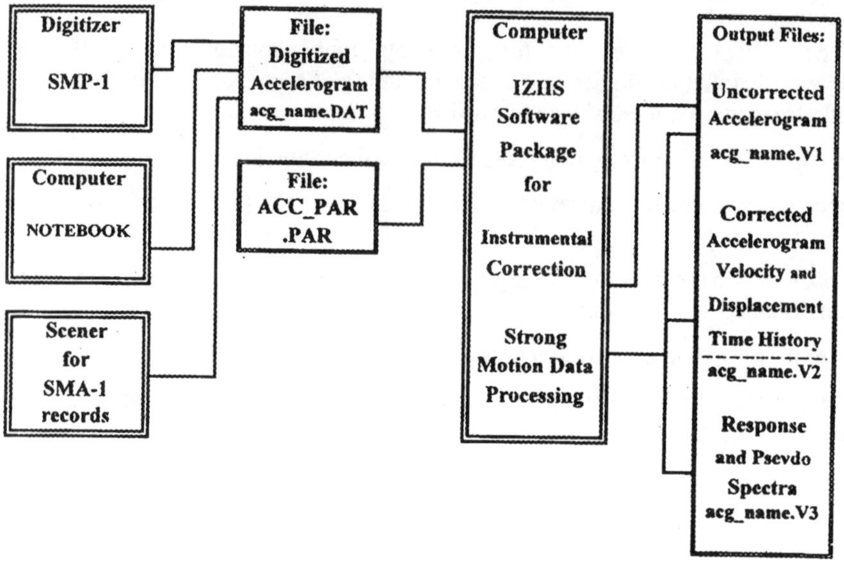

Figure 2: Functional block diagram of the IZIIS programme package for instrumental correction and processing of accelerogrammes.

Comparative processing of accelerogrammes has been done on two types of records:

- Earthquake record obtained on a seismic shaking table (comparative tests of different types of accelerographs), and
- Blast record ("in situ" experiments made at the NPP "Kozloduy" site).

The results from the comparative processing of an accelerogramme and discussion as to essential correlation and differences between the IZIIS programme package and the SWS "Kinemetrics" programme package are presented in Chapter 3 of this study.

2.1. STANDARD PROCEDURE OF PROCESSING OF RECORDS

The Institute of Earthquake Engineering and Engineering Seismology has developed its own procedure for processing of strong motion accelerogrammes. Used as a basis for development of this procedure is the standard procedure of the Californian Institute of Technology -CALTEH, Pasadena, in which certain modifications have been made because of use of different equipment for digitization and data processing as well as certain improvements through the latest investigations in IZIIS. A detailed description of the procedure is given in IZIIS publication IZIIS 66-1977 [2].

The whole processing process can be divided into several phases:

- Digitization;
- Obtaining of uncorrected data;
- Instrumental correction;
- Computation of accelerations, velocities and displacements;
- Computation of response spectra and the Fourier amplitude spectrum.

3. Comparison Between Sws "Kinemetrics" And Iziis Procedures For Processing Of Accelerogrammes

Elaborated in this part, separately and comparatively, are two concepts for processing of accelerogrammes as follows:

- Programme package developed by "Kinemetrics" from Pasadena, USA, known as SWS - "Seismic Works Station";
- Programme package for processing of accelerogrames applied in the Institute of Earthquake Engineering and Engineering Seismology (IZIIS) in Skopje for all scientific and applicative research programmes in the field of earthquake engineering realized at IZIIS.

Comparative processing by use of both procedures have been made on two types of records:

- Earthquake record obtained on three different types of accelerographs (SMA-1, SMA-2 and SSA-2) generated by the biaxial shaking table in IZIIS from the longitudinal component of the El Centro earthquake, and,
- "In situ" record of a blast in the immediate vicinity of NPP "Kozloduy" recorded at level 61.7 on block 5 by SMA-2 instrument with serial number 263.

The main goals of these comparative analyses were:

- ➢ To analyze the effect of input parameters (characteristics of digital band-pass filters and frequencies for computation of response spectra),
- ➢ To make comparative analysis of both procedures by application of the same input parameters;
- ➢ To analyze the effect of time interval of digitization.

To gain an insight into the effect of the parameters of filters used in the process of processing of records, analysis of the effect of characteristics of filters given according to Table 1 was made. From the presented results in the same table, it is obvious that their effect is the greatest in computation of displacement (there is a difference of up to several centimetres).

Table 1. Computations of acceleration, velocity and displacement by the same values of filter parameters obtained by SMA-2 accelereograph

Band-pass filter		"Kinemetrics"- SWS procedure			"IZIIS" - procedure		
Lower boundary freq. (Hz)	Upper boundary freq. (Hz)	Acceleration (cm/s/s)	Velocity (cm/s)	Displacement (cm)	Acceleration (cm/s/s)	Velocity (cm/s)	Displacement (cm)
0.03-0.05	33 - 35	281.24	31.56	12.34	281.0	31.253	11.715
0.075-0.1	33 - 35	280.26	30.42	9.94	280.0	30.597	9.864
0.05-0.10	33 - 35	280.65	30.32	9.05	280.3	30.430	9.113
0.10-0.20	33 - 35	280.72	29.80	8.07	280.5	29.785	8.046

The selection of frequencies for computation of response spectra has been done based on the regulations for aseismic design of nuclear power plants Regulatory Guide 1.60 [10], Regulatory Guide 1.61 [11] and design spectra of acceleration from design and local earthquake for free field, horizontal and vertical component, used in design and construction of NPP "Kozloduy". Selected are a total of 100 values of natural frequencies for a single-degree of freedom system (Table 2) for which the maximum values of spectral quantities shall be defined in the "Kinematrics" and "IZIIS" procedure.

Table 2. Interval for computation of Ground and Floor Response Spectra

Ground Response Spectra		Floor Response Spectra	
Frequency range (Hz)	Increment (Hz)	Frequency range (Hz)	Increment (Hz)
0.5-3.0	0.10	0.5-1.6	0.10
3.0-3.6	0.15	1.6-2.8	0.20
3.6-5.0	0.20	2.8-4.0	0.30
5.0-8.0	0.25	4.0-9.0	0.50
8.0-15.0	0.50	9.0-17.0	1.00
15.0-18.0	1.00	17.0-22.0	2.00
18.0-22.0	2.00	22.0-34.0	3.00
22.0-34.0	3.00		

To compare the difference between the two procedures, the El Centro earthquake record - longitudinal component obtained by SMA-2 accelerograph has been used as an input data. The record was processed by use of the Kinemetrics and IZIIS procedure for the same boundary frequencies for the band-pass filter and the same frequencies for computation of response spectra. Selected as boundary frequencies of the band-pass filter were: (0.05 - 0.10) Hz and (33 - 35) Hz, in accordance with the theoretical criteria for use of filters in processing of accelerogrammes.

The values of acceleration, velocity and displacement obtained by use of the "Kinemetrics" and "IZIIS" procedure are presented in Table 3 .

Table 3. Acceleration, velocity and displacement by the El Centro earthquake record - longitudinal component obtained by SMA-2 accelerograph with same values of filter parameters

Procedure	Band-pass filter		Acceleration	Velocity	Displacement
	Lower boundary freq. (Hz)	Upper boundary freq. (Hz)	(cm/s/s)	(cm/s)	(cm)
"Kinemetrics"	0.05 - 0.10	33 - 35	280.65	30.32	9.050
"IZIIS"	0.05 - 0.10	33 - 35	280.30	30.43	9.113

The effect of the digitization interval, i.e., the "sample rate" has been analysed based on accelerogrammes obtained by three different types of accelerographs: SMA-1, SMA-2 and

SSA-2. The accelerogrammes have been obtained by the shaking table in IZIIS - Skopje excited by the "El Centro" longitudinal component.

For that purpose, the accelerogramme obtained by SMA-1 has been digitized on a digital table of the type of Drawing Board II, Calcomp digitizer, products division Anahein, USA, with a specially prepared software for that purpose and processed by the IZIIS procedure with a sampling rate of 100 samples/sec.

The accelerogrammes obtained by accelerographs type SMA-2 have been digitized with a standard speed of 256 samples/sec., while the SSA-2 accelerogrammes have been recorded by 200 samples/sec.

While processing the accelerogrammes, the lower boundary frequency of the band-pass filter has been 0.05 to 0.10 Hz, while the upper boundary frequency of the band-pass filter has been selected depending on the performances of the recording system, i.e., (25 to 27) Hz in SMA-1 and (33 - 35) Hz for the SMA-2 and SSA-2 instruments.

Table 4 shows the obtained values of acceleration, velocity and displacement. Figs. 3 and 4 show the comparison between both the procedures for response spectra of absolute acceleration and Fouier spectrum for damping of 5% of the critical. As clear from the figures, there are differences but they are not so great.

Table 4. Acceleration, velocity and displacement by the El Centro earthquake record - longitudinal component obtained by SMA-2 accelerograph with same values of filter parameters

Accelero graphs type	Band-pass filter		Acceleration (cm/s/s)	Velocity (cm/s)	Displacement (cm)
	Lower boundary freq. (Hz)	Upper boundary freq. (Hz)			
"SMA-2"	0.05 - 0.10	33 - 35	280.65	30.32	9.05
"SSA-2"	0.05 - 0.10	33 - 35	-282.76	-29.93	-8.46
"SMA-1"	0.05 - 0.10	25 - 27	278.50	30.18	10.18

In July 1996, in the course of realization of this project, "in situ" investigation was performed in the immediate surrounding of NPP "Kozloduy", by use of explosives in bore holes specially prepared for that purpose. Although the explosions were of a low intensity, they still were recorded by some instruments from the system of seismic monitoring installed on the nuclear power plant.

To get a better insight into some advantages and disadvantages of one or the other procedure, a real record obtained at level 61.7 on block 5 by the SMA-2 instrument with a serial number 263, was processed for longitudinal (LONG), transverse (TRAN) and vertical (VERT) direction. From the recorded record, a record with a duration of 7.6 sec referring to the explosion with the greatest charge was processed.

The processing was done by the two processing packages packages - Kinemetrics and IZIIS. It should be mentioned that the algorithm used in processing of the accelerogrammes by use of the Kinemetrics procedure is not known to us.

While processing the record by use of the IZIIS procedure, it was concluded that the "DC" (direct component) of the signal is eliminated with this procedure. This is presented in Table 5 and figure 5 (VERT).

It was concluded that the Kinemetrics procedure does not involve elimination of DC of the signal which is present in the very record which results in non-real values of acceleration, velocity and displacement which is presented in Table 5. and figure 6 (VERT).

By elimination of the DC component from the input values for processing by the Kinemetrics procedure, more realistic values are obtained for acceleration and velocity, but those for the displacement still deviate from the real values presented in Table 5. Even after such additional correction, the long period components are still present in the results obtained by use of the "Kinemetrics" procedure. To eliminate them and for the purpose of obtained more realistic values of acceleration, velocity and displacement and their corresponding graphic presentations, a band-pass filter with lower boundary frequencies of 0.3 - 0.4 Hz was used, in compliance with regulations 10, 11 and 12. The obtained values given in Table 5.

Table 5. Acceleration, velocity and displacement by obtained by SMA-2 accelereograph under blast efect

Procedure	Compo-nent	Band-pass filter		Acceleration (cm/s/s)	Velocity (cm/s)	Displacement (cm)
		Lower boundary freq. (Hz)	Upper boundary freq. (Hz)			
"IZIIS"	LONG. TRAN. VERT.	0.05 - 0.10	33 - 35	6.2 3.5 16.0	0.118 0.169 0.249	0.033 0.071 0.037
"Kinemetrics"	LONG. TRAN. VERT.	0.05 - 0.10	33 - 35	-9.92 25.09 15.98	9.41 36.36 2.59	-5.23 -20.10 -1.45
"Kinemetrics" (remove DC offsets in data)	LONG. TRAN. VERT.	0.05 - 0.10	33 - 35	-6.32 -3.82 17.28	0.23 0.33 -0.46	-0.14 -0.18 -0.23
"Kinemetrics" (remove DC offsets in data)	LONG. TRAN. VERT.	0.30 - 0.40	33 - 35	6.26 3.80 17.06	-0.11 0.15 0.25	0.01 -0.02 -0.01

For the purpose of a better insight between the IZIIS procedure and the Kinemetrics procedure (with and without "DC" components in the signals) presented is a comparison done for the response spectra of absolute acceleration and Fourier spectrum for damping of 5 percent of the critical. In figures 7 and 8 for the longitudinal component, transverse component in figures 9 and 10 and for the vertical component - in figures 11 and 12.

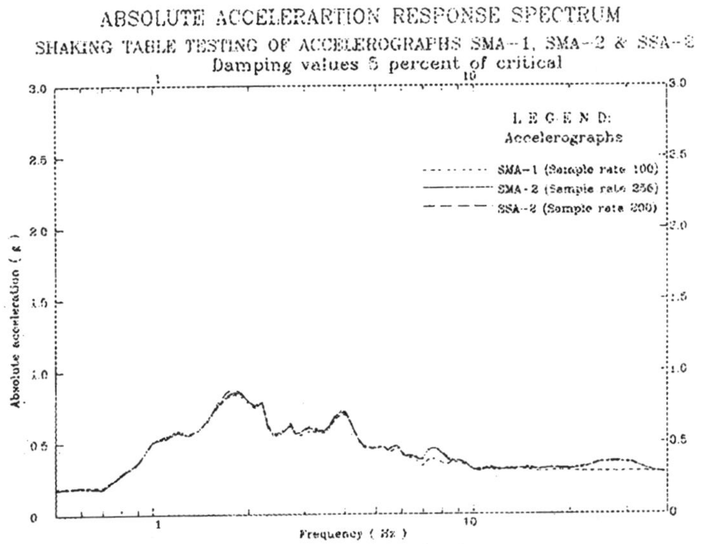

Figure 3. *Response spectra of absolute acceleration graph, El Centro, Comp. LONG*
SMA-1, Band pass filter limits (0.05 ÷ 0.10) Hz and (25 ÷ 27) Hz
SMA-2 and SSA-2, Band pass filter limits (0.05 ÷ 0.10) Hz and (33 ÷ 35) Hz.

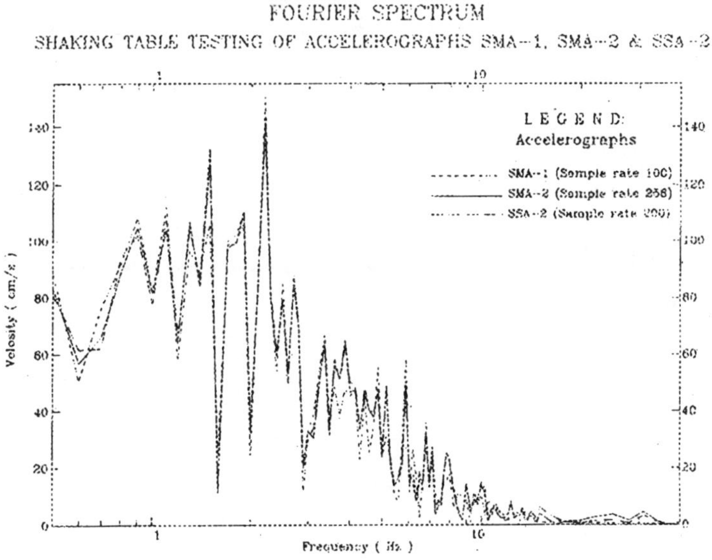

Figure 4. *Fourier spectrum graph, El Centro, Comp. LONG*
SMA-1, Band pass filter limits (0.05 ÷ 0.10) Hz and (25 ÷ 27) Hz
SMA-2 and SSA-2, Band pass filter limits (0.05 ÷ 0.10) Hz and (33 ÷ 35) Hz..

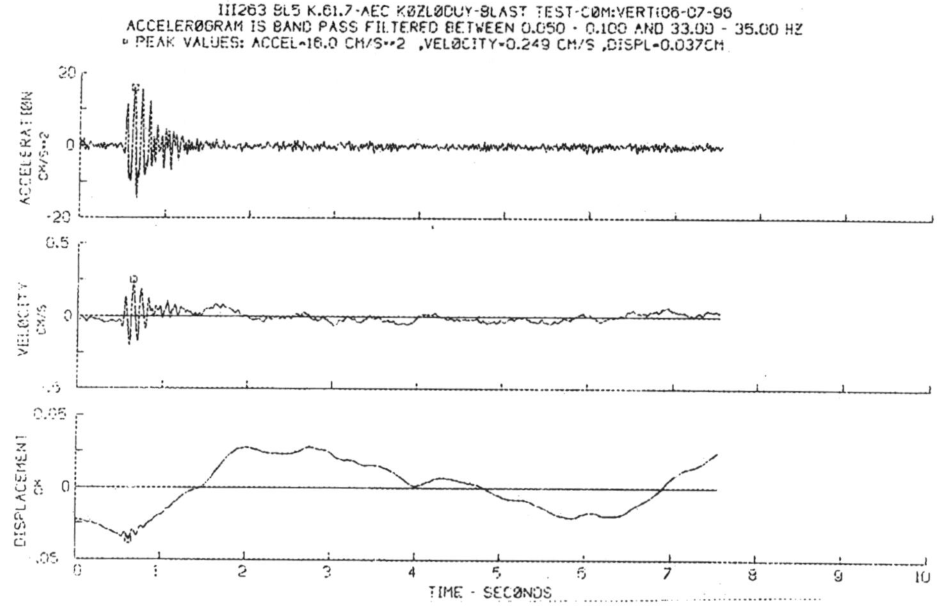

Figure 5. *Acceleration, Velocity and Displacement graph obtained by using the IZIIS procedure BLAST-TEST, Comp. VERT; Band-pass filter limits: (0.03 ÷ 0.05) Hz and (33 ÷ 35) Hz.*

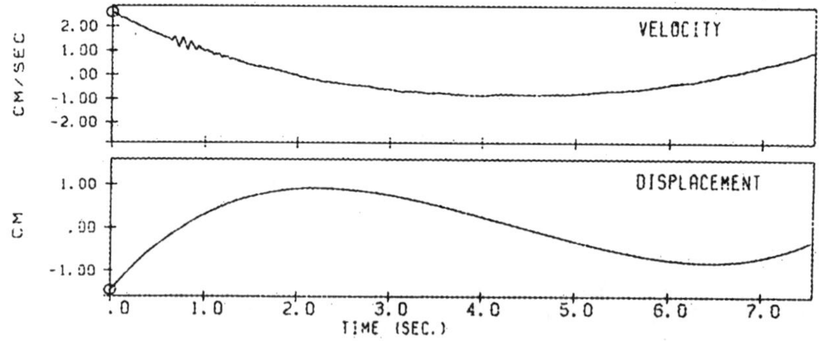

Figure 6. *Acceleration, velocity and displacement graph obtained by using the Kinemetrics procedure BLAST-TEST, Comp. VERT; Band-pass limits: (0.03 ÷ 0.05)Hz and (33 ÷ 35) Hz.*

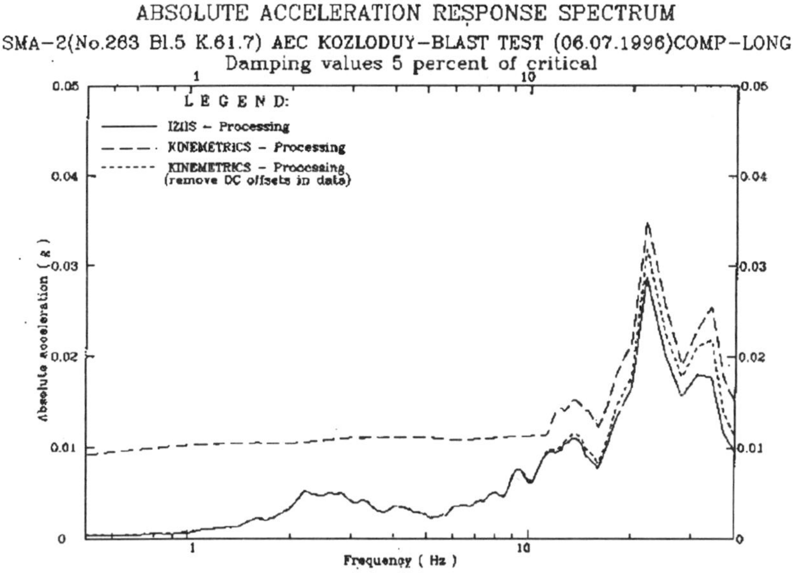

Figure 7. *Response spectra of absolute acceleration graph, BLAST-TEST, Comp. LONG; Band pass filter limits (0.05 ÷ 0.10) Hz and (33 ÷ 35) Hz*

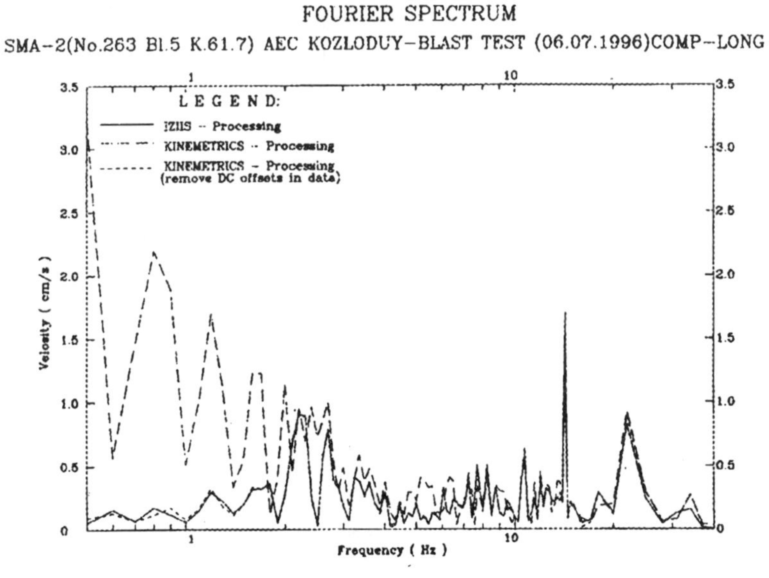

Figure 8. *Fourier spectrum graph, BLAST-TEST, Comp. LONG Band pass filter limits (0.05 ÷ 0.10) Hz and (33 ÷ 35) Hz*

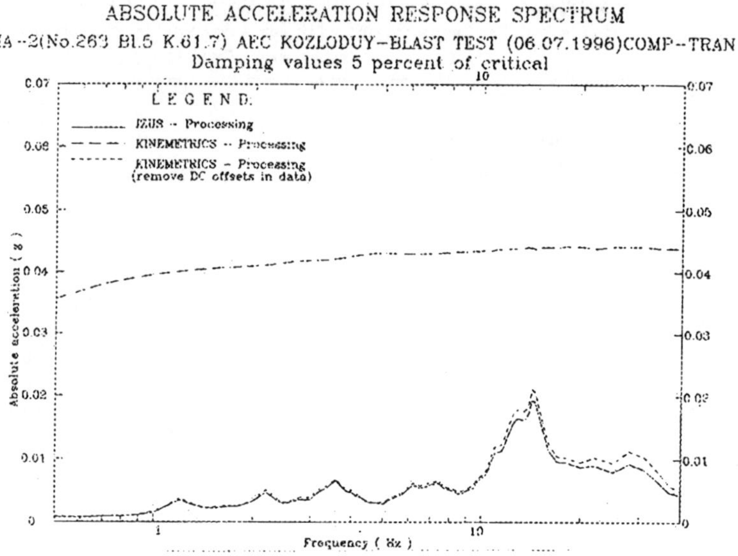

Figure 9. *Response spectra of absolute acceleration graph, BLAST-TEST, Comp. TRAN; Band pass filter limits (0.05 ÷ 0.10) Hz and (33 ÷ 35) Hz*

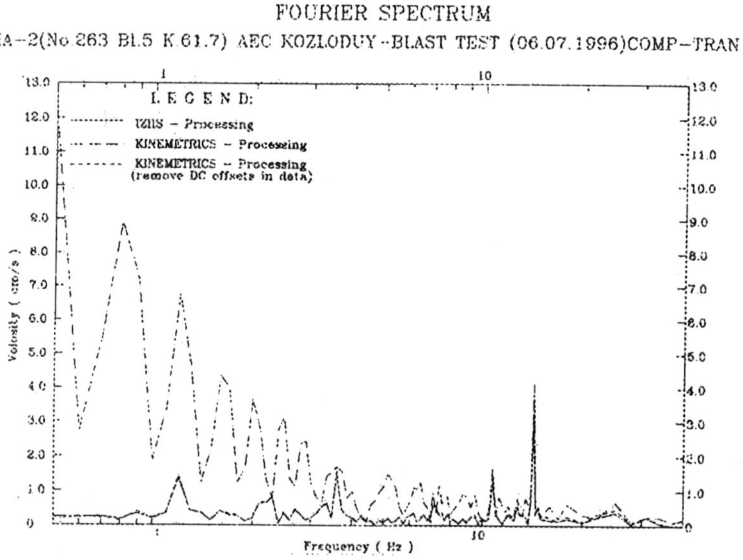

Figure 10. *Fourier spectrum graph, BLAST-TEST, Comp. TRAN Band pass filter limits (0.05 ÷ 0.10) Hz and (33 ÷ 35) Hz*

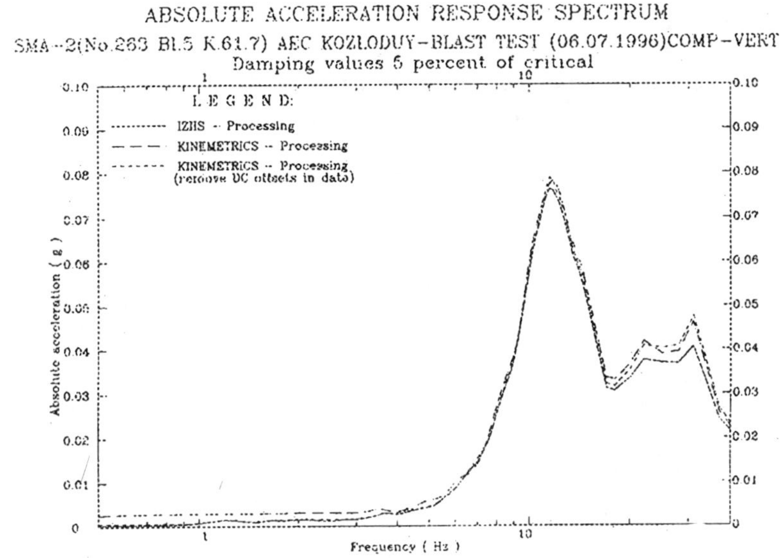

Figure 11. *Response spectra of absolute acceleration graph, BLAST-TEST, Comp. VERT; Band pass filter limits (0.05 ÷ 0.10) Hz and (33 ÷ 35) Hz*

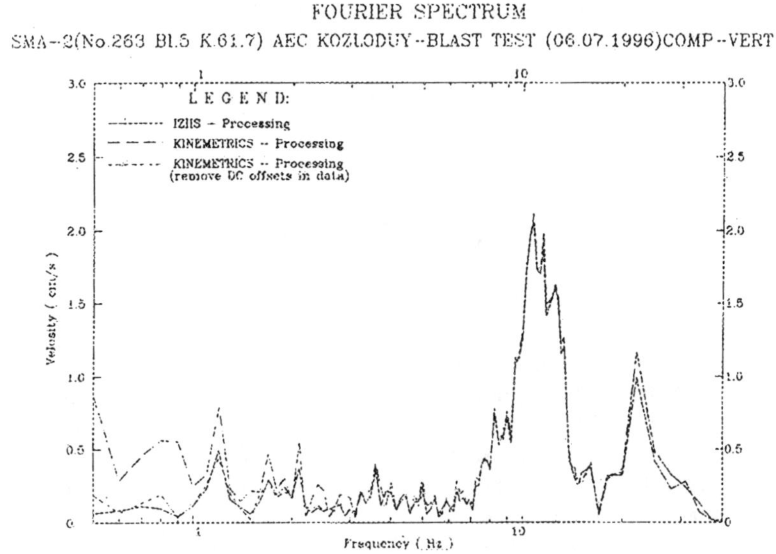

Figure 12. *Fourier spectrum graph, BLAST-TEST, Comp. VERT Band pass filter limits (0.05 ÷ 0.10) Hz and (33 ÷ 35) Hz*

4. Conclusions And Recommendations

Based on the previously presented theoretical bases of the procedure for processing of accelerogrammes applied in IZIIS and the performed processing and results obtained by analysis of a record from a real earthquake and record obtained from explosion, the following conclusions and recommendations are arrived at:

4.1 CHARACTERISTICS OF FILTERS

The theoretical conditions are satisfied if the lower boundary frequencies of the band-pass filter of 0.05 to 0.1 Hz are selected, however on the basis of regulations, it is correct to work also with lower boundary frequencies of a band-pass filter of 0.3 to 0.5 Hz.

Based on the regulations, for accelerogrammes obtained on SMA-2, SMA-3 and SSA-2, it is correct to select upper boundary frequencies of band-pass filter of 33 to 35 Hz, whereas for accelerogrammes obtained on SMA-1, these frequencies should be 25 - 27 Hz (due to the dynamic characteristics of the accelerograph).

4.2 FREQUENCIES FOR COMPUTATION OF RESPONSE SPECTRA

The values of the frequencies for computation of response spectra should be defined according to the regulations. Use of identical values of frequencies is recommended for all the accelerogrammes. It was concluded that these frequencies and the frequencies from the design spectra of acceleration of NPP Kozloduy are compatible.

4.3 DIGITIZATION INTERVAL

The technical characteristics of equipment imposes the use of the following digitization intervals: for SSA-2 $\Delta t = 1/200s$, for SMA-2 and SMA-3 $\Delta t=1/256s$ and for SMA-1 $\Delta t = 1/100s$ or $\Delta t = 1/50s$ (depending on the performances of the digitizing equipment.).

4.4 SIMILARITIES AND DIFFERENCES BETWEEN PROCEDURES FOR PROCESSING OF ACCELEROGRAMMES USED BY “IZIIS” AND “KINEMETRICS” SWS

- It has been concluded that in processing of identical accelerogrammes obtained from a record of an actual earthquake, both procedures (the one used by IZIIS and the one used by Kinemetrics SWS) yield approximately the same results. Presented in this report is the theoretical basis of the procedure for processing of accelerogrammes used by IZIIS which is probably the same or similar to SWS. From the results, it may be concluded that different algorithms and programme solutions are most probably used. These generate the differences. This is particularly manifested in instrumental correction of accelerogrammes where different algorithms for numerical filtration are most probably used.
- While computing the response spectra and other spectra, almost identical results are obtained for the same frequencies which points to the fact that the same algorithms are most probably used for computation in both procedures.
- While processing a record from a blast recorded on the dome of Block 5, it has been concluded that the IZIIS procedure correctly computes all the time histories and spectra.

- While using the Kinemetrics procedure, it has been concluded that the algorithms for instrumental correction of accelerogrammes obtained on SMA-2 (and certainly on SMA-3 since these have the same technological solutions), and digitized on SMP-1 Digitizer, could not eliminate the DC direct components of the accelerogramme signals. For this purpose, a programme has been elaborated for elimination of the present direct components which should be used prior to instrumental correction of accelerogrammes obtained on SMA-2 and SMA-3.

References

1. "Kinemetrics - Operating and Instructions Manuals" for: - Accelerografs type: SMA-1, SMA-2, SMA-3, DSA-1, DSA-3, SSA-2; - Playback Devices type: SMP-1, DSP-2; - Digitizing System SMP-1; - Data Retrieval PC based Software Package QTSSA; - Data Processing Software Package SWS (Seismic workstation)
2. Dimitar Petrovski, Nove Naumovski, "Processing of Strong Motion Accelerogramms", Part -1, "Analytical Methods", Publication No. 66, IZIIS, Skopje, Macedonia, 1979
3. Mamucevski, D. and Dojcinovski D. (1996) Seismic Monitoring System of NPP Kozloduy, Report IZIIS 96-74
4. Mamucevski. D., Dimitrovski, T. and Vitanovski N. (1995), "NPP Kozloduy, Section HTS and SK, 1. Repair and Calibration of Three Accelerographs type SMA-2, 2. Comparative Shaking Table Testing of SSA-2, SMA-2 and SMA-1" IZIIS 95-04, Skopje, Macedonia
5. Mihailov, V. and Celebi M., (1999) Seismic Monitoring of Structures - An Important Element in Seismic Hazard Mitigation for Urban Media, Mac – U.S. Research Project – Progress Report 1998/99
6. Mihailov, V. Dojcinovski, D. and D. Mamucevski, (1994), "Seismic monitoring on structures: Some experience and related problems", 10th European Conference on Earthquake Engineering, Vienna, Austria,
7. Mihaiulov, V. and Dojcinovski D. (1998) Analysis of Records Taken by SMA-1 acclegraph Installed on HPP "Moste", Earthquake of April 12, 1998 12^{h} 55' and M=5.5
8. Mihaiulov, V. and Dojcinovski D.(1995) Processing of Records from the earthquake of May 1, 1995 obtained on HPP "Strezevo" dam; Report IZIIS 95-24
9. Regulatory Guide 1.12, 1974, "Instrumentation for Earthquakes", U.S. Atomic Energy Commission, 1974
10. Regulatory Guide 1.60, 1973, "Design Response Spectra for Seismic Design of Nuclear Power Plants", American Nuclear Society, 1973
11. Regulatory Guide, 1.61, 1973, "Damping Values for Seismic Design of Nuclear Power Plants", American Nuclear Society, 1973
12. Standard ANSI/ANS - 2.2 – 1978, "Earthquake Instrumentation Criteria for Nuclear Power Plants", American Nuclear Society, 1978
13. Standard ANSI/ANS - 2.10 – 1979, "Guidelines for Retrieval, Review, Processing and Evaluation of Records, obtained from Seismic Instrumentation", American Nuclear Society, 1979
14. Standard IEEE 344-1987, "Recommended Practices for Seismic Qualification of Class 1E Equipment for Nuclear Power Generating Stations", The Institute of Electrical and Electronics

ANALYSIS OF DATA FROM STRONG MOTION NETWORK IN FRASER DELTA, BRITISH COLUMBIA, CANADA

W.D.L. FINN
Anabuki Chair of Foundation Geodynamics
Department of Safety Systems Construction Eng., Kagawa University
1-1 Saiwai-cho, Takamatsu City, Kagawa, 760-08526 Japan, and
Department of Civil Engineering, University of British Columbia

E. ZHAI, T. THAVARAJ, X-S. HAO and C.E. VENTURA
Department of Civil Engineering, University of British Columbia
2324 Main Mall, Vancouver, B.C. Canada V6T 1Z4

ABSTRACT: The 1996 Duvall earthquake in Washington State triggered ground motion stations in the Fraser Delta, British Columbia, Canada. Recorded ground motions were used to examine the applicability of 1-D and 2-D site response analyses for amplification studies in the Delta. 1-D response analysis gave a good indication of the period of peak response, but response spectra were generally not satisfactory for engineering purposes. The use of 2-D analyses which included buried topography generally improved the predictions of site response spectra.

1 Introduction

The Geological Survey of Canada (GSC) has been concerned about the potential for large amplification of ground motions due to the soft sediments in the Fraser Delta in South-Western British Columbia. In the deeper sediments, the long period motions are expected to be amplified considerably. In 1994, the GSC initiated a project to model the seismic response of the thicker deposits of soft sediments. Site response analyses were conducted based on data from two 300 m deep boreholes drilled in 1994 [5]. The Harris et al. [5] study also benefitted from industry seismic reflection profiles which established the depth to bedrock over the Delta [1]. Similar response studies had been conducted previously by Byrne and Anderson [2], Finn and Nichols [4], and Sy et al. [12]. However, these earlier studies had to make assumptions about depth to bedrock and the boundary between Pleistocene and Holocene deposits. Also, soil property information, especially at depth, was poor.

The Harris study reached the conclusion that significant amplification of long period motions ($3.5s \leq T \leq 5.0s$) is likely in the deeper sediments of the Delta. These amplifications pose increased hazards for tall buildings, bridges, industrial facilities and

M. Erdik et al. (eds.), Strong Motion Instrumentation for Civil Engineering Structures, 433–446.

pipelines. Harris et al. [5] questioned whether 1-D site response analyses would be adequate at such long periods, quoting Sykora et al. [13], "...for periods greater than 4 s, motions are likely to be affected by 2-D effects and surface wave energy." Harris et al. [5] also expressed concern about the topographic effects of the buried structures revealed by the seismic reflection studies.

Further field studies by the GSC to 1997 revealed a complex buried topography with rapidly varying thicknesses of Holocene and Pleistocene deposits over irregular bedrock. This new data increased concern about the applicability of 1-D analysis. Therefore, the GSC commissioned the present study to expand the range of the Harris report. The present study is based on a greatly expanded data base of soil properties from additional deep boreholes, extensive geophysical work giving shear wave velocities in the Holocene and Pleistocene sediments and better definition of the Holocene/Pleistocene boundary.

In the present study, site response analyses were conducted at strong motion stations in the Delta which were triggered during the 1996 Duvall earthquake. Computed and recorded responses were compared to assess whether 1-D analyses are adequate for evaluating potential seismic response in the Delta. A well-defined cross-section of the Delta was also analysed using 2-D models to see what the effects of buried topography may be and to investigate whether 2-D analysis gives a better definition of site response. This paper focusses on the most important issue for design, how well computed response spectra compare with measured spectra.

2 Geology of the Delta

The Fraser Delta of south-western British Columbia is located just south of Vancouver below the North Arm of the Fraser River. The fast growing City of Richmond is located in the Delta. The sediments above the Tertiary bedrock have been described by Rogers et al. [10] as follows. The sediments beneath the Delta are Holocene-age deltaic deposits and Pleistocene-age glacial and interglacial deposits. The Holocene sediments are up to about 300 m thick [7] and are mainly silts and sands. The Pleistocene sediments are mostly ice-compacted tills and glaciomarine silts and sands that overlie tertiary bedrock. Shear wave velocity in the Holocene sequence increases with depth [6] with typical average values in the range of 200 m/s to 300 m/s, but about 100 m/s near the surface in many places. The Pleistocene sediments typically have an average shear wave velocity near 500 m/s but with considerable variability. The shear wave velocity of the underlying tertiary bedrock is 1500 m/s or greater. The surface of the tertiary bedrock has been mapped by Britton et al. [1], as shown in Fig. 1.

Figure 2 shows the thickness of Holocene sediments in the study area. The Holocene sediments thin rapidly to the north at the edge of the basin. They are about 300 m thick in the basin centre near strong motion instrument sites ARN and RHA and relatively shallow on the north side of the Fraser River near strong motion station MNY. In contrast, the Pleistocene sediments thin to the north more gradually and extend beyond the Fraser Delta and mantle most of the Greater Vancouver area. Depth to the tertiary

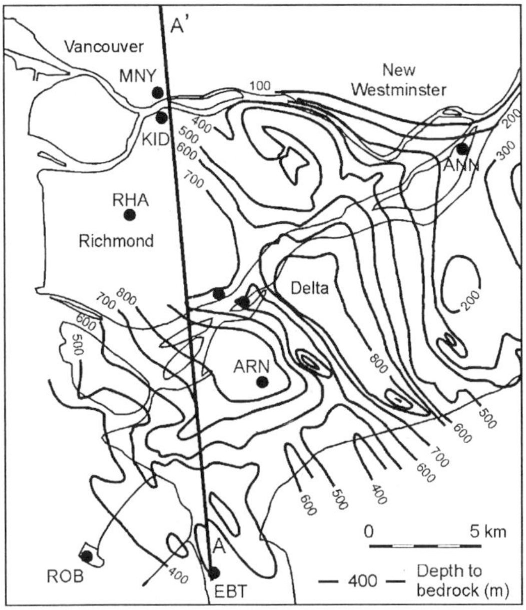

Figure 1. Interpreted depth-to-bedrock (tertiary) beneath the Fraser Delta based on industry seismic reflection data [1].

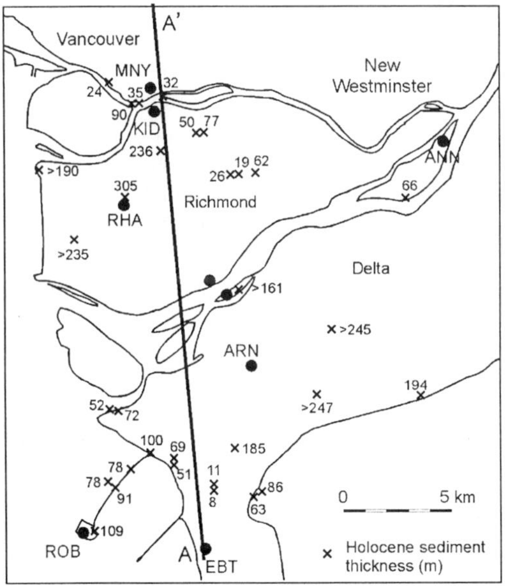

Figure 2. Thickness of Holocene sediments in the Fraser Delta [10].

bedrock is about 800 m in the centre of the Delta (strong motion stations ARN, RHA) and about 200-400 m near the North Arm of the Fraser River (strong motion stations KID, MNY, ANN), where the Holocene sediments pinch out. There are some localised highs on the buried bedrock and Pleistocene surfaces, such as one near strong motion station EBT. Figures 1 and 2 provide the basis for constructing a roughly NS cross-section through the Delta along the A-A′. This cross-section, giving the approximate locations of the Holocene/Pleistocene boundary and the top of the tertiary bedrock, is shown in Fig. 3.

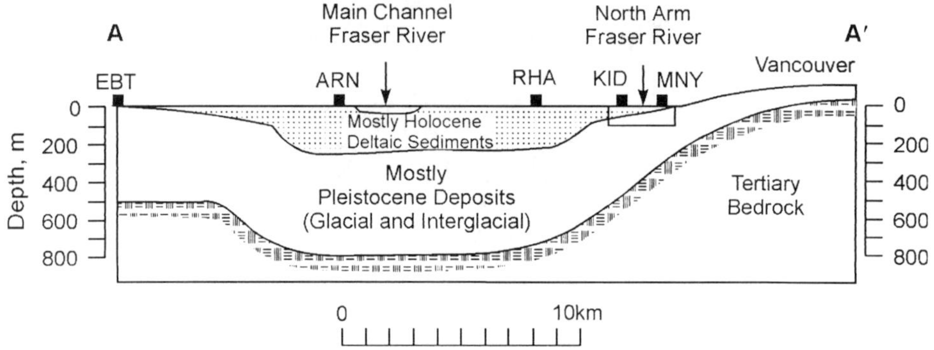

Figure 3. Cross-section of Delta along line A-A′ in Figure 1 [10].

3 Strong Motion Data

The strong motion stations in the Fraser Delta which were triggered during the Duvall earthquake, M=5.1, which occurred in Washington State, USA, are shown in Fig. 1. All the instruments that triggered were digital recording and were set to trigger between 0.4% to 1.0% g. The digital data gave high resolution which was essential for interpreting the low amplitude records. Details of the strong motion stations and the peak accelerations recorded during the earthquake are given in Table 1 for the stations studied in this report.

The only strong motion recording on rock is from strong motion station PGC which is located at the Pacific Geoscience Centre on Vancouver Island near Sidney, B.C. This instrument is located on granitic bedrock. The range in distance from the epicentre to the strong motion stations considered in this study varies between 160 and 180 km, and the azimuth is roughly the same for all the stations. Therefore, travel path effects are not expected to be significant. In addition, the stations considered in this study are away from the SH, SV radiation minima and, therefore, source effects are not considered to be significant [10].

Filtered peak accelerations for the stations that triggered during the earthquake, are given in Table 2. Two sets of components are given; L and T, which are defined by clockwise rotation from north, and Rad and Tan which refer to radial and transverse components of

Table 1. Information on strong motion stations in the Fraser Delta used in this study including the as-recorded peak accelerations.

Station	Distance (km)	Soil Conditions	Peak Acceleration (cm/s^2)		
EBT	166.3	500 m till	6.0	3.2	1.5
KID	185.3	45 m (H) >200 m (P)	12.8	13.6	2.9
MNY	186.1	~3 m (H) 200 m (P)	15.0	6.9	3.7
RHA	182.7	300 m (H) 500 m (P)	9.0	10.7	2.3

Note: Distance is the epicentral distance; (H) indicates Holocene sediments (mainly silts and sands), and (P) indicates Pleistocene sediments (mainly tills).

Table 2. Filtered peak accelerations for various components of the recorded Duvall earthquake ground motions.

Site	L	Dir.	T	Dir.	V	Rad	Tan
EBT	4.6	0°	-4.5	90°	1.5	3.23	5.93
KID	11.9	180°	-14	270°	2.9	13.61	9.31
PGC	4.6	0°	3.5	90°	1.8	3.38	4.54
MNY	-6.2	0°	15.5	90°	-3.7		
RHA	11.1	270°	-8.2	0°	2.4	10.65	8.87

the accelerations. The radial motions are along the line between the epicentre and the strong motion station. The L and T components for all the different stations are in the same directions whereas the Rad and Tan components are all in slightly different directions. Therefore, the L and T components form the best basis for evaluating the predictive capability of different methods of analysis and results are reported here only for these directions.

The reliability of the low level motions recorded on the strong motion instruments was confirmed by comparing the spectra of the recorded motions with the spectra of motions from the broad band instruments of the seismic network operated by the Pacific Geoscience Centre [9].

4 Results of 1-D Analyses

4.1 KID STATION

Seismic response analyses were conducted using the SHAKE program [11] with the L and T components of the PGC filtered data from Table 2 as input motions. Accelerations, Fourier spectra and acceleration spectra were computed for each component. The acceleration response spectrum is the crucial information for design engineers. Therefore, comparisons between computed and recorded motions will be confined to acceleration response spectra.

The computed response spectrum for the 180° component is shown in Fig. 4. It can be seen that for periods greater than about 0.4 s, it agrees reasonably well with the recorded spectrum and shows a peak response at a period of about 0.35 s. However, for periods shorter than 0.4 s the comparison is not good. The computed spectrum substantially exceeded the recorded spectrum in the period range 0.2 to 0.4 s. The computed motions show strong response peaks at 0.1 s and 0.35 g, which greatly exceed the spectrum of the recorded motions. At periods between 0.1 s and 0.2 s, the spectrum of the recorded motion shows a range of strong response with a peak at about 0.13 s. In this short period range, the computed spectrum considerably substantially underestimates the recorded response.

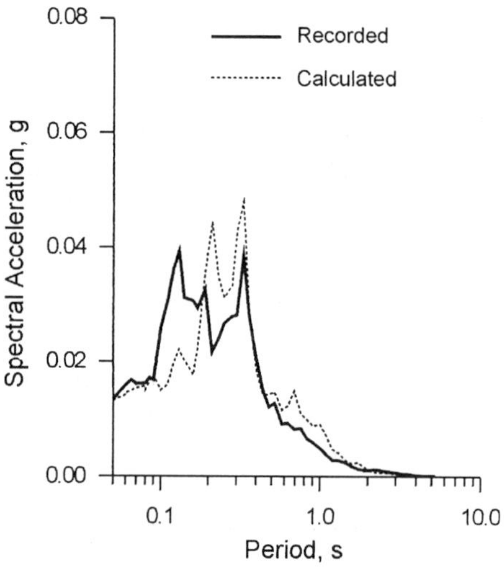

Figure 4. Comparisons of recorded and calculated spectra at KID (180°).

The pattern is very similar for the 270° component shown in Fig. 5. In this case, there is reasonable agreement for periods greater than about 0.3 s, and the peak of strongest response agrees closely in magnitude and frequency with that of the recorded motion. However, at periods below 0.3 s, once again the computed motions underestimate the recorded motions.

The dynamic response analyses correctly identify the periods of peak response at the KID station for both L and T components of motion. The response spectrum for periods greater than about 0.4 s are generally quite good, but at shorter periods, the response is badly underestimated. The response spectra of the PGC rock motions indicate

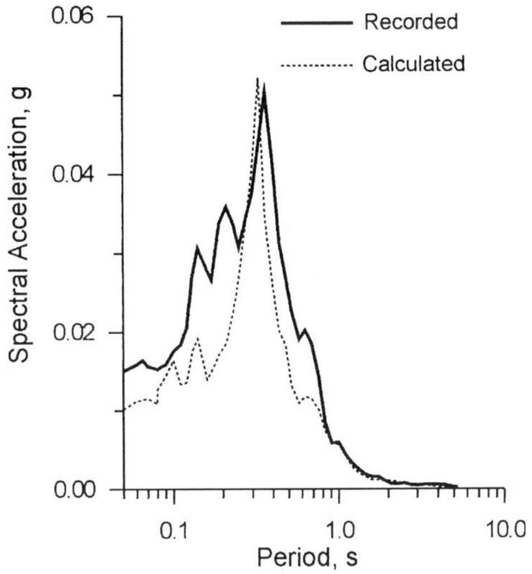

Figure 5. Comparisons of recorded and calculated spectra at KID (270°).

amplification for periods less than about 0.4 s. This was confirmed recently by microtremor measurements at the site which suggested a peak spectral response at 0.1 s. This would be consistent with some weathering at the site. The PGC site is under further investigation to clarify these issues and to determine the potential effects on site response.

4.2 RHA STATION NEAR RICHMOND CITY HALL

The depth of tertiary bedrock at RHA is 800 m and the depth of Holocene deposits is 300 m. This is a site where significant amplification of long period motions would be expected. The response spectra of the computed and recorded motions for the 0° and 270° components are compared in Figs. 6 and 7, respectively.

This station showed the best comparison between computed and recorded motions. The period of peak response was fairly accurately determined.

For the L and T components, there was reasonable agreement over the whole range of frequencies. The recorded and computed responses both were very narrow banded with most of the strong response concentrated at one period. This period is a surprising 0.3 s, considering that at this site there were 300 m of Holocene sediments. It would seem that the dominant response here is due to a form of resonance in one of the upper strata, or the excitation of one of the higher periods of the site.

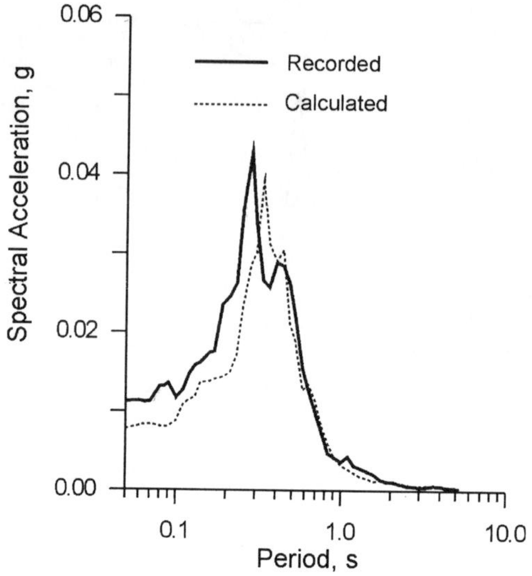

Figure 6. Comparisons of recorded and calculated spectra at RHA (0°).

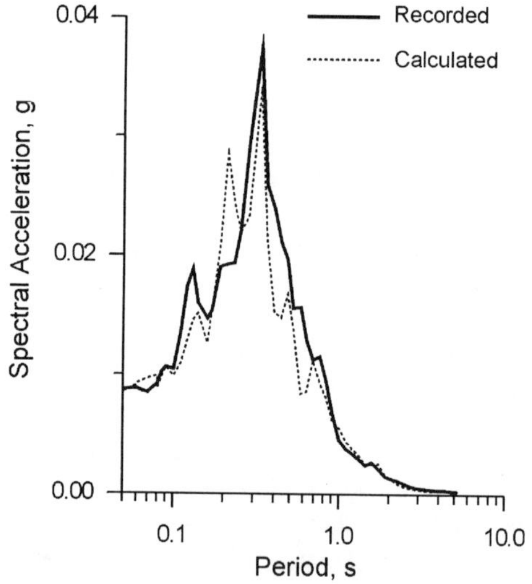

Figure 7. Comparisons of recorded and calculated spectra at RHA (270°).

At this deep site, it would seem that 1-D response analysis is adequate. This is probably due to the fact that the Holocene/Pleistocene boundary is at great depth and the basin edge is far removed.

4.3 MNY STATION

There is some uncertainty about the thickness of Holocene deposits at Station MNY as no borehole information is available. The thicknesses derived from forward and reverse reflection surveys were 29.5 m and 49.2 m, respectively. Dynamic response analyses were conducted for each model using the L and T components of the PGC motions as input and shear wave velocities derived from the reflection surveys.

4.3.1 *Reflection Forward Model*

The response spectrum of the recorded motions for the 0° component is broad band with a region of peak response spanning the period range, 0.2 to 0.6 s. The computed response is a very narrow band spectrum (Fig. 8), centered on 0.3 s, and amplified about 3 times compared to the recorded spectrum at the same site. For the 90° component (Fig. 9), the agreement between the spectra is also bad. In this case, the computed period of peak response is about 0.1 s shorter than the period of the recorded response, and the amplitude is about 40% less. In general, the spectra compare better at both long and short periods outside the period range of the peak response.

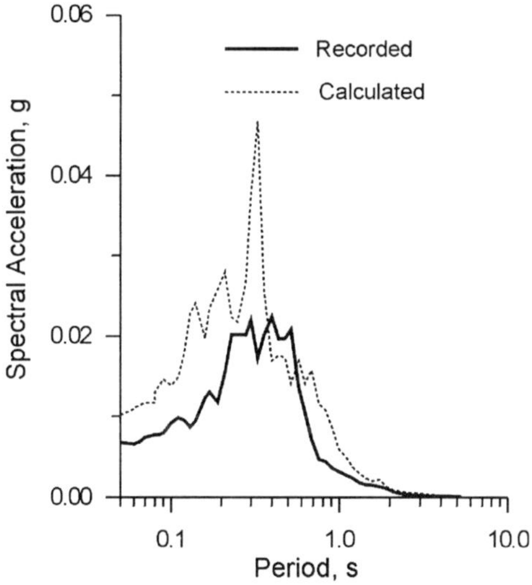

Figure 8. Comparisons of recorded and calculated spectra at MNY (0°).

4.3.2 *Reflection Reverse Model*

The calculated response spectrum for the 0° component badly underestimates the recorded spectrum in the period range of peak response at about 0.4 to 0.6 s. For the 90° component, the computed spectrum is again a narrow band spectrum amplified about 2.5

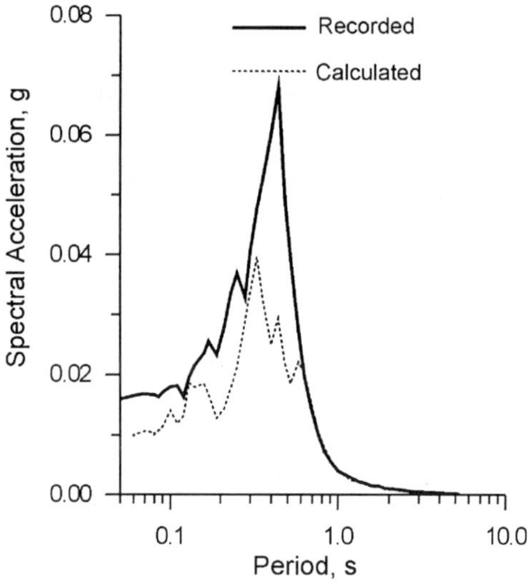

Figure 9. Comparisons of recorded and calculated spectra at MNY (90°).

times compared to wider band recorded response spectra. In essence, the response of the reverse model is the reverse of the forward model.

The disparity between recorded and computed motions for the MNY models cannot be attributed to the selected damping ratio because for both one model overestimated and the other model underestimated the recorded response. It is likely that the true depth lies between the estimated values and would lead to better results. Cone penetration tests are planned for this site to resolve the depth issue.

5 2-D Response Analyses

2-D response analyses were conducted using the program FLUSH [8] on a segment of cross-section A-A' shown in Fig. 3. After detailed studies of the requirements of extensive 2-D analyses, it was decided that the best strategy for available resources would be to choose a section which included the two strong motion stations KID and MNY. The stations are relatively close together and the Holocene sediments are less than 50 m deep. To further simplify the computations, the depth of the analysis was limited to 100 m. This ensured that the horizontal base of the model was entirely in the Pleistocene. The extent of the model is shown by the rectangle surrounding KID and MNY, in Fig. 3.

For input motions at the 100 m depth, the motions recorded in rock at PGC were scaled by a factor of 0.5 to remove the free surface magnification effect. It was assumed that these motions would not be significantly amplified during transmission through the Pleistocene. The motions at 100 m below the surface were not calculated by deconvolution using SHAKE because comparisons between computed and recorded motions at the surface at these locations as described earlier showed that 1-D analyses near the edge of the basin did not simulate the ground motions very well.

A severe limitation to analysis is the allowable layer thickness in the geometrical model of the section to ensure the passage of motions with appropriate frequency content. The limiting thickness depends on the desired high cut-off frequency and the shear wave velocity in the layer. For example, near the surface, the thickness is limited to a metre or two, if one wants to pass waves with frequencies up to 10 Hz near the surface. Even with some compromises on layer thickness, the 2-D model has 2048 elements and 2145 nodes. A comparison of the area of the restricted model with that of the entire basin cross-section clearly indicates the magnitude of the job in analysing the entire cross-section.

Each model was analysed with and without the inclusion of the channel of the North Arm of the Fraser River which flows between the stations but is much closer to KID. It was expected that these analyses would indicate whether the river needs to be included in future analyses.

6 Results of 2-D Analyses

6.1 MNY STATION

The analysis was first conducted, including the North Arm of the Fraser River. The acceleration spectrum for MNY (0°) is given in Fig. 10. The response spectrum of the motions computed by FLUSH gives a much better approximation of the spectrum of recorded motions than does the spectrum of the SHAKE motions. In particular, the strong narrow band response at a period of about 0.3 s predicted by SHAKE is not evident in the FLUSH spectrum.

Modelling the river in the analysis leads only to a minor adjustment of periods of peak response in the FLUSH analysis.

6.2 KID STATION

The analysis was first conducted including the North Arm of the Fraser River. The acceleration spectrum for KID (180°) is given in Fig. 11. The response spectrum computed by FLUSH is a somewhat better approximation to the spectrum of the recorded motions. The strong spectral amplification at a period of 0.2 s predicted by SHAKE analysis does not appear in the FLUSH analysis. However, the strong recorded response in the period range 0.1 s to 0.25 s is still underestimated.

If the river is not included in the model, the response is very similar except that the response at T = 1.05 is amplified about 2 times compared to when the river is included. In addition, the response at T = 0.35 s is strongly amplified. The rest of the spectrum is

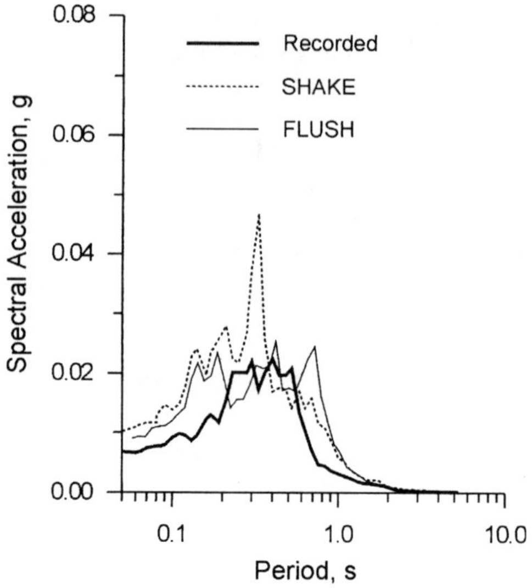

Figure 10. Comparison of response spectra for model 1 at MNY Station.

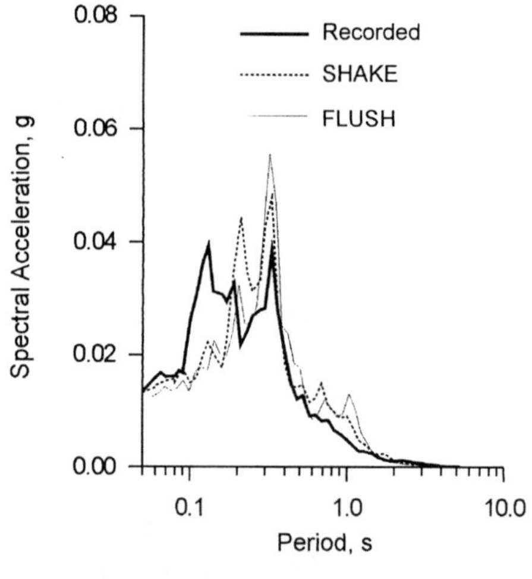

Figure 11. Comparison of response spectra for model 1 at KID Station.

approximately the same. The spectrum at periods below 0.2 s is still badly underestimated in the 2-D FLUSH analysis.

7 Conclusions

1-D response analysis in all cases gave a good indication of the period of peak response. It is interesting to note that the period is associated with the higher modes of the deeper sites. Frequently, such modes are closely associated with soft surface layer at the site.

At some sites such as KID, the recorded motions showed strong spectral response at short periods to the range 0.1 s to 0.25 s. Neither 1-D nor 2-D analysis predicted this response. 1-D analysis provided a good representation of the recorded notion at the very deep site at RHA near Richmond City Hall which is more removed from the basin edge, and where the Holocene/Pleistocene boundary is at great depth.

The use of 2-D analyses improved somewhat the predictions of site response spectra. These analyses reflected the effects of only one buried topographic feature, the variation in the depth of the boundary between Holocene and Pleistocene sediments. However, one would expect the irregular Holocene/ Pleistocene boundary to be the dominant influence where the depth of Pleistocene sediments is significant.

Other features such as the topography of the bedrock also affects the wave transmission to the surface. External and buried topography can have a significant impact on response and ideally this should be captured by 3-D dynamic response analyses. The feasibility of conducting such analyses in the Delta is currently under review.

Ideally, more analyses should be conducted for a range of ground motions sufficient to provide statistically significant response parameters for different locations in the delta.

Such analyses should take into account the data from the 1997 Georgia Strait earthquake. This earthquake occurred much closer to the delta and the source and path effects may be quite different. These new data would be crucial in providing more robust validation for the reliability and potential of dynamic response analyses for Delta sites.

8 Acknowledgements

The study benefited greatly from close cooperation with Dr. John Cassidy of the Pacific Geoscience Centre, Sidney, B.C., Dr. James Hunter of the Terrain Sciences Division, GSC Ottawa, and H. Christian of GSC Atlantic Canada, in Dartmouth, Nova Scotia. The study relied very heavily on the geological structure of the Fraser Delta as delineated by the continuing research program of the Geological Survey of Canada.

9 References

1. Britton, J.R., Harris, J.B., Hunter, J.A. and Luternauer, J.L. (1995) The bedrock surface beneath the Fraser River delta from seismic measurements, Current Research 1995-E, Geol. Survey of Canada.

2. Byrne, P.M. and Anderson, D.L. (1987) Earthquake design in Richmond, B.C., version II, Soil Mechanics Series No. 109, Department of Civil Engineering, University of British Columbia, Vancouver, B.C.
3. Cassidy, J.F., Rogers, G.C. and Weichert, D.H. (1997) Soil response on the Fraser Delta to the M_w=5.1 Duvall, Washington, earthquake, accepted for publication, Bulletin of the Seism. Soc. of Amer., April.
4. Finn, W.D.L. and Nichols, A.M. (1988) Seismic response of long-period sites: lessons from the September 19, 1985 Mexican earthquake, Canadian Geotech. Journal 25, 128-137.
5. Harris, J.B., Hunter, J.A., Luternauer, J.L. and Finn, W.D.L. (1995) Site amplification modelling of the Fraser delta, British Columbia, Proc., 48th Canadian Geotechnical Conference, Vancouver, B.C., Sept. 25-27, Vol. 2, 947-954.
6. Hunter, J.A. (1995) Shear wave velocities of holocene sediments, Fraser river delta, British Columbia, in Current Research 1995-A, Geological Survey of Canada, 29-32.
7. Luternauer, J.L. and Hunter, J.A. (1996) Mapping Pleistocene deposits beneath the Fraser river delta: preliminary geological and geophysical results, in Current Research 1996-E, Geol. Survey of Canada, 41-48.
8. Lysmer, J., Udaka, T., Tsai, C.F. and Seed, H.B. (1975) FLUSH: A computer program for approximate 3-D analysis of soil-structure interaction problems, Report No. EERC-75/30, Earthq. Eng. Research Center, Univ. of California, Berkeley, California.
9. Rogers, G.C. (1997) Private Communication.
10. Rogers, G.C., Cassidy, J.F. and Weichert, D.H. (1997) Variation in earthquake ground motion on the Fraser delta from strong motion seismograph records.
11. Schnabel, P.B., Lysmer, J. and Seed, H.B. (1972) SHAKE: a computer program for earthquake response analysis of horizontally layered sites, Report EERC-71/12, Earthq. Eng. Center, Berkeley, California.
12. Sy, A., Henderson, P.W., Lo, R.C., Siu, D.Y., Finn, W.D.L. and Heidebrecht, A.C. (1991) Ground motion response for the Fraser delta, British Columbia, Proc., 4th Int. Conf. on Seismic Zonation, Stanford University, Palo Alto, California, 2, 343-350.
13. Sykora, D.W., Wahl, R.E. and Wallace, D.C. (1991) WESHAKE for personal computers, US Army Corps of Engineers Geotechnical Earthquake Engineering Software, Instructional Report GL-91-1.

ON IDENTIFICATION OF DAMAGE IN STRUCTURES VIA WAVE TRAVEL TIMES

S.S. IVANOVIC[1], M.D. TRIFUNAC[2] and M.D. TODOROVSKA[2]
[1)] University of Monte Negro, Department of Civil Engineering
Cetinjski Put BB, Podgorica 81000, Monte Negro, Yugoslavia
[2)] University of Southern California, Department of Civil Engineering Los Angeles, CA 90089-2531, U.S.A.
e-mail: trifunac@mizar.usc.edu

Abstract

The traditional identification methods for detection of damage and for health monitoring of structures are based on detection of changes in the system natural frequencies. These changes tend to be small in the early stages of damage, and therefore may be difficult to quantify, even from accurately processed recorded motions. Other difficulties arise from the nonuniqueness in the model representation. Unless the model accounts for the soil-structure interaction, and it has been carefully validated and calibrated, it is very difficult to identify the true causes and sources of observed nonlinearities in the response. In this paper, it is suggested that the formation of damaged zones in structures could be monitored (identified) via the delays in travel times of seismic waves through these zones. This approach needs further development and testing. A preliminary analysis presented in this paper (of a building damaged by the 1994 Northridge, California, earthquake) shows that this method (1) can lead to detectable changes in the travel times of the waves passing through the areas known to have experienced damage, and (2) in its simplest form does not to require detailed modeling or analysis of soil-structure interaction.

Key Words: damage identiiication, wave propagation, travel time.

M. Erdik et al. (eds.), Strong Motion Instrumentation for Civil Engineering Structures, 447–467.

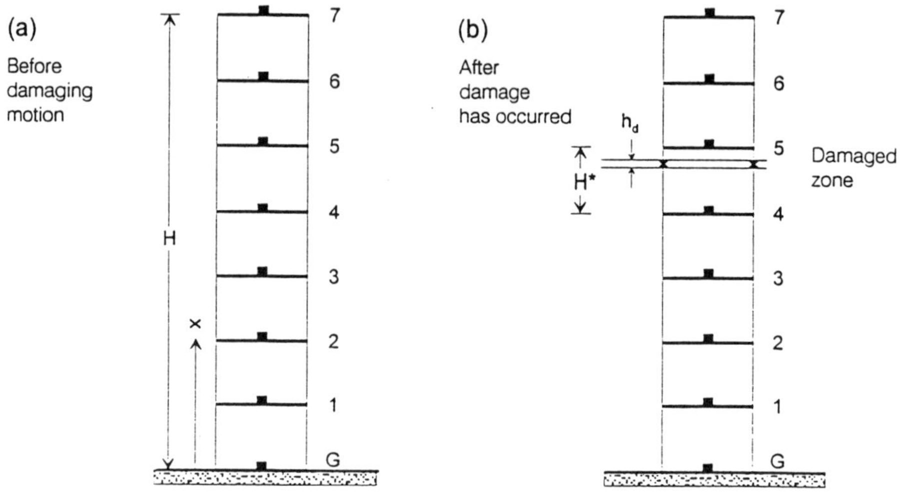

Figure 1. A multi degree-of-freedom system (a) before and (b) after localized damage has occurred (e.g. in the columns below the 5th floor). The solid squares indicate locations of the strong motion instruments.

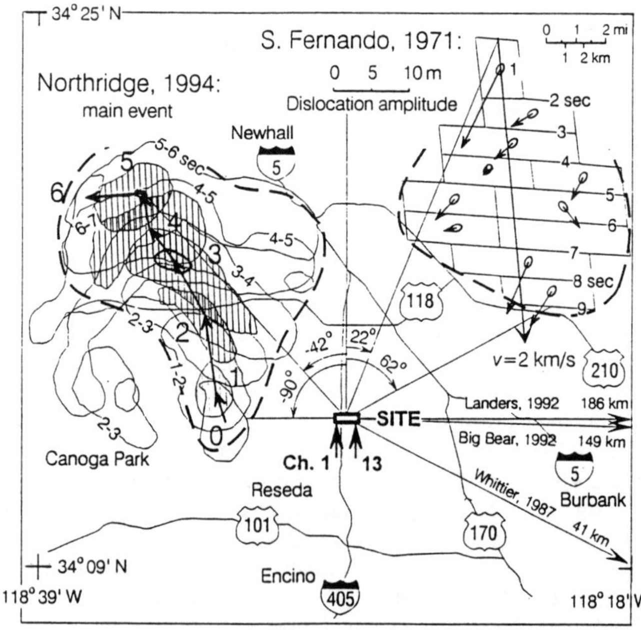

Figure 2. San Fernando Valley area showing the site of the 7-storey hotel relative to the fault planes of 1971 San Fernando and 1994 Northridge earthquakes. Directions toward the 1987 Whittier-Narrows, 1992 Landers and 1992 Big Bear earthquakes are shown by arrows.

1. INTRODUCTION

One of the purposes for testing full scale structures, before during and after earthquakes, is to detect damages caused by severe earthquake shaking [3, 15]. In ideal setting, the measurements should identify the location, evolution and the extent of damage. For example, the recorded data would show the time history of reduction of stiffness in the damaged member(s), and would identify the damaged member(s). Minor damage, which weakens the structural members but does not alter the form of their participation in the overall stiffness matrix, is expected to be reflected only in changes of the corresponding terms of the system stiffness matrix, and this will result in changes of the corresponding mode shapes and natural periods of vibration [19]. Consequently, a partially damaged member would reduce the overall stiffness of the system, and would cause the natural periods of vibration to lengthen. A simple approach to structural health monitoring has been to measure these changes in the natural periods (usually the first period, T_1) before and after strong shaking [5]. However, there are at least two problems with this approach. The first problem is that such period changes are usually small, and therefore are difficult to measure accurately [7]. The second problem is that the apparent system period, T, which is the quantity usually measured, depends also on the properties of the foundation soil, i.e.

$$T^2 = T_1^2 + T_r^2 + T_h^2 \tag{1}$$

where T_1–first fixed-base building period, T_r–period of the building rocking as a rigid body on flexible soil, and T_h–period of the building translating horizontally as a rigid body on flexible soil. The apparent system period, T, can and often does change appreciably during strong shaking, by factors which can approach two [20]. These changes are caused mainly by nonlinear response of the foundation soils, and appear to be self-healing, probably due to dynamic settlement and compaction during aftershocks and small earthquakes [5, 17]. To detect changes in T_1 only, special purpose instrumentation must be installed in structures. With the currently available instrumentation in various buildings in California, one can evaluate changes in T, but separate contributions from T_r, T_h and T_1 cannot be detected [6].

For periods shorter than T (this corresponds to short wave lengths and to higher modes of building vibration), the soil-structure interaction effects become more complex and must be analyzed by wave propagation methods [12, 13]. In principle this higher complexity may offer good resolution for the purposes of identification of the soil-structure parameters, but this depends also on our ability to model the system realistically and calls for far more detailed full scale tests. Therefore most studies consider measured data only in the vicinity of T.

To illustrate the order of magnitude of the changes in T_1, we consider the model shown in Fig. 1. Let us assume that this model deforms in shear only, and let the period of the first mode of vibration be equal to T_1. Since the mode-shapes represent interference patterns of shear waves propagating up and down the structure [9, 11], it can be shown that the period T_1 is proportional to the travel time H/β ($T_1 = 4H/\beta$ and $\beta =$ shear-wave velocity in this structure), before any damage has occurred. After strong shaking, suppose that some columns have been damaged at a particular floor. Let the "length" of this damaged zone be h_d, and the reduced velocity of shear waves within this damaged zone, β_d. Then the period of the first mode will be proportional to $T_d \sim (H-h_d) / \beta + h_d / \beta_d$, and the percentage increase in T_d, relative to T_1, will be $p = \frac{100h_d}{H}\left(\frac{\beta}{\beta_d} - 1\right)$. For example, for H = 20 m, h_d = 2 m, β = 100 m/s and β_d = 50 m/s, $p = \frac{100 \times 2.}{20}\left(\frac{100}{50} - 1\right) = 10$ present.

In the following, we explore whether simple measurements of wave velocity in structures during strong shaking can be carried out, and whether the location of the observed changes will coincide with the areas of observed damage. For this purpose, we analyze strong motion recordings in a 7-storey reinforced concrete hotel building in Van Nuys, California, severely damaged by the 1994 Northridge earthquake. We show that this task appears to be feasible, and suggest that accurate digitization of accelerograms recorded in buildings is essential, before this type of analysis can be further developed and refined.

2. METHODOLOGY

Let it be assumed that recordings of strong motion are available at two adjacent floors, and that it is possible to measure the velocity of shear waves propagating in the structure. Before damage has occurred, the travel time between two adjacent floors, i and j, would be

$$t_{i,j} = H^* / \beta \tag{2}$$

and after damage has occurred

$$t_{i,j}^d = \left(H^* - h_d\right) / \beta + h_d / \beta_d, \tag{3}$$

where H^* = H/N, N = the number of stories (in our example, N = 7), and H^* = story height. The percent change from $t_{i,j}$ to $t_{i,j}^d$ is then $p = \frac{100h_d}{H^*}\left(\frac{\beta}{\beta_d} - 1\right)$. For h_d = 2 m, $\beta_d = 0.5\beta$, and H^* = 20/7 m, $p = \frac{100 \times 2 \times 7}{20}\left(\frac{1}{5.0} - 1\right) = 70$ percent. Note that this is N times larger than the percent change in T_1, because the observation "length" has been reduced N times.

A typical value of H^* is 3 m and of β is 100 m/sec; then $t_{i,j} \sim 0.03$ sec. Old data processing of strong motion acceleration provided equally spaced data at 50 points/s. Since the early 1990's, most data is processed with time step $\Delta t = 0.01$ s or 100 points/s. Clearly, to detect time delays of the order of 0.03, the accuracy of origin time and the accuracy of the time coordinates in digitized and processed data must be better than 0.03 s.

3. RESULTS

3.1 Building Description

The building we study is a 7-story reinforced concrete hotel building in Van Nuys, central San Fernando Valley, California (Fig. 2). It was designed in 1965 [1] and constructed in 1966. Its plan dimensions are approximately 63 by 160 feet (Fig. 3a; 1 foot=0.3048 m)). The typical framing consists of columns spaced at 20 foot centers in the transverse direction (Fig. 3b) and 19 foot centers in the longitudinal direction. Spandrel beams surround the perimeter of the structure. Lateral forces in each direction are resisted by interior column-slab frames and exterior column spandrel beam frames. Due to the stiffness added by the spandrel beams, the exterior frames are approximately twice as stiff as the interior frames. With the exception of some light framing members supporting the stairway and elevator openings, the structure is essentially symmetric. The participation in the response of the nonstructural brick filler walls and some exterior cement plaster could cause some asymmetry for lateral motion in the longitudinal direction, expected to be minor.

The first floor is a slab on grade over about 2 feet of compacted fill. Except for two small areas at the ground floor, which are covered by one-story canopies, the plan configurations of each floor, including the roof, are the same. The floor system is reinforced concrete flat slab with thickness 10 inches at 2nd floor, 8.5 inches at 3rd to 7th floors and 8 inches at the roof (1 inch=2.54 cm). A penthouse, with mechanical equipment, covers approximately 10 percent of the roof area.

The interior partitions, in general, are gypsum wallboard on metal studs. Cement plaster, 1 inch thick, is used for exterior facing at each end of the building and at the stair and elevator bays on the long side of the building. Double 16 gauge metal studs support the cement plaster. Some additional cement plaster walls are located on the south side of the building at the first floor. The north side of the building, along column line D, has four bays of brick masonry walls located between the ground and the second floor at the east end of the structure. Nominal 1-inch expansion joints, separate the walls from the underside of the second floor spandrels. Although none of the wall elements described are designed as a part of the lateral force-resisting system, they do contribute in varying degrees to the stiffness of the structure.

Geological data indicate that the site lies on recent alluvium. A typical soil boring shows the underlying soil to be primarily fine sandy silts and silty fine sands. The foundation system consists of 38 inch deep pile caps, supported by groups of two to four poured-in-place 24-inch-diameter reinforced concrete friction piles. These are centered under the main building columns. All pile caps are connected by a grid of beams. Each pile is roughly 40 feet long and has design capacity of over 100 kips (444.8 KN) vertical load and up to 20 kips (89 KN) lateral load.

3.2 Strong Motion Acceleragrams

The first known strong motion records in the building date back to the 1971 San Fernando earthquake (Fig. 2). Then, the building had only three self contained triaxial AR-240

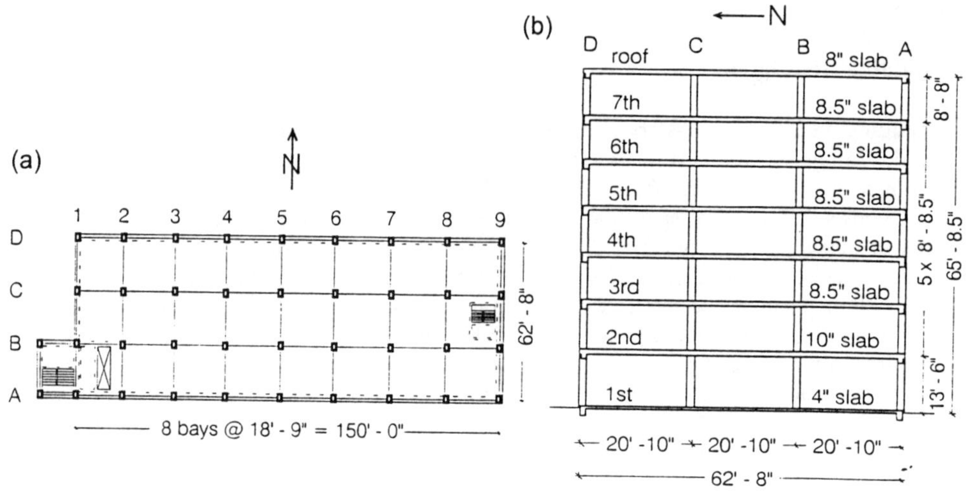

Figure 3 (a) A typical floor plan, and (b) a typical transverse section for the 7-storey hotel.

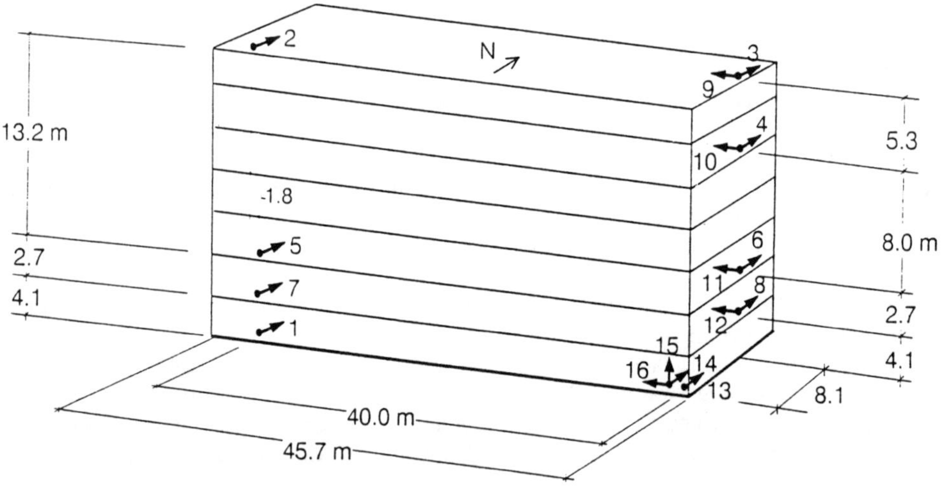

Figure 4. Sensor locations and orientations for the CR-1 recording system (channels 1–13) and for the SMA-1 accelerograph (channels 14–16).

accelerographs, one in the south-eastern corner at ground level, one in the center of the fourth floor, and one in the south-western corner; on the roof.

The central recording system (CR-1 accelerograph; Fig. 4), was installed in the building prior to the 1987 Whittier-Narrows earthquake. It is operated by California Department of Mines and Geology (CDMG). Between 1987 and 1994, the CR-1 system was triggered by many local and distant larger events [2]. Here, we analyze recorded accelerations only from the 1987 Whittier-Narrows (M_L=5.9), 1992 Landers (M_S=7.5), 1992 Big Bear (M_L=6.5) and 1994 Northridge (M_L=6.4) events, at epicentral distances 41, 186, 149 and 1.5 km. The direction of strong motion arrivals from the Whittier-Narrows, Landers and Big Bear earthquakes are E 27°S, East, and E 1.5°S. The largest accelerations were those recorded during the 1994 Northridge earthquake. Table 1 summarizes selected parameters of the above earthquakes and their recordings.

Table 1 Selected earthquake and accelerogram Parameters describing the data used in this paper

Earthquake	Date	M_L	R [km]	φ [deg]	L_{REC} [s]	L_{PROC} [s]	$a_{H,MX}$ [g]	$a_{V,MX}$ [g]
San Fernando	02/09/71	6.6	22	22-62	59.5	59.5	0.25	0.17
Whittier-Narrows	10/01/87	5.9	41	117	43	22.7* 43	0.16	
Landers	06/28/92	7.5 (M_S)	186	90	79	43.0* 79	0.04	0.007
Big Bear	06/28/92	6.5	149	91.5	43	22.6* 43	0.01	0.007
Northridge	01/17/94	6.4	1.5	270-318 240-350	60	60.0	0.44	0.27

* data digitized by M.D. Trifunac from xerox copy of one page of film record, presented in CDMG data reports

M_S– surface wave magnitude
M_L– local wave magnitude
R– epicentral distance
φ– azimuth of wave arrival
L_{REC} – length of recorded accelerogram
L_{PROC} – length of processed accelerogram
$a_{H,MX}$ – peak horizontal acceleration
$a_{V,MX}$ – peak vertical acceleration

3.3 Description of Damage

Between 1972 and 1993, the building was shaken many times, but suffered no visible damage. During the San Fernando earthquake, the peak horizontal acceleration at the foundation level was 0.25g. Between 1972 and 1993, the largest peak acceleration (0.16g) was recorded during the 1987 Whittier-Narrows earthquake (see Table 1). During the1994 Northridge earthquake, the peak ground acceleration was 0.44g and the building suffered considerable damage.

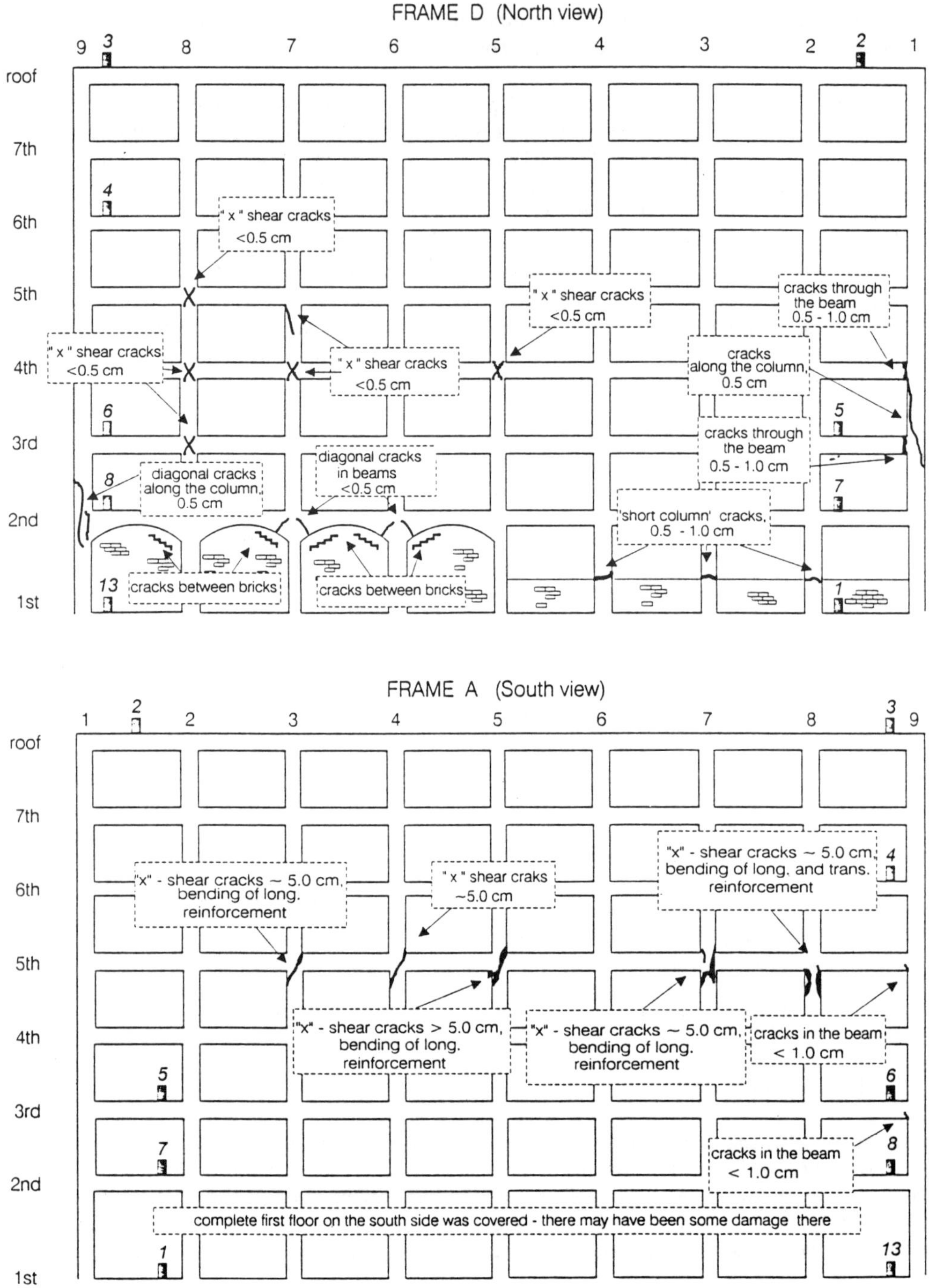

Figure 5. Schematic representation of damage: (top) frame D (North view), and (bottom) frame A (South view). The sensor locations for channels 1–8 and 13 (oriented towards North), are also shown (see also Figure 4).

3.3.1 *Damage Caused by the 1971 San Fernando Earthquake*

Structural damage from the San Fernando earthquake was minor. Epoxy was used to repair spalled concrete of the second floor beam column joints on the north side and east end of the building. The nonstructural damage was extensive, and about 80 percent of all repair cost was used to fix drywall partitions, bathroom tiles and plumbing fixtures. The damage was most extensive on the second and third floors and minimal at the sixth and seventh floors [1].

3.3.2 *Damage Caused by the 1994 Northridge Earthquake and its Early Aftershocks*

During the Northridge Earthquake, the building experienced significant damage. Serious structural damage occurred in the exterior longitudinal, south (A) and north (D) frames. Those were designed to take most of the lateral loads in the EW direction.

In Frame A (Fig. 5 bottom, south side), wide shear cracks appeared in columns A3, A4, A5, A7 and A8, just below the contact with the spandrel beam of the 5th floor. The cracks width was 5–10 cm on the surface of the column at A5F5 and A7F5, and more than 10 cm at A8F5 (A5F5 means frame A, column 5, spandrel beam of 5th floor. At all these locations, deformation (buckling) of the longitudinal bars (due to large motions) was evident. Large deformation of the longitudinal bars at A8F5 also caused deformation of the transverse reinforcement. There was no visible damage along column A6 of the frame A, although both adjacent columns, A5 and A7, were seriously damaged. The 1st floor of this frame was covered from the exterior, possibly hiding some minor damage. Minor cracks (width <1 cm), occurred at A9F3 and A9F5.

In exterior frame D, the width of shear cracks on the surface of the columns was moderate (0.2–1 cm; Fig. 5 top). These cracks had clearly a visible "x" shape at locations D5F4, D7F4, D7F5, D8F3, D8F4, D8F5. Column D1 cracked (width ~1 cm) vertically along the 3rd floor. Those cracks extended also into the spandrel beams at 3rd and 4th floors in the form of vertical cracks (width < 1 cm).

At the 1st floor of frame D, at columns D2, D3 and D4, the cracks appeared to have been caused by a "short column" effect, due to the brick wall, built up to 1/3 of the height of the columns in the first four bays. These cracks were 0.5–1 cm wide (Fig. 5 bottom). A diagonal crack (about 0.5 cm wide) occurred at D9F2. Small diagonal cracks (width less than 0.5 cm) were noticed at locations D6F2 and D7F2. In the brick walls, in the four eastern bays of the first floor, cracks occurred between the bricks (through mortar), at the upper corners of the walls (Fig. 5 top).

There was no visible damage to the interior longitudinal frames (B and C), although some cracks could have been hidden by large furniture (especially on the first floor), wallpaper and carpets. No damage was observed in the reinforced concrete slabs. Small cracks (width < 0.2 cm) were noticed in the slab around the central columns on 5th and 6th floors. There were no signs of large deformations in the foundation, neither at the 1st floor slab nor in the pavement around the building.

The nonstructural damage was extensive. Every guest room suffered some type of nonstructural damage. Furniture was overturned at the upper floors (above 3rd). Due to large relative motions

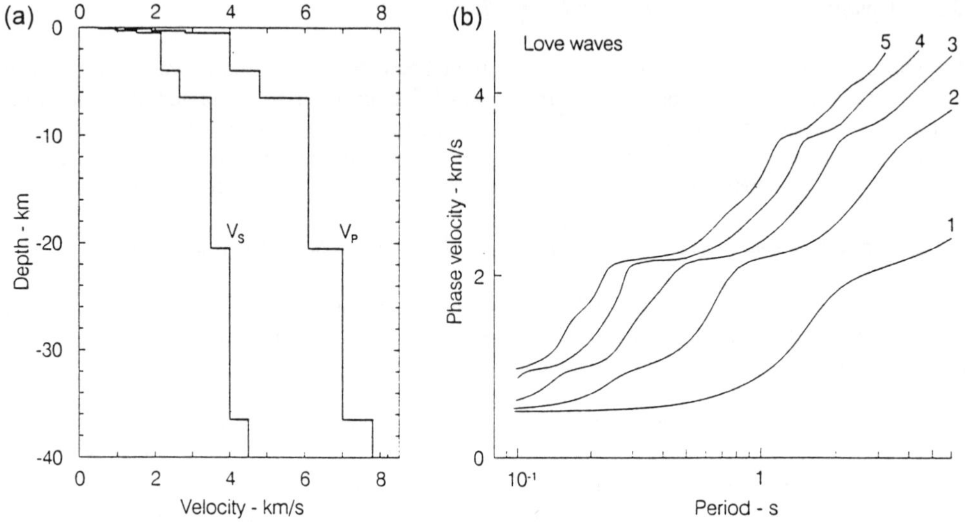

Figure 6 (a) Typical S- and P-wave velocity profiles (V_s and V_p) for San Fernando Valley. (b) Phase velocities of the first five Love wave modes for a parallel layer earth model with profile as in part (a).

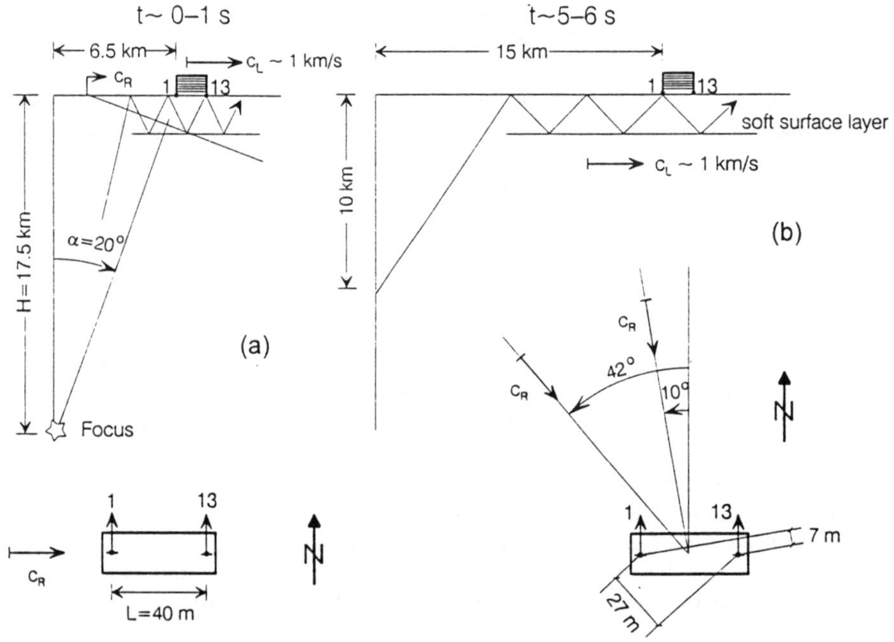

Figure 7 An illustration of predominant wave arrivals from the rupture of the 1994 Northridge earthquake (Figure 2). Parts (a) and (b) show vertical cross-sections through the building site, at the beginning (~0–1 s) and towards the end (~5–6 s) of the main dislocation sequence. Note the propagation of energy predominantly towards east in part (a), and predominantly towards south-east in part (b).

and deformation of the interior walls, the wallpaper was distorted or torn off. The relative displacements also caused extensive damage to the brittle ceramic tile covers in the bathrooms. There were numerous cracks in the bathtubs, and many ceramic tiles fell off.

3.4 Wave Propagation In and Around the Building

In the following, we analyze the accelerograms recorded in the building, using simple approximate representation based on wave propagation. Figure 2 shows the setting of the building, in central San Fernando Valley, and the horizontal projections of the 1971 San Fernando and 1994 Northridge earthquake faults. During the San Fernando earthquake, the faulting started at depth, ~ 9 km below surface, and propagated up and towards south with average dislocation velocity of ~2 km/s, along the fault plane dipping 40° and striking N72W [16]. The consequence of this propagation is that the waves arrived at the building site first at angle φ = 22°, and then φ increased towards 62°, during the following 9 s (Fig. 2). The strongest motion arrived from the deep part of the dislocation surface, during the first 4 seconds, and then from the shallow part of the fault during the last two seconds of faulting.

The fault motion for the Northridge earthquake [21] is shown by snapshots of fault dislocations with time increments of one second. The open empty ovals, the darker vertically hatched zones and the dark horizontally hatched zones show respectively areas with dislocation amplitudes >0.5, 1.0 and 1.5 m. It is seen that the largest dislocations propagated towards NW, during the first 5 seconds, and then, only during the last second, towards West, with dislocation velocity 2.8–3 km/s. Assuming that most of the strong motion energy arrived from the area with largest dislocation amplitudes, implies wave arrivals with azimuths φ =270–318°. However, because the entire fault surface radiated energy, the range of azimuths for wave arrivals was actually broader, φ= 240–350°.

During the Whittier-Narrows, Landers and Big Bear earthquakes, the building was at epicentral distances 41, 186 and 149 km. Because of the small fault dimensions relative to the corresponding epicentral distances, in laterally homogeneous layered earth (Fig. 6a,b), all the wave energy would have been arriving along essentially same respective paths. However, the complex geologic structure (the Santa Monica and Verdugo Mountains, and the Elysian and Repetto Hills are between San Gabrial and Los Angeles Valleys) contributed to lateral scattering resulting in more complicated multiple direction arrivals of wave energy towards the building.

Figure 7 shows a schematic vertical cross-section through the focus of the Northridge earthquake and the building site (see Fig. 2). It is seen that for the waves arriving first towards channel 1 and then towards channel 13 (40 m apart, at the western and eastern ends of the building), the SH- and Love-wave components of the early motions could be delayed for these two channels by $\Delta t \leq 0.05$ sec. Direct body waves would be delayed less (due to large radial phase velocity C_R), while the waves traveling through shallow low velocity layers would be delayed at most by $\Delta t \approx 0.05$ sec. About 5 to 6 s later, the energy originates near the NW end of the fault, at depth ≈10 km, and at distance ≈5 km NW from the building site. For azimuths φ = 318–350°, the "separation distance" for channels 1 and 13 decreases to ≈ 27 and 7 m respectively, and this reduces the relative delay of wave arrivals at channels 13 and 1 to Δt

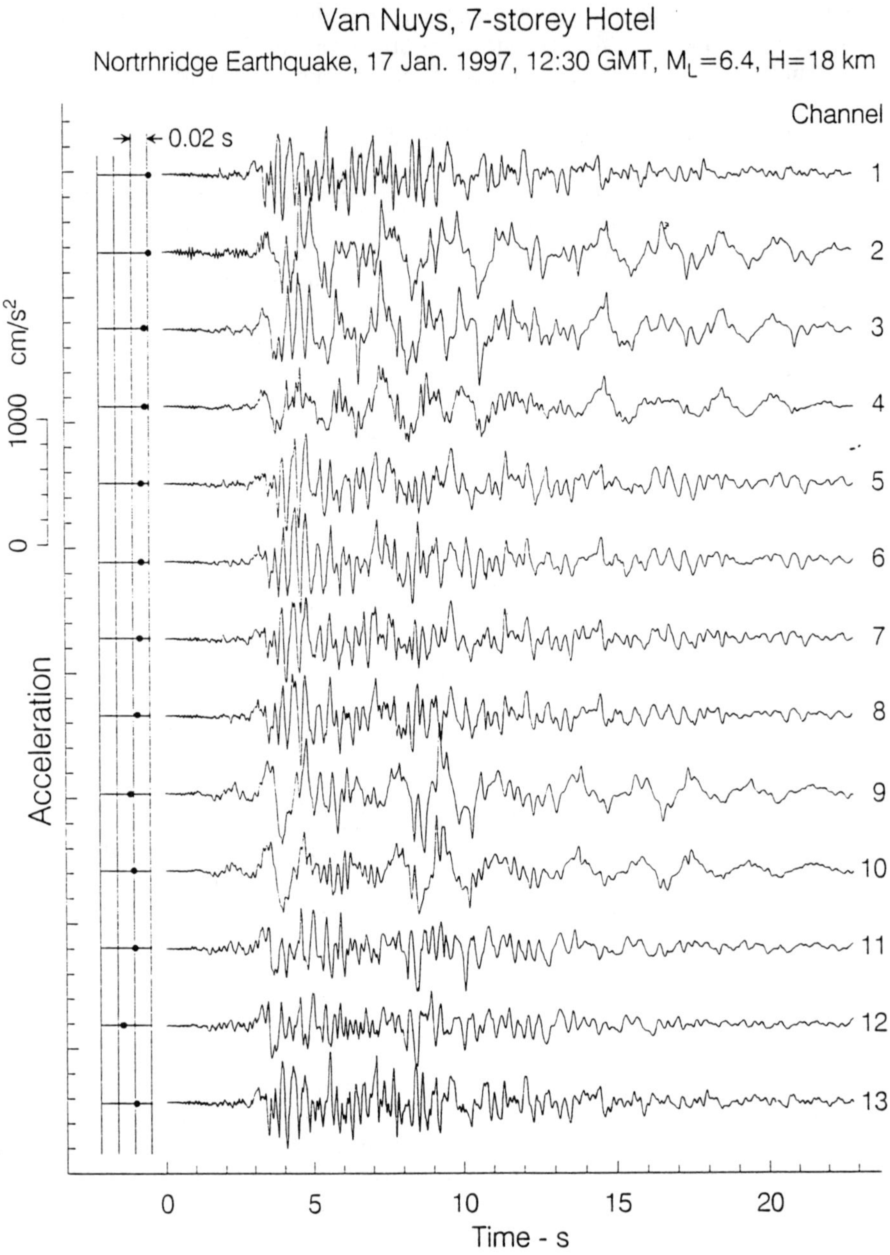

Figure 8. Recorded strong motion accelerations and time delays for the 13 CR-1 channels digitized by CDMG, relative to the same data digitized by USC. The delay is the largest for Channel 12 (~0.03 s).

≈0.008 s or less. Following the direct arrivals from the fault plane, the surface and coda waves of strong motion may arrive from all directions, scattered and reflected from the edges of sediments in San Fernando Valley.

3.5 Required Accuracy in Data Processing

For the above delays to be seen in recorded data, it is necessary to have accurate relative timing of the data recorded by different channels of the CR-1 recording system. This requires (1) accurate digitization of the beginning of all traces, and (2) accurate corrections for variations in film speed. Our routine accelerogram processing software produced corrected data (for baseline and instrument response) equally spaced at 50 points/s until the early 1990's, and currently at 100 points/s. It is seen that the expected "delays" between motions at channels 1 and 13 due to wave propagation (see Fig. 4) are ≈ 0.01–0.03 s ≈ 1 to several intervals of equally spaced points at Δt=0.01 s.

To evaluate the relative accuracy of time coordinates of the Northridge earthquake record digitized by CDMG, we redigitized independently the same record, from a reproduction of the film record in reference [8]. By visual comparison of the first several seconds of digitized data, and by cross-correlation analysis, we found that CDMG digitization "omits" (starts late) up to ~0.03 s at the beginning of some traces. In Fig. 8, we document these "omitted" intervals for all 13 channels for this record. It is seen that the origin times of both digitizations agree for channels 1 and 2. The CDMG data begins progressively late with increasing channel number, and the delay is ≈ 0.02 s for channels 8 through 13. For channel 12, the delay is the largest (≈ 0.03 s).

Figure 9 shows similar comparison of initial delays for three other earthquakes (Whittier-Narrows, Big Bear and Landers). It is seen that the differences are of systematic and repetitive nature. Though there are several simple and plausible explanations for these differences, we cannot state their cause with certainty. As an illustration only, we suggest that one possible explanation can be related to different optical densities of different traces on the film, when digitized with same threshold level of gray [4, 18].

3.6 Rough Estimation of Apparent Travel Times

An approach to investigating the wave propagation inside the building may be formulated starting with the assumption that the incident energy consists of pulses, which then propagate from one recording point to another. To find the time delay representing the average difference at stations i and j of all arrival times collectively, we can evaluate cross-correlation functions of the recorded motions, $R_{ij}(\tau)$, where τ represents the average delay or "travel time". To emphasize the local pulses and to avoid randomization caused by numerous pulses during the entire excitation history, we multiply recordings at stations i and j by windows, $W(t)$, as shown in Fig. 10. These windows have unit amplitude in a 2 s interval, 0.5 s ramps on each side, and zero amplitudes otherwise. Then we compute

$$R_{i,j}(t_0,\tau)=\int_{t_0}^{t_0+\tau+3} f_i(t)W(t-t_0)f_j(t+\tau)W(t-t_0-\tau)dt \qquad (4)$$

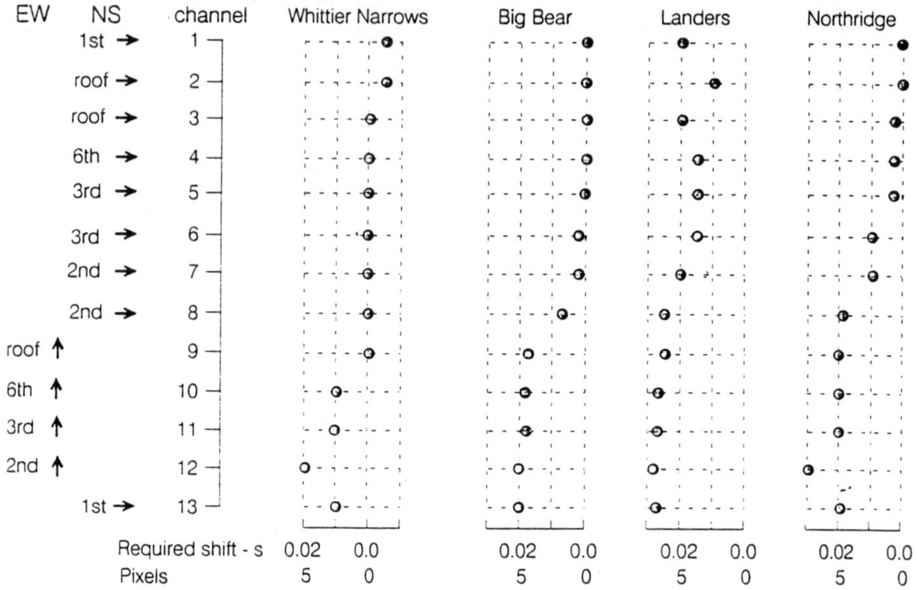

Figure 9 Time delays (in seconds and pixels) for the CDMG-digitized data of the 13 CR-1 channels, relative to the USC-digitized data, for the records of the 1987 Whittier-Narrows, 1992 Big Bear, 1992 Landers and 1994 Northridge earthquakes.

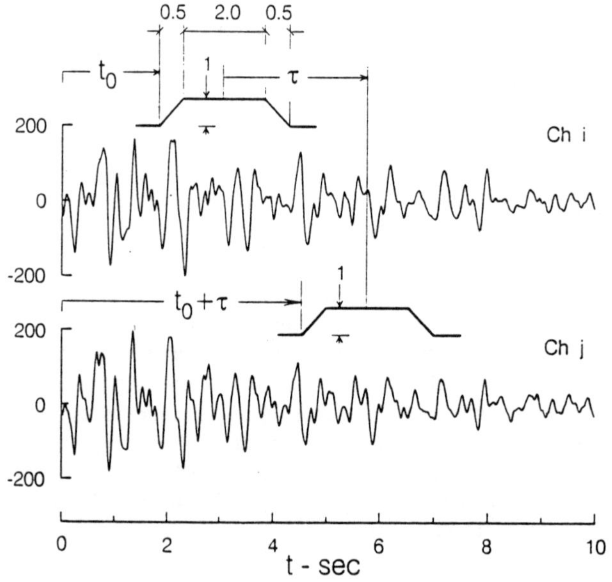

Figure 10 An illustration of the time windows applied to a pair of signals (channels i and j), and the delay, τ, as used in computing the cross-correlation function of the two acceleration signals, $R_{ij}(t_0, \tau)$ (see Figures 11 and 12).

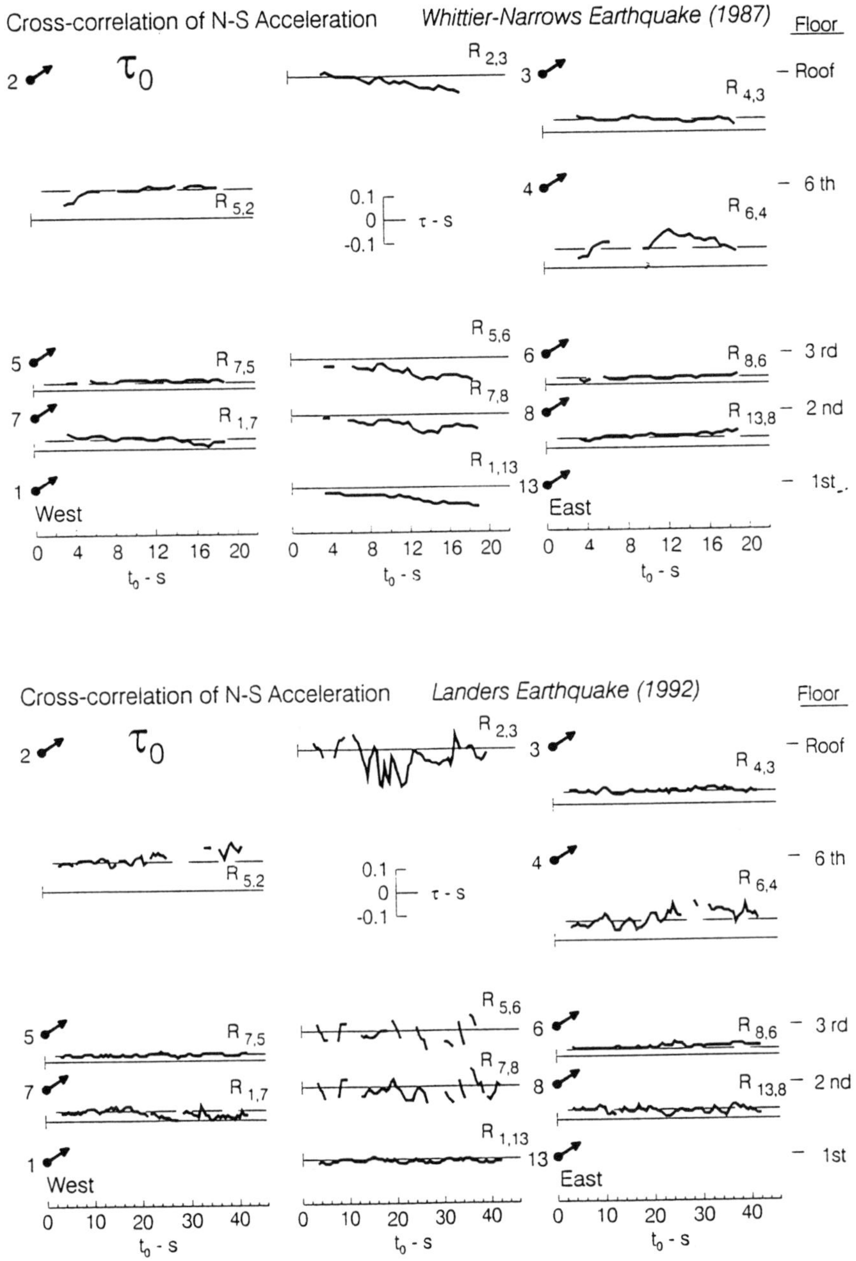

Figure 11 Delay, $\tau=\tau_0$, for maximum cross-correlation, $R_{ij}(t_0, \tau)$, versus time t_0 (see Figure 10), for recorded N-S accelerations during the 1987 Whittier-Narrows (top) and 1992 Landers (bottom) earthquakes.

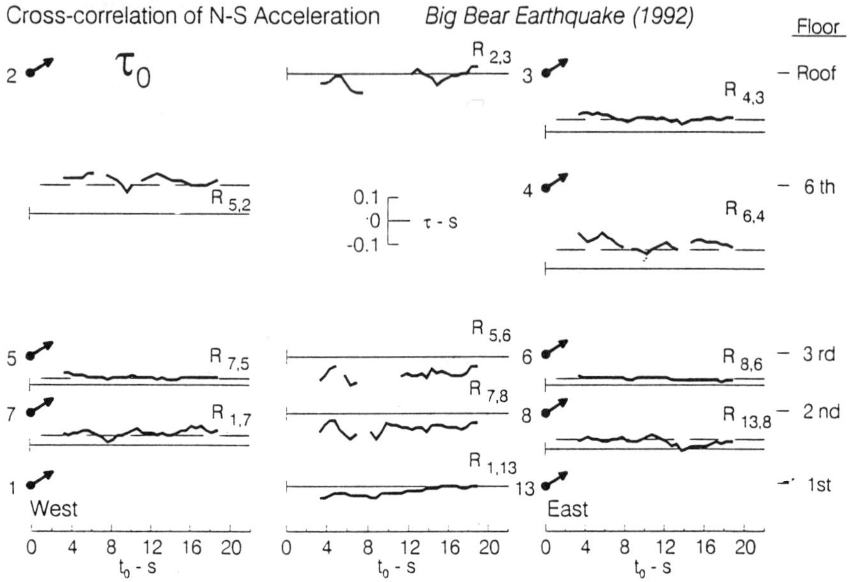

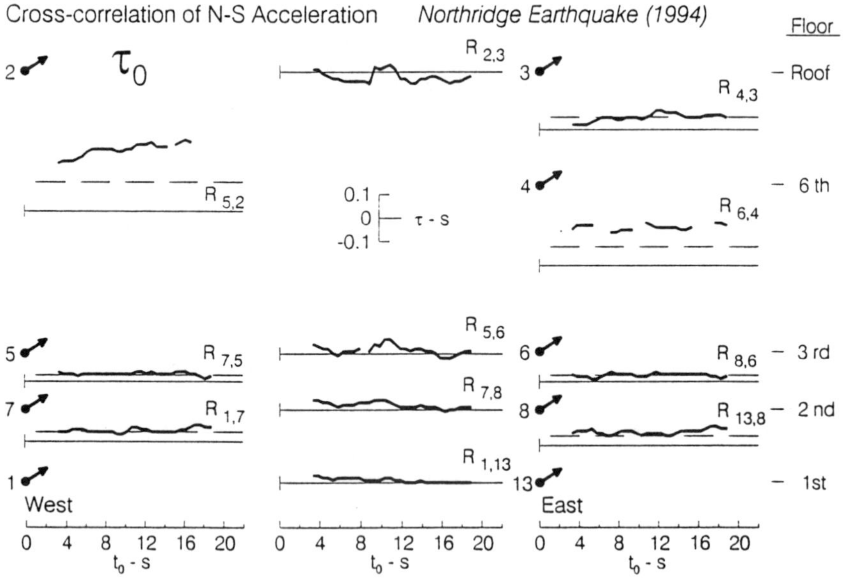

Figure 12 Delay, $\tau=\tau_0$, for maximum cross-correlation, $R_{ij}(t_0, \tau)$, versus time t_0 (see Figure 10), for recorded N-S accelerations during the 1992 Big Bear (top) and 1994 Northridge (bottom) earthquakes.

and find the delay τ_0 for which the amplitude of R_{ij} (t_0, τ) is maximum. The physical interpretation of the result and the estimates of τ_0 depend on the assumed properties of the window function $W(t)$. Analysis of these properties is beyond the scope of this paper. Here, for simplicity and to illustrate the general trends, we adopt only one window function $W(t)$, as shown in Fig. 10. Figures 11 through 12 show plots of τ_0 versus t_0 only for the NS accelerations in the building recorded during the Whittier-Narrows, Landers, Big Bear and Northridge earthquakes.

Forward evaluation of the average shear-wave velocity in the building requires representation of the structure by at least two types of anisotropic layers with different properties [14], one for the floor slabs and one for the columns. It also requires estimation of the relative fixity of the columns. Details of this modeling are beyond the scope of this paper. Here, we only state that such analysis shows that the velocity of vertically propagating shear-waves in this building is ≈100 m/s, and that the velocity of SH- waves along the floor slabs is ≈2000 m/s. These estimates are consistent with the experimental results presented here, and with other independent analyses of wave motion in this building, which will be presented in future papers.

Assuming plane vertically propagating waves, with average shear-wave velocity in the vertical direction β=100 m/s would imply $\tau_0 = h_{ij} /100$, where h_{ij} is the vertical separation of stations i and j (in meters). In Figs 11 and 12 (left and right columns), these propagation times are shown by thin dashed lines. The computed τ_0 delays are shown by heavy interrupted lines. Occasionally, reflections off the top or off the sides of the building may cause peaks of R_{ij} (t_0, τ_0) to occur at much longer or shorter τ_0 than what we assume corresponds to the direct waves. To simplify this presentation, we chose not to show those values.

The central columns in Figs 11 and 12, show time delays, τ_0, of pulses propagating along the longitudinal axis of the building, in the EW direction. Complete interpretation and modeling of what those delays represent is a complex problem whose analysis is beyond the scope of this paper. Assuming that the building has to follow the ground motion [14] and ignoring soil structure interaction, would lead to τ_0 for $R_{1,13}$ to depend on the average travel time of shear-waves and of horizontal components of P, SV and Rayleigh waves in the soil between locations of channels 1 and 13. For EW propagation, these two locations are 40 m apart, but for other directions of wave approach, this distance can be anything between 0 and 40 m. For the Whittier-Narrows, Landers and Big Bear earthquakes, most of the energy traveled from East to West (see Fig. 2) and so τ_0 must be negative (see eqn 4). Between 5.5 and 7.0 s after trigger, the records of Whittier-Narrows earthquake show shear-wave arrivals which should be associated with "high" values of phase velocities, and τ_0 should be "small" and negative (Fig. 13). Twelve seconds after trigger, the recorded motions show arrival of surface waves and τ_0 for $R_{1,13}$ is in the range from −0.04 to −0.08 s, suggesting phase velocity between 0.5 and 0.95 km/s. These values are "reasonable" for intermediate and high frequency (f > 1 Hz), first and second modes of surface Love waves in San Fernando Valley (e.g. Fig. 6b).

For the Landers earthquake, the largest earthquake of the four considered here, τ_0 for $R_{1,13}$ is negative and "small", and is constant throughout the entire duration of wave arrival (Fig. 13). This suggests predominantly surface waves with strong participation of long periods and higher

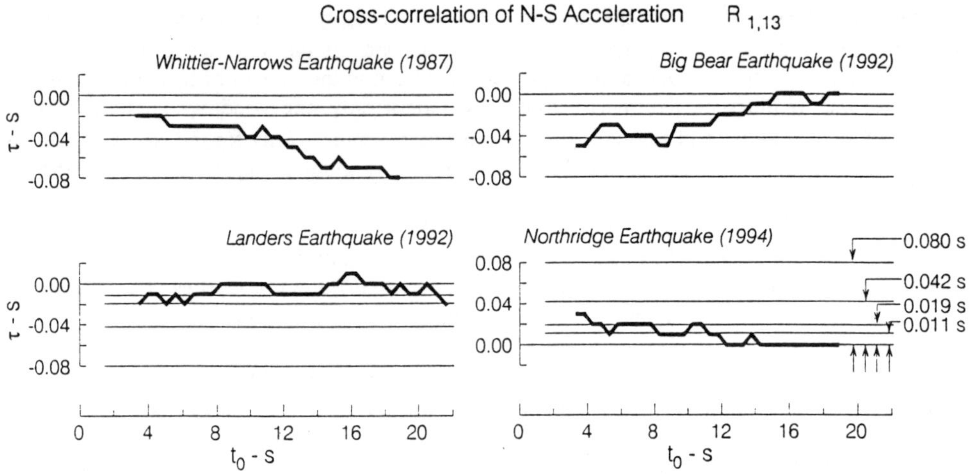

Figure 13 Delay, $\tau=\tau_0$, for maximum cross-correlation $R_{1,13}(t_0, \tau)$ of recorded N-S accelerations at channels 1 and 13 (1st floor), versus time t_0, for three earthquakes East of the site (1987 Whittier-Narrows, 1992 Landers and 1992 Big Bear) and one earthquake West of the site (1994 Northridge).

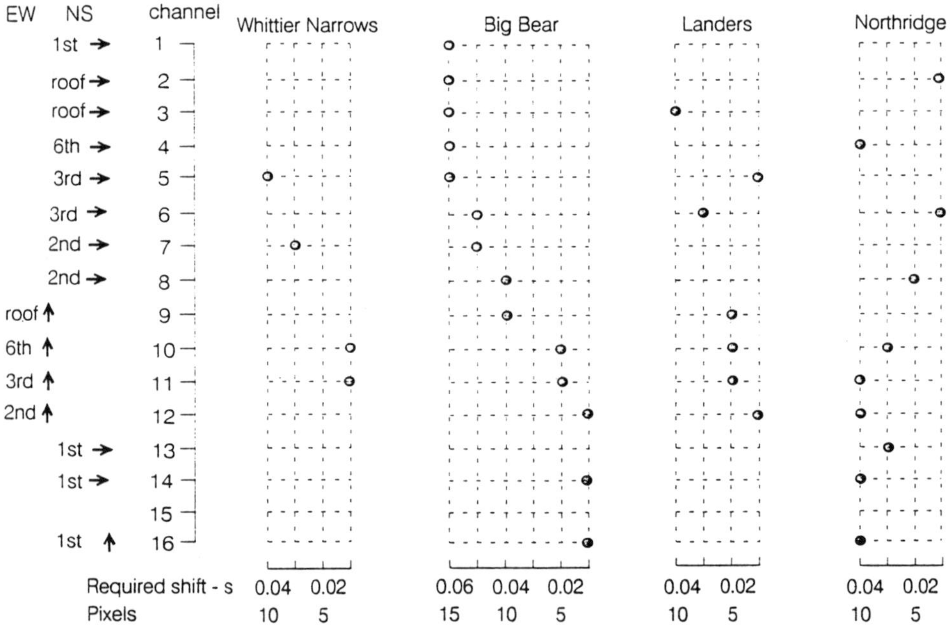

Figure 14 Time shifts for USC digitalization of the CR-1 (channels 1–13) and SMA-1 (channels 14–16) records of four earthquakes, adopted for the analysis of delays in Figures 11–13 (1 pixel ~ 0.004 cm (s)). E.g., shift of 0.04 s means padding four zero-amplitude points (equally spaced at 0.01 s) at the beginning of the record, to delay the recorded motions (i.e. advance the origin time) by 0.04 s.

modes, contributing with phase velocities $c = 2.15$ to 3.5 km/s. During the Big Bear earthquake, the accelerations at the building site were small, and the instruments triggered late, apparently after the arrival of shear waves. The motion appears to be dominated by high frequency surface waves with phase velocities $c = 0.5$ to 0.95 km/s, during the first 12 s of recorded motion (Fig. 13). Between 12 and 20 s, the waves appear to be associated with larger EW and NS motions, suggesting arrivals from all directions and SH components arriving from North or South. This would reduce the apparent separation distance between the recording sites for channels 1 and 13, resulting in small values of τ_0.

During the first 12 s of the Northridge earthquake, the waves arrived first from west and then later from north-west (see Fig. 2 and 7) resulting in positive τ_0 in $R_{1,13}$. With time, as the predominant direction of wave approach changed from West to NW, the effective separation of stations 1 and 13 decreased, thus leading to gradual decrease of τ_0. Beyond 12 to 15 s, the motion most probably arrived from "all" directions representing scattered and reflected waves from irregular geologic boundaries of San Fernando Valley, and this resulted in small values of τ_0. These trends and variations of τ_0 versus t_0 (Fig. 13) can be seen also at the 2nd ($R_{7,8}$), 3rd ($R_{5,6}$) and 8th ($R_{2,3}$) floors of the building, but other effects caused by wave reflections off the East and West ends of the building are superimposed (see Fig. 11 and 12). We show these results for completeness of this presentation, but leave the detailed analyses and interpretation for future work.

The above results on τ_0 in R_{ij} (t_0, τ_0) are based on USC digitalization and processing of CR-1 and SMA-1 records (using xerox copies of the records published in CDMG reports). After digitization and processing, to achieve internally consistent and physically plausible results, in Fig. 11 and 12, it was necessary to shift (advance) the origin time of different components (channels) as shown in Fig. 14. These shifts are small relative to the significant prolongations of τ_0 for $R_{5,2}$ and $R_{6,4}$, seen only for the Northridge data (Fig. 12 bottom). The required shifts for the Big Bear earthquake however are surprisingly large.

The Big Bear acceleration records began with a late trigger (during arrival of surface waves and after the S-wave arrival). The amplitudes are small and the traces achieved full optical density after more than about 0.1 s. It is possible that the unexpected pattern of "required" shifts is caused by artificial reduction of trace width (and thus loss of the trace beginning) by the high contrast characteristics of xerox copying. This effect may apply to both USC and CDMG digitalizations, to a similar degree, such that a relative starting time between these two digitalizations (see Fig. 9) still applies. Further analysis and interpretation of this interesting and important data processing problem [18] is however beyond the scope of this work.

Large reduction of "propagation velocities" is seen only in the values of τ_0 for $R_{5,2}$ and $R_{6,4}$, during the Northridge earthquake. As Fig. 12 (bottom) shows, the computed τ_0 is roughly two times longer than expected. With reference to Fig. 5, it is possible to speculate that this is caused by damage of the columns just below the spandrel beams on the 5th floor.

4. DISCUSSION AND CONCLUSIONS

We explored the possibility of measuring propagation velocity of waves in a simple symmetric structure, and interpreting the observed delays in wave arrivals as indicators of damage [7]. We computed the propagation "time" from a pair of stations by using windowed cross-correlation function. This rough method ignores wave dispersion and the presence of "other" reflecting boundaries, and therefore is contaminated and can give wrong answers in the presence of many reflections. Nevertheless, the results we obtained show promise and appear to give significantly long travel times in the areas where damage did occur.

The first requirement for the success of this method is that the time coordinates of processed strong motion data must be (1) as accurate as possible and (2) their accuracy must be analyzed, interpreted and quantified. Since we are able to identify the first digitization point for each acceleration trace with accuracy of about one pixel (e.g. for scanners with 600 dpi the width of one pixel is 1/236 cm, that is 0.004 sec), it is necessary only to calibrate the accuracy of electrical signals creating two pulses per second (2 PPS) time trace, which is usually written on the edge of the film. For digital instruments, in principle, neither the starting nor relative timing of samples from different channels should pose problems, but the actual speed of multiplexing, and the frequency of sampling must be accurately measured and should be documented.

In this study we use the concept of equivalent continuous medium to represent a three-dimensional structure, but for accurate identification of damaged zones the modeling of structures may have to be more detailed. The difficult and challenging task will be to develop digital filters which can account for various scattered waves and for reflections in a three-dimensional matrix of structural members, which will then be used to measure the "wave" velocity between the two points.

Finally, we mention the obvious limitation of the above method. It has to do with its ability to resolve "small" and concentrated damage zones. The method can offer only an order of magnitude (~ N) improvement over measurements of changes in natural frequencies. It is, of course, possible in principle to saturate buildings with transducers, densely distributed, on all structural members, but this is obviously not a practical alternative. The best we can expect, at present, is to have one instrument per principal direction per floor. This will correspond to approximately three times better resolution then in the example presented in this paper.

5. REFERENCES

1. Blume, J.A. and Assoc. (1973). Holiday Inn, Chapter 29 in "San Fernando, California Earthquake of February 9, 1971," Volume I, Part A, U.S. Dept. of Commerce, National Oceanic and Atmospheric Administration, Washington, D.C.

2. Graizer, V.M. (1997). Personal Communication.

3. Ivanović, S. and M.D. Trifunac (1995). Ambient vibration survey of full scale structures using personal computers (with examples in Kaprielian Hall), Dept. of Civil Engrg, Rep. No. 95-05, Univ. of Southern California, Los Angeles, California.

4. Lee, V.W. and M.D. Trifunac (1990). Automatic digitization and processing of accelerograms using PC, Dept. of Civil Engrg, Rep. No. 90-03, Univ. of Southern California, Los Angeles, California.

5. Luco, J.E., H.L. Wong and M.D. Trifunac (1986). Soil-structure interaction effects on forced vibration tests, Dept. of Civil Engrg, Report No. 86-05, Univ. of Southern California, Los Angeles, California.

6. Moslem, K. and M.D. Trifunac (1986). Effects of soil structure interaction on the response of buildings during strong earthquake ground motions, Dept. of Civil Engrg, Report No. 86-04, Univ. of Southern California, Los Angeles, California.

7. Safak, E. (1998). Detection of seismic damage in multi-story buildings by using wave propagation analysis, Proc. Sixth U.S. National Conf. on Earthquake Eng., Seattle, Washington.

8. Shakal, A., M. Huang, R. Darragh, T. Cao, R. Sherburne, P. Malhotra, C. Cramer, R. Syndov, V. Graizer, G. Maldonado, C. Peterson and J. Wimpole (1994). CSMIP strong motion records from the Northridge, California, Earthquake of 17 January 1994, Report No. OSMS 94-07, Calif. Dept. of Conservation, Div. Of Mines and Geology, Sacramento, California.

9. Todorovska, M.I. and M.D. Trifuanc (1989). Anti-plane earthquake waves in long structures, J. Engrg Mechanics, ASCE, 115(12), 2687-2708.

10. Todorovska, M.I. and M.D. Trifunac (1990). A note on excitation of long structures by ground waves, J. Engrg Mechanics, ASCE, 116(4), 952-964.

11. Todorovska, M.I. and M.D. Trifunac (1990). Propagation of earthquake waves in buildings with soft first floor, J. Engrg Mechanics, ASCE, 116(4), 892-900.

12. Todorovska, M.I. and M.D. Trifunac (1990). Analytical model for the in-plane building-foundation-soil interaction: incident P-, SV-, and Rayleigh waves, Dept. of Civil Engrg, Report No. 90-01, Univ. of Southern California, Los Angeles, California.

13. Todorovska, M.I. and M.D. Trifunac (1991). Radiation damping during two-dimensional in-plane building-soil interaction, Dept. of Civil Engrg, Report No. 91-01, Univ. of Southern California, Los Angeles, California.

14. Todorovska, M.I., V.W. Lee and M.D. Trifunac (1988). Investigation of earthquake response of long buildings, Dept. of Civil Engrg, Report No. 88-02, Univ. of Southern Calif., of Los Angeles, California.

15. Trifunac, M.D. (1972). Comparison between ambient and forced vibration experiments, Earthquake Enrg and Struc. Dynam., 1, 133-150.

16. Trifunac, M.D. (1974). A three-dimensional dislocation model for the San Fernando, California, Earthquake of February 9, 1971, Bull. Seism. Soc. Amer., 64, 149-172.

17. Trifunac, M.D., S.S. Ivanović and M.I. Todorovska (1999). Experimental evidence for flexibility of foundation supported by concrete friction piles, Soil Dynam. and Earthquake Engrg, 18(3), 169-187.

18. Trifunac, M.D., M.I. Todorovska and V.W. Lee (1999). Selected common problems with automatic digitization of strong motion accelerograms (submitted for publication).

19. Udwadia, F.E. (1977). Uniqueness problems in structural identifications from strong motion records, Proc. Sixth World Conf. On Earthquake Engrg, Sarita Prakasan, Meerut, India, Vol. II, 1010-1015.

20. Udwadia, F.E. and M.D. Trifunac (1974). Time and amplitude dependent response of structures, Earthquake Engrg and Struct. Dynam., 2, 359-378.

21. Wald, D.J. and T.H. Heaton (1996). The slip history of the 1994 Northridge, California, earthquake determined from strong-motion, teleseismic, GPS and leveling data, Bull Seism. Soc. Amer., 86(1B), S49-S70.

EARTHQUAKE MOTION OBSERVATION IN AND AROUND 8-STORY SRC BUILDING

TOSHIHIDE KASHIMA, IZURU OKAWA and SHIN KOYAMA
Building Research Institute, Ministry of Construction
Tatehara 1, Tsukuba, Ibaraki 305-0802, Japan

1. Introduction

Building Research Institute (BRI), Ministry of Construction, Japan, started the strong motion observation project in 1957. The observation network has been enriched and enlarged in the past forty years. BRI is now operating several networks with a number of strong motion instruments. In this paper, we are introducing a new system that is installed to buildings and underground at BRI, and some observation results.

2. Outline of Strong Motion Observation at BRI Annex

BRI has started earthquake motion observation with the dense instrumentation to buildings and surrounding ground including deep boreholes in 1998. We set up this system in a newly constructed 8-story steel reinforced concrete (SRC) building next to the main building in the BRI premises. Totally 66-channel accelerograms are recorded when certain level of motion is triggered. We hope the recording system of this kind will be very directly useful for evaluating the soil-structure interaction (SSI) phenomena.

2.1. GEOLOGICAL CONDITIONS

BRI is located with approximately 60 kilometers to the northeast from downtown Tokyo. The site is situated 30 meter above sea level on the diluvial heights between Sakuragawa River flowing into Kasumigaura Lake and Kokaigawa River, a branch flow in the greater Tonegawa River water system.

The geological investigation shows that clay and fine sand are main contents up to 90 meters depth underground, inserting sandy gravel. Table 1 indicates surface soil structure at BRI. We confirmed a sandy gravel layer at 88 meters in depth. Its shear wave velocity, however, is not obtained. Transfer functions for shear waves between ground surface and depths of 42 meters, 68 meters and 88 meters are shown in Figure 1. We assume here that the shear wave velocity of the layer 88 meters deep is about 500 m/s. The figure shows common predominant periods of 2 to 4 Hz and 9 Hz. Moreover, the transfer function of 88-meter layers includes the peak at 1 Hz. This might be the

M. Erdik et al. (eds.), Strong Motion Instrumentation for Civil Engineering Structures, 469–479.

peak relevant to the fundamental frequency of the surface soil layers. In addition, the layer between 88 meters deep and approximately 250 meters deep is called upper Kazusa formation group, and the underlying layer is called lower Kazusa formation group, the Tertiary layers.

Table 1. Structure of surface soil layers

No.	H (m)	D (m)	V_P (m/s)	V_S (m/s)	ρ (t/m^3)	Soil Type
1	2.0	2.0	170	110	1.30	Loam
2	6.0	8.0	1430	200	1.30	Sandy Clay & Clayey Sand
3	6.0	14.0		160	1.50	Sandy Clay & Clay
4	8.0	22.0	1630	260	1.80	Fire Sand & Clayey Fine Sand
5	6.0	28.0	1500	200	1.75	Sandy Clay & Clay
6	14.0	42.0	1570	270		
7	6.0	48.0	1880	460	1.90	Gravel
8	8.0	56.0	1780	340	1.75	Sandy Clay & Clay
9	12.0	68.0	1690	290		
10	12.0	80.0	1790	380	1.95	Gravel & Fine Sand
11	8.0	88.0	1600	280	1.75	Sandy Clay & Clay
12				*500*	*2.00*	Gravel

H: Thickness, D: Depth, V_P: P-wave Velocity, V_S: S-wave Velocity, ρ: Density

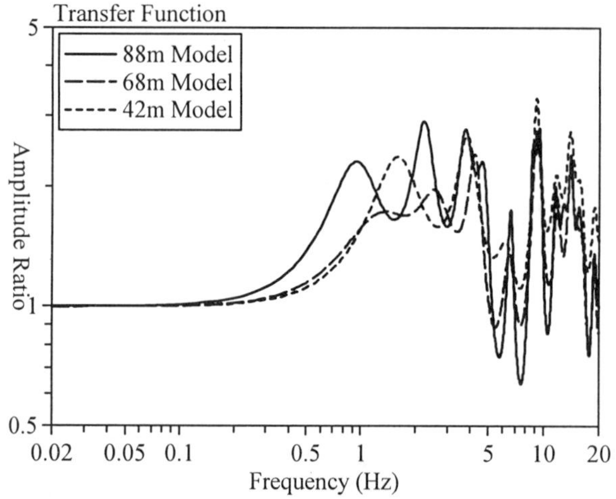

Figure 1. Transfer function of surface soil layers

2.2. BUILDING CHARACTERISTICS

The building for accelerometer installation is a newly constructed Urban Disaster Prevention Research Center building (annex) that was completed in March 1998. The annex building has eight stories with single basement floor. Total building area is approximately 5,000 m^2 and supported by mat foundation on the clayey layer of 8.2 meters underground. The annex is connected to the main building with passages, but is structurally separated.

2.3. RECORDING SYSTEM

The sensors are installed with 11 locations (33 channels) in the new building, 7 locations in the surrounding ground, and 4 locations (12 channels) in the main (older) building. Sensor configuration is shown in Figures 2 and 3. The farthest sensor on the ground is 100 meters away from the annex building, and the deepest sensor is set up 89 meters in depth. Three sensors are set up at the basement floor and the eighth floor in order to investigate torsional vibration.

All sensors are connected to the recording equipment at the observation room in the annex. The specifications of the recording system are shown in Table 2. The system has 66-channel 24-bit A-D converters, a digital processing unit and 40 MB flash memory storage. PC cards are used as storage device and are directly processed by PC. Broad dynamic range and certain operation are ensured by the most up-to-date facilities.

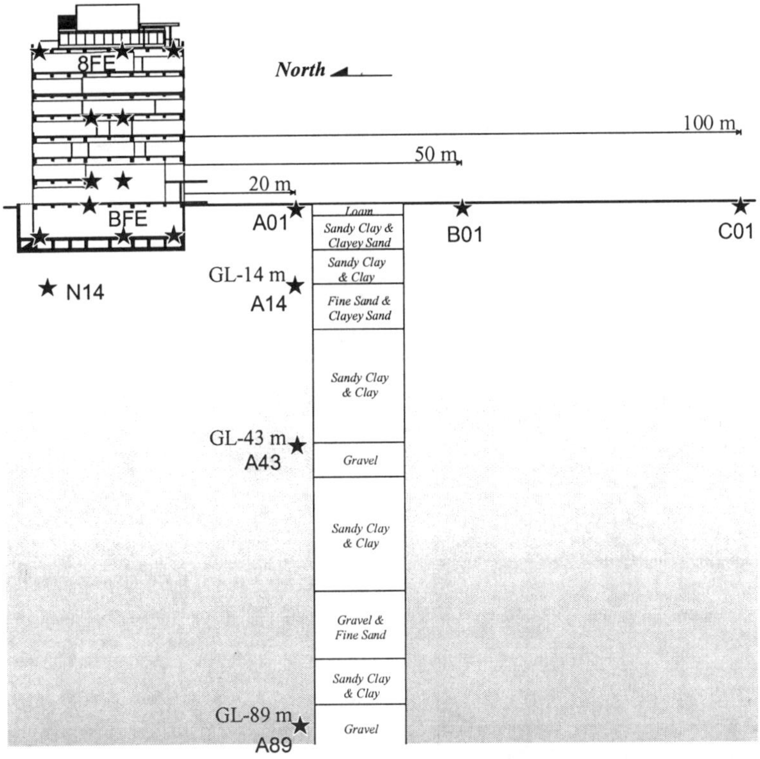

Figure 2. Vertical sensor configuration

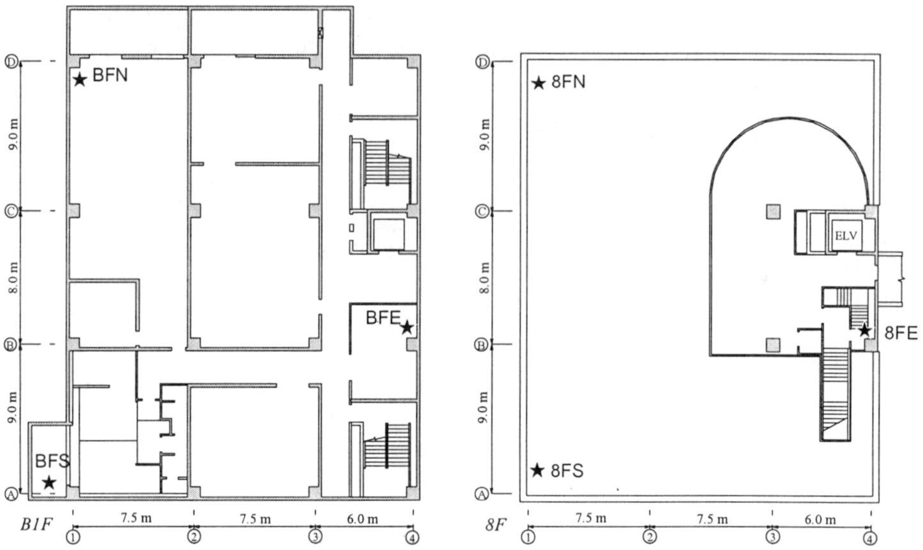

Figure 3. Sensor configuration at basement floor and 8th floor

Table 2. Specifications of recording system

Model	AJE-8200, Akashi Corp.
Sensor	V403BT (negative feedback servo), Akashi Corp.
Number of Channels	66
Frequency Range	DC ~ 30 Hz
Acceleration Range	± 2 G
A/D Converter	24-bit, (Delta-Sigma)
Dynamic Range	114 dB
Recording Medium	ATA Flash Memory Card
Triggering Logic	Disjunction (OR) of Specified Three Components
Functions	Peaks, JMA Seismic Intensity, Spectral Intensity (SI)

3. Observed Records and Discussion

3.1. OBSERVED RECORDS

In a past year, we obtained some records with moderate acceleration levels. Observed earthquakes and peak accelerations on the ground (A01) and at the eighth floor (8FE) are listed in Table 3.

Horizontal acceleration records on the ground (A01), 89 meters below the ground (A89), at the basement floor (BFE), and at the eighth floor (8FE) for the eathquke#4 are plotted in Figure 4. Dynamic characteristics of ground motions and building response are discussed through the analysis using these records in the following paragraphs.

Table 3. List of observed records

No.	Date	Time	Latitude	Longitude	h (km)	M	Δ (km)	Peak Acc. (cm/s^2)	
								A01	8FE
1	1998/04/27	07:06	036°05.1'N	139°53.2'E	50	4.0	18	14	14
2	1998/06/24	23:52	036°07.3'N	140°06.4'E	73	4.6	2	19	53
3	1998/08/29	08:46	035°36.2'N	140°02.7'E	67	5.1	58	10	17
4	1999/03/26	08:31	036.5°N	140.6°E	50	5.1	63	46	61
5	1999/04/25	21:27	036.5°N	140.5°E	50	5.2	56	37	67

h: Focal Depth, *M*: JMA Magnitude, *Δ*: Epicentral Distance

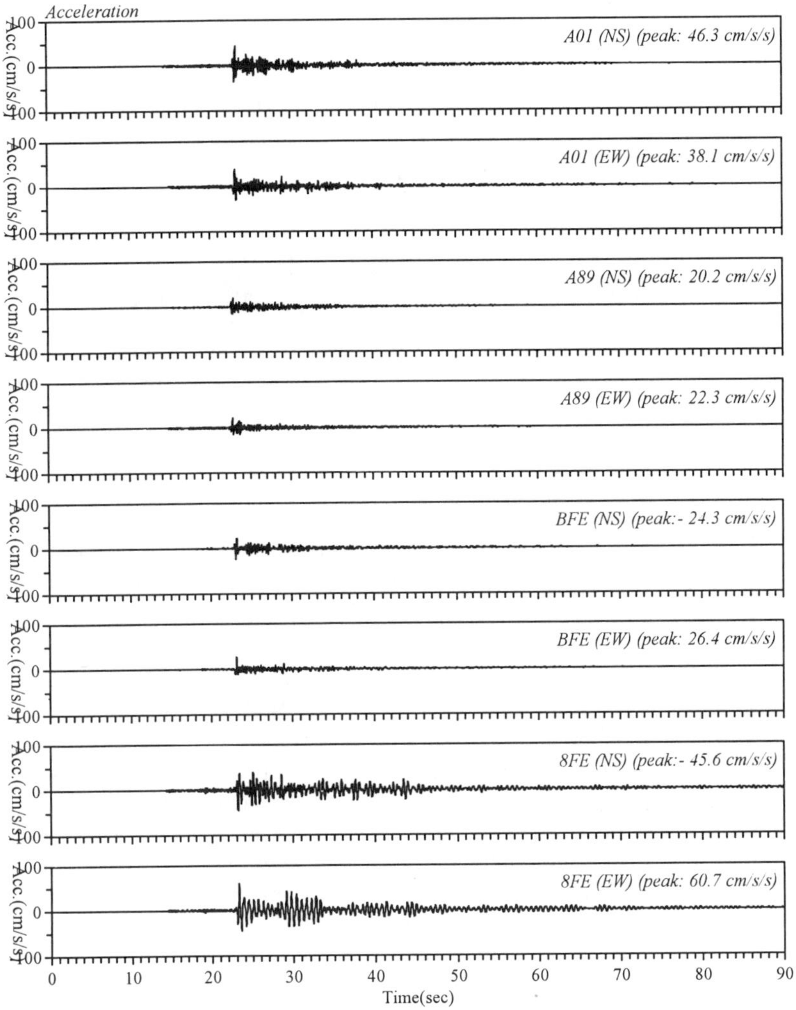

Figure 4. Acceleration records of the earthquake on March 26, 1999

3.2. AMPLIFICATION OF SURFACE SOIL LAYERS

Figure 5 shows Fourier amplitude spectral ratios of earthquake motions on the ground (A01) to ones at 89 meters below the ground (A89). Theoretical ratio calculated from the soil structure shown in Table 1 is also indicated by the dotted line in the figure. Both spectral ratios, N-S component and E-W components, have common peaks at 0.8 to 1.0 Hz, 2.2 Hz and 3 to 4 Hz. Observed results show good agreement with the theoretical ratio upon the location of peaks. However, there are differences in heights of peaks. We assumed damping ratio of 2% for each soil layer in the theoretical calculation. Reevaluation of physical constants, especially damping property, of soils using observed data is necessary for further discussion.

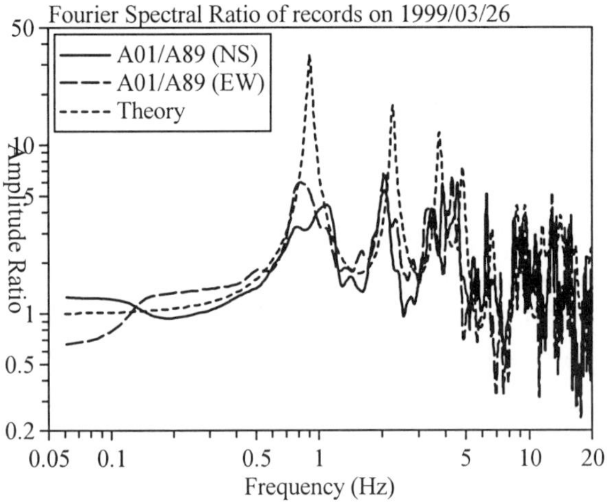

Figure 5. Fourier spectral ratios of records on the ground to ones at 88 meters below the ground

3.3. DYNAMIC CHARACTERISTICS OF ANNEX BUILDING

Fourier amplitude spectral ratios of records at BFE and 8FE to ones at A01 are plotted in Figure 6. The spectral ratio of 8FE/A01 in the N-S component has clear peaks of 1.7 Hz and 2.3 Hz. Natural frequency of the annex building in the E-W direction can be estimated at 1.7 Hz. The peak of 2.3 Hz originated from the torsional vibration judging from phase differences of Fourier spectra. The spectral ratio of 8FE/A01 in the E-W component also has a peak of 1.7 Hz, but doesn't have a peak of 2.3 Hz. First natural frequency in the N-S direction is very close to the E-W direction. The vibration system discussed here includes effect of the soil-structure interaction. Incidentally, damping ratios of the first vibration modes in N-S direction and E-W direction are estimated at 5% and 7%, respectively.

In comparison of records at the basement floor with ones on the ground, differences appear in the high frequency range. The spectral ratio of BFE to A00 falls from the frequency of 1.7 Hz that is the first natural frequency of the building. The spectral ratios become 1/2 in frequency range higher than 4 Hz. SSI causes such a phenomenon and reduces peak accelerations at the basement floor as shown in Figure 4.

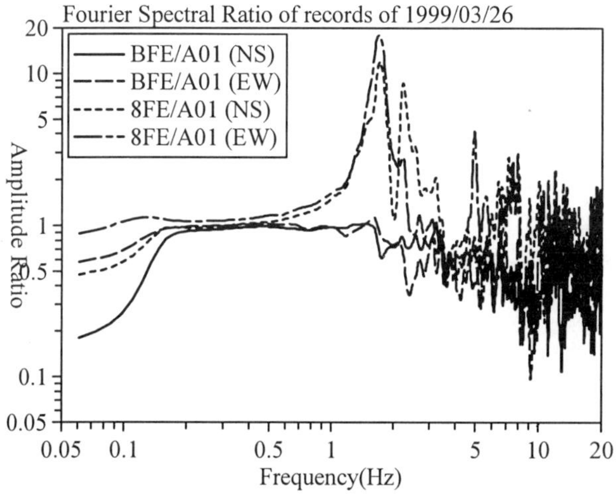

Figure 6. Fourier spectral ratios of records in the building to ones on the ground surface

4. Other BRI Activities on Strong Motion Observation

In addition, BRI is conducting three strong motion instrument networks recently. Each project has its own aim and background. Those projects are briefly introduced in this chapter.

4.1. NATIONWIDE STRONG MOTION OBSERVATION IN BUILDINGS

BRI has installed strong-motion instruments in major cities throughout Japan as shown in Figure 7. There are now 49 observation sites in operation using the digital strong-motion instrument. The objects of observation are mainly buildings, and the measuring point is usually placed both on the top and in the foundations of the building. Every observation site is connected to BRI via telephone line in order to collect data immediately and to mitigate maintenance work.
The observation network has obtained many noteworthy records. For example, in the 1993 Kushiro-oki (Off Kushiro) Earthquake, 711 cm/s^2 was recorded as the peak acceleration on the ground surface at Kushiro Local Meteorological Observatory as cited in [1]. In addition, in the 1994 Sanriku-haruka-oki (Far off Sanriku) Earthquake, a large acceleration amplitude record was obtained in the building next to the severely damaged old Hachinohe municipal office building as reported in [2].

The network is still being enlarged year by year. We are proceeding to instrument buildings that have up-to-date seismic control systems recently.
A large number of records have been obtained from this network up to now. Part of strong motion database is open to public through the Internet.

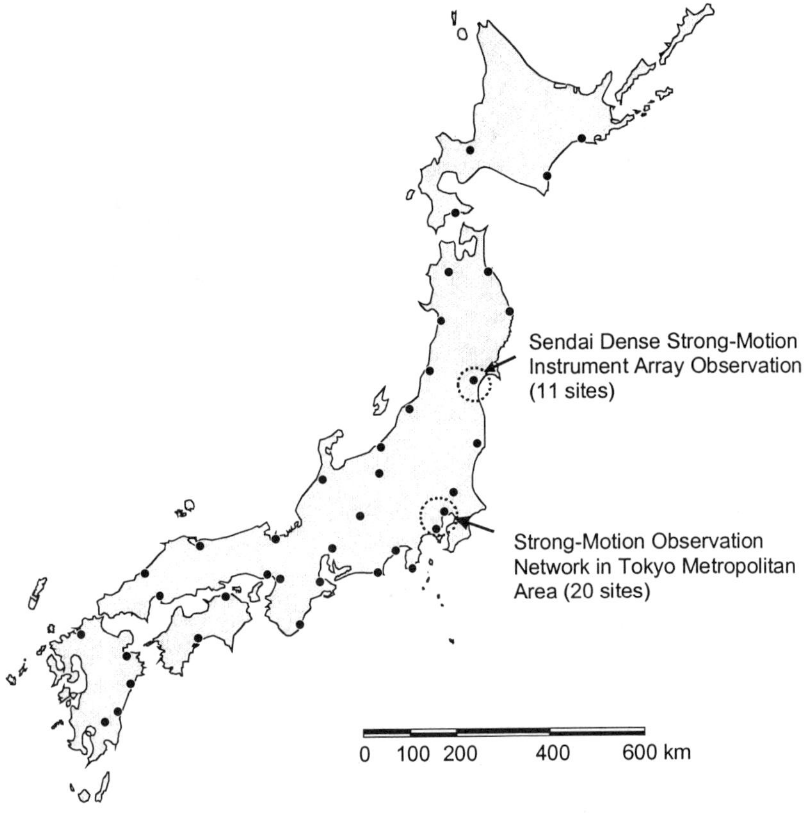

Figure 7. Sites of nationwide strong motion observation

4.2. DENSE STRONG-MOTION INSTRUMENT ARRAY IN SENDAI

We also installed seismographs in grounds as introduced in [3]. Eleven recording sites are deployed around Sendai area as shown by solid circles in Figure 8. Solid squares show sites of the Nationwide Network. Eleven sites were selected so that typical surface soil conditions of urban areas in Japan are contained. Sendai area was severely damaged by the 1978 Miyagi-ken-oki (Off Miyagi Pref.) Earthquake. The result of damage investigation concluded that the damage distribution had clear correlation with the soil conditions as reported in [4]. Therefore, we will get the difference of strong motion characteristics when large earthquake database is completed. Each site has three tri-axial sensors. They are installed from surface to so-called engineering bedrock with shear wave velocities of 660 cm/s^2 to 1400 cm/s^2. These records have provided data to evaluate the effect of surface geology on seismic motions. This project had been implemented as a cooperative research project between BRI and the Association for Promotion of Building Research (KKSK). A number of earthquake motions have been recorded for these 12 years, and the final report including strong motion data will be published in autumn, 1999. Although the cooperative research has completed in March 1999, the observation will be reduced and continued in another five years as the individual project of BRI.

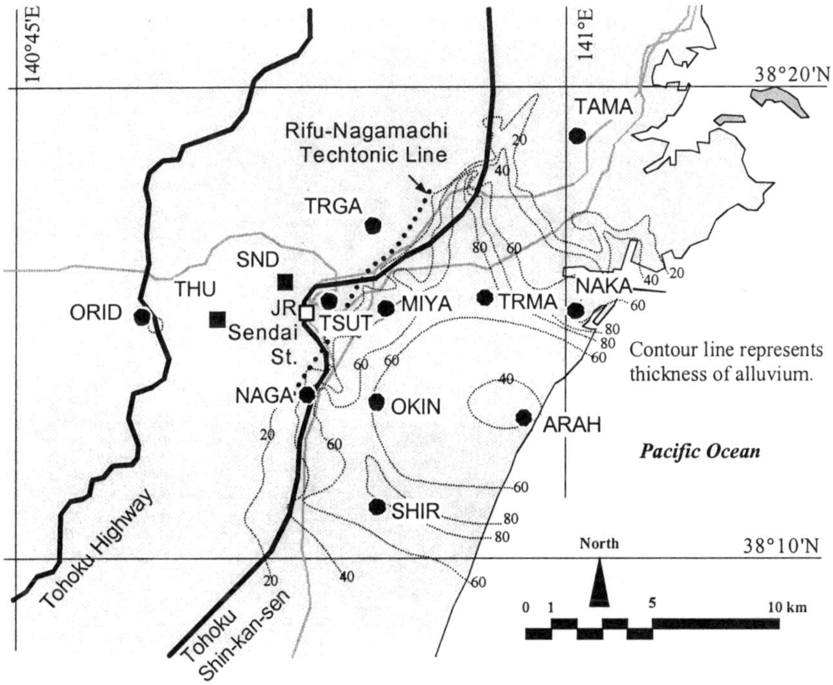

Figure 8. Dense strong motion instrument array in Sendai

4.3. STRONG-MOTION INSTRUMENT NETWORK IN THE METROPOLITAN AREA

The 1995 Hyogo-ken-nanbu Earthquake (Kobe Earthquake) awakened us again to the importance of disaster prevention measures for large-scale urban areas. It is important to predict the probability of a future earthquake and its impact, and make as many preparations as possible in anticipation of such an event. It is also very essential to grasp the damage situation immediately to put in effect the necessary countermeasures.

BRI has established twenty new observation sites placed radially in the Tokyo metropolitan area. Figure 9 indicates site configuration of this network with site locations of other projects. Objects of instrumentation are buildings. Sensors were installed on the top and at the basement of a building, and a sensor was added on the ground if possible. This project also aims to investigate the characteristics of the seismic motion affecting the whole Kanto Plain through observation records. The system immediately collects information on the seismic intensity at the time of an earthquake occurrence.

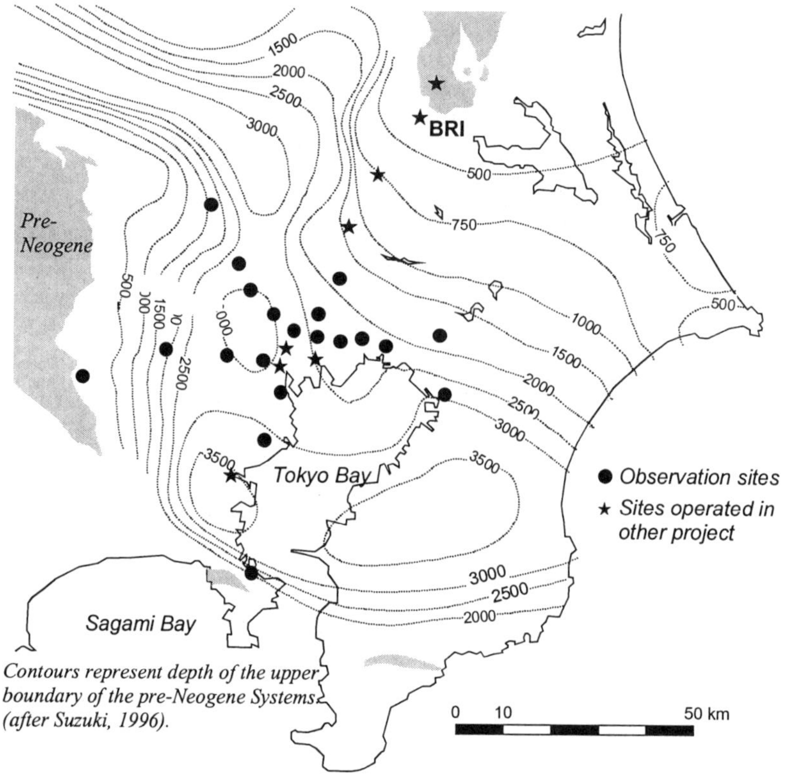

Figure 9. Strong motion instrument network in the metropolitan area

5. Conclusions

BRI has started new earthquake motion observation with the dense instrumentation. Amplification effect of surface geology was examined with good agreement between observed and the theoretical result. Dynamic characteristics of the annex building are also discussed through analyses of earthquake motion records. First natural frequencies of annex building in the N-S direction and the E-W direction can be clearly distinguished from the spectral analysis. Results here will be fundamental information for following studies.

We intend to proceed with studies on reevaluation of soil properties, examination of detailed vibration modes, investigation of soil-structure interaction, and so on. Non-linearity of those dynamic characteristics is also an interesting research subject.

The damping ratio of the annex building was estimated by the study meeting of young researchers, chaired by Dr. T. Saito, BRI, in the Building Research Institute Council. Authors deeply appreciate their study work.

6. References

1. Kashima, T. and Y. Kitagawa (1993), The 1993 Kushiro-oki Earthquake, Prompt Report on Strong Motion Records, Building Research Institute
2. Kashima, T. and Y. Kitagawa (1994), The 1994 Sanriku-haruka-oki Earthquake, Prompt Report on Strong Motion Records Vol.3, Building Research Institute
3. Kitagawa Y., I. Okawa and T. Kashima (1994), Dense Array Observation and Analyses of Strong Ground Motions at Sites with Different Geological Conditions in Sendai, BRI Research paper, Building Research Institute
4. Building Research Institute (1978) Report on the Damage by 1978 Off-Miyagi Prefecture Earthquake (in Japanese), Research Report of BRI, No.86, Building Research Institute

INSTRUMENTATION, MEASUREMENTS AND NUMERICAL ANALYSIS OF BRIDGES: AN EXAMPLE OF THE CABLE-STAYED BRIDGE ON EVRIPOS CHANNEL, GREECE

V. A. LEKIDIS[1] , C. Z. KARAKOSTAS[1] ,D.G.TALASLİDİS[2]
[1]*Institute of Engineering Seismology and Earthquake Engineering (ITSAK),46 Georgikis Scholis Str.,*
P.O. Box 53, 551 02, Finikas, Thessaloniki, Greece
[2]*Dept. Civil Engineering., School of Technology,*
Aristotle University of Thessaloniki,
540 06 Thessaloniki, Greece
e-mail:lekidis@itsak.gr

Abstract

The instrumentation of special structures such as large bridges, tunnels, monuments etc., contributes towards a better understanding of their dynamic performance, as well as a more accurate and reliable prediction of the earthquake resistance of such large-scale structures. In experimental analysis the classical ways to estimate the modal parameters of the whole structural system, as well as of its structural elements, is either to excite the structure artificially (using, e.g. vibrators, heavy vehicles) or to evaluate the recordings obtained from a weak or strong ground motion. In modeling the behaviour of a civil engineering structure in a realistic way, among the important parameters to be defined are the mass distribution, the damping characteristics, the stiffness of the main load resisting system, the influence of secondary elements and interaction phenomena. Large vibration tests provide reliable data for the evaluation of the influence of such modal parameters.

The effective analytical evaluation of the bridge through reliable and effective numerical models is necessary in order to verify the experimental data. By comparing the results of both experimental and analytical approaches and adequately updating the analytical model, a more realistic modeling of the bridge can be obtained. Changes in the modal parameters during the time-life of the bridge can contribute towards detection of damages, such as crackings or any other reasons reducing the stiffness of the structural system. The cable-stayed bridge in Evripos channel (Greece) was selected to apply the aforementioned methodology. The main ambient vibrations recorded are due to traffic, wind and moderate earthquakes at small distances from the bridge. Three-dimensional finite element models of the bridge were created and the dynamic behaviour of the bridge was analysed for both a small and a strong earthquake with high frequency content. Results of the experimental estimation of eigenvalues of the

M. Erdik et al. (eds.), Strong Motion Instrumentation for Civil Engineering Structures, 481–493.

vibration modes are also presented. Finally, comparisons between experimental and analytical values have been performed.

1. Introduction

One of the main objectives in earthquake engineering is the realistic prediction of the dynamic behaviour of a structural system due to strong ground motion. From the analytical point of view, the dynamic behaviour of each structural system can be investigated by means of different numerical (mostly finite element) models. The complexity of the model depends on the level of sophistication used in the analysis, the number, arrangement and description of masses, the number of degrees of freedom of the finite element model and the seismic input. Apart from the analytical approach, recent investigations of the seismic effects upon structures are also based on experimental research methods. As such, one can mention testing of models on shaking tables, and definition of the dynamic response of structures using records obtained from installed structural arrays, when the latter are subjected either to environmental (e.g. earthquakes, wind, traffic loads) or artificial (e.g. blasts, controlled vibrations) excitations. The instrumentation of special structures such as large bridges, tunnels, monuments etc., contributes towards a better understanding of their dynamic performance as well as a more accurate and reliable prediction of the earthquake resistance of such large-scale structures. In modeling the dynamic behaviour of a civil engineering structure in a realistic manner, among the important parameters to be defined are the mass distribution, the damping characteristics, the stiffness of the main load resisting system, the influence of secondary elements and interaction phenomena. Large vibration tests provide reliable data for the evaluation of the influence of such structural parameters. A considerable amount of research has been carried out in the field of experimental and analytical study of various structures -and especially bridges- in the last decade (Ewins & Griffin [1], Gates & Smith [2], Haibach [3], Imai et al [4], Karabinis [5], Lee & Yun [6], Hong & Yun [7], Johanson [8], Kiousis & Karabinis [9], Paultre et al. [10], Ewins [11], Broquet & Bruhwiler [12]). Dynamic bridge testing has been used on many occasions as a means of evaluating various static and dynamic properties of modern highway bridges as well as suspension and cabled- stayed bridges. Many of these tests were carried out to obtain vibration frequencies and mode shapes, and were based on ambient wind-induced or traffic-induced vibrations (Paultre et al. [10], Yun & Shinozuka [13], Paultre et al. [14]).

This paper presents testing procedures used at the Institute of Earthquake Engineering and Engineering Seismology (ITSAK). These procedures were applied to the cable-stayed bridge on Evripos channel (Halkis bridge), which is up till now the largest cable-stayed one in Greece. The main objectives of the corresponding research program carried out at ITSAK are: (a) to evaluate the dynamic properties of the bridge; (b) to calibrate finite-element models of the bridge developed at the Institute; (c) to study the influence of temporal changes in the stiffness of main and secondary structural elements on the dynamic response of the bridge; and (d) to predict the response of the structure due to strong earthquakes. In this paper recent results of the research project, which is still under way, are presented. Results of previous investigations can be found in [15], [16]. The bridge over Evripos channel has a total length of 694.5 meters and a width of 13.5 meters (Stathopoulos [17], [18]). The central part of the bridge is of a cable-stayed type with a total length of 395 meters, the middle span of which is 215 meters long. The transversely pretensioned deck (with a thickness of 45cm) is suspended through two levels of cables from the two piers which are of a height of about 90 meters each. The rest of the bridge consists of 8 single-span simply supported parts of lengths of 36 - 39 meters, each made up of 4 prefabricated pretensioned I - beams seated through elastomeric bearings on piers. In the framework of the research project, only the behaviour of the central -cable stayed- part of the bridge is studied.

Table 1. Technical specifications of the Kinematics PX-23 Structural Array

- 12 bit resolution analog-digital converter
- Full scale ± 2g
- Output voltage ± 2.5 V
- Frequency response DC-50 Hz within -3 dB
- Damping 0.7 nominal
- Natural frequency 600 Hz
- Dynamic range > 72 dB
- Sampling rate 200 sps
- Built-in anti-alias filter : 2-pole, 50 Hz Butterworth response
- Automatic self identification after sensing an event
- Common triggering, common timing and external timing device (GPS) facilities
- Internal timing 2pps
- Remote communication, data retrieval and maintenance facilities
- Operating temperature - 20° C to + 85° C
- Humidity 100 %
- Full DC power (by external battery box)
- Four days autonomy in standby position

A Kinemetrics data acquisition system and control unit is used to record the response (Lekidis [19]). The structural array purchased in 1994 was the Kinemetrics PX-23 model -of SSA2 type-and was installed the same year by the Kinemetrics representative in Greece, in collaboration with the electronic laboratory personnel of ITSAK. The array consists of 42 sensors - of FBA11 type -, ending up in sixteen modules situated at the deck-level rooms in the eastern pier of the bridge. The positions of the sensors were carefully selected so as a complete recording of the three-dimensional dynamic behaviour of the bridge is possible. Six vertical and two transverse sensors record the response at the middle span of the bridge, whereas at each one of the two piers six sensors are installed. There are also four triaxial sensors, two located at the base of each pier and two free-field. The layout of the structural array is presented in Figure 1. The system is equipped with common time, common triggering and GPS facilities. The array is triggered and starts recording when the threshold value of certain sensors is exceeded. It is possible to predetermine explicitly the sensors that are capable of

triggering the system. The technical specifications of the recording system are summarized in Table 1.
The system became fully functional by the end of 1994. Since then, a satisfactory number of ambient vibrations has been recorded, as well as certain small earthquakes at distances of 20 - 70 km. The recorded data are stored in the system's memory and can be retrieved either in situ or through modem from the Institute's computational centre at Thessaloniki. An evaluation of the data so far recorded was conducted, in order to determine the dynamic behaviour of the bridge. The results obtained were compared with those of analytical models of the structure, also developed at the Institute.

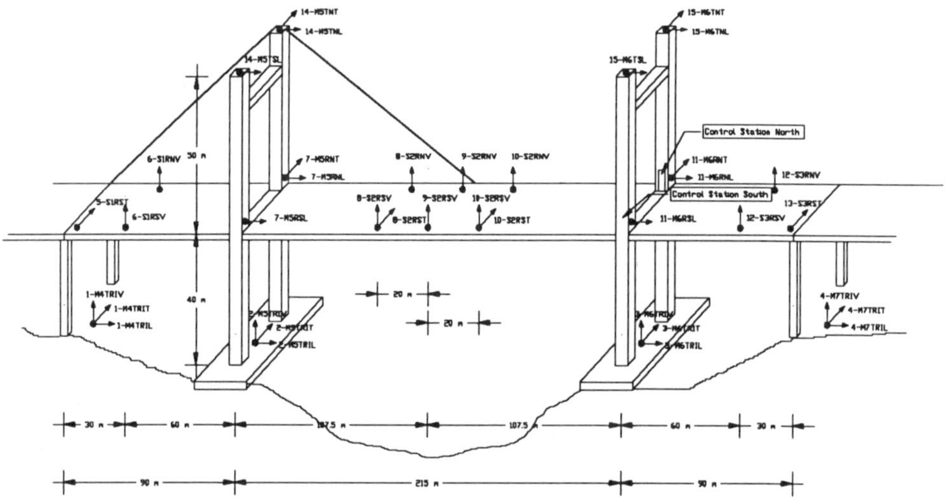

Figure 1. Central part of the bridge and layout of the structural array.

2. Analytical and Experimental Investigation of the Bridge

A three-dimensional finite element model was developed using the program SAP2000 [20] for the analytical evaluation of the bridge, in order to verify the experimental data. The dynamic behaviour of the bridge has been numerically investigated first by performing an eigenvalue analysis and then by using time-history analysis to predict the response at specific positions of the bridge due to recorded excitations.
Some of the results of the eigenvalue analysis are shown in Figure 2 and Table 2. A more detailed presentation of the analysis can be found in [16]. A time-history analysis was performed, using as input motion the longitudinal component of a small earthquake at a distance of 23.2 km Northeast of the bridge site. The earthquake, of magnitude M_w=4.3 R, was recorded on 22/5/98 (20.56 GT). The input motion is presented in Figure 3. SAP2000 performs a modal solution of the time-history analysis. Response in the time domain was calculated at the top of the pier and the middle point of the central deck of the bridge. As it can be seen from Figure 3, the peak value of the input

longitudinal record is 44.65 cm/sec^2 at the base of pier M5 and the peak value of the corresponding response at the top of the pier is 43.9 cm/sec^2. The duration of the vibration (> 0.008g) is about 3.0 sec both for the input motion and the response at the top of the pier. At the middle of the bridge the peak value of the horizontal response (Figure 3) is 46.28 cm/sec^2. The analysis, as well as the recorded response (see, e.g., Fig.9) show that the characteristics of the structural response are very close to those of the input motion.

In Figure 4 the response spectrum of the input motion is presented. It can be seen from the response spectrum, which has a peak value at 0.20 sec, that in the period range corresponding to the first modes of the bridge the pseudoacceleration values are significantly lower than the peak ground value. This confirms the well known fact [21] that the seismic loading of a structure with a high (>1 sec) fundamental period can not be accurately predicted from the response spectrum of the input motion. In deriving design spectra, modifications have been proposed in the range of long periods, due to uncertainties involved in the combination of modal responses and because the spectral values are not conservative enough for the lower modes of vibration.

A similar time-history analysis was performed for the vertical component of the seismic event. In Figure 5, both the input motion and the vertical response at the middle of the central deck are presented. In this case, an amplification factor of ca. 1.33 is observed.

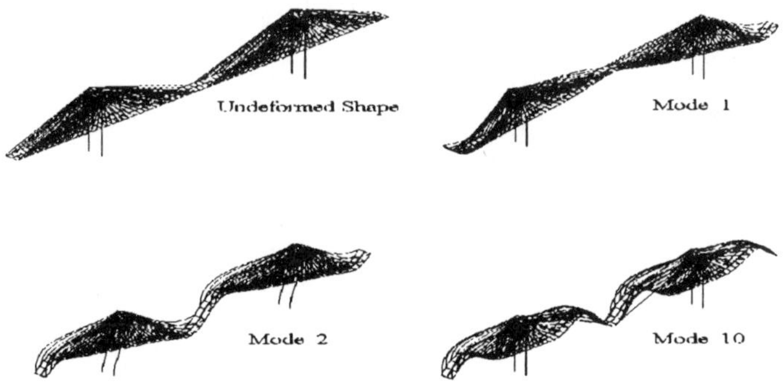

Figure 2. Undeformed shape and numerically obtained eigenmodes of the bridge.

Table 2. Periods and modal participation factors from eigenvalue analysis of the bridge.

Mode	Period	UX	UY	UZ
1	2.763790	-0.027249	4.53E-05	-22.688358
2	2.600826	72.302830	0.000239	-0.006605
3	2.282575	-2.51E-06	-89.182061	2.53E-05
4	1.811659	-83.045079	6.11E-05	0.003249
5	1.556144	8.83E-05	7.826857	-0.000183
6	1.420366	1.11E-05	-0.443822	0.001196
7	1.306252	0.006232	-0.000695	24.748342
8	1.242427	0.000100	25.815171	0.001081
9	1.117267	6.592468	0.001409	7.35E-05
10	1.065136	5.40E-05	0.000643	57.695471

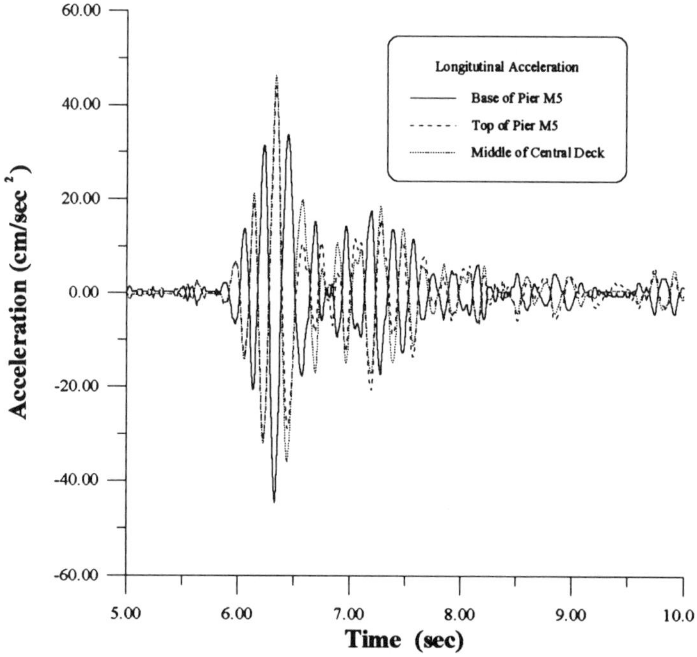

Figure 3. Earthquake of 22/5/98: Computed longitudinal acceleration time histories.

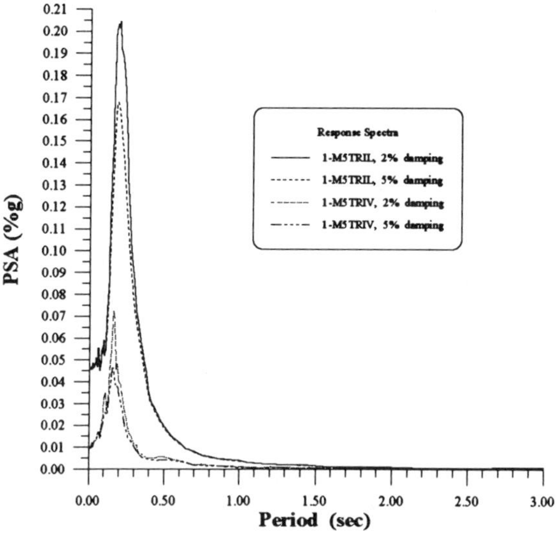

Figure 4. Response spectra of the recorded earthquake of 22/5/98.

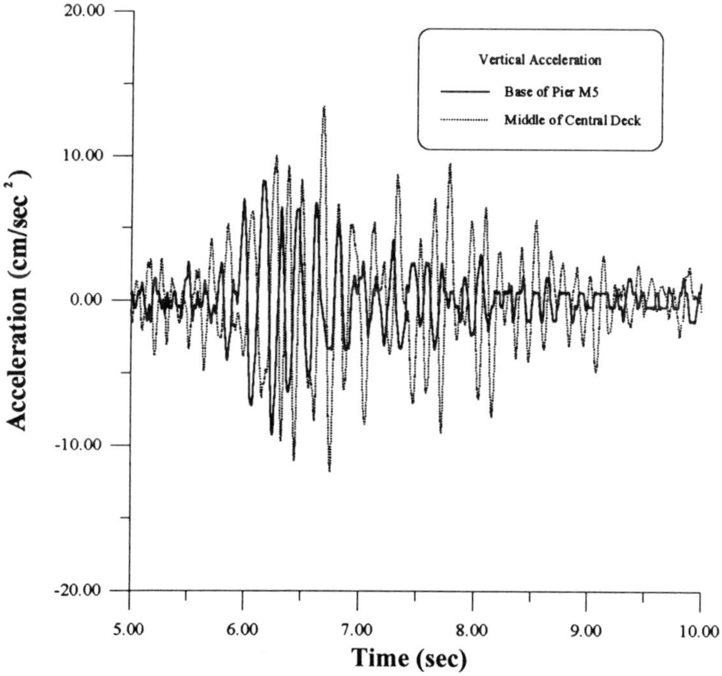

Figure 5. Earthquake of 22/5/98: Computed vertical acceleration time histories.

It is obvious that this earthquake did not affect the structural system of the bridge, because of its high (>3 Hz) frequency content. In Figure 6 the Fourier transforms of the recordings used in the previous analysis are presented. For most moderate distance earthquakes the frequencies contained are expected to be high, far from the first modes of the bridge. On the other hand, a previously recorded small earthquake at a distance of 50 km had affected the first modes of the superstructure and the amplification factors of the superstructure response were much higher (Lekidis et al., [16]).

During its operational time the special array didn't record any major strong ground motion. For this reason, a similar major strong ground motion was selected for a further investigation of the dynamic behaviour of the bridge. It is presumed that a very strong ground motion expected from the seismic fault of Atalanti (an important seismic fault 55 km long, about 30 km to the Northwest of the bridge) might be similar to the recording of the Kozani earthquake caused by the Rimnio (30 km long) fault. Both faults are of the normal (extensional) type, and in both cases the seismotectonic environment is similar. This earthquake occurred in West Macedonia-Greece on 13/5/ 1995 and caused very heavy damage to towns and villages into a zone 30-40 km far from the epicentre. The input signal used for the analysis in the longitudinal direction is shown in Figure 7, together with the corresponding responses at the top of the pier and the middle of the deck.. A similar analysis was conducted in the vertical direction. The input motion and the corresponding response at the middle of the deck are shown in Figure 8. The results of the analyses show a behaviour of the bridge similar to that of the event of 22/5/1998.

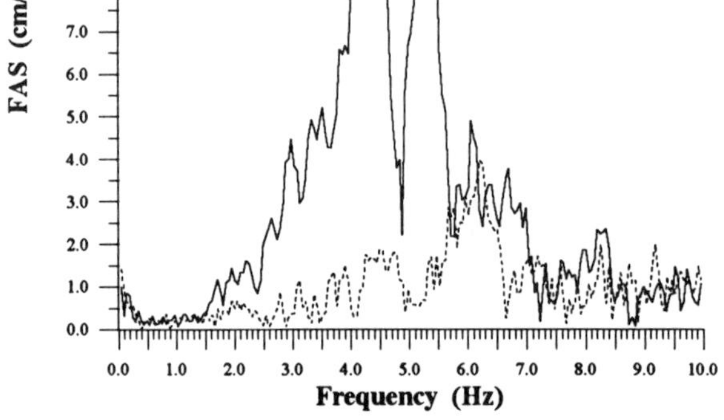

Figure 6. Fourier Amplitude Spectra of the earthquake of 22/5/98 for station 2-M5.

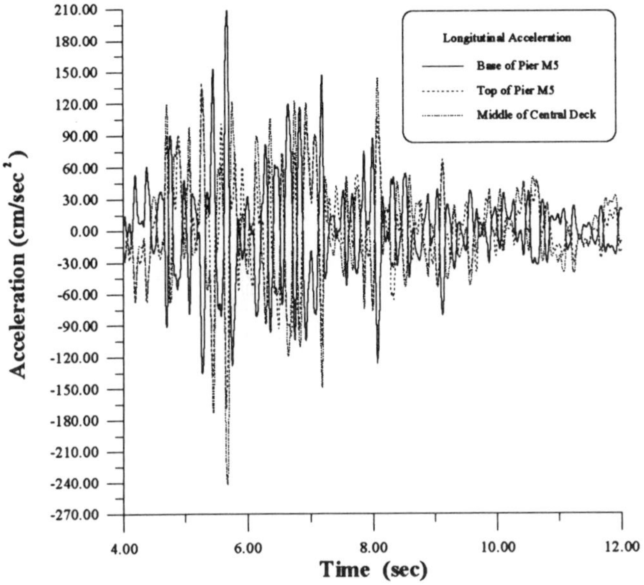

Figure 7. Kozani Earthquake: Computed longitudinal acceleration time histories.

This, in combination with previous tests show that the superstructure is mostly vulnerable to far field earthquakes and some near-field ones, with energy released predominantly in the low frequency range.

In Figure 9, the recorded longitudinal response at the top of pier M5 for the event of 22/5/1998 is presented. A comparison with the analytical results (Figure 3) shows a good agreement for the peak value of acceleration, the duration of the vibration and the time of the maximum response. It is obvious that such recordings of the response of a flexible structure subjected to actual earthquakes are very useful towards a better definition of design spectra at least for Greece.

Similarly, a 50% increase of the pseudoacceleration at the 2.0 sec period was recommended by Newmark & Hall [21], (and adapted by seismic codes, e.g. ATC-3-06 [22]), for the uniform hazard spectra for stiff soil conditions proposed by Seed et al. [23]. The Fourier transforms of the recorded responses at stations 14-M5TNL, 8-S2RNV and 9-S2RNV (see Figure 1) are presented in Figure 10. It is seen that the event of 22/5/1998 excited higher modes (at 3.5 and 5.5 Hz), associated mainly with the logidutinal direction.

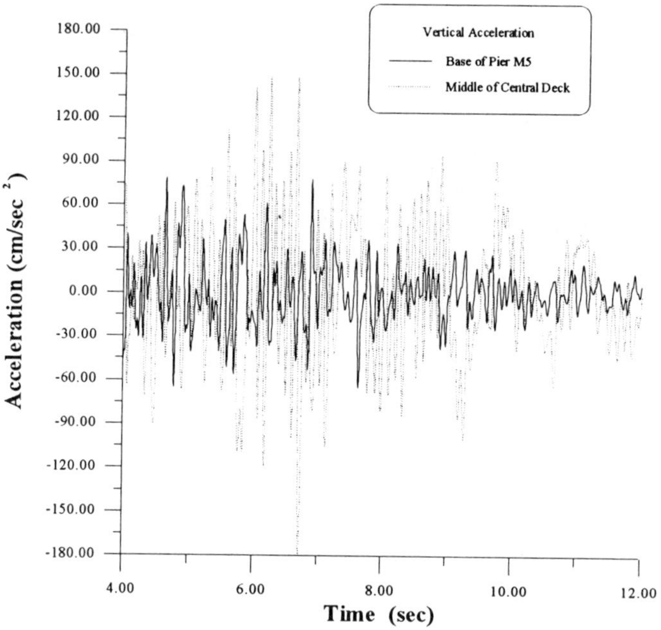

Figure 8. Kozani earthquake: Computed vertical acceleration time histories.

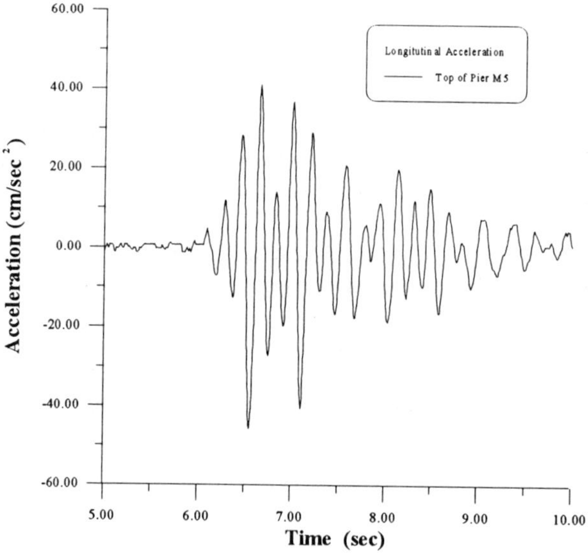

Figure 9. Earthquake of 22/5/98: Recorded longitudinal acceleration at top of pier M5.

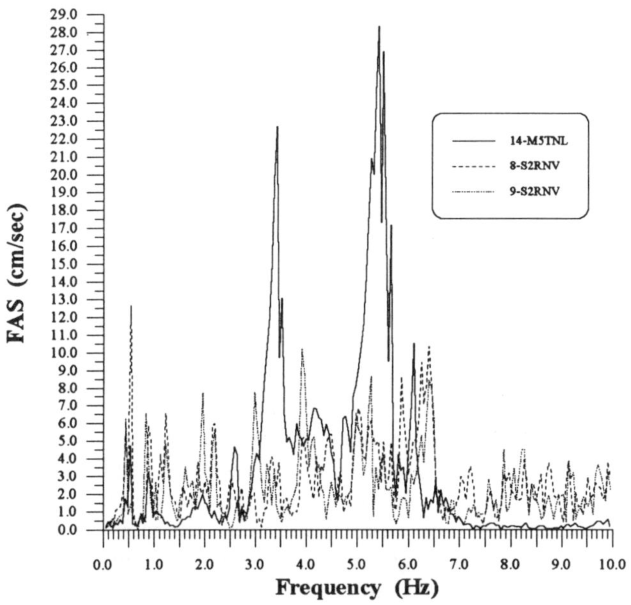

Figure 10. Fourier Amplitude Spectra of the earthquake of 22/5/98.

Apart from the previous comparison between the analytical and the experimental data for the acceleration response at specific points of the superstructure, it is also interesting to examine the variation of the input motion across the bridge.

An important variation of the strong ground motion along the longitudinal axis of the main bridge has also been observed. In Table 3 the peak values of the recorded motion at various free-field stations along the axis of the bridge are shown.

Table 3. Recorded Peak Acceleration-Velocity Values (cm/sec^2-cm/sec) at various free–field stations

	Station (1-M4)	Station (2-M5)	Station (3-M6)	Station (4-M7)
Longitud. Recording	A=19.74 V=0.854	44.65 1.585	12.58 0.541	9.44 0.488
Transv. Recording	A=14.73 V=0.484	35.35 1.099	19.94 0.652	7.19 0.436
Vertical Recording	A=6.41 V=0.286	9.31 0.347	4.63 0.216	4.90 0.213

The distance between station 1-M4 (see Figure 1) and 2-M5 is 90m, as it is also the distance between stations 3-M6 and 4-M7, but the distance between 2-M5 and 3-M6 is 215 m. Although a further investigation of the space variability of the input motion might be necessary, it is noted that recent analyses [24] seem to indicate that neglecting such phenomena in the design leads to more conservative results.

3. Conclusions

In this work recent results of investigations concerning the dynamic behaviour of the bridge on Evripos channel are presented. The numerical results of a time-history analysis for an actual small earthquake at a moderate distance are compared with the recorded response. There seems to be good agreement in the response duration, the peak values of acceleration at several locations of the superstructure and the time of maximum response. A further numerical analysis for a major earthquake, with characteristics similar to those expected at the bridge site, has also been carried out.
From the investigations performed so far, the Halkis bridge does not seem to be vulnerable to moderate distance earthquakes, since in most of them the contained frequencies are expected to be high, not exciting the first modes of the bridge. This preliminary conclusion is of significant practical importance, due to the adjacency of the bridge site to the well-known Atalanti fault, a potential source of big earthquakes.

Both recorded and numerical time-history computations of the bridge response show that it is not possible to accurately predict the seismic loading simply from response spectra of the input motion for structures in the high (> 1 sec) period range, as is the case for the Halkis bridge. This conclusion conforms with the practice followed for the design spectra of seismic codes, in which an increase of the expected acceleration values is imposed in the high period range. In this context, it should be mentioned that special attention should be taken in the seismic design of flexible structures, in case that only response spectrum analysis is used. It is recommended that as additional knowledge is gained and especially for the design of special structures it may be desirable to develop a set of response spectra representative of a near-field source as well as moderate and far distant sources.

4. Acknowledgments

The authors would like to thank graduate students of the Civil Engineering Department of Aristotle University of Thessaloniki, Florentin Ilias and Hatzakos Giannis, for their help in the numerical analyses of the Halkis Bridge. Thanks are also due to the technical personnel of ITSAK for the highly efficient maintenance of the structural array of the bridge. Finally the authors wish to thank the designer of the bridge Dr. S. Stathopoulos for providing data for structural elements and other useful information.

References

1. Ewins, D.J. & Griffin, J. (1981) A State-of-the-art assessment of mobility measurement techniques-results for the mid-range structures, J. of Sound and Vibration 78 (2), 197-222.
2. Gates, J. & Smith, M. (1984) Results of ambient vibration testing of bridges, Proceedings of the 8th Word Conference on Earthquake Engineering, San Francisco , IV, 873-880.
3. Haibach, E. (1986) Measurement and interpretation of dynamic loads on bridges, Technical Steel Research , Synthesis Report to C.E.C.
4. Imai , H., Yun, C.B., Maruyama, O. & Shinozuka , M. (1989) Fundamentals of system identification in structural dynamics, Technical Report NCEER -89-0008.
5. Karabinis, A.I. (1991) Dynamic characteristics of a typical prestressed highway bridge, International Conference on Dimensions in Bridges and Flyovers, Singapore, 143-149.
6. Lee, C.G. & Yun, C.B. (1991) Parameter identification of linear structural dynamic systems, Computers and Structures 40(6), 1475-1487.
7. Hong, K.S. & Yun, C.B. (1993) Improved method for frequency domain identifications of structures, J. of Engineering Structures 15(3), 179-188.
8. Johanson, R. (1993), System modeling and identification, Prentice-Hall.
9. Kiousis, P. & Karabinis, A. (1994) Dynamic characteristics of bridges. Proc.of the 11th Greek Concrete Conference, Corfu, Greece, Tech. Chamber of Greece (in Greek).
10. Paultre, P., Proulx, J. & Talbot, M. (1995) Dynamic testing procedures for highway bridges using traffic loads, Journal of Structural Engineering, American Society of Civil Engineers (ASCE) 121(2), 362-376.
11. Ewins, D.J. (1995) Modal Testing: Theory and Practice, Research Studies Press Ltd, England.
12. Broquet, B.,Bruhwiler, E. (1999) In situ and model tests and numerical analysis of a curved cable-stayed bridge, Int. Journal of Structural Engineering, 1, pp 57-62.
13. Yun, C.B. & Shinozuka, M. (1990) Program LINEARID for identification of linear structural dynamic systems, Technical Report NCEER-90-0011.
14. Paultre, P., Chaallal, O. & Proulx, J. (1992) Dynamics and dynamic amplification factors- A review of analytical and experimental findings, Canadian Journal of Civil Engineering 19(2), 260-278.
15. Lekidis, V.A., Karakostas, C.Z., Papazachos, C.B., Margaris B.N.and Talaslidis D.G. (1997) Special strong motion array in Evripos bridge : Networking, data aquisition and processing", Abstract volume of the 29th General Assembly of the International Association of Seismology and Physics of the Earth's Interior, Geophysical Laboratory, University of Thessaloniki (Ed.), P. Ziti & Co, Thessaloniki, Greece, 352.
16. Lekidis, V. A., Karakostas, C.Z. and Talaslidis, D.G. (1998) Dynamic Characteristics of the Cable-Stayed Bridge on Evripos Channel, Greece, Proceedings 11th European Conference on Earthquake Engineering, CNIT, Paris, France, September 6-11 1998 (in CD-ROM)
17. Stathopoulos, S. (1987) High bridge over Evripos channel, Proceedings of the 8th Greek Congress of Reinforced Concrete, Xanthi-Kavala 27-29 May 1987, 215-236.
18. Stathopoulos, S. (1994) Construction of the high bridge over Evripos channel, Proceedings of the 11th Greek Congress of Reinforced Concrete, Corfu, 18-20 May 1994, 326-342.
19. Lekidis , V.A. (1988) In situ vibration testing for structures under earthquake excitation, Report ITSAK 88-05, Thessaloniki (in Greek).
20. Sap2000 Nonlinear, v. 6.0 (1996) Computers and structures Inc., Berkeley, California.
21. Newmark, N. M., Hall W. J. (1987) Earthquake spectra and design, Dpt. of Civil Engineering, University of Illinois at Urbana-Champaign,Earthquake Engineering Research Institute.
22. ATC-3-06 (1984) Tentantive Provisions for the development of seismic regulations for buildings, Applied Technological Council (second printing)
23. Seed, H. B., Ugas, C. and Lysmer, J. (1976) Site-dependent Spectra for earthquake-resistant design, Bull.Seism. Soc. Am., 66, 221-243.
24. Vanmarrcke, E., (1999), Spatial variation of earthquake ground motion, invited lecture at the Dpt. Civil Engng., 21/4/1999, Aristotle University

DAMAGE DETECTION IN SEMI-RIGID JOINT, RC FRAMES SUBJECTED TO STRONG MOTION EXCITATION

J. LEONOV*
Seismology Division, The Geophysical Institute of Israel, P.O. Box 2286, Holon 58122, Israel,
e-mail: jossleon@iprg.energy.gov.il

A new approach to fault detection and prediction in RC frame resistant structures subject to strong ground excitation, considering both input uncertainty and material non-linearity, is presented. It could also serve as a basis for developing a near-real-time structure health assessment procedure.

Problem Statement

The general task of this proposal may be stated as follows:

To develop a robust rule/decision criterion for the purposes of damage detection of RC frame structures due to strong motion excitations. The decisions are to be based on building response monitoring at a number of locations during a particular event and should account for uncertainty sources such as:

- meagreness or lack of data regarding excitation. i.e. **load uncertainty**;
- incompleteness of the measurements with regard to the locations at which the response is being observed, i.e. **spatial uncertainty**.

The solution to the proposed task lies in checking whether a set of design codes provide criteria are held or violated as the adopted structural model, being subjected to the actual excitation, responds according to the recorded response. This based on the assumption that the structural model is correct.

* This proposal has been developed in the framework of author's Ph.D. thesis and is soon to be realised numerically. The thesis is supervised by Dr. Y. Frostig and Dr. Y. Ben-Haim, professors at the Haifa Technion, the Israeli Institute of Technology.

M. Erdik et al. (eds.), Strong Motion Instrumentation for Civil Engineering Structures, 495–504.

Structure - Modelling Assumptions and Limitations

The structures under consideration are skeletal RC buildings whose floors are assumed rigid diaphragms in their own plane with out-of-plane flexibility. Accordingly, the structure is considered as a series of planar, lateral loading resisting, frames connected at each floor level by rigid diaphragms, ensuring their equivalent transversal deformations. It is assumed that the frames are designed according to the conventional *strong column - week beam* design philosophy. In such structures, the input earthquake energy is released through inelastic deformations developed mainly in the beams/girders. Thus, the columns retain their elastic flexural behaviour whilst, at the beam, the formation of two zones is encountered. The first is that of residual deformations, located at both beam-ends close to the joints, and the second is the intermediate fraction still behaving elastically. This behaviour leads us to adopt a model known as a frame with a *semi-rigid* connection. This is composed of an inelastic rotational spring lumped at both girder ends and flexural, elastically responding, elements (Figure 1).

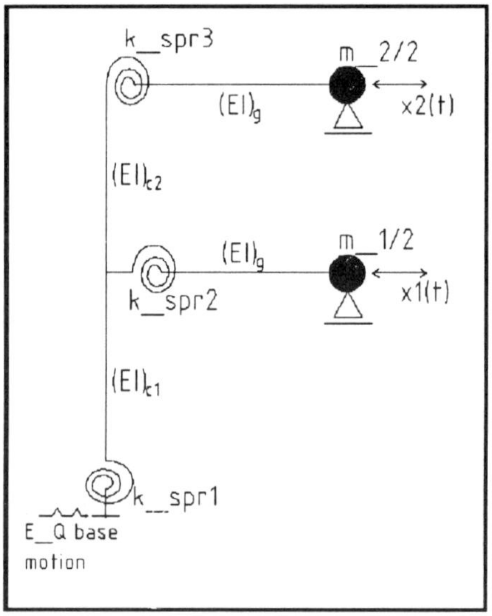

Figure 1

The inelastic spring is an energy-dissipated device represented by a phenomenological hysteretic model. The model is a set of rules governing the moment- rotation relationship. The so-called *Q-hyst* model – [8] is chosen here. This model is controlled by three parameters: the *initial stiffness*, the *yield point* and the *post-yielding stiffness*. In spite of the small parameter number, it is able to reproduce adequately the typical concrete hysteretic behaviour due to alternate cyclic loading. The small number of parameters is also an advantage in the subsequent damage detection procedure.The elastic portion of the beams and the columns is modelled by one-dimensional *beam* element. Analytically, it is described by polynomial function of the transversal displacements. This approach facilitates versatility in different model utilisation for the beam and columns when the $P-\Delta$ effect could be significant and so the *column-beam* element is more eligible. The elements assemblage into a plane frame model, which can be analysed dynamically, is accomplished by satisfying the compatibility and the force equilibrium demands at the joint connections. The phenomenological model adopted for the inelastic concrete behaviour causes the stiffness of the compound model to be a piece-wise function of its response. With regard to dynamic analysis, this means that a sequence of linear problems has to be solved with initial conditions those at the end of the predecessor functional branch.

Treatment Of The Excitation Uncertainty

It is widely acknowledged that environmental events such as earthquakes are accompanied by considerable uncertainty because of their random nature. They are unpredictable and any attempt to enclose them in analytical expression results in simplifications, approximations and subsequent decision limitations. In general, there are two approaches to handling the uncertainty: probabilistic and non-probabilistic. The more common and, thus, conventional is the probabilistic.

"Any probabilistic theory contains two main components: sets of events, and a measure function defined on these sets", [2] The measure function, otherwise known as distribution, is a parametric model which is tuned on the basis of available information. Typically, the data at one's disposal contains more frequent rather than rare events, thus the tail of the calibrated model may differ considerably from the actual distribution. That is, the model-based conclusions will be valid for the bulk of occurrences rather than for the exceptions. However, it is the exceptions which are of great concern to the analyst. A wrongly conceived model could induce either over-conservative or underestimated decisions.

The non–probabilistic conception is also based on sets of events, but no measure of occurrences frequency need be defined. Instead, the available data "is invested in the **structure** of the event sets" [2] Herein the *convex*-set model of uncertainty is chosen as a non-probabilistic alternative. The convex model is a convex set of functions each standing for possible event realisation and assembled according to similar attributes. For instance, the *energy bound convex models* represent sets of functions with a given bound on the energy, *envelope-bound convex set* consists of functions bounded into given linear intervals and the *spectral convex model* defines sets of all functions with limitations in their spectral content.

In the approach to damage detection described in the following, one of the spectral convex models is chosen known as the *Fourier bounded convex* model:

$$U(\alpha)=\left\{\vec{\beta}\middle|\left(\vec{\beta}-\vec{\beta}_0\right)^T\cdot W\cdot\left(\vec{\beta}-\vec{\beta}_0\right)\leq\alpha^2\right\} \tag{1}$$

where: $\vec{\beta}$ is a vector of Fourier spectrum coefficients of the unknown signal, $\vec{\beta}_0$ - an a priori information for them if any.

Implement Of The Idea For Damage Detection And Solution Procedure Development.

Recalling that the essence of the idea consists of:

1. A choice of a set of relevant design codes provide criteria ensuring both structure serviceability and keeping the limited state restrictions

$$\bigcup_k \{\Theta_k\}, k=s_1\ldots s_p; \tag{2}$$

These could be limitations on the lateral story displacements - s_1 or on the relative story drifts - s_2, comparison of the actual against the allowable compress members stresses - s_k, provided versus demanded local or overall ductility - s_r, dissipated energy as a ratio of the input energy – s_q, tracking for the values of some damage index - s_p).

2. Each one of the above selected criteria could serve as an indicator for damage, considering its violation at the most critical of the locations (the *semi-rigid* joints) to be a signal of damage/failure. The violation of any of the 'critical' criteria is verified by comparing its threshold value - $\hat{\Theta}_k$ with the extremum of the following performance index:

$$J_k=\max\Theta_k \tag{3}$$

3. The response of the real building is measured at several aptly selected locations:

$$\tilde{y}_m(t_l), l=1\ldots L, m=1\ldots M; \tag{4}$$

The fact that the stiffness of the model is a piece-wise function makes it possible to present each one of the design code criteria. Θ_k as a linear combination of the state vector components - $\vec{x}$:

$$\Theta_k=\vec{h}_k^T\cdot\vec{x}, k=s_1\ldots s_p; \tag{5}$$

However, it is to be aware for the pseudo-linearity of the above relations since its coefficients - $\{\hbar_i\}_k$ vary in accordance with the rules of the moment-rotation law of the joint-lumped springs.

As previously mentioned, the failure examination could be carried out by checking whether any of the chosen criteria at any of the likely critical places has been violated during the entire period of observation. However, neither excitation nor the subsequent model response is known in order to estimate the structural condition! The only possible and apparently most natural approach to this problem is to use the Optimisation theory, i.e. to find the extremum value of each of the members in the set (2), taking into account the constraints imposed by:

- the system dynamics described by the following non-linear differential equations

$$\dot{\vec{x}} = \vec{f}\left[\vec{x}(t), u(t,\vec{\beta})\right], \tag{6}$$

with $\vec{x}$ being an n dimensional displacements vector, $u(t,\vec{\beta})$ - the excitation scalar function.

- the excitation uncertainty, analytically expressed by *convex model*

$$U(\alpha) = \left\{\vec{\beta} \middle| \vec{\beta}^T \cdot W \cdot \vec{\beta} \le \alpha^2\right\}^*, \tag{7}$$

with W - $q \times q$ weighting matrix and α - scalar termed *uncertainty index* and $\vec{\beta}$ - a q dimensional vector of unknown Fourier decomposition coefficients in the notation:

$$u(t,\vec{\beta}) = \sum_{i=n_1}^{n_q} \beta_i \cdot \cos(2\pi \cdot f_i \cdot t) = \vec{\beta}^T \cdot \sigma(t); \tag{8}$$

- the presumption that the real structure response - $\vec{y}_m(t_l)$, at one or several locations - $(m = 1 \ldots M)$ observed at even time instance $(l = 1 \ldots L)$, match exactly that of the structural model - $\vec{x}_m(t_l,\vec{\beta})$ at corresponding positions and time intervals that is:

$$\vec{x}_m(\vec{\beta}, t_l) = \vec{y}_m(t_l) \tag{9}$$

---

[*] For the sake of simplicity, vector $\vec{\beta}_0$,in expression (1), will not be further considered

Where the extremum values exist, they can be compared with the lower and upper thresholds of the comparative design criterion - $\left\{\hat{\Theta}^{(-)},\hat{\Theta}^{(+)}\right\}_k$ and thus checking which of the p-th criteria are held or violated.

Obviously, the proposition as stated is an optimisation problem and can be identified as the Bolza problem from the Optimal Control Theory, namely:

Find the q *-dimension vector -* $\vec{\beta}$ *which minimises (maximises) the performance index –*

$$J_{kl}(\vec{\beta}) = \vec{h}_k\left[\vec{x}(\vec{\beta},t_l)\right]\cdot\vec{x}(\vec{\beta},t_l),\ k = s_1\ldots s_p,\ l = 1\ldots L, \tag{10}$$

subject to the following equality constraint:

$$\left|\begin{array}{l}\dot{\vec{x}} = \vec{f}\left[\vec{x}(t),u(t,\vec{\beta})\right]\\ \vec{\beta}^T\cdot W\cdot\vec{\beta} = \alpha^2\\ \vec{x}_m(t_l,\vec{\beta}) = \vec{y}_m(t_l)\end{array}\right. \tag{11}$$

Typically, a solution is sought adjoining the equalities (11) with undermined (*Lagrange*) multipliers- $\mu,\vec{\upsilon},\vec{\lambda}(t)$ to the performance index - J_{kl}. In such a way, the problem is transformed into unconstrained one:

$$\bar{J}_{kl} = \vec{h}_k\cdot\vec{x}(\vec{\beta}_{kl},t_{ll}) + \vec{\upsilon}_{kl}^T\cdot\left(\vec{x}_m(\vec{\beta}_{kl},t_l) - y_m\right) + \mu_{kl}\cdot\left(\vec{\beta}_{kl}^T W\vec{\beta}_{kl} - \alpha^2\right) + \int_{t_0}^{t_l}\left\{\lambda_{kl}(t)^T\cdot\left[\vec{f}(\vec{x},\vec{\beta}) - \dot{\vec{x}}\right]\right\}dt \tag{12}$$

with: $\vec{\lambda}(t)$ an n dimensional vector termed co-state or influence vector, μ - scalar, υ - an M scalar vector and $t_l, l = 1\ldots L$ the l-th observation time instant.

The l-th time instant, in the stated formulation, serves as the final time and could be any of the measurement instances, or a multiple of the sampling interval up to the terminal time - T. In other words, the problem has to be solved as many times as the product between considered intermediate instances L and the number of the design code criteria - p.

After defining a scalar function H (the *Hamiltonian*) for convenience:

$$H\left[\vec{x}(t),\vec{\beta},\vec{\lambda}(t)\right] = \vec{\lambda}(t)^T\cdot\vec{f}(\vec{x},\vec{\beta}); \tag{13}$$

the augmented performance index (11) becomes:

$$\bar{J}_{kl} = \vec{h}_k \cdot \vec{x}(\vec{\beta}, t_l) + \vec{\upsilon}_{kl}^T \cdot [\vec{x}_m(\vec{\beta}, t_l) - \vec{y}_m] + \mu_{kl} \cdot (\vec{\beta}_{kl}^T \cdot W \cdot \vec{\beta}_{kl} - \alpha^2) + \\ + \int_{t_0}^{t_l} \{H[\vec{x}(t), \vec{\beta}, \vec{\lambda}_{kl}(t)] - \vec{\lambda}_{kl}(t)^T \dot{\vec{x}}\} dt \tag{14}$$

It is also to account for the particular non-linearity form of $\vec{f}(\vec{x}, \vec{\beta})$, expressing in slope discontinuities each time when any of the nodal rotations reaches a new curve branch or its velocity vanishes. This constrains some of the state vector components to prescribed values at unknown intermediate time instances. The new **internal** constraints require extra lagrangian multipliers - $\vec{\pi}$. These interior constraints are also functions of the state variables:

$$\vec{\zeta}_i[\vec{x}(t = \vartheta_i)] = 0, i = 1,2 \ldots s \tag{15}$$

with ϑ_i - intermediate, unknown apriori time instant, s is the overall number of discontinuities within $[t_0, t_l]$, and $\vec{\zeta}_i$ is an n dimensional vector function. The modified augmented performance index becomes:

$$\bar{\bar{J}}_{kl} = \vec{h}_k^T \cdot \vec{x} + \vec{\upsilon}_{kl}^T \cdot [\vec{x}_m(\vec{\beta}_{kl}, t_l) - \vec{y}_m] + \mu_{kl} \cdot (\vec{\beta}_{kl}^T \cdot W \cdot \vec{\beta}_{kl} - \alpha^2) + \sum_{i}^{s} \vec{\pi}_i^T \cdot \vec{\zeta}_i[\vec{x}(t = \vartheta_i)] + \\ + \int_{t_0}^{t_l} \{H[\vec{x}(t), \vec{\beta}, \vec{\lambda}(t)] - \vec{\lambda}(t)^T \cdot \dot{\vec{x}}\} dt \tag{16}$$

To find the solution, the stationary value of the performance index $\bar{\bar{J}}_{kl}$, its variation $\delta\bar{\bar{J}}_{kl}$, due to variation in each of the unknowns, namely $\delta\{\vec{x}, \vec{\beta}_{kl}, \vec{\lambda}_{kl}, \mu_{kl}, \vec{\pi}_{i(kl)}, \vec{\upsilon}_{kl}\}$, has to vanish. Let us now consider the variation - $\delta\bar{\bar{J}}_{kl}$. Splitting the integral into $\int_{t_0}^{\vartheta_1^-} ..dt + \int_{\vartheta_1^+}^{\vartheta_2^-} ..dt + \ldots + \int_{\vartheta_s}^{t_l} ..dt$ and integrating by parts, allowing possible discontinuities in $\vec{\lambda}$ at $t = \vartheta_1, \ldots, \vartheta_s$, results in:

$$\delta\bar{\bar{J}}_{kl} = \left[\left(\frac{\partial}{\partial\vec{x}}\left(\vec{h}_k^T\vec{x}\right)\right)^T + \left(\frac{\partial}{\partial\vec{x}}\left(\upsilon_{kl}^T\vec{x}_m\right)\right)^T\right]^T d\vec{x}\Big|_{t_f} + \sum_i \vec{\pi}_{i(kl)}^T \frac{\partial\zeta_i}{\partial t}dt\Big|_{t=\vartheta_i} + \sum_i \vec{\pi}_{i(kl)}^T \frac{\partial\zeta_i}{\partial\vec{x}}d\vec{x}\Big|_{t=\vartheta_i} -$$
$$-\vec{\lambda}_{kl}^T\delta\vec{x}\Big|_{t_0}^{\vartheta_1^-} - \vec{\lambda}_{kl}^T\delta\vec{x}\Big|_{\vartheta_1^+}^{\vartheta_2^-} - \ldots - \vec{\lambda}_{kl}^T\delta\vec{x}\Big|_{\vartheta_s^+}^{t_{obs}} + \left(H - \vec{\lambda}_{kl}^T\dot{\vec{x}}\right)dt\Big|_{\vartheta_1^-} - \left(H - \vec{\lambda}_{kl}^T\dot{\vec{x}}\right)dt\Big|_{\vartheta_1^+} +$$
$$+\left(H - \vec{\lambda}_{kl}^T\dot{\vec{x}}\right)dt\Big|_{\vartheta_2^-} + \ldots + \left(H - \vec{\lambda}_{kl}^T\dot{\vec{x}}\right)dt\Big|_{\vartheta_s^-} - \left(H - \vec{\lambda}_{kl}^T\dot{\vec{x}}\right)dt\Big|_{\vartheta_s^+} + 2\mu\left(W\vec{\beta}_{kl}\right)^T d\vec{\beta} +$$
$$+\left(\vec{\beta}_{kl}^T W\vec{\beta}_{kl} - \alpha^2\right)d\mu + \left(\vec{x}_m - \vec{y}_m\right)^T d\vec{\upsilon} + \int_{t_0}^{t_f}\left[\left(\vec{H}_x + \vec{\lambda}_{kl}\right)^T\delta\ \vec{x} + \left(H_\lambda - \dot{x}\right)^T\delta\vec{\lambda} + H_\beta^T\delta\vec{\beta}\right]dt \tag{17}$$

Making use of the following analytical relations

$$d\vec{x}\Big|_{t=\vartheta_i} = \begin{cases}\delta\,\vec{x}\Big|_{t=\vartheta_i^-} + \dot{\vec{x}}\Big|_{t=\vartheta_i^-} dt, \\ \delta\,\vec{x}\Big|_{t=\vartheta_i^+} + \dot{\vec{x}}\Big|_{t=\vartheta_i^+} dt\end{cases} \tag{18}$$

to eliminate $\delta\vec{x}(\vartheta_i^-)$ and $\delta\vec{x}(\vartheta_i^+)$, noting that $\vec{\beta}_{kl}$ are not time dependent, and also that the $dt = 0$ at the final time, the expression (17) becomes:

$$\delta\bar{\bar{J}}_{kl} = \left[\vec{h}_k^T + \vec{\upsilon}_{kl}^T I - \vec{\lambda}_{kl}\right]^T d\vec{x}\Big|_{t_f} + \left[2\mu W\vec{\beta}_{kl} + \int_{t_o}^{t_f} H\vec{\lambda}_{kl}dt\right]^T d\vec{\beta} + \sum_i\left[\left(\vec{\pi}_{i(kl)}^T\zeta_{i,x} - \vec{\lambda}^T(\vartheta_i^-) + \lambda^T(\vartheta_i^+)\right)d\vec{x}\Big|_{t=\vartheta_i}\right] +$$
$$+\sum_i\left[\left(\vec{\pi}_i^T\zeta_{i,t} + H(\vartheta_i^-) - H(\vartheta_i^+)\right)dt\right] + \left(\vec{\beta}_{kl}^T W\vec{\beta}_{kl} - \alpha^2\right)d\mu + \left(x_m - y_m\right)^T d\vec{\upsilon} +$$
$$+\int_{t_0}^{t_f}\left[\left(H_x^T + \vec{\lambda}\right)^T\delta\,\vec{x} + \left(H_\lambda - \dot{\vec{x}}\right)^T\delta\vec{\lambda}\right]dt \tag{19}$$

with: $\mathbf{I}$ an $N \times N$ dimensional diagonal matrix, having all its elements zero but 1 on the main diagonal just at the position of the observed state counterparts.

The condition that the variation of $\delta\bar{\bar{J}}_{kl}$ has to vanish results in the following simultaneous differential (*Euler-Lagrange*) equations:

1. $H_x^T + \dot{\vec{\lambda}}_{kl} = 0 \Rightarrow \vec{f}_x^T\vec{\lambda}_{kl} + \dot{\vec{\lambda}}_{kl} = 0$ (*N differential equations*)
2. $H_\lambda - \dot{\vec{x}} = 0 \Rightarrow \vec{f}\left(\vec{x},\vec{\beta}\right) - \dot{\vec{x}} = 0$ (*N differential equations*)
3. $2\mu_{kl}W\vec{\beta}_{kl} + \int_{t_o}^{t_f} H_\beta^T dt = 0 \Rightarrow 2\mu_{kl}W\vec{\beta}_{kl} + \int_{t_o}^{t_f}\vec{f}_\beta^T\vec{\lambda}_{kl}dt = 0$ (*q integral equations)*

(20)

4. $\vec{\beta}_{kl}^T W\vec{\beta}_{kl} - \alpha^2 = 0$ (*1 algebraic equation*),
5. $\vec{x}_m(t_i) - \vec{y}_m = 0$ (*1 algebraic equation*);

together with the following constraints on the multiple inner boundaries:

$$\vec{\lambda}\left(\vartheta_1^{(-)}\right)=\vec{\lambda}\left(\vartheta_1^{(+)}\right)+\vec{\zeta}_{1,x}^{T}\left(\vartheta_1\right)\vec{\pi}_1 \rightarrow N\,eq-s, \qquad H^{(-)}\left(\vartheta_1^{(-)}\right)=H^{(+)}\left(\vartheta_1^{(+)}\right)-\vec{\pi}_1^T\vec{\zeta}_{1,t}\left(\vartheta_1\right)\rightarrow 1eq.,$$

$$\vec{\lambda}\left(\vartheta_2^{(-)}\right)=\vec{\lambda}\left(\vartheta_2^{(+)}\right)+\vec{\zeta}_{2,x}^{T}\left(\vartheta_2\right)\vec{\pi}_2 \rightarrow N\,eq-s, \cup \; H^{(-)}\left(\vartheta_2^{(-)}\right)=H^{(+)}\left(\vartheta_2^{(+)}\right)-\vec{\pi}_2^T\vec{\zeta}_{2,t}\left(\vartheta_2\right)\rightarrow 1eq., \tag{21}$$

$$\vdots \qquad\qquad \vdots$$

$$\vec{\lambda}\left(\vartheta_s^{(-)}\right)=\vec{\lambda}\left(\vartheta_s^{(+)}\right)+\vec{\zeta}_{s,x}^{T}\left(\vartheta_s\right)\vec{\pi}_s \rightarrow N\,eq-s, \qquad H^{(-)}\left(\vartheta_s^{-}\right)=H^{(+)}\left(\vartheta_s^{(+)}\right)-\vec{\pi}_s^T\vec{\zeta}_{s,t}\left(\vartheta_s\right)\rightarrow 1eq.;$$

$$\left[\vec{\hbar}_x+\vec{\upsilon}I-\vec{\lambda}\right]_{t=t_f}=0\rightarrow N\,eq-s,$$

$$\vec{x}\left(t_0\right)=\vec{x}_0\rightarrow N\,eq-s;$$

as well as the state vector continuity at the interior points:

$$\vec{x}\left(\vartheta_i^{(-)}\right)=\vec{x}\left(\vartheta_i^{(+)}\right), \tag{22}$$

where $t_0<\vartheta_1<\ldots<\vartheta_s<t_l$, $H^{(-)}\left(\vartheta_i^{(-)}\right)=\vec{\lambda}^T\left(\vartheta_i^{(-)}\right)\cdot\vec{f}\left(\vartheta_i^{(-)}\right), H^{(+)}\left(\vartheta_i^{(+)}\right)=\vec{\lambda}^T\left(\vartheta_i^{(+)}\right)\cdot\vec{f}\left(\vartheta_i^{(+)}\right)$

The number - s of the internal discontinuity points ϑ_i is not known and depends upon the 'final' time - $t_l, l=1\ldots L$. It could be zero and increases with the time progression.
Evidently, the formula just developed applies to a *Multiple Points Boundary Value Problem* (MBVP) since the response/state vector ($\vec{x}$) is known as the process start, t_0, while the co-state vector ($\vec{\lambda}$) is given at several internal time instances, up to the final - T.
It is worth noting one more detail. In general, the analytical definition of a convex model is an inequality (7). However, for the sake of avoiding complexity in the formulated optimisation problem it is used as equality. To overcome the fact that the uncertainty index - α is actually unknown, essentially it is the uncertainty descriptor, α will be given different reasonable values. The smaller the value of α, the more confident the conclusion for damage or, respectively, lack of damage. In the case when do not exist values of α for which the optimisation problem can be solved the amount of uncertainty either in the excitation or in the measurement completeness is too much to be tolerable.

Conclusions and Suggestions

The theoretically developed solution is not yet realised numerically. Therefore, the difficulties which could arise are unknown. In principal, a non-linear BVP has not always solution. This fact could lead to an erroneous conclusion regarding the level of uncertainty. Hence, criteria to distinguish this from other possibilities must be formulated.

The approach discussed in this paper offers two methods for strong motion data utilization:

as a priori data for the Convex model adjustments; and
as 'on-line' information via the $\vec{y}_m$ monitoring;

References

1. Ben – Haim Y., and Ellishakoff I. (1990) Convex models of Uncertainty in Applied Mechanics, Elsevier, Amsterdam.
2. Ben – Haim Y. (1994) A non – probabilistic concept of reliability, Structural Safety 14, 227-245, Elsevier, Amsterdam.
3. Biddah A. and Heidebrect C. (1998) Seismic performance of moment – resisting steel frame structures designed for different levels of seismic hazard Vol.14, No. 4, 597 – 628.
4. Bryson,A. E. Jr. and Ho Y.-C. (1975) Applied optimal control, Taylor & Francis, USA.
5. Drenick, R. F. (1970) Model – free design of aseismic structures, Journal of the Engineering Mechanics Divisions, ASCE, Vol96, No. EM4, 483 – 738.
6. Ghobarah, A., Aziz, T. and Abou – Elfath, H. (1999) Softening effects on seismic response of non – ductile concrete frames, Earthquake Engineering Vol. 3, No. 1, 59 – 82.
7. Aschheim, M. and Black, E. (1999) Effects of prioir earthquake damage on response of simple stiffness – degrading structures, Earthquake Spectra Vol. 15, No. 1, 1-24.
8. Saiidi, M. (1982) Hysteresis model for reinforced concrete, Journal of the Structural division, ASCE, Vol. 108.
9. Westermo, B. D. (1985) The critical excitation and response of simple dynamic systems, Journal of sound and vibration, 100(2), 233

ASSESSMENT OF SEISMIC HAZARD IN ROMANIA BASED ON 25 YEARS OF STRONG GROUND MOTION INSTRUMENTATION

D. LUNGU, C. ARION, A. ALDEA, S. DEMETRIU
Technical University of Civil Engineering – Bucharest
124, Bd. Lacul Tei, 72302, Bucharest, Romania
e-mail: lungud@hidro.utcb.ro

1. Romania Seismic Networks And Available Records

The first and the most important free-field strong ground motion in Romania was recorded in Bucharest, at INCERC seismic station, during March 4, 1977 Vrancea earthquake (M_w=7.5, h=109 km, epicentral distance to Bucharest 105 km).

The accelerogram (*0.2g* peak ground acceleration and *1.6s* long predominant period of soil vibration) was recorded on a *SMAC-B* Japanese instrument and further digitized and processed by the *Observational Committee of Strong Motion Earthquake* of the *Building Research Institute, Ministry of Construction of Japan*, January 1978 (254 pages Report).
In the decade following the 1977 strong earthquake, on the territory of Romania were installed 3 strong motion networks: *INCERC (Building Research Institute)*, *INFP (National Institute of Earth Physics)* and *GEOTEC (Institute for Geotechnical and Geophysical Studies)*.

The free-field instrumentation at *INCERC* was developed in connection with building instrumentation. The present *INCERC* network have more than 70 analog *SMA-1 Kinemetrics* instruments as well as 15 digital *ADS* Romanian instruments. It is still the largest and the richest (as number of existing strong accelerograms) seismic network of Romania.

The free-field strong motion of *INFP* was developed from a *PNUD/UNESCO* support and now it has about 25 *SMA-1 Kinemetrics* instruments. In the last decade the network was developed in cooperation with *Geophysical Institute of Karlsruhe University*. The two institutes recently installed about 30 *K2 Kinemetrics* accelerometers as a one of the components of the *SFB 461* German-Romanian project on Vrancea earthquakes.

The *GEOTEC* seismic network is the network that monitorizes dams and hydraulic structures. It has about 20 instruments (3/4 are *SMA-1* accelerometers).The presently in use strong motion accelerographs of Romania seismic networks are indicated on the map in Figure 1 and in Table 1. It is emphasized that the main target of the net is the Vrancea seismic zone in Carpatians, as well as Bucharest and densely urbanized areas of the South-East of Romania.

M. Erdik et al. (eds.), Strong Motion Instrumentation for Civil Engineering Structures, 505–518.

The envelope of the peak ground acceleration (*PGA*, cm/s^2) and the effective peak velocity (*PGV*, cm/s) recorded in Romania during the last 3 strongest Vrancea earthquakes are interpolated in Figure 2 and Figure 3. They clearly show directivity (N45^{O}E) of the subcrustal Vrancea sources activity, as well as the phenomenon of large velocities on the soft soil of Romanian Plain, in the South-East of Romania.

2. Recurrence Of The Vrancea Subcrustal Earthquakes

The Vrancea region, located when the Carpatians Mountains Arch bents, is the source of subcrustal (60-170km) seismic activity, which affects more than 2/3 of the territory of Romania and an important part of the territories of Republic of Moldova, Bulgaria and Ukraine

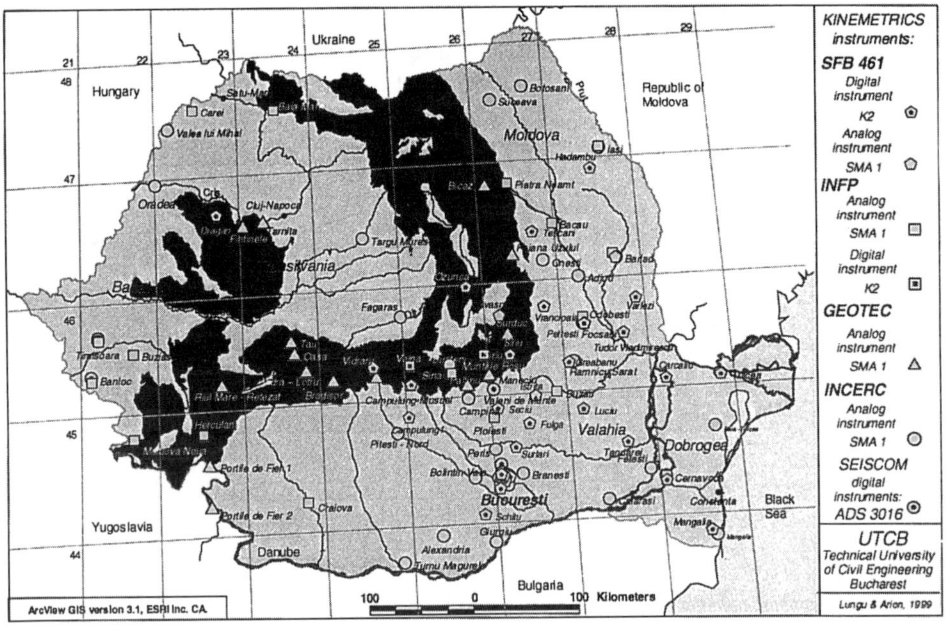

Figure 1. ROMANIA. Strong Motion Networks

Table 1. Free field strong motion instruments targeting Vrancea earthquakes in Romania

Seismic network	*Bucharest*				*Vrancea influence area in Romania*				*Total*			
	Digital instruments		Analog Instr.	*Total*	Digital Instruments		Analog instr.	*Total*	Digital Instruments		Analog instr.	*Total*
	K2	*ADS*	*SMA-1*		*K2*	*ADS*	*SMA-1*		*K2*	*ADS*	*SMA-1*	
INCERC	-	8	5	13	-	2 3 3	10[1)] 20[2)] 27[3)]	12 23 30	-	10 11 11	15 25 32	25[1)] 36[2)] 43[3)]
INFP	2	-	5	7	2	-	13	15	4	-	18	22
GEOTEC	-	-	-	-	-	2	6	8	-	2	6	8
SFB461	4	-	-	4	21	-	-	21	28	-	-	25
Total	6	8	10	24	23	4[1)] 5[2)] 5[3)]	29 39 46	56 67 74	29	12 13 13	39 56 49	80[1)] 91[2)] 98[3)]
	14				27 28 28				41[1)] 42[2)] 42[3)]			

1) Installed in buildings having only Ground floor (GF)
2) Installed in buildings having GF+1Story
3) Installed in buildings having GF+2Story

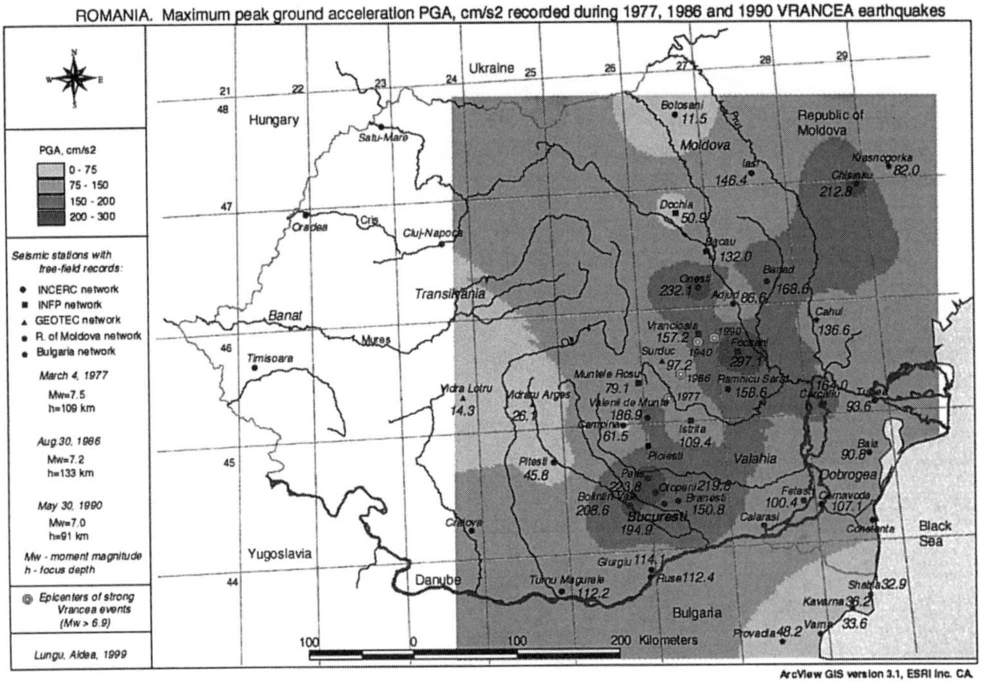

Figure 2. Maximum recorded peak ground acceleration during the last Vrancea strong events

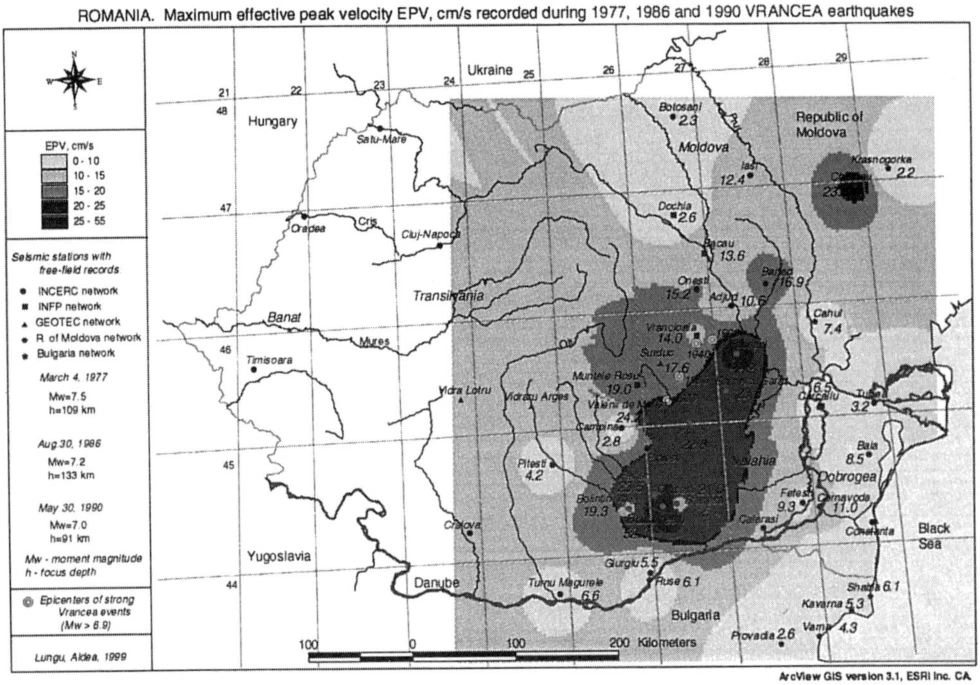

Figure 3. Computed effective peak velocity during the last Vrancea strong events

According to the 20th century seismicity, the epicentral Vrancea area is confined to a rectangle of 40x80km^2 having the long axis oriented N45E and being centered at about 45.6° Lat.N 26.6° and Long. E.

The Vrancea source induces a high seismic risk in the densely built zones of the South-East of Romania. On March 4, 1977, during the most severe Vrancea earthquake of this century, in the capital city of Romania, Bucharest, more than 1570 casualties were registered and 29 fragile reinforced concrete high-rise buildings built before 1945 completely collapsed.

Two catalogues of earthquakes occurred on the territory of Romania were compiled, more or less independently, by Radu (1974, 1980, 1995) and by Constantinescu and Marza (1980, 1995) Table 2. The Radu's catalogues are more complete, even the majority of significant events are also included in the Constantinescu and Marza catalogue. The magnitude in Radu catalogue is the Gutenberg-Richter (1936) magnitude, M_{GR}. The magnitude in Constantinescu & Marza catalogue was the surface magnitude, M_S. Tacitly, that magnitude was later assimilated as Gutenberg-Richter magnitude (Marza, 1995).

Table 2. Catalogue of subcrustal Vrancea earthquakes ($M_w \geq 6.3$) occurred during the 20th century

Date	Time (GMT) h:m:s	Lat. N□	Long. E□	RADU Catalogue, 1994				MARZA Catalogue, 1980		www.infp.ro Catalogue, 1998
				h, km	I_0	M_{GR}	M_w	I_0	M_s	M_w
1903 13 Sept	08:02:7	45.7	26.6	>60	7	6.3	-	6.5	5.7	6.3
1904 6 Feb	02:49:00	45.7	26.6	75	6	5.7	-	6	6.3	6.6
1908 6 Oct	21:39:8	45.7	26.5	150	8	6.8	-	8	6.8	7.1
1912 25 May	18:01:7	45.7	27.2	80	7	6.0	-	7	6.4	6.7
1934 29 March	20:06:51	45.8	26.5	90	7	6.3	-	8	6.3	6.6
1939 5 Sept	06:02:00	45.9	26.7	120	6	5.3	-	6	6.1	6.2
1940 22 Oct	06:37:00	45.8	26.4	122	7 / 8	6.5	-	7	6.2	6.5
1940 10 Nov	01:39:07	45.8	26.7	150[1)]	9	7.4	-	9	7.4	7.7
1945 7 Sept	15:48:26	45.9	26.5	75	7 / 8	6.5	-	7.5	6.5	6.8
1945 9 Dec	06:08:45	45.7	26.8	80	7	6.0	-	7	6.2	6.5
1948 29 May	04:48:55	45.8	26.5	130	6 / 7	5.8	-	6.5	6.0	6.3
1977 4 March[2)]	19:22:15	45.34	26.30	109	8 / 9	7.2	7.5	9	7.2	7.4
1986 30 Aug	21:28:37	45.53	26.47	133	8	7.0	7.2	-	-	7.1
1990 30 May	10:40:06	45.82	26.90	91	8	6.7	7.0	-	-	6.9
1990 31 May	00:17:49	45.83	26.89	79	7	6.1	6.4	-	-	6.4

[1)] Demetrescu's original (1941) estimation: 150Km; Radu's initial estimation (1974) was 133 km
[2)] Main shock

As a systematisation requirement for seismic hazard assessment, usually it is recommended the use of the moment magnitude, M_w. For Vrancea subcrustal events the relation between two magnitudes can be simply obtained from recent events data given in Table 2:

$$M_w \cong M_{GR} + 0.3 \qquad 6.0 < M_{GR} < 7.7 \tag{1}$$

Even the available catalogues of Vrancea events were prepared using the Gutenberg-Richter magnitude M_{GR}, the recurrence-magnitude relationship was herein newly determined using the moment magnitude M_w. The relationship is determined from Radu's 20th century catalogue of subcrustal magnitudes with threshold lower magnitude M_w=6.3.

The average number per year of Vrancea subcrustal earthquakes with magnitude equal to and greater than M_w, as resulting also from Figure 4, is:

$$\log n(\geq M_w) = 3.76 - 0.73\, M_w \tag{2}$$

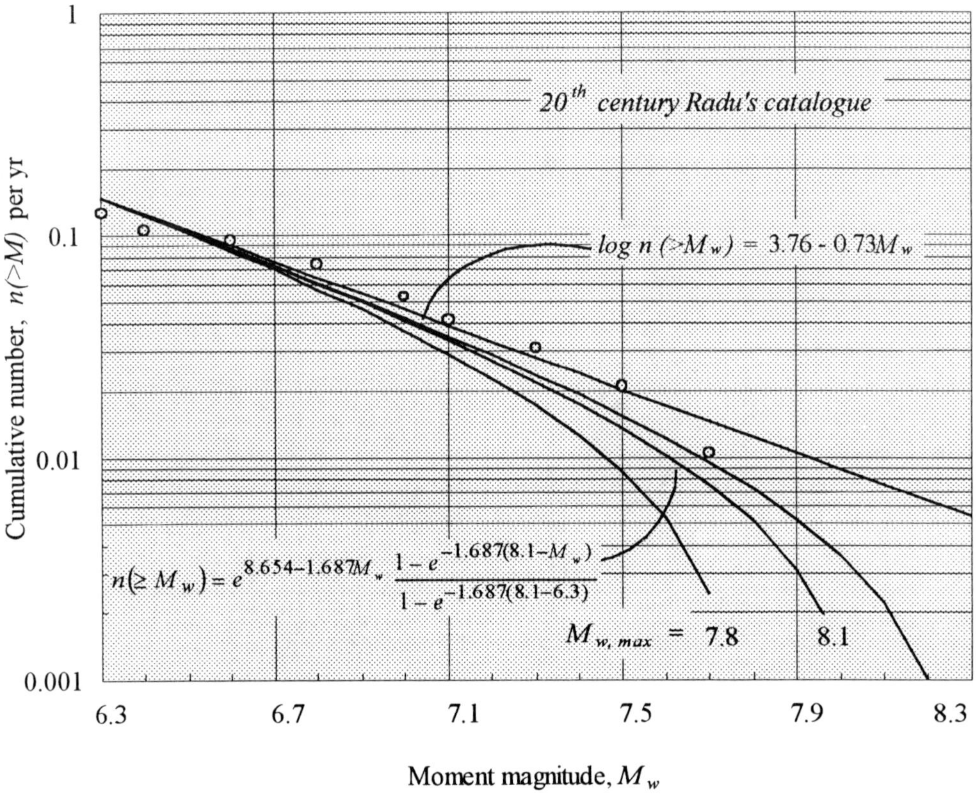

Figure 4. Magnitude recurrence relation for the subcrustal Vrancea source ($M_w \geq 6.3$)

The values of surface rupture area (*SRA*) and surface rupture length (*SRL*) from Wells and Coppersmith (1994) equations for "strike slip" rupture were used to estimate maximum credible Vrancea magnitude. According to Romanian geologists Sandulescu & Dinu, in Vrancea subduction zone: $SRL \leq 150 \div 200\ km$, $SRA < 8000\ km^2$. Based on this estimation, from Table 3one gets:

$$M_{w,max} = 8.1. \tag{3}$$

Table 3. Application of Wells and Coppersmith equations to the Vrancea source (mean values)

M	M_w	Event	Experienced SRA, km^2	Wells & Coppersmith equations	
				$logSRA=-3.42+0.90M_w$	$logSRL=-3.55+0.74M_w$
				SRA, km^2	SRL, km
6.7	7.0	May 30,1990	1100[1)]	759	43
7.0	7.2	Aug. 30,1986	1400[1)]	1148	60
7.2	7.5	March 4, 1977	63 x 37=2331[2)]	2138	100
	8.1	Max. credible	-	7413	278

[1)]As cited in Tavera (1991) [2)] Enescu *et al.* (1982)

If the source magnitude is limited by an upper bound magnitude $M_{w,max}$, the recurrence relationship can be modified in order to satisfy the property of a probability distribution (Hwang and Huo 1994, Elnashai and Lungu 1995):

$$n(\geq M_w) = e^{\alpha - \beta M_w} \frac{1 - e^{-\beta(M_{w,\max} - M_w)}}{1 - e^{-\beta(M_{w,\max} - M_{w0})}} \tag{4}$$

and, in the case of Vrancea source:

$$n(\geq M_w) = e^{8.654 - 1.687 M_w} \frac{1 - e^{-1.687(8.1 - M_w)}}{1 - e^{-1.687(8.1 - 6.3)}} \tag{5}$$

In Eq.(5), the threshold lower magnitude is $M_{w0}=6.3$, the maximum credible magnitude of the source is $M_{w,max}=8.1$, and $\alpha = 3.76\ ln10 = 8.654$, $\beta = 0.73\ ln10 = 1.687$.

The maximum credible magnitude of the source governs the prediction of source magnitudes in the range of large recurrence intervals, where classical relationship (2) does not apply, Table 4.

Table 4. Mean recurrence interval *(MRI)* of Vrancea magnitudes, $T\ (\geq M_w)=1/n(\geq M_w)$, in years, as function of the maximum credible magnitude of the source

Date	Gutenberg-Richter magnitude, M_{GR}	Moment magnitude, M_w	*MRI* from Eq. (5), years	*MRI* from Eq. (2), Years
		8.1	-	150
		8.0	475 → 778	127
		7.9	356	107
		7.8	217	91
10 Nov. 1940	7.4	7.7	100 → 148	76
		7.6	108	65
4 March 1977	7.2	7.5	82	55
		7.4	50 → 63	46
30 Aug. 1986		7.3	50	37
	7.0	7.2	40	33
30 May 1990	6.7	7.0	26	23

The depth of the Vrancea foci has a great influence on the experienced seismic intensity. The damage intensity of the Vrancea strong earthquakes is the combined result of both magnitude and location of the focus inside the earth.

The relationship between the magnitude of a destructive Vrancea earthquake $(M_w \geq 6.3)$ and the corresponding focal depth shows that higher the magnitude, deeper the focus:

$$\ln h = -0.866 + 2.846 \ln M_w - 0.18\,P \qquad (6)$$

where P is a binary variable: $P=0$ for the mean relationship and $P=1.0$ for mean minus one standard deviation relationship. The mean minus one standard deviation curve in Figure 5 should be used in hazard analysis as pessimistic correlation of Vrancea magnitude with focus depth. The earthquakes with $M_w \leq 6.3$ display non-correlation between h and M_w.

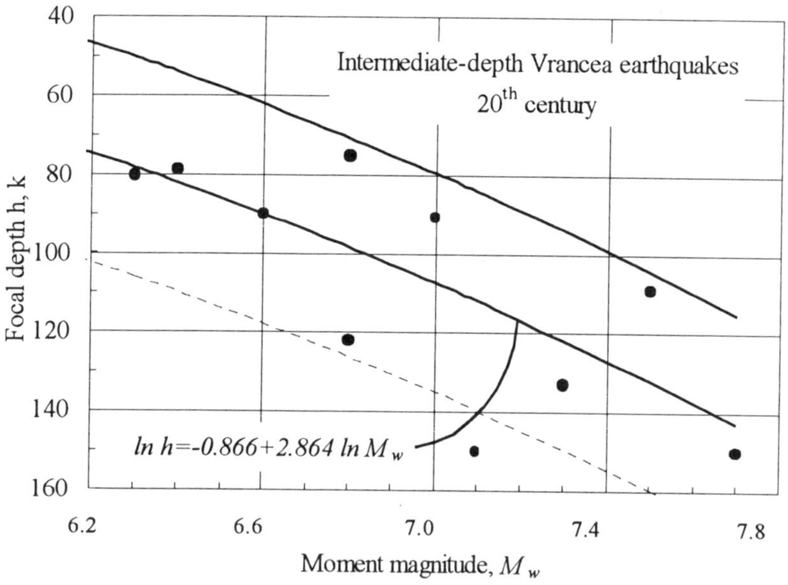

Figure 5. Correlation of Vrancea moment magnitude with focus depth

3. Seismic Macrozonation Using Gis Technology

The research devoted to the prediction of attenuation of Vrancea ground motions started in 1994. An increased accuracy in selection of the free-field accelerograms and the redigitisation (in 1997) of the accelerograms of *INCERC* seismic network allow the refinement of the previous predictions (Lungu *et. al*, 1995-1999).

The database used for the present analysis of the Vrancea ground motions attenuation contains 71 triaxial or biaxial accelerograms recorded at 47 free-field stations in Romania distributed on networks and events as in Table 5. The free field accelerograms were obtained at the ground level or the basement of 1 - 2 storey buildings. The accelerograms recorded on instruments installed at the basement of mid-rise and tall buildings (3 ÷ 12 storeys) were not used in the attenuation analysis. The database still have 9 accelerograms recorded in Republic of Moldova and Bulgaria.

Table 5. Distribution of the free field accelerograms used in the present attenuation analysis

Seismic network	Romania			Republic of Moldova	Bulgaria	Total
Earthquake	*INCERC*[1)]	*INFP*[2)]	*GEOTEC*[3)]	*IGG*[4)]		
March 4, 1977	1	-	-	-	-	1
Aug. 30, 1986	24	8	3	2	-	37
May 30, 1990	23	10	2	2	5	42
Total	48	18	5	4	5	80

[1)] *INCERC, Building Research Institute, Bucharest*
[2)] *INFP, National Institute for Earth Physics, Bucharest-Magurele*
[3)] *GEOTEC, Institute for Geotechnical and Geophysical Studies, Bucharest*
[4)] *IGG, Institute of Geophysics and Geology, Moldavian Academy of Science, Chisinau*

Taking into account the N45°E directionality of the seismic field produced by the Vrancea source, the attenuation analysis was performed on two orthogonal directions (the average direction of the rupture surface N45°E and the normal to this direction E45°S) and on 3 circular sectors (of 90° each) centred on these directions: Bucharest sector, Moldova sector and Cernavoda sector. The distribution of the recorded accelerograms within the 3 sectors is given in Table 6.

Table 6. Distribution of the free field accelerograms used in the present attenuation analysis

	Epicentral area	Bucharest Sector	Moldova Sector	Cernavoda sector	Total
March 4, 1977	-	1	-	-	1
Aug. 30, 1986	9	15	8	5	37
May 30, 1990	10	15	7	10	42
Total	19	31	15	15	80

The following Joyner-Boore type model was selected for the analysis of attenuation:

$$ln\ PGA = c_0 + c_1 M_w + c_2 \ lnR + c_3 R + c_4 h + \varepsilon \qquad (7)$$

where: *PGA* is peak ground acceleration at the site, M_w- moment magnitude, *R* - hypocentral distance to the site, *h* - focal depth, c_0, c_1, c_2, c_3, c_4 - data dependent coefficients and ε - random variable with zero mean and standard deviation $\sigma_\varepsilon = \sigma_{ln\,PGA}$, Table 7.

Table 7. Regression coefficients inferred for horizontal components of peak ground acceleration during Vrancea subcrustal earthquakes, Equation (7)

Sector	c_0	c_1	c_2	c_3	c_4	σ_{lnPGA}
All data	3.098	1.053	-1.000	-0.0005	-0.006	0.502
Bucharest sector	1.685	1.181	-1.000	0.002	-0.005	0.461
Moldova sector	0.144	1.102	-1.000	0.0008	-0.003	0.588
Cernavoda sector	1.565	1.252	-1.000	-0.001	-0.003	0.411

It is emphasized that the extrapolation of data in the range of large magnitudes $(M_w \geq 7.5)$ is entirely based on the peak ground acceleration recorded in Bucharest during March 4, 1977 event. That is why in the attenuation analysis that peak was also added to the data set recorded in 1986 and 1990 in each of the three sectors.

Based on *EUROCODE 8* and American *ASCE 7-95* or *UBC-1997* codes hazard level requirements, predictive GIS mapping of peak ground acceleration *PGA* produced by the subcrustal Vrancea earthquakes is herein illustrated for *50yr* MRI and for 10% probability of exceedance in 50 yr (i.e. $\overline{T} = 475\ yr$ MRI). The values on the *50 yr* map represent $(m+\sigma)$ values and the values on the *475 yr* map represent $(m+1.5\sigma)$ attenuated values. That gives for Bucharest $PGA_{50yr} \cong 0.2g$ and $PGA_{475yr} \cong 0.4g$.

The maps in Figure 6 accounts for all available free-field ground motions recorded in Romania, Republic of Moldova and Bulgaria during the 1977, 1986 and 1990 Vrancea earthquakes.

In Figure 7 is illustrated the present seismic zonation of Romania, Ukraine and Republic of Moldova to the Vrancea earthquakes in Romania. The general pattern of the map (which is based on deterministic macroseismic observations) as well as the content of the table describing the conversion of seismic intensity into ground acceleration are self-explanatory. They clearly suggest the need for regional harmonization of seismic macrozonation maps and prove the need for a joint zonation of the seismic hazard in the influence area of the Vrancea source.

Most probable seismic scenario, MRI=50yr.
Subcrustal Vrancea source in Carpatians.

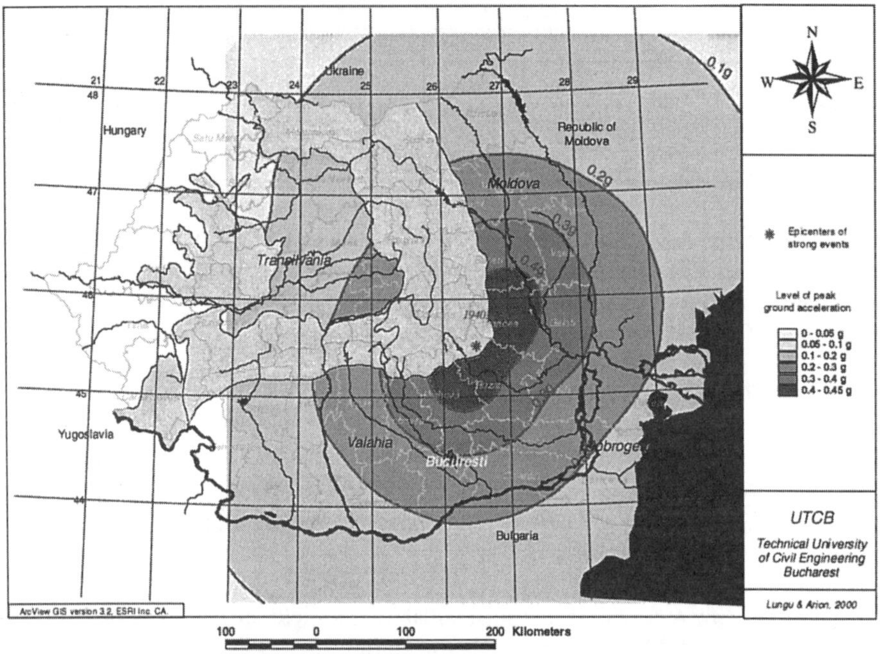

Worst case seismic scenario, MRI=475yr.
Subcrustal Vrancea source in Carpatians

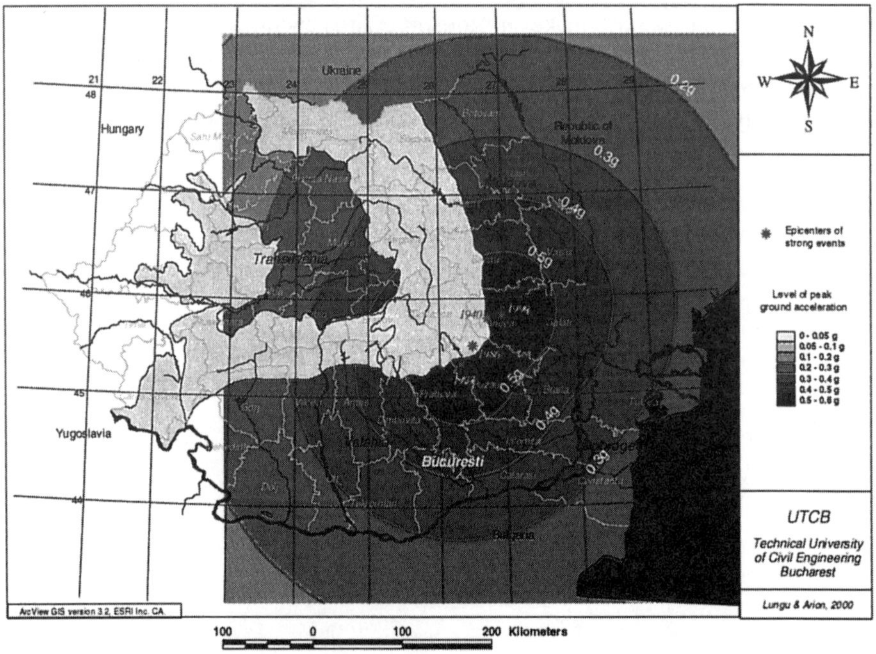

Figure 6. Predictive contour maps for peak ground acceleration from subcrustal Vrancea source

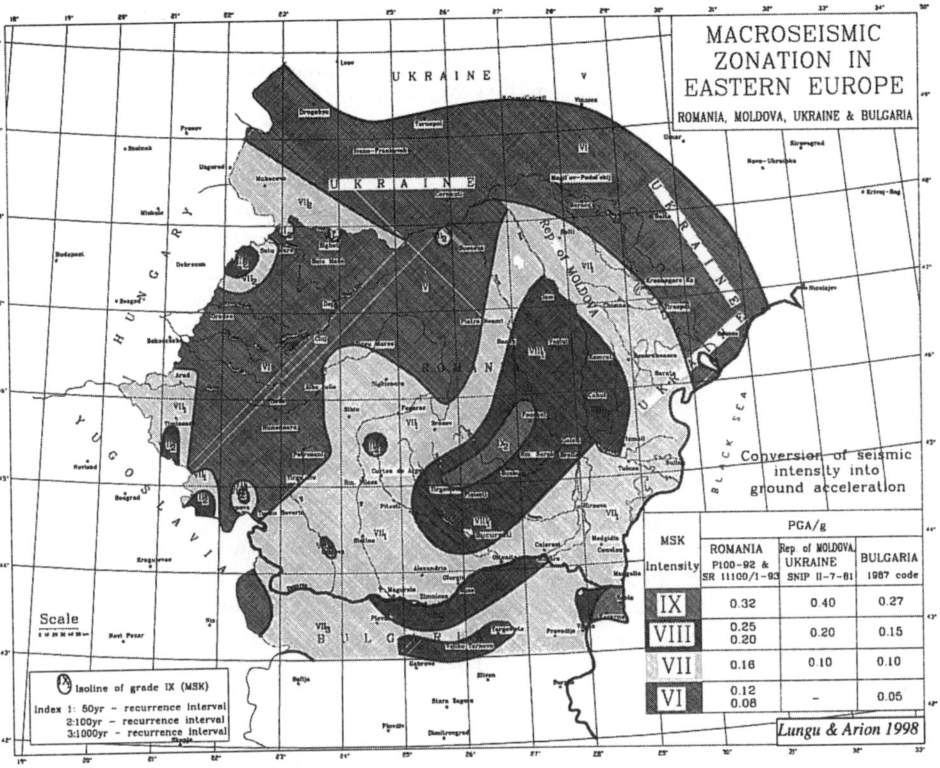

Figure 7. Macroseismic zonation for Vrancea source

4.Conclusions

Predictive *Geographic Information System (GIS)* maps delivered herein are consistent with *ASCE 7 & EUROCODE 8* hazard format: 10% probability of exceedance in 50 yr. It represents provision to be considered for the new Romanian and regional codes for earthquake resistance of buildings and others structures.

5.Acknowledgements

The ground motion data were provided by the late Dr. C. Radu, *INFP*, by Mr. S.Borcia *INCERC*-Bucharest and Dr. T. Moldoveanu, *GEOTEC*. We are indebted to all of them. The authors acknowledge with thanks the generous access to *ArcView Spatial Analyst, ESRI Inc.*, provided by Mr. C. Vasile from *GEOSYSTEMS*, Romania.

6. References

1. Constantinescu, L, Marza, V., 1980, A computer-compiled and computer-oriented catalogue of Romania's earthquakes during a millennium (984-1979), *Géophysique*, Tome 24, No2, p.193-206.
2. Elnashai A., Lungu D., 1995. Zonation as a tool for retrofit and design of new facilities. Report of the Session A1.2 - *5th International Conference on Seismic Zonation*, Nice, France, Oct. 16-19. Proceedings Vol.3. Quest Editions, Presses Academiques, Nantes, p. 2057-2082.
3. Hwang H.H.M., Huo J.R., 1994. Generation of hazard-consistent fragility curves for seismic loss estimation studies. Technical Report NCEER-94-0015. National Center for Earthquake Engineering Research, State University of New York at Buffalo, Aug.
4. Lungu, D., Aldea, A., Arion, C., 2000 "Engineering, state & insurance efforts for reduction of seismic risk in Romania" In: Proceedings of the *12th World Conference on Earthquake Engineering,* Auckland, New Zealand, Jan/Feb.
5. Lungu, D., Aldea, A., Arion, C., Cornea, T. 1998, "Seismic hazard zonation in Eastern Europe", *Second International Conference on Earthquake Hazard and Seismic Risk Reduction*, September 15-21, Yerevan, Armenia (Proceedings will be published by Kluwer Academic Publisher, 1999).
6. Lungu D., Cornea T., Nedelcu C., 1998. Probabilistic hazard assessment and site-dependent response for Vrancea earthquakes. In *Vrancea Earthquakes. Tectonics, Hazard and Risk Mitigation*, Kluwer Academic publishers b.v. Wenzel F., Lungu D., editors, p.251-268
7. Lungu D., Cornea T., Coman O., 1996 and 1995. Probabilistic hazard analysis to the Vrancea earthquakes in Romania. Part I in "Experience database of Romanian facilities subjected to the last three Vrancea earthquakes", *Research Report to the International Atomic Energy Agency*, Vienna, Contract No. 8233/EN R1 and Contract No. 8223/EN , Stevenson & Assoc. Office in Bucharest (57p+41 p. and 75 p.).
8. Marza V., 1995. Romania's seismicity file: 1. Pre-instrumental data. Special publications of the *Geological Society of Greece*, 1996.
9. Radu C. manuscripts, 1994. Catalogues of earthquakes occurred on Romanian territory during the periods 984-1990 and 1901-1994.
10. Tavera J., 1991. *Etude des mécanismes focaux de gros séismes et sismicité dans la région Vrancea-Roumanie.* DEA de Géophysique Interne, Institut de Physique du Globe de Paris, Université de Paris 7.
11. Wells D.L., Coppersmith K.J., 1994. New empirical relations among magnitude, rupture length, rupture width, rupture area, and surface displacement. *Bulletin of the Seismological Society of America, Vol.84, No 4*, p. 974-1002.

MEASUREMENT OF LATERAL EARTH PRESSURES ON AN EMBEDDED FOUNDATION DURING EARTHQUAKES

CHIKAHIRO MINOWA[1], MICHIO IGUCHI[2] AND MASANORI IIBA[3]

[1]*National Research Institute for Earth Science and Disaster Prevention, Tsukuba City 305, Japan.*

[2]*Science University of Tokyo, Faculty of Science and Engineering, Noda City 278-8510, Japan.*

[3]*Building Research Institute, Ministry of Construction, Tsukuba City 305, Japan.*

e-mail: iguchi@rs.noda.sut.ac.jp; minowa@eq-eng.bosai.go.jp

Abstract

The dynamic earth pressures during earthquakes have been observed on both sides of an embedded foundation for about 30 events. Some characteristics of dynamic earth pressures were extracted from the observations. The main findings obtained in this study can be summarized as follows. (1) The phase characteristics of earth pressures between both sides of the foundation show remarkable dependence on the frequency components of acceleration motions. (2) The earth pressures on lateral sides of the foundation are strongly related with horizontal velocity motions of the foundation. (3) The peak values of earth pressures have shown a tendency to increase with increase of lower frequencies contained in accelerograms on the ground surface. (4) The fact that the earth pressure has strong relation with the velocity motions of the foundation may be roughly explained by means of a simplified one-dimensional wave propagation theory for a soil rod model.

l. Introduction

In seismic design of embedded foundations or underground stuctures, it becomes important to estimate earth pressures on foundations or underground walls during earthquakes. The observations of earth pressures during earthquakes, however, have been scarcely conducted [l, 2]. Instead, there have been presented some experimental researches conducted in laboratories using small models [3, 4].

M. Erdik et al. (eds.), Strong Motion Instrumentation for Civil Engineering Structures, 519–532.

The earth pressure would be affected by various factors such as frequency components of earthquake ground motions, the motions of foundations or underground structures, composition of soil and its properties around the structures, types of earthquake waves, incident angle of seismic waves and so on. It is desirable to examine the effects of these factors on the basis of earthquake observations.

The recordings of dynamic earth pressures using a large scale shaking table foundation have been made on both sides of the embedded foundation. In addition to the earth pressures, acceleration and velocity motions of the foundation as well as the free-field motions at the site have been also observed. The objective of this paper is to discuss the dynamic characteristics of lateral earth pressures on the rigid embedded foundation based on earthquake observations. Special emphases are placed in this study on the investigations about effects of the frequency components of earthquake ground motions and motions of the foundation on dynamic characteristics of the earth pressures.

2. Outlines of Foundation and Earthquake Observation

2.1 OUT LINE OF FOUNDATION

The size of the foundation is 25m x 39m in plane and is used as a shaking table foundation in Tsukuba. The base of the foundation is directly supported on firm sand at a depth of 8.2m. The outlines of the foundation and soil profile are shown in figures 1(a), (b) and 2, respectively. The weight of the foundation itself and the shaking table is about 11,600 tf and 180 tf, respectively. The total weight corresponds almost to the excavated soil for the foundation. The fundamental frequency of the soil-foundation system is about 4.1 Hz in EW direction. It is confirmed that the foundation behaves as a rigid body within frequencies less than 10 Hz.

The detailed soil constants at the site were measured to a depth of about 40m and may be found elsewhere [5].

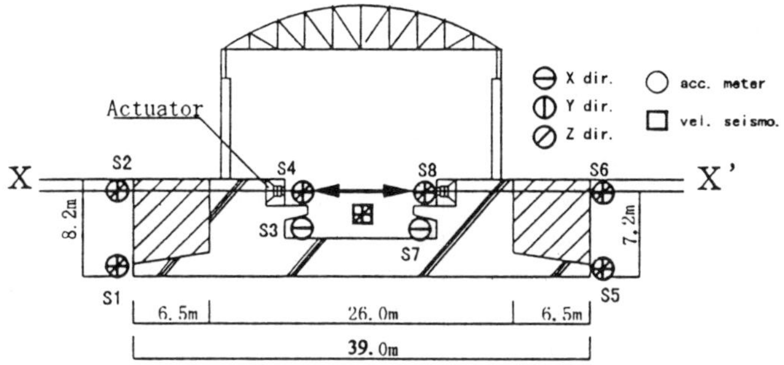

(a) Section of Foundation and Location of Seismographs.

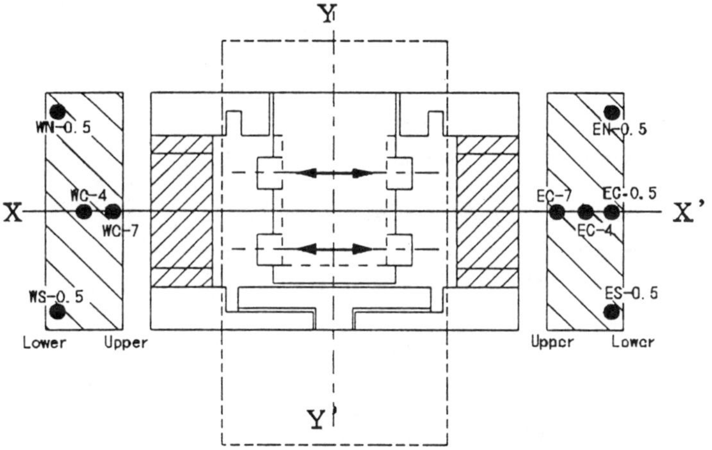

(b) Location of Earth Pressure Gauges.

Fig. 1 Outline of Foundation and Observation.

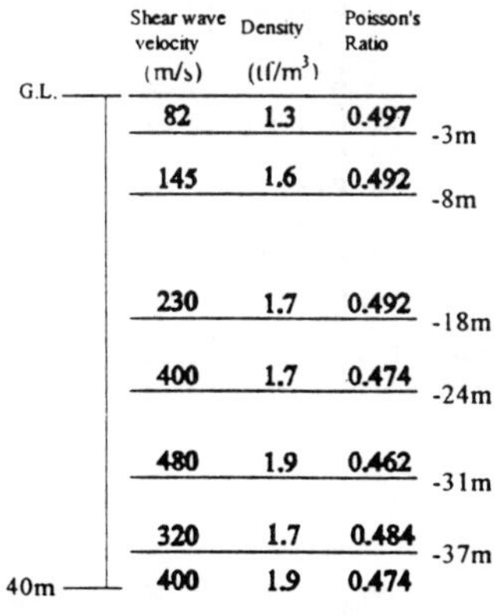

Fig. 2 Soil Profile

2.2 EARTHQUAKE OBSERVATION

Figure 1 (b) shows the location of earth pressure gauges on both sides of the foundation.On each side of the foundation, five earth pressure gauges (BE-2KR of Kyowa-Dengyo)have been instrumented at different depth of the embedded foundation, but one installedon the west side became out of function and is omitted from the figure. The observation of free-field ground motions has also been performed on the soil surface and at a depth of 40m. As for the foundation, the earthquake motions of the foundation have been observed at several points with accelerographs and velocity seismographs as shown in figure 1(a).The motions of the foundation are represented by accelerograms recorded at the point S3,which is located almost at the center of the foundation.

The records of dynamic earth pressures for about 30 earthquakes have been recorded for six years from 1991 to 1996. It includes incomplete records and they were omitted from analyses. The locations of epicenters of 20 earthquakes which were used for the analyses are illustrated in figure 3 and the earthquake parameters are summarized in table 1. The observed peak accelerations in the EW direction on the free surface have been less than about 30 gals except the event of No. 22 as shown in table 1.

Table 1. Earthquake Parameters

Eq.No	Date	Time	Epicenter (N:N.L., E:E.L.)	Depth (km)	Mag. (M)	Max. of Acc (gal) (EW)	Group
1	1991. 6.25	12:48	N36.64, E140.97	49	5.1	31.96	C
2	8. 6	23:49	N35.87, E141.16	26	5.8	8.92	A
3	10.19	8:31	N36.08, E139.92	59	4.3	31.92	C
4	11.19	17:19	N35.60, E140.02	81	4.9	12.60	C
5	12.12	11:22	N36.46, E140.66	48	4.6	16.21	C
6	1992. 2. 2	4:04	N35.23, E139.79	92	5.9	18.71	B
9	5.11	19:04	N36.53, E140.54	56	5.6	29.28	B
10	6. 1	22:51	N36.67, E141.27	44	5.7	20.48	B
11	8.30	4:19	N33.20, E138.34	325	6.6	16.30	B
14	1993. 6. 7	16:47	N36.02, E141.76	28	5.9	9.32	B
16	9.18	11:18	N36.18, E140.88	35	5.0	16.21	B
17	10.12	0:55	N32.02, E138.24	390	7.0	24.96	B
21	1995. 1. 7	7:39	N40.18, E142.32	30	6.9	7.35	A
22	1. 7	21:34	N36.17, E139.59	70	5.4	100.47	B
23	1. 8	4:38	N36.19, E139.58	72	4.6	17.01	C
24	1.10	3:00	N35.56, E141.26	45	6.2	9.90	A
27	4.12	14:20	N36.27, E140.37	52	4.6	13.23	B
28	7. 3	8:53	N35.06, E139.30	120	5.6	5.80	A
29	7.30	3:11	N35.54, E140.36	50	5.0	20.90	B
30	1996. 9.11	11:37	N35.07, E141.03	30	6.6	26.52	A

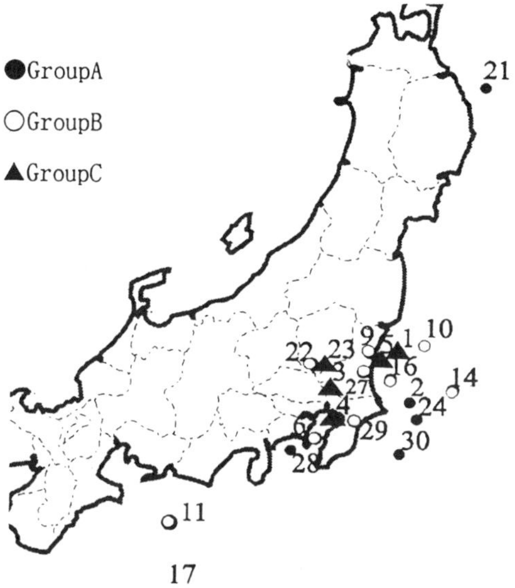

Fig. 3 Location of Epicenters of Earthquakes

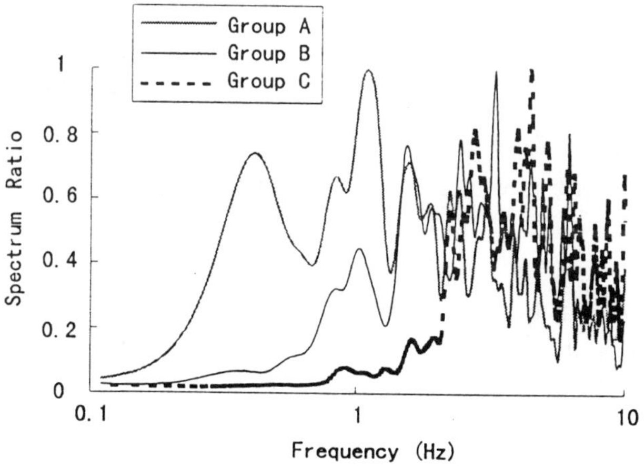

Fig. 4 Fourier Spectra of Accelerograms Chosen Each Group

2.3 GROUPING OF EARTHQUAKE MOTIONS

In order to examine the effects of frequency component of earthquake ground motions on the earth pressures, all records were categorized into three groups (groups A, B and C) according to the frequency components of accelerograms recorded on the soil surface. Figure 4 shows Fourier spectra of the representative motions chosen from each group. The spectra shown in figure 4 were smoothed by use of Parzen window and normalized by the maximum values. As seen from this ,iigure, group A is characterized by a spectrum with predominant frequencies less than 1 Hz, and groups B and C, on the other hand, include less lower frequencies and more higher frequencies. In table 1 the categorized group symbols are listed for each earthquake.

3. Characteristics of Observed Earth Pressure

3.1 PHASE CHARACTERISTICS OF EARTH PRESSURES

Phase characteristics of earth pressures on both sides of the foundation can be extracted by calculating motion products of the earth pressures (product of two motions at every instance). T'he computed results are shown in figures 5(a), (b) and (c) for the representative motions of each group. The positive values of the motion product can be understood as the earth pressures on both sides to be in phase, and, on the contrary, the negative values imply that the two motions are out of phase 180 degrees. Inspection of the results shown in figure 5 (a) indicates that for the motions belonging to group A earth pressures on both sides are almost in phase throughout the record. The phenomenon of in phase cannot be explained by an analysis assuming the vertical incidence of seismic waves [2].

As for group B shown in figure 5(b), the earth pressures on both sides of foundation are showing opposite phase in primary portion of the motions and tend to shift into in phase with time. With respect to motions of group C which contain higher frequencies, the opposite and in phase of earth pressures appear alternately throughout the whole motions as shown in figure 5(c).

Summarizing the results shown in figure 5, it may be said that the phase characteristics of earth pressures between both sides of the foundation are strongly affected by frequency component included in the ground motions and tend to be in phase for the motions with lower frequencies.

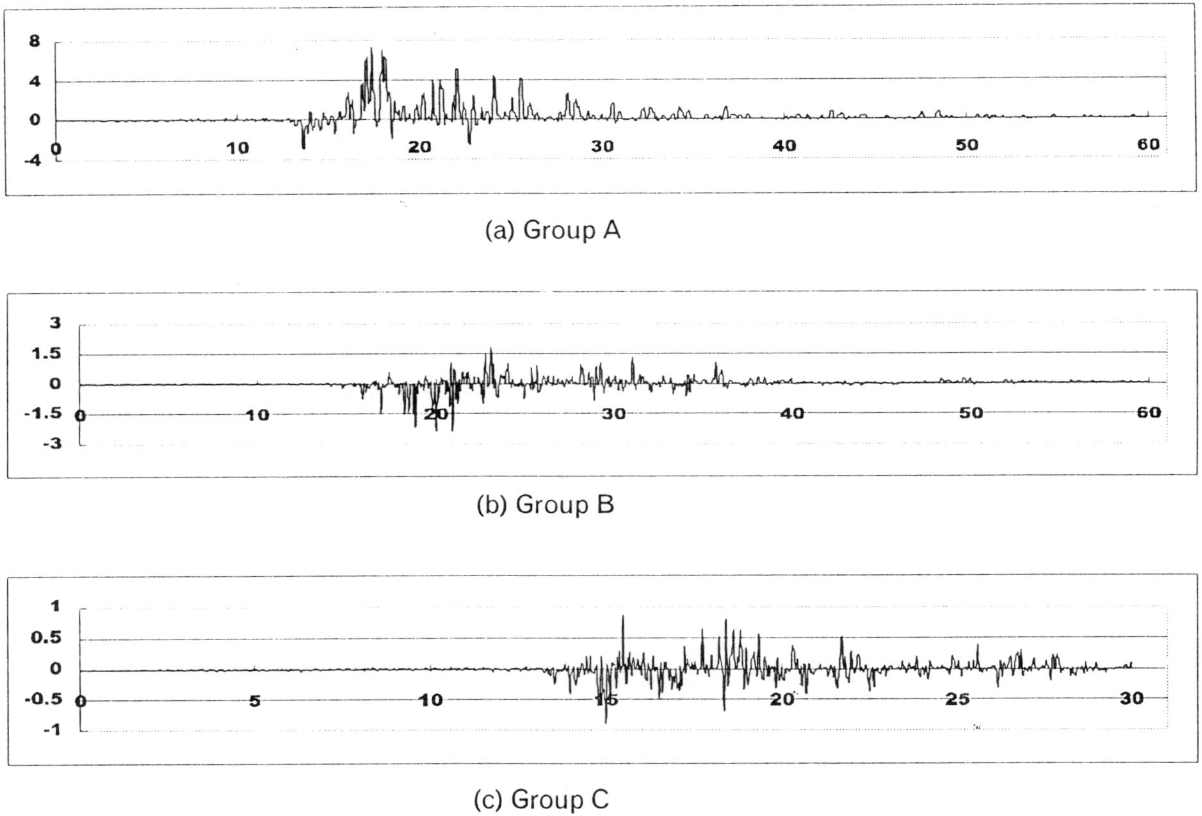

Fig. 5 Motion Products of Earth Pressures on Both Sides of Foundation

3.2 FOUNDATION MOTIONS AND EARTH PRESSURES

In order to study what motions of the foundation is related with the earth pressures, the displacement, velocity and acceleration motions of the foundation were compared with the records of earth pressures. The displacement and velocity motions of the foundation were evaluated by a numerical integration of acceleration motions observed at the point S3. It is noted that the computed velocity motions have shown almost the same velocity motions recorded on the foundation. In figures 6(a), (b) and (c), the velocity motions of the foundation are compared with those of earth pressures, which have shown, among three, the strongest relation with earth pressures. The waveforms shown in figure 6 are normalized by corresponding maximum value.

In summary of this section, it was found on the basis of observations that the earth pressures on both sides of the foundation are strongly related with the horizontal velocity motions of the foundation.

3.3 MAXIMUM GROUND MOTIONS AND EARTH PRESSURES

It is important to estimate the peaks of earth pressures when subjected to earthquake excitations. Figures 7(a) and (b) show the relationships between the peak ground accelerations on the soil surface and maximum eartti pressures at points WC-4 and EC-7. The results are plotted with different marks for each group. It will be noticed that the correlation between the peak accelerograms and peak earth pressures can be clearly captured by categorizing the recordings into three groups. The slope for each group was drawn by means of the least-squares method. It is noticed that the slopes clearly differ with different groups, and tend to increase in order of groups C, B and A. This fact suggests that the earth pressures tends to increase for the ground motions with increasing of lower frequencies. Taking an example for the earth pressures at point WC-4, the relationship between the maximum earth pressures $p_{o\,max}$ (kPa) and the peak surface accelerations $\ddot{z}_{max}$ (gal) can be approximated by following equations;

Group A ..... $p_{omax} \approx 0.22\, \ddot{z}_{max}$ (1-a)

Group B .....$p_{omax} \approx 0.085\, \ddot{z}_{max}$ (1-b)

Group C .....$p_{o\,max} \approx 0.046\, \ddot{z}_{max}$ (1-c)

Similarly, figure 8 indicates the relationship between the maximum ground velocities on the soil surface and the peak values of earth pressures at point WC-4. It mav be noticed from the figure that the relationship between peak ground velocities and maximum values of earth pressures almost identical regardless of groups. This fact apparently indicates that the earth pressures are less frequency dependent on velocity ground motions. An approximate expression correlating the maximum earth pressures $P_{o\,max}$ (kPa) with the peak surface velocities $\dot{z}_{max}$ (cm/sec) for a point WC-4 can be written as,

$$p_{omax} \approx 1.8\, \dot{z}_{max} \quad (2)$$

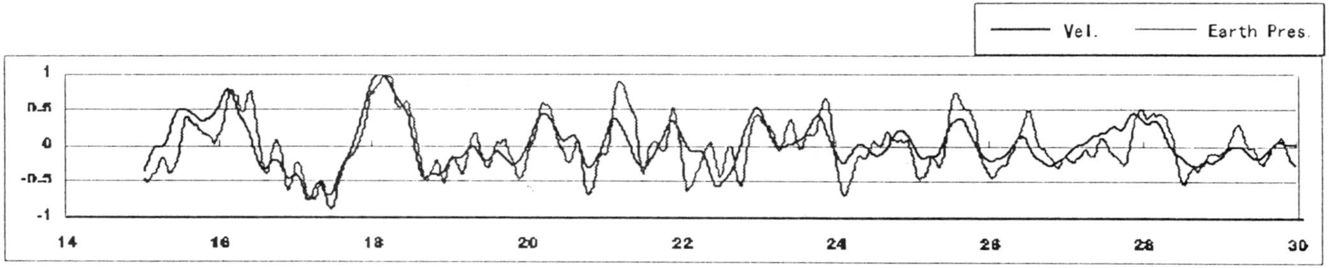

(a) Group A (EC-7)

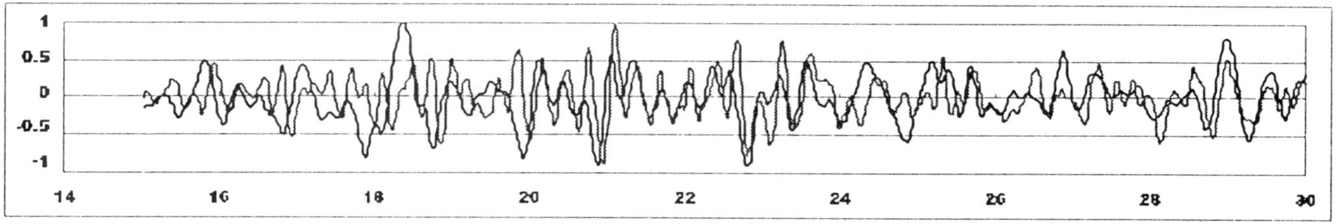

(b) Group B (EC-4)

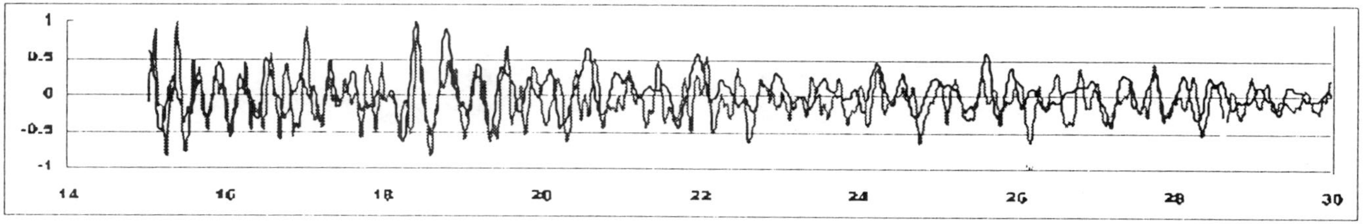

(c) Group C (EC-7)

Fig. 6 Comparison between Velocity Motions of Foundation and Earth Pressures

The coefficients of above expressions might be subjected to modifications with change of the observation points and of the soil properties around foundation, and therefore it is needed to be careful in applying the equations to others. More general expressions for estimation of the maximum earth pressures are left for future studies.

4. Simplified Method to Predict Relationship between Velocity Motions of Foundation and Earth Pressures

4.1 SIMPLIFIED SOIL MODEL

As stated in the section 3.2, the earth pressures on the lateral sides of the foundation are strongly related with the horizontal velocity motions of the foundation. We try in this section to explain the fact with a simplified method and introducing an assumption that a one-dimensional soil rod can be used to estimate lateral soil resistant against horizontal motions of the foundation.

Figure 9(a) and (b) show a foundation-soil model and a soil rod extracted from the lateral soil surrounding the foundation. It is assumed here that the soil rod is extending infmitely in the lateral direction and having a uniform section and homogeneous soil constants. It is also assumed that the rod is free from constraint along sides of the rod.

The equation of motion with respect to an axial displacement of the rod, $u(x, t)$, can be expressed by a well-known wave equation as follows;

$$\frac{\delta^2 u}{\delta x^2} = \frac{\rho\, \delta^2 u}{E \delta t^2} \qquad (3)$$

where x is the axial coordinate, and ρ, E are mass density and Young's modulus of the soil rod, respectively. Considering only a wave travelling in the direction of x-axis, the solution of equation (3) can be expressed as follows,

$$u(x, t) = f\,(t - x/c\,) \qquad (4)$$

where $f(\ )$ is an arbitrary function of the argument and $c = \sqrt{E / \rho}$, which represents a longitudinal wave velocity of the soil rod.

Supposing the end of the rod is subjected to longitudinal excitation, we will have following equation at the boundary $x = 0$.

$$\left. \frac{E\, \delta u}{\delta x} \right|_{x=0} = -p_0\,(t) \qquad (5)$$

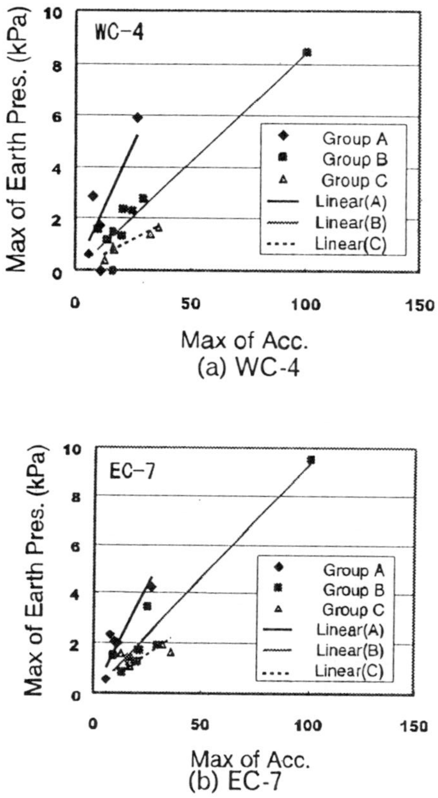

Fig. 7. Peak Ground Acceleration vs. Max of Earth Pressures

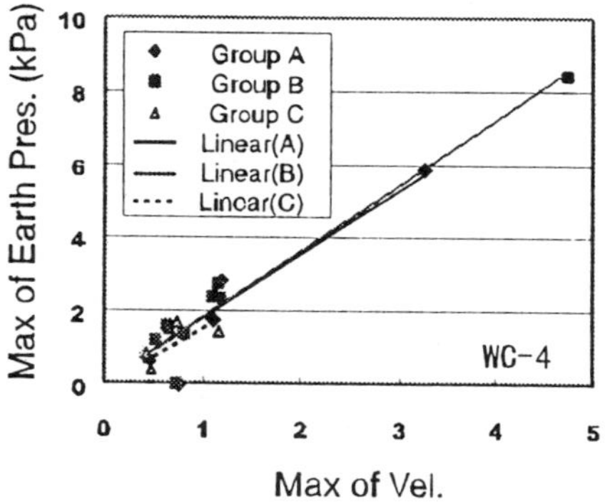

Fig. 8 Peak Ground Velocities vs. Max of Earth Pressures (WC-4)

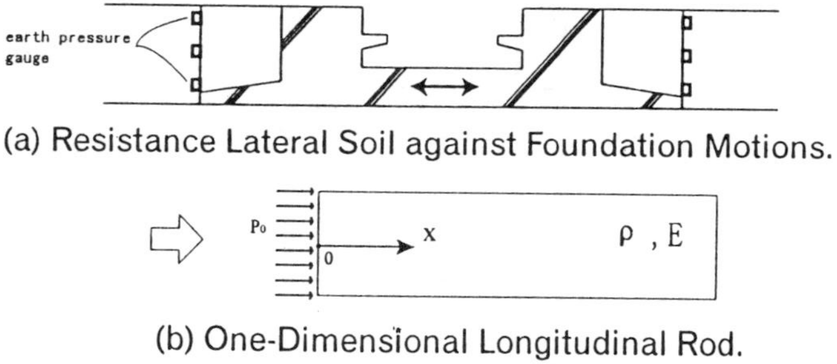

Fig. 9 Simplified Foundation-Soil Model

where $p_o(t)$ is an excitation per unit area. Applying the forward difference formulas for $\delta u / \delta x|_{x=0}$ in equation (5), we obtain an approximate expression for $p_o(t)$ as shown below.

$$p_o(t) \approx E \frac{u_0 - u_1}{\Delta x} \tag{6}$$

where $u_o(t)$ corresponds to lateral displacement at the end of rod and $u_1(t)$ is displacement in the soil rod at a small distance Δx from the end. Equation (6) suggests the validity of a method to estimate the earth pressure based on a relative displacement between the foundation and the soil. Equation (1) is one possibility to estimate the earth pressure. Another form to estimate the earth pressure can be obtained from equation (5), by making use of a relation $\delta u / \delta x|_{x=0} = -\frac{1}{c} \dot{u}|_{x=0}$
and substitution from the relation into equation (5), and we obtain,

$$p_o(t) = \frac{E}{c} v_o(t) \tag{7}$$

in which $v_o = \dot{u}|_{x=0}$ corresponds to velocity motions at the end of the soil rod. The $p_o(t)$ in equation (7) corresponds to the earth pressure acting on sides of the foundation, and therefore the equation indicates that the earth pressure is proportional to the velocity motion of foundation.

4.2 NUMERICAL EXAMINATION

In order to validate equation (7) numerically, we introduce soil constants for this equation. As for the value of a longitudinal wave velocity c, it might be considered to use the Lysmer's analog velocity V_{La}, which has been used successfully to estimate soil reaction against piles. The Lysmer's analog velocity is defmed as follows [6].

$$C = V_{La} = \frac{2.4}{\pi(1-\nu)} V_s \tag{8}$$

where ν and V_s are the Poisson's ratio and shear wave velocity of a soil. Furthermore, the corresponding Young's modulus E_a may be evaluated by

$$E_a = \rho V_{La}^2 \tag{9}$$

The value E_a is to be used for the Young's modulus of the soil rod E. Thus, the ratio between the peak values of velocity motions of the foundation and earth pressures (p_{omax} / v_{omax}) may be expressed using new symbols as follows;

$$(p_{omax} / v_{omax}) = E_a / V_{La} \tag{10}$$

In equation (10), it is assumed that the peak values occur simultaneously. The observed peak ratios shown in the left side of above equation were compared with the results of $E_a\ /V_{La}$ evaluated by use of the measured soil constants. The lateral soil of the foundation is composed of two layers and the shear wave velocities (V_s), the Poisson's ratios (ν) and soil densities (γ) have been obtained as shown in table 2. This table also shows the compared results between the computed values of $E_a\ /\ V_{La}$, for each layer and the ratios of observed peak values $p_{omax}\ /\ v_{omax}$ for each group. Inspection of the results shown in table 2 reveals that the ratios of $p_{omax}\ /\ v_{omax}$ tend to increase in order of group A, B and C. The fact is indicating that the ratios are frequency dependent. It will be also noticed that though the ratios of observed peak values differ depending on the group and with varying the observation points the computed results of $E_a\ /\ V_{La}$ based on the actual soil constants correspond almost to the ratios of observed peak values $p_{omax}\ /\ v_{omax}$. This fact perhaps suggests the validity of the hypothesis that the lateral soil of the foundation is approximately reacting as one-dimensional uniform soil rod. And, it should be noted that a part of dynamic characteristics of the earth pressures on the foundation may be explained roughly by use of a simplified model considered above.

Table2: Surface soil constants and Ea/V_{La} vs. $P_{omax}\ /\ V_{omax}$

Surface Soils	Soil const.				Value of $E_a\ /\ V_{La}$			Soil press.Per velocity ($P_{omax}\ /\ V_{omax}$)		
	Vs	ν	γ	Thick	V_{La}	Ea	Ea/V_{La}	GroupA	GroupB	GroupC
1	82	0.497	1.3	3.0	176	4120	23.4	5 ~ 21	10 ~ 38	19 ~ 52
2	145	0.492	1.6	5.2	309	15560	50.4			
	m/s		tf/m^3	m	m/s	tf/m^2	gf/m^2/cm/s	gf/m^2/cm/s		

5. Conclusions

In this paper, characteristics of dynamic earth pressures were examined on the basis of observations for about 30 events recorded on the lateral sides of an embedded foundation. It was revealed that the lateral earth pressures were strongly related to the horizontal velocity motions of the foundation, and the phase characteristics of lateral earth pressures on both sides of the foundation were inlluenced greatly by frequency components of earthquake motions. It was shown that a part of these characteristics could be approximately explained by using a simplified one-dimensional soil model.

6. Acknowledgement

The authors wold like to express their gratitude to Ms. Aya Shigetoh and Mr. Takayoshi Mutoh for their help in preparation of this paper.

7. References

1. Onimaru, S., Sugawara, R., Uetake, T., Sugimoto, M. and Ohmiya, Y (1994) Dynamic characteristics of earth pressure on seismic array observation (in Japanese), Proc. 9'" Japan Erthq. Engng, Symp., pp1051-1056.
2. Uchiyama, S. and Yamashita, T. (1999) Experimental and analytical studies of lateral seismic earth pressure on a deeply embedded building model (in Japanese), J. of truct. and Constr. Engng., Trans. ofAIJ, No. 516, pp105-112.
3. Kazama, M. and Inatomi, T. (1988) A study on seismic stability of large scale embedded rigid structures, Proc. 9'" WCEE (Tokyo-Kyoto), Vol. III, III-629-III-634.
4. Watanabe, K., Maeda, T., Kobayashi, Y and Towhata, I. (1998) Shaking table tests on seismic earth pressure exerted on retaining wall model, Proc. the 3rd Symp. on Urban Earthq. Hazard to Near-Field Earthquakes, National Science Fund of Ministry of Educa- tion of Japan, pp219-222.
5. Minowa, C., Ohyagi, N., Ogawa, N., and Ohtani, K. (1991) Response data of large shaking table foundation during improvement works (in Japanese), Tech. Note of the Na- tional Research Institute for Earth Science and Disaster Prevention, No.151.
6. Gazetas, Cz and Dobry, R. (1984) Horizontal response of piles in layered soils, J. Geotech. Engng, ASCE, Vol. 110, No. 1, pp20-40.

SEISMIC INSTRUMENTATION OF THE I-40 MISSISSIPPI RIVER BRIDGE IN MEMPHIS, TENNESSEE

SHAHRAM PEZESHK [1], MEHMET CELEBI [2], GREG STEINER [3], CHARLES V. CAMP [1], HOWARD HWANG [3]
[1] *The University of Memphis, Department of Civil Engineering The University of Memphis, Memphis, TN 38152*
[2] *U.S. Geological Survey, USGS (MS977), 345 Middlefield Road Menlo Park, Ca. 94025*
[3] *The University of Memphis, Center for Earthquake Research and Information, The University of Memphis, Memphis, TN 38152*
e-mail: spezeshk@memphis.edu

1. Background

The purpose of this paper is to describe installation of a seismic instrumentation system which is proposed to be deployed on and in the vicinity of the I-40 Hernando DeSoto Mississippi River Bridge in Memphis, Tennessee. This bridge is planned to be retrofitted to withstand a magnitude (mb) 7 event at 65 km distance from the site with a depth of 20 km. The goal of the retrofit is to have this bridge fully operational following the maximum probable earthquake (2500 year return period). As part of the I-40 bridge retrofit, Friction Pendulum™ Isolation Bearings will be used to insure the integrity of the main spans of the bridge.The I-40 bridge is scheduled to be retrofitted in several phases. The first phase of the retrofit is scheduled to begin in year 2000. Therefore, there is a window of opportunity to install an integrated system of seismic strong-motion instruments on and around the vicinity of the bridge during the retrofit construction phase. A strong-motion instrumentation with 65 data channels at 25 different locations on the bridge and at the free-field in the vicinity of the bridge is proposed to be installed. This will provide needed data to better understand the ground motion and ground failure and the response of the retrofitted bridge in a strong earthquake. In addition, the instrumentation will strengthen the free-field strong-motion network in the metropolitan Memphis area.

Currently, in the United States and elsewhere in the world, there is very little data available on the response of long-span bridges during seismic events. Since such data are scarce, our ability to understand the behavior of such structures and to verify dynamic analyses of such structures during design/analyses/retrofit phases is limited. Currently, there are no long-span bridge structures instrumented in the New Madrid seismic zone. Therefore, data collected from instrumentation of the I-40 bridge will be an invaluable asset in evaluating the structure. The

M. Erdik et al. (eds.), Strong Motion Instrumentation for Civil Engineering Structures, 533–544.

data will be used to assess the performance of the bridge following the retrofit and in particular for the assessment of the performance of the base-isolation system used in the retrofit project. The proposed I-40 bridge strong-motion instrumentation system with 65 data channels at 25 different locations will provide a sufficient number of sensors to reconstruct the behavior of the structure in sufficient detail to verify the response predicted by mathematical models. Using data collected from smaller earthquakes an improved mathematical model of the bridge can be developed. Furthermore, a well-instrumented structure for which a complete set of recordings has been obtained, should provide useful information to:

(1) check the appropriateness of the dynamic model in the elastic range,
(2) determine the importance of nonlinear behavior on the overall and local response of the structure,
(3) follow the spreading nonlinear behavior throughout the structure as the response increases and determine the effect of this nonlinear behavior on frequency and damping,
(4) correlate the damage with inelastic behavior models,
(5) determine the ground-motion parameters that correlate well with bridge response and/or damage,
(6) quantify the interaction of soil and structure (this is particularly important for the I-40 bridge which is located on 3000 feet of soil above bedrock), and
(7) make recommendations to improve seismic codes and/or future bridge designs.

An improved model of the I-40 bridge can be used to predict potential damage/failure that the structure may experience during larger seismic events. An accurate bridge model will be a cost-effective approach in evaluating the retrofit scheme, investigating ways to improve bridge performance, and reducing the possibility of catastrophic failure during a large seismic event. A good example is the data recorded from the Vincent-Thomas Bridge in Los Angeles Harbor. Figure 1 shows the vertical acceleration record obtained from center of side span of the bridge during the Northridge earthquake. This location recorded the largest peak acceleration. The amplitude spectra for two 40-second bands (20-60 second and 80-120 second) exhibit two distinctive frequencies suggesting that some form of deformation occurred in the mid-span to alter its frequency (Celebi, paper in preparation). As a result of studies of the bridge, including those that utilized this record, deficiencies in the bridge were realized and a retrofit program was developed that will eliminate possible failure during a larger event.

The I-40 bridge instrumentation, when complete, will be used to develop a rapid warning system. The system will be triggered when certain acceleration thresholds at designated locations on the bridge are exceeded. Key officials at the Tennessee Department of Transportation will be automatically notified of the exceedance of a threshold earthquake within minutes over standard communication devices. This information will help determine the forces that the bridge underwent in a large earthquake and should aid engineers in determining if the bridge should be closed or/and what precautions rescuers and engineers should take. This type of monitoring and the associated rapid response may save lives.

The instrumentation system will be implemented in stages to assess the behavior and performance of the bridge before, during, and after the retrofit. To this end, the first step will be deployment of free-field stations on each side of the bridge. The motions recorded from free-field stations will be used as input to mathematical models of the bridge.

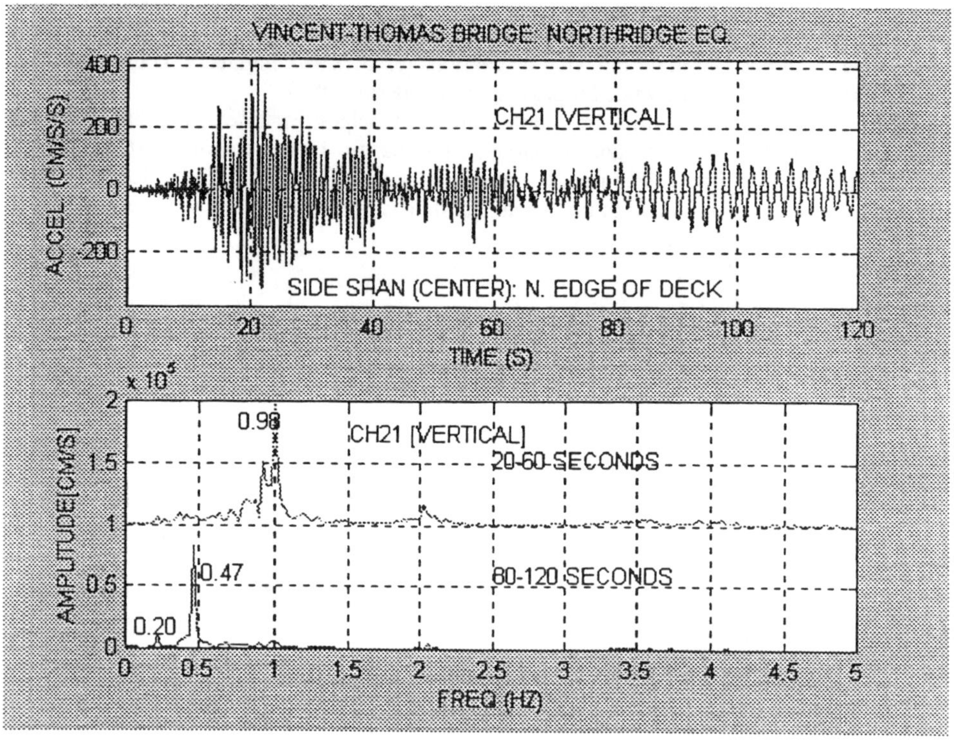

Figure 1. The recorded acceleration time-history of vertical motions at center of side-span exhibited the largest peak acceleration during the 1994 Northridge earthquake. This record also shows that the frequency of the bridge changed drastically (to 0.47 Hz [~ 2 sec]) during the last 40 seconds of the record as compared to the 20-60 second band (0.98Hz [~ 1 sec]).

2. Project Plan

In this project, we will install 65 sensors on the I-40 Hernando Desoto Mississippi River Bridge in Memphis, Tennessee to fully characterize the response of the bridge to strong ground motion. The location of the I-40 bridge is shown in Figure 2. Figures 3 and 4 show the bridge's main two-span tied arch bridge. The general location and sensing direction of the sensors are marked in Figures 3 and 4. The final locations of the sensors may be adjusted based on field investigations. Figure 5 shows the Arkansas-side approach spans, while Figure 6 shows the east main approach structure and connecting ramps on the Memphis side. Figures 5 and 6 also show the general location and sensing direction of the sensors on the approach spans. The information collected from the sensors will include: (1) free-field ground motion near the instrumented bridge; (2) motion of the bridge foundation; (3) motion of the bridge below the isolation bearings; (4) motion of the bridge above the isolation bearings; (5) the spatial variation of ground motion along the total span; and (6) the lateral and torsional motion of the bridge.

2.1 MEASUREMENT OF FREE-FIELD MOTION

The free-field ground motion is the basic required information for a seismic response analysis of the bridge. It can be measured by a triaxial accelerometer located on Mud Island near the main span of the bridge (see Figure 4). These sensors record the ground motion in two horizontal directions (sensors 44 and 45) and a vertical direction (sensor 46). Placing the free-field sensors on Mud Island will help avoid any effects induced by structural response (soil-structure interaction) on the recorded ground motion; thus, the free-field sensors should not be placed too close to the instrumented bridge.

2.2 MEASUREMENT OF BRIDGE FOUNDATION MOTION

The motion at the bridge foundation can be measured by a triaxial accelerometer placed on the pile cap (sensors 41, 42, and 43 indicated in Figure 3). These sensors will record the foundation motion, including the effect of soil-structure interaction, in two horizontal directions and a vertical direction. An estimate of the soil-structure interaction effect may be obtained by comparing the motion recorded at the bridge foundation to the free-field motion.

2.3 MEASUREMENT OF BRIDGE MOTION

The bridge motion can be measured by sensors 1 through 40 in transverse, longitudinal, and vertical directions as shown in Figure 3. These sensors will record the motion of bridge piers above and below their isolation bearings and of the mid-span of each the main spans. Sensors 47 through 65 will measure transverse, longitudinal, and vertical motions of approach spans. Using data recorded by these sensors, we can establish the dynamic characteristics of the bridge, such as vibrational mode shapes, structural periods, and main span deflections in the longitudinal, the transverse, and the vertical direction. The torsional response will be estimated from the motion recorded by pairs of sensors placed on opposite sides of the bridge deck. Furthermore, the effect of base-isolation will be estimated by comparing the motion recorded above and below the isolation bearings.

2.4 INSTRUMENTATION

The Desoto bridge strong motion monitoring array consists of a large number of sensors installed over the full length of the bridge and its approaches (1.8Km). Installation economics and reliability concerns dictate that the array be broken down into smaller standalone sub-array systems; while recorder economics favor a single common recording and archiving facility. A synergistic relationship can be exploited by upgrading an existing strong motion recording facility at the nearby Autozone building on the Memphis river front.

Sensors will be wide temperature range accelerometers with direct to digital outputs yielding a wide dynamic range with 16 bit resolution and 2G full scale sensitivity. Sensor bandwidth will be from 0.1 to at least 50 Hz. All sensors and associated electronics will be housed in NEMA 4X style enclosures, which will be configured to facilitate mounting and leveling on varied bridge surfaces. These enclosures also resist the extreme environmental conditions found in and around the bridge. Multilevel transient suppression will be used on all signal and power cables because of the frequent severe thunderstorms, which plague this region. UV resistant, flooded, direct burial, shielded telephone cable will be used to connect the sensors to the telemetry system nodes thus minimizing installation costs while maintaining high levels of signal integrity and reliability.

Six to twelve sensors will be connected and concentrated at each radio telemetry system node, thus necessitating approximately 11 radio telemetry systems to fully monitor the bridge. Because of the short radio path length (2Km), low power radio equipment and small low gain antennas can be used to implement the telemetry portion, which will also facilitate the use of solar derived battery backed up power systems. Solar power also minimizes failures from power line induced transients, while providing power during power outages that are likely to occur during seismic events. Each telemetry node will be totally independent of the others resulting in a greater degree of total system reliability.

The recording site for the bridge system will be co-located with an existing strong motion array located in the only, seismically resistant constructed, building in downtown Memphis. Radio signals originating from each telemetry node on the bridge will be received on the roof of the building and travel on existing CAT 5 twisted pair cables to the recording room located on the first floor of the building. These signals will be combined in the data acquisition system with those generated by the internal building-monitoring array already installed. A new USGS Earthworm PC based recording system will be installed as part of the upgrades to this facility as well as a newer GPS based timing system. Data will be recorded and archived at the site for redundancy while simultaneously being telemetered over a high speed (2Mb) microwave Ethernet directly to CERI for real time processing and analysis. UPS and Extended generator back up power are already provided by the building owners as well as high reliability building infrastructure services.

Long term maintenance of this system will be facilitated by a real time calibration system which will exercise each sensor on a weekly (or daily) basis in order to assess total acquisition system performance. Sufficient spare components will be purchased initially to support operation and maintenance of the system for its proposed 25 year life span. The only renewable items required for maintenance of the system will be backup batteries, which have a limited life cycle and solar panels, which may be subject to human vandalism. Operational expenses will arise from recording and archiving media and computer hardware and software system upgrades and replacements.

This system will provide valuable data on engineering performance of the bridge during a felt earthquake. In addition, this system will be able to provide immediate data on the dynamic characteristics of the bridge excited by high winds. Both kinds of information are significant in evaluating the possible applications of bridge design and retrofit concepts in the Eastern United States. In addition, the I-40 bridge instrumentation will provide a downtown Memphis free–field strong-motion station, which will strengthen the four-station Memphis–Shelby County network. In view of the scarcity of seismic strong–motion instrumentation of bridges in this region, the I-40 bridge system may become the single most important strong-motion installation in the Eastern United States.

Data that can be collected from a felt earthquake will be invaluable in a realistic, quantitative assessment of earthquake hazards of this region. Valuable lessons have been learned from the study of data obtained from well-instrumented structures. For example, the Imperial County Buildings, during the moderate-sized Imperial Valley earthquake (*Ms* 6.5) of October 15, 1979 (Cassaro et al. 1987 and Rojahn and Mork, 1982) and many buildings in San Francisco, during the Loma Prieta of October 18, 1989 (Krawinker et al. 1991) provided invaluable data about seismicity and structural performance. The collected data can be used in future code modifications and improvement of existing seismic codes.

3. Acknowledgments

This work is partly funded by the Mid-America Earthquake Center.

4. References

1. Cassaro, M., Celebi, M. (Coordinator), W. Durbin, P. Gould, W. Grahm, A. Johnston, H. Karabinis, A. Lin, R. Maley, O. Nuttli, G. Schwable, J. Theiss, M. Walsh, T. Winstead. (1987). "Report on Recommended List of Structures for Seismic Instrumentation in San Bernardino County, California; USGS Open-File, Report 85-583.
2. Krawinkler, H., A. Nassar, and M. Rahnama. (1991). "Damage Potential of Loma Prieta Ground Motions." Bulletin of the Seismological Society of America, Vol. 81, No. 5, pp. 2048-2069.
3. Rojahn, C. and P.N. Mork. (1982). "An Analysis of strong-motion Data from a Severely Damaged Structure - The Imperial County Services, El Centro, California, in The Imperial Valley, California, Earthquake October 15, 1979."USGS Professional Paper. 1254.

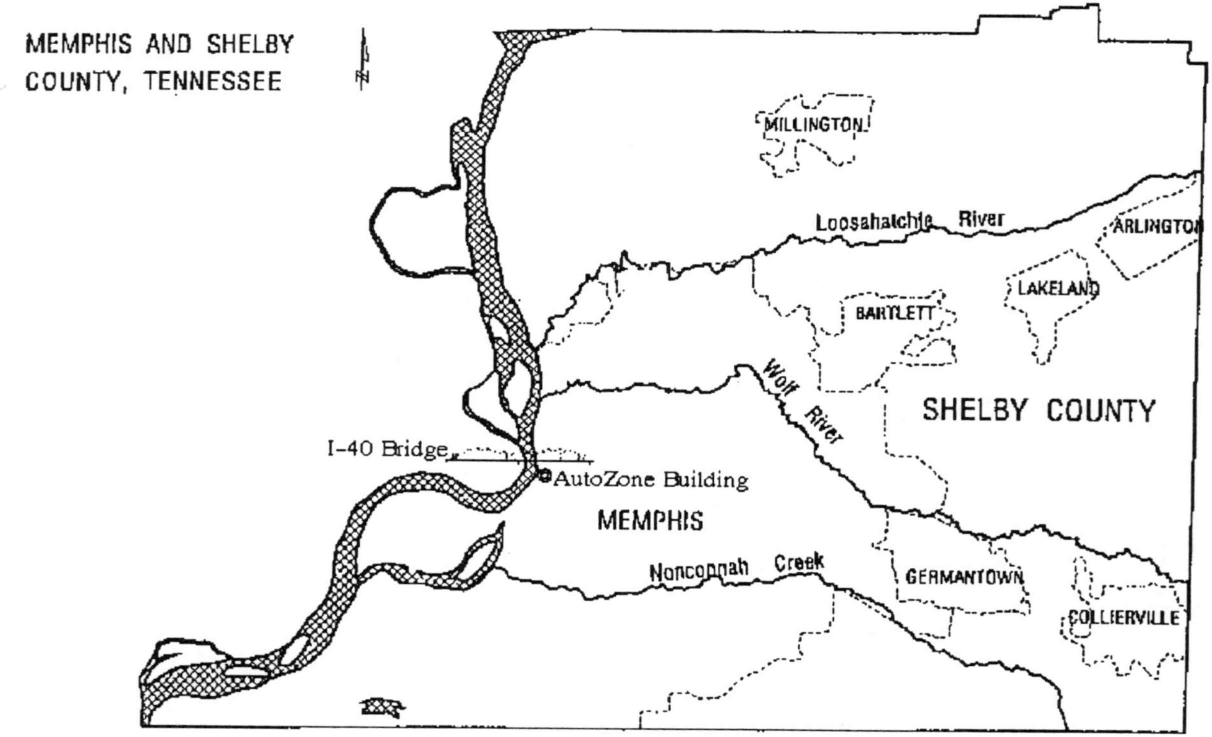

Figure 2. Location of the I-40 Hernando Desoto Mississippi River Bridge.

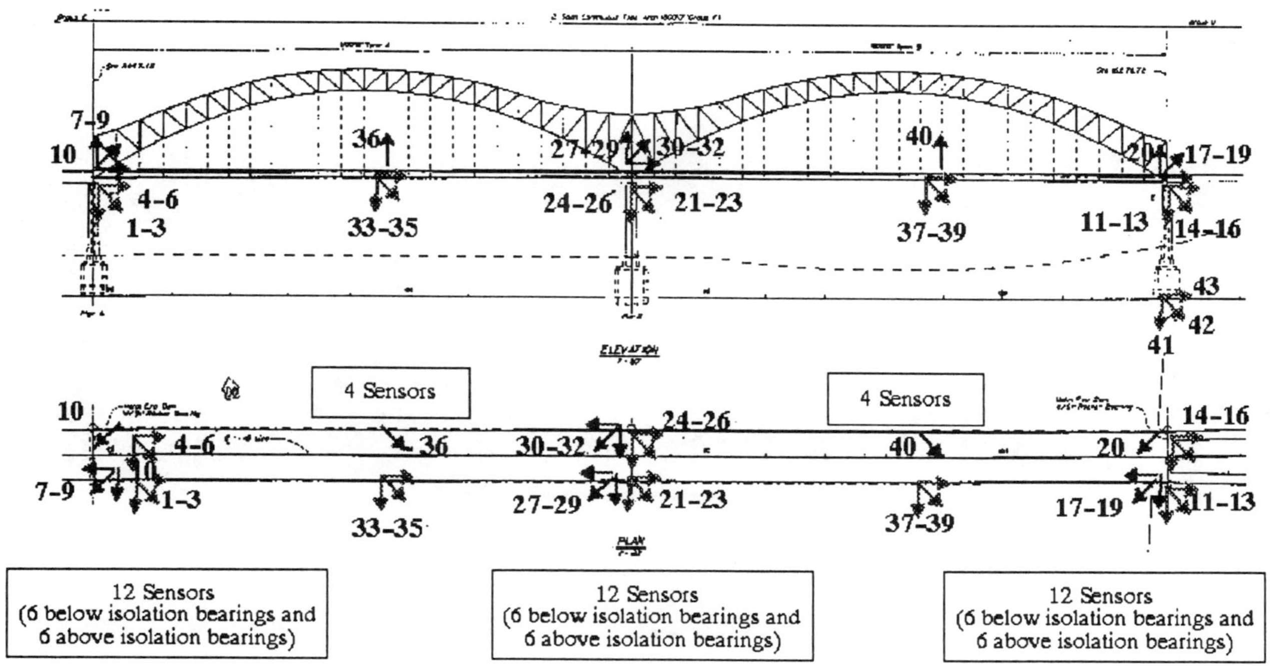

Figure 3. Main Two-Span Tied Arch of the I-40 Bridge and Sensor Locations.

Figure 4. Mud Island and the Location of the Free -Field Sensors.

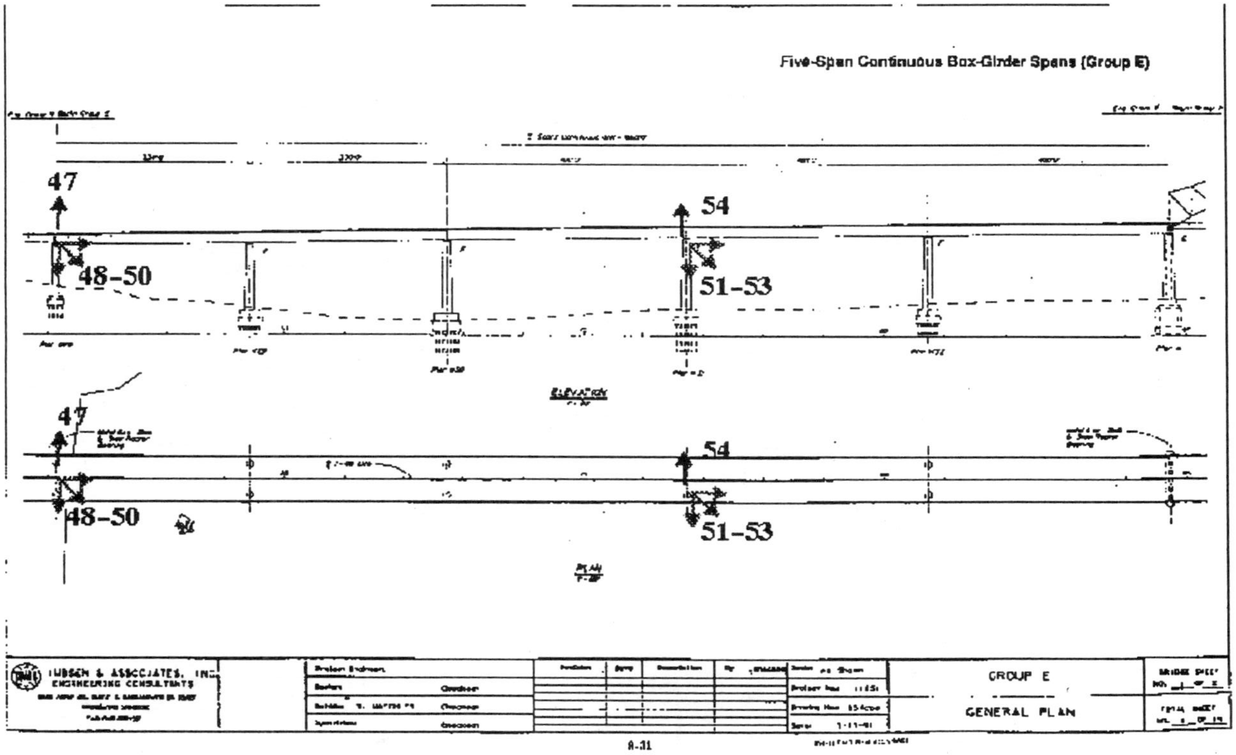

Figure 5. West Approach to the I-40 Bridge.

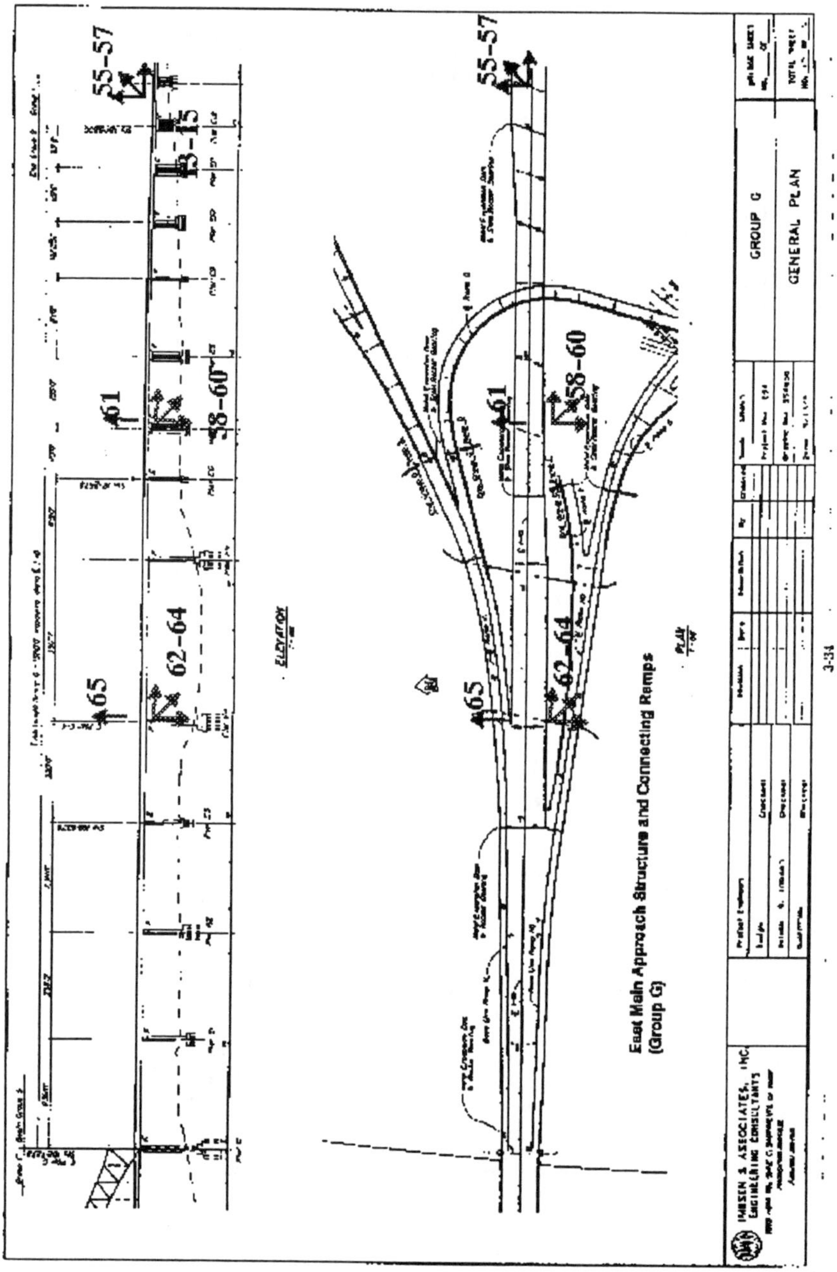

Figure 6. East Approach to the I-40 Bridge.

ENGINEERING AND SEISMOMETRIC SERVICE IN THE BUILDINGS OF TASHKENT AND TASHKENT REGIONS

T.RASHIDOV
Institute of Mechanics and Seismic Stability of Structures of the Uzbekistan Academy of Sciences. Akademgorodok, 700143 Tashkent, Uzbekistan
e-mail: iskan@seismo.com.uz

Abstract

Tashkent city is located in the zone where I=8-9 (Mercalli intensity scale) and some faults are in the adjoining regions. Microseismozoning carried out with due regard for the soil conditions showed that there are areas on the city territory where I=7,8,9 and areas exposed to landslides and dilution of soil under earthquakes. The buildings constructed in Tashkent are of different constructive schemes. Stations of engineering and seismometric service (SESS) are located in the 20-storeyed panel administrative building, in the 9-storeyed dwelling house of a special construction, in the 4-storeyed brick building of the Hotel Tashkent, in the 5-storeyed brick buildings of research institutes, in the 17-storeyed building of the Hotel Uzbekistan. Seismodetectors are installed in ground and on different levels inside these constructions.

The stations record ground motions caused by earthquakes if I>4. Equipment of these stations was constructed in the former Soviet Union and is obsolete at present. Besides that tracking of underground communications and the Metro where SESS has been installed is also carried out. The Charvak reservoir whose water storage reach 2 milliard cubic metres is at 60 km' distance from Tashkent. Dam of the reservoir is in the vicinity of a seismic fault. If the dam whose altitude is 167 m is destroyed the Tashkent oasis will be flooded. Detectors have been incorporated into the dam and seismometric service was organised. At present equipment of the station is obsolete but is in a serviceable condition.

Ground motions recorded by these SESS were used for improvement of the map of microseismozoning, Constructive Standards and Regularities and for calculation of city buildings for seismic stability. Now organisation of common network for seismometric control in Tashkent and Tashkent regions, use of new-type machinery for registration of strong ground motion caused by earthquakes are actual.

1. Charackteristics Of Seismicity Of Tashkent Region

Tashkent is situated in the transition zone from Tien-Shan mountains to Turan plate, altitude 480 m above sea level. Several active faults crossed the territory of Tashkent city and its surrounding. The length of those faults is 60-220 km, width is 10-15 km and seismic potential is estimated within magnitudes 5.0-7.3. These faults can produce earthquakes in Tashkent with intensity I=8. According to the map of General seismic

M. Erdik et al. (eds.), Strong Motion Instrumentation for Civil Engineering Structures, 545–550.

zoning Tashkent is situated in the zone of intensity 8 (MSK scale). Ground conditions of Tashkent city territory are presented mainly by loess soils macroporous, dusty. By seismic properties this soils divided into 3 categories. More than 50% of the territory is covered with soils of the third category, which are characterised as weak, with high possibility of settlement and high level of underground water. Because of these conditions on the most part of the territory earthquakes producing intensity 8 lead to effect of intensity 9, which is taken into account on the seismic microzoning map of the city. Information about destructive earthquakes is available since 1494. Total number of earthquakes with magnitude 4.0 < M < 7.5 and macroseismic intensity 6-7 on Tashkent territory since 1868 is 14. Recurrence period of shocks with macroseismic effect for Tashkent: intensity 6 - 20, 7 - 50, 8 - 100 years. For the period 1981 to 1995 years on the territory of the city 186 earthquakes were registered of intensity 2 to 6. For the last 60 years twice (in 1946 and 1966) Tashkent suffered from earthquakes of intensity 8. In 1966 the epicentre was located within the territory of the city. Consequences of this earthquake are studied in details, materials collected, analysed and published [1].

2. Engineering And Seismometric Service In The Buildings

Financial difficulties of the last years do not allow modernising the SESS in Tashkent with new-type machinery. At present in the Tashkent buildings functionate only three SESS of six existing. These SESS are installed in the building of the Institute of Mechanics and Seismic Stability of Structures (IMSSS), in the 20-storeyed administrative building located on the Mustakillik square and in the building of the Hotel Uzbekistan (Table 1).

All SESS intended for strong ground motion registration and recording. Strong ground motions caused by the Kirghiz earthquake of August 19, 1992 (I=8 in the epicentral zone) are among the records obtained. Instrumental data from the SESS/2 show that intensity of the earthquake in Tashkent was I=6 [1].

The administrative building where the SESS/2 has been installed is situated on the site of possible seismicity I=9 (Mercalli intensity scale). The engineering and geological conditions of the cite are listed in the Table 1. Foundation of the building has been constructed in the form of a monolithic reinforced concrete plate 6-7 m high. Its dimensions are 50 x 45 m. Distance between columns in the surface part is 6.0 x 5.2 x 4.5 m. The monolithic stiffening diaphragms provide spatial stiffness of the reinforced concrete frame.

The station includes 9 measuring points, 5 of them for vertical axis to determine form and regularities of the building motion (Table 1). Twin observation points are organised on the ground (+3.3 m) and 19th technical (62.7 m) floors to register relative motion of the ground part of the building. All seismodetectors are fixed on the metal support and record longitudinal, transversal and vertical ground displacements and accelerations caused by earthquakes with I=4÷9 on the basis of galvanometric registration method. 15 records from 5 points of observation placed on the following levels: -12.3 m, +10.8 m, +30.6 m, +50.4 m, +67.2 m were chosen to analyse how the earthquake of August 19,

1992 affects the 20-stireyed administrative building. Duration of the records is 13 seconds, numbering was carried out with constant subinterval t=0.05 seconds. Motion of ground and overlap was used to determine dynamic parameters of the buildings (periods of damping forms and decrement). Autocorrelation function of the ground motion and covariational function between the ground motion and overlap are constructed by the correlation method. Weight function for the building is obtained from the Wiener-Hopf equation. Correlation spectral properties of the weight functions are studied to determine some dynamic parameters of the building.

Table 1. Characteristics of SESS in the buildings of Tashkent city.

Number of Station	Construction of building and dimensions, m	Geological section	Observation points	Parameters of registration
1	5-storeyed brick building of IMSSS 45.4 x 15.8 x 18.8	Ground water level - 1.7 - 2.5; Subsidence loess-like loam - up to 2 m; Not subsidence ground sand loam - 15-16 m Below - thick layer of gravel	On the ground in bunker 35 m distance; Ground floor +0.00 m; 2nd floor +7.20 m; 3rd floor +10.80 m; Garret +18.80 m.	Displacements, accelerations.
2	20-storeyed administ-rative building of frame-work constructive system with stiffening diaphragms 46.0 x 40.0 x 67.2 depth of the underground part - 12.3 m	Ground water level - 14.0 - 17.0 m; Fill-up soil - 1.5 m; Sand loam - 7.5 m; Loam - 17.0 m; Stone loess - 30.0 m	Lower level of basement - 12.30 m; 3rd floor +10.8 m; 9th floor +30.60 m; 15th floor +50.40 m; Roof +67.2 m.	Displacements, accelerations.
3	17-storeyed building of the Hotel Uzbekistan of framework metal construc-tive system consisting of central part, right and left wings and underground part 8.60 m deep	Ground water level - 14-17 m; Fill-up soil - 2.5 m; Loess-like loam - 27.5 m; Gravel - 50.0 m; Below - stone loess +25.80 m; Roof of the central part, right and left Wings +53.70 m.	Lower level of basement of the central part, right and left wings - 8.60 m; Eighth floor of the central part	Displacements, accelerations.

The spectrum of the weight functions contains information about period of the motion and the autocorrelation function - about decrement of the motion. In the present case the periods of four forms of motion: T = 1,1; 0,65; 0,5; 0,35sec, and decrement of motion of the building equal to δ = 0,32; 0,28 are determined. Use of the results obtained in the calculation shows the sufficient seismic stability of the building, which is in a good agreement with the designed data.

3. Engineering And Seismometric Service In Tashkent Metro

Functioning of the Tashkent Metro whose lines are in the region of high seismicity and subsidence grounds and whose construction is notable for originality and novelty is of interest for instrumental control of constructions and their interaction with ground during earthquakes. Method for SESS organisation in the Metro has some features consisting in the fact that its span tunnels and stations located in different soil conditions are of different constructive design.

The parts of the Metro complex where few constructions (platforms, escalators, station premises and other part of the station complex, span tunnels adjoining the station) adjoin each other on the relatively small area are of the greatest interest. It is also expedient to install measuring points of SESS in constructions located in the zones of sharp change of the Metro depth or physical and mechanical properties of soil and in the places of span outcrop and metro trestle.

Two SESS have been organised in the first line of the Tashkent Metro as a result of analysis of its engineering and geological conditions and constructive design. The regional complex of the first SESS is located near the station Hkamza. Measuring points are installed in the station lining of single-arched type made of monolithic reinforced concrete, in span tunnel of solid sectional lining and in the Metro trestle over the Ak-Tepe canal. Measuring points of the second SESS located at the station Pakhtakor in zone of the first and the second metro lines crossing are in the constructions of columnar type made of prefabricated reinforced concrete elements and in span tunnels of circular shape including span tunnels under the Ankhor canal.

Experiment seismostation for registration of instrumental observations from 13 bases installed in construction of the station and in adjoining span tunnels has been constructed. Seismodetectors of the experiment station are installed in constructions of the right span tunnel as moving from the station Mustakillik Maidoni to the station Pakhtakor.

The span tunnel of rectangular shape made of five solid sectional lining adjoins the station tunnel being the constructive extension of the station through the functional joint. Further towards the station Pakhtakor the span tunnel 22 m long of rectangular shape, two sections of circular tunnel made of prefabricated reinforced concrete blocks with antiseismic joints adjoin each other through the functional antiseismic joints. The measuring points of the experiment station are located in the ends of the sections under investigation from both sides of the functional antiseismic joints.

They are completed with detectors of two types (for displacement and acceleration recording) oriented along the longitudinal and transversal axes of the tunnel. To avoid influence of the electromagnetic field caused by the rails on the engineering seismometric registration channel the measuring points are placed on the side of the station and span tunnel walls.

The detectors are installed on special horizontal platforms on the level of the rail head or rail base. Depth of the tunnels is 18-20 m. The experiment station recorded actions of few earthquakes occurs in the last years including the Nazarbek earthquake on tunnel constructions [2,3]. Values of intensities in the region of seismostation caused by earthquakes of December 11, 1980, December 30, 1980 and January 1, 1981 were 6, 4 and 3 respectively.

The instrumentally obtained records of displacements and accelerations were processed according to the method of spectral analysis i.e. they were approximated by the Fourier series. Comparison of amplitude-frequency spectra of longitudinal and transversal displacements and accelerations recorded in different points was carried out. The maximum values of longitudinal displacements caused by the earthquakes of December 11, 1980 and January 1, 1981 refer to the frequencies f=0.3 Hz and f=0.7 Hz respectively. The maximum longitudinal acceleration refers to the frequencies 20-25 Hz.

4. Engineering And Seismometric Service In Charvak Dam

The Charvak reservoir whose water storage reaches 2 milliard cubic metres is at 60 km' distance from Tashkent. Dam of the storage is in the vicinity of a seismic fault. It relates to the rock filling type of constructions, has transversal profile with the central core of loam, lateral prisms of limestone and two-layered conversion zone of sandstone and gravel. If the dam whose altitude is 167 m is destroyed the Tashkent oasis will be flooded. Control system of the dam includes 21 points of registration situated in the most typical point of the dam and canyon: four points are situated along the dam crest, two points - along the canyon side (in adits), four - along the foundation contour, six - in the alignments, four - in the core (inspection shaft) and one point - outside the dam. Registration includes 3-component records along the main axes of the construction. Number of registering channels is 105, the amplitude dynamic diapason of seismic motion registration is 120 dB (0,000002÷2 m/sec), frequency band - 0.5-20 Hz, duration of the record for any seismic event - 0.5-1 minute [4].

5. Conclusion

Ground motion recorded by these SESS was used for improvement of the map of microseismozoning and the Constructive Standards and Regularities, for calculation of seismic stability of buildings constructed on the city territory. At present one of the actual questions is the question of a common network for seismometric control in Tashkent and Tashkent region organisation and its supply with new-type machinery providing registration and recording of strong motions caused by earthquakes.

References

1. (Collective of the authors) Tashkent earthquake of April 26 1966, (1967), FAN Academic publishing house, Tashkent.
2. Report on scientific and research work "Study of the Strain State of transport and power construction interacting with ground under seismic loads", (1996), G.R./01940002967 Institute of Mechanics and Seismic Stability of Structures, Tashkent.
3. Rashidov, T.R., Ishankhojaev, A.A. (1993) Seismic Stability of Shallow Tunnel Constructions of the Metro, FAN Academic publishing house, Tashkent.
4. Report "Development of the automated system for seismometric control of the Charvak dam motion under earthquakes"(1986), G.R./01825000398, Institute of Seismology, Tashkent

STRESS WAVE PROPAGATION DUE TO A MOVING FORCE: COMPARISON OF FEM AND BEM SOLUTIONS

K. M. RASMUSSEN, S. R. K. NIELSEN , P. H. KIRKEGAARD
Department of Building Technology and Structural Engineering,
Aalborg University, Sohngaardsholmsvej 57, DK9000 Aalborq, Denmark
e-mail: soren.nielsen@civil.auc.dk

Abstract.

In this paper the performance of two numerical methods of solving the problem of a time dependent moving force on the surface of an elastic continuum will be evaluated. One method is the finite element method (FEM) formulated in convected coordinates coupled with an absorbing boundary condition of the impedance type. The other method to be considered is the boundary element method (BEM), where a new formulation using Green's functions transformed to a moving coordinate system is introduced. The methods are tested by the classic wave propagation problem of a Ricker pulse propagating from the surface of an elastic halfspace. The time integral net impulse of the considered loading must be null for the considered FEM to work. Further, the FEM is unable to absorb Rayleigh waves, since the considered impedance condition has been tuned to P- and S-waves. By contrast the BEM is able to handle also these cases. The disadvantage of the BEM is the increased calculation time.

Keywords: Boundary element method, finite element method, impedance, absorbing boundaries, moving force.

1. Introduction

Traffic induced vibrations are a known source of discomfort and damage problems in nearby buildings, either directly in the form of vibrations of the buildings or through structure borne noise. During the last couple of decades the general amount of traffic has increased considerably and along with heavier trucks this has emphasized the problems of traffic induced vibrations. But also train induced vibrations have been given some attention in recent years, mainly due to the dramatic increase in speed, (Krylov, 1995). Especially the greater number of underground railway systems in urban areas have resulted in severe noise and vibration problems. A considerable effort has been done to minimize or soften the efiects of the trad'ic induced vibrations by developing e.g. vibration and noise barriers and construction of vibration rails and roads, see e.g. (Nelson, 1996). With respect to analysis, semi-empirical methods for estimating the vibrations have been introduced and some numerical methods used to estimate the vibrations have been introduced, but mainly in the frequency domain.

M. Erdik et al. (eds.), Strong Motion Instrumentation for Civil Engineering Structures, 551–560.

Simulation of traffic induced vibrations in an elastic media can be achieved as a sum of moving concentrated forces due to the superposition principle. Thus, the basic problem that must be solved to simulate traffic induced vibrations is the problem of a moving time dependent concentrated force on the surface of an elastic half-space. When using the FEM or BEM to solve this problem an important problem arises. With an element mesh of finite size the moving force will soon move beyond the boundary of the mesh. Therefore, it will be preferable to formulate the problem in a coordinate system moving along with the force.

In this paper both FEM and BEM formulations for moving coordinate systems are presented and the obtained results for a classic wave propagation problem are compared.

2. Convected coordinates

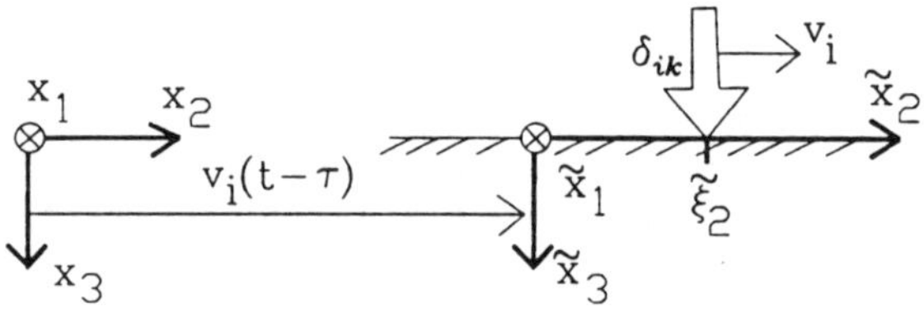

Figure 1. Moving force with velocity v_i in fixed and moving coordinate systems.

A Cartesian $(\tilde{x}_1, \tilde{x}_2, \tilde{x}_3)$-coordinate system is introduced, which follows the moving force, with velocity components v_i, see figure 1. $\tilde{\xi}_i$ signifies the coordinates of the force in the moving coordinate system. The two coordinate systems coalesce at the time $t = \tau$ which corresponds to the Galilean transformation

$$x_i = \tilde{x}_i + v_i(t - \tau) \tag{1}$$

The equations of motion in the fixed coordinate system are then given as

$$\frac{\partial \sigma_{ij}}{\partial x_j} - \rho \frac{\partial^2 u_i}{\partial t^2} = 0 \tag{2}$$

σ_{ij} is the Cauchy stress tensor, u_i is the displacement vector and ρ is the mass density. In (2) the summation convention of Cartesian tensors has been applied. (2) is transformed to the moving coordinate system by introducing the partial differentiation operators

$$\left.\begin{array}{rcl}\frac{\partial}{\partial x_j} & = & \frac{\partial}{\partial \tilde{x}_j} \\ \left.\frac{\partial}{\partial t}\right|_{x_j} & = & \left.\frac{\partial}{\partial t}\right|_{\tilde{x}_j} - v_j \frac{\partial}{\partial \tilde{x}_j} \\ \left.\frac{\partial^2}{\partial t^2}\right|_{x_j} & = & \left.\frac{\partial^2}{\partial t^2}\right|_{\tilde{x}_j} - 2v_j \left.\frac{\partial^2}{\partial \tilde{x}_j \partial t}\right|_{\tilde{x}_j} + v_j v_k \frac{\partial^2}{\partial \tilde{x}_j \partial \tilde{x}_k}\end{array}\right\} \tag{3}$$

Hereby the equation of motion in the moving coordinate system reads

$$\frac{\partial \tilde{\sigma}_{ij}}{\partial \tilde{x}_j} - \rho \left(\frac{\partial^2 \tilde{u}_i}{\partial t^2} - 2v_j \frac{\partial^2 \tilde{u}_i}{\partial \tilde{x}_j \partial t} + v_j v_k \frac{\partial^2 \tilde{u}_i}{\partial \tilde{x}_j \partial \tilde{x}_k} \right) = 0 \tag{4}$$

$\tilde{u}_i(\tilde{\mathbf{x}}, t)$ is the displacement field in the moving coordinate system. The coordinate transformation (1) and the equation of motion in the moving coordinate system (4) are the basis of both the FEM and BEM formulations in convected coordinates.

3. FEM formulation

(4) forms the basis of the finite element formulation in convected coordinates. The equation of motion in convected coordinates, (4), is brought to the weak formulation by multiplication with a virtual displacement field $\hat{u}_i$, followed by integration over the volume and use of the divergence theorem. Hereby the following variational equations may be obtained, (Krenk et al., 1999)

$$\begin{aligned}&\int_V \left(\frac{\partial \hat{u}_i}{\partial \tilde{x}_j} D_{ijkl} \frac{\partial \tilde{u}_k}{\partial \tilde{x}_l} - \rho v_j v_k \frac{\partial \hat{u}_i}{\partial \tilde{x}_j} \frac{\partial \tilde{u}_i}{\partial \tilde{x}_k} \right) dV - \\ &\int_S \hat{u}_i \left(\tilde{\sigma}_{ij} - \rho v_j v_k \frac{\partial \tilde{u}_i}{\partial \tilde{x}_k} \right) n_j dS + \\ &\int_V \rho v_j \left(-\hat{u}_i \frac{\partial \dot{\tilde{u}}_i}{\partial \tilde{x}_j} + \frac{\partial \hat{u}_i}{\partial \tilde{x}_j} \dot{\tilde{u}}_i \right) dV - \int_S \rho v_j \hat{u}_i \dot{\tilde{u}}_i n_j dS + \\ &\int_V \rho \hat{u}_i \ddot{\tilde{u}}_i dV = 0\end{aligned} \tag{5}$$

where D_{ijkl} is the elasticity tensor. The variational and displacement fields are interpolated in the following way

$$\left.\begin{array}{rcl}\tilde{u}_i(\tilde{\mathbf{x}}, t) & = & \mathbf{N}_i(\tilde{\mathbf{x}}) \mathbf{u}(t) \\ \hat{u}_i(\tilde{\mathbf{x}}, t) & = & \hat{\mathbf{N}}_i(\tilde{\mathbf{x}}) \hat{\mathbf{u}}(t)\end{array}\right\} \tag{6}$$

where $\mathbf{N}_i(\tilde{x}_j)$ and $\hat{\mathbf{N}}_i(\tilde{x}_j)$ are the shape functions of the displacement field and the variational field and $\tilde{u}_i(\tilde{\mathbf{x}}, t)$ and $\hat{u}_i(\tilde{\mathbf{x}}, t)$ are the nodal values of the displacement and variational fields.

4. Impedance absorbing boundary condition

The surface traction $\tau_i = \sigma_{ij} n_j$ on the artificial boundary expresses the reaction force from the surrounding elastic media on the modelled field. For a class of problems it can be shown that τ_i is proportional to the velocity on the boundary as follows

$$\tau_i = -Z_{ij}\dot{u}_j \tag{7}$$

A boundary condition of the type (7) is known as an impedance boundary condition, a designation borrowed from acoustics, where an expression of the type (7) is formulated between the pressure and the velocity at the artificial boundary. For plane P- and S-waves in a homogenous isotropic medium, an expression for the impedance tensor Z_{ij} in moving coordinates was derived by (Krenk et al., 1999)

$$\tau_i = -Z_{ij}\left(\dot{\tilde{u}}_j - v_k\frac{\partial \tilde{u}_j}{\partial \tilde{x}_k}\right) \tag{8}$$

$$\begin{aligned} Z_{ij} &= \frac{2\mu}{c_P}\left(n_k r_k^P\right) r_i^P r_j^P + \frac{\lambda}{c_P} n_i r_j^P + \\ &\frac{\mu}{c_S}\left(n_k r_k^S\right)\left(\delta_{ij} - 2 r_i^S r_j^S\right) + \frac{\mu}{c_S} r_i^S n_j \end{aligned} \tag{9}$$

μ and λ are the Lamé material parameters, $c_P = \sqrt{\frac{\lambda+2\mu}{\rho}}$ and $c_S = \sqrt{\frac{\mu}{\rho}}$ are the P- and S-wave phase velocities. r_i^P and r_i^S are the propagation directions of P- and S- waves in the moving coordinate system, which are different, even if the source is common, see (Krenk et al., 1999). The boundary condition (8) will be perfectly transparent for plane P- and S-waves. All other wave components will be more or less reflected.

Insertion of (8) into (5) and use of (6) provide the following system of discretized FEM differential equations

$$\mathbf{M}\ddot{\mathbf{u}}(t) + \mathbf{C}\dot{\mathbf{u}}(t) + \mathbf{K}\mathbf{u}(t) = \mathbf{P}(t) \tag{10}$$

where $\mathbf{P}(t)$ is the load vector $\mathbf{K}$, $\mathbf{C}$ and $\mathbf{M}$ signify the stiffness-, damping- and mass matrices given as

$$\begin{aligned} \mathbf{K} \;=\; &\int_V \left(\hat{\mathbf{N}}_{i,j}^T D_{ijkl}\mathbf{N}_{k,l} - \rho v_j v_k \hat{\mathbf{N}}_{i,j}^T \mathbf{N}_{i,k}\right) dV - \\ &\int_S \hat{\mathbf{N}}_i^T \left(Z_{ij} - \rho n_l v_l \delta_{ij}\right) v_k \mathbf{N}_{j,k} dS \end{aligned} \tag{11}$$

$$\mathbf{C} = \int_V \rho v_j \left(\hat{\mathbf{N}}_{i,j}^T \mathbf{N}_i - \hat{\mathbf{N}}_i^T \mathbf{N}_{i,j}\right) dV + \int_S \hat{\mathbf{N}}_i^T \left(Z_{ij} - \rho n_k v_k \delta_{ij}\right) \mathbf{N}_j dS \tag{12}$$

$$\mathbf{M} = \int_V \rho \hat{\mathbf{N}}_i^T \mathbf{N}_i dV \tag{13}$$

$$\tag{14}$$

where $\mathbf{N}_{i,j}(\tilde{\mathbf{x}}) = \frac{\partial}{\partial \tilde{x}_i}\mathbf{N}_i(\tilde{\mathbf{x}})$. The impedance boundary condition of the type (7) and (8) is not able to sustain static loading. Further permanent displacements occur, if the net impulse $\int_{-\infty}^{\infty} \mathbf{P}(t)dt \neq \mathbf{0}$.

5. BEM formulation

(2) is the basis for the differential equations for Green's functions, therefore (4) gives the differential equations for Green's functions in a moving coordinate system

$$\frac{\partial \tilde{\sigma}_{ijk}}{\partial \tilde{x}_j} - \rho\left(\frac{\partial^2 \tilde{g}_{ik}}{\partial t^2} - 2v_j \frac{\partial^2 \tilde{g}_{ik}}{\partial \tilde{x}_j \partial t} + v_j v_l \frac{\partial^2 \tilde{g}_{ik}}{\partial \tilde{x}_j \partial \tilde{x}_l}\right) + \delta_{ik}\delta(\tilde{\mathbf{x}} - \tilde{\boldsymbol{\xi}})\delta(t-\tau) = 0 \tag{15}$$

$\tilde{g}_{ik}(\tilde{\mathbf{x}}, t; \tilde{\boldsymbol{\xi}}, \tau)$ is Green's function for the displacement field $\tilde{u}_i(\tilde{\mathbf{x}}, t)$ in the moving coordinate system and similarly for Green's function for the surface traction $\tilde{\sigma}_{ijk}(\tilde{\mathbf{x}}, t; \tilde{\boldsymbol{\xi}}, \tau)$. Introducing the coordinate transformation (1) in (15) gives the following equations for $\tilde{g}_{ij}$ as functions of the fixed coordinates x_i

$$\frac{\partial \tilde{\sigma}_{ijk}}{\partial x_j} - \rho \frac{\partial^2 \tilde{g}_{ik}}{\partial t^2} + \delta_{ik}\delta(\mathbf{x} - \mathbf{v}(t-\tau) - \tilde{\boldsymbol{\xi}})\delta(t-\tau) = 0 \tag{16}$$

Since $\delta(\mathbf{x} - \mathbf{v}(t-\tau) - \tilde{\boldsymbol{\xi}})\delta(t-\tau) = \delta(\mathbf{x} - \tilde{\boldsymbol{\xi}})\delta(t-\tau)$, (16) reduces to the normal differential equations for Green's functions. Hence

$$\tilde{g}_{ik}(\mathbf{x} - \mathbf{v}(t-\tau), t; \tilde{\boldsymbol{\xi}}, \tau) = g_{ik}(\mathbf{x}, t-\tau; \tilde{\boldsymbol{\xi}}, 0) \Rightarrow \tilde{g}_{ik}(\tilde{\mathbf{x}}, t; \tilde{\boldsymbol{\xi}}, \tau) = g_{ik}(\tilde{\mathbf{x}} + \mathbf{v}(t-\tau), t-\tau; \tilde{\boldsymbol{\xi}}, 0) \tag{17}$$

Green's function for Cauchys stress tensor $\tilde{\sigma}_{ijk}(\tilde{\mathbf{x}}, t; \tilde{\boldsymbol{\xi}}, \tau)$ is obtained upon insertion of (17) into the constitutive equations for the considered linear elastic material. Since these only involve spatial differential operations with respect to $\mathbf{x}$, the result is seen to be

$$\tilde{\sigma}_{ijk}(\tilde{\mathbf{x}},t;\tilde{\boldsymbol{\xi}},\tau) = \sigma_{ijk}(\tilde{\mathbf{x}}+\mathbf{v}(t-\tau),t-\tau;\tilde{\boldsymbol{\xi}},0) \tag{18}$$

Hence, a BEM formulation in the moving coordinate system that calculates the displacement field following the force can be established by using the Green's functions $\tilde{g}_{ik}(\tilde{\mathbf{x}},t;\tilde{\boldsymbol{\xi}},\tau)$ and $\tilde{\sigma}_{ijk}(\tilde{\mathbf{x}},t;\tilde{\boldsymbol{\xi}},\tau)$ as given by (17) and (18) instead of the original ones. Notice that this result is valid even for non-homogeneous or anisotropic linear materials. Using the Betti reciprocal theorem and the symmetry properties of Green's functions Somigliana's identity in 2D can then be written, see e.g. (Banerjee et al., 1987)

$$\begin{aligned} &C^{+}(\tilde{\mathbf{x}})u_{\alpha}^{+}(\tilde{\mathbf{x}},t) = \\ &\int_S \int_0^t g_{\alpha\beta}(\tilde{\mathbf{x}},t-\tau;\tilde{\boldsymbol{\xi}},0)t_{\beta}^{+}(\tilde{\boldsymbol{\xi}},\tau)d\tau dS_{(\tilde{\boldsymbol{\xi}})} - \\ &\fint_S \int_0^t \sigma_{\alpha\gamma\beta}(\tilde{\mathbf{x}},t;\tilde{\boldsymbol{\xi}},0)n_{\gamma}(\tilde{\mathbf{x}})u_{\beta}^{+}(\tilde{\boldsymbol{\xi}},\tau)d\tau dS_{(\tilde{\boldsymbol{\xi}})} \end{aligned} \tag{19}$$

where

$$C^{+}(\tilde{\mathbf{x}}) = \begin{cases} 1 \ , & \tilde{\mathbf{x}} \in A^{+} \\ 0 \ , & \tilde{\mathbf{x}} \in A^{-} \\ \frac{1}{2} \ , & \tilde{\mathbf{x}} \in S \end{cases} \tag{20}$$

A^{+} signifies the media, A^{-} the surrounding continuum and S the boundary. The last singular integral in (19) should be interpreted in the sense of a Cauchy principal value. Discretization of (19) in time and space leads to the BEM in time domain.

6. Numerical results

The problem of a Ricker impulse on an isotropic, homogenous linear elastic 2D plane strain halfspace is analyzed with both methods. The medium is subjected to a moving time dependent surface loading per unit length $P(t)$ (Ricker impulse). The time variation of the loading is given as

$$\left. \begin{array}{l} P(t) = p_0\tau(1-\tau^2)^2 \\ \tau = 2t/T - 1 \quad , \quad -1 < \tau < 1 \end{array} \right\} \tag{21}$$

This pulse has vanishing derivatives at $t = 0$ and $t = T$.
The example is performed with the following parameters : Intensity of force $p_0 = 1$ MN, $\mu = \lambda = 100 \cdot 10^6$ N/m^2 and $\rho = 2.0 \cdot 10^3$ kg/m^3. This

gives the wave velocities for the P-, S- and Rayleigh waves, respectively, $c_P = 387$ m/s, $c_S = 224$ m/s and $c_R = 206$ m/s.
The force is distributed in triangular shape over an area of 16 m. 160 m of the surface is discretized and the force is applied at the centre. The applied FEM mesh is rectangular with depth 80 m. The BEM mesh consists of 4 m 2 node line elements with linear interpolation and similarly the FEM mesh consists of 4 m by 4 m, 4 node quadrilateral elements, see figure 2.

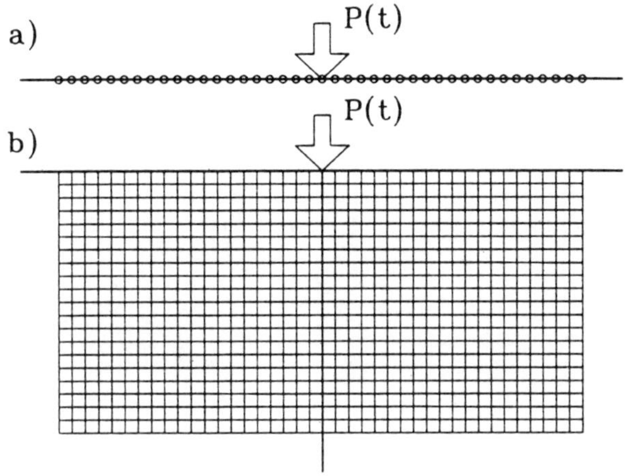

Figure 2. a) BEM and b) FEM meshes.

The horizontal, u_1 and vertical, u_2 displacement histories at (x_1, x_2, x_3) = (+20 m, 0 m, 0 m), 'in front of the force' and at (x_1, x_2, x_3) = (-20 m, 0 m, 0 m), 'behind the force' are presented in the following figures.

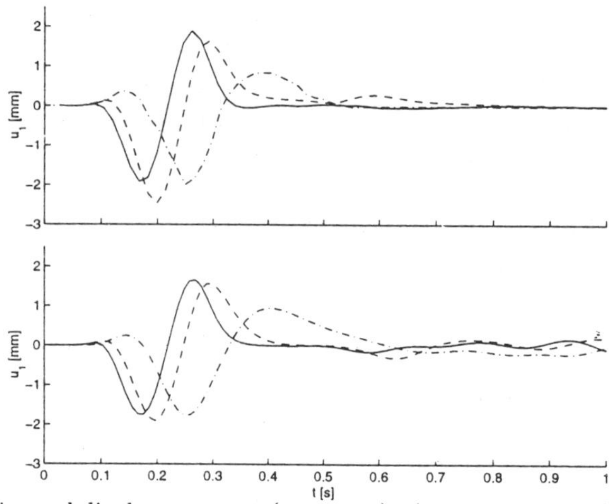

Figure 3. Horizontal displacements at (x_1, x_2, x_3)=(+20 m, 0 m, 0 m) obtained using BEM (top) and FEM (bottom) for $v_1 = 0 \cdot c_S$ $-$, $v_1 = 0.2 \cdot c_S$ $--$ and $v_1 = 0.5 \cdot c_S$ $- \cdot -$.

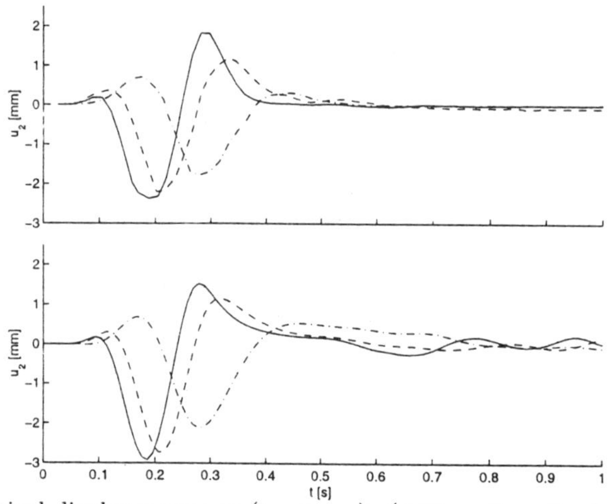

Figure 4. Vertical displacements at (x_1, x_2, x_3)=(+20 m, 0 m, 0 m) obtained using BEM (top) and FEM (bottom) for $v_1 = 0 \cdot c_S$ $-$, $v_1 = 0.2 \cdot c_S$ $--$ and $v_1 = 0.5 \cdot c_S$ $- \cdot -$.

Comparison between figures 3 and 4 shows that the minimum displacement increases with greater speed and that the maximum displacement decreases. However, for $v_1 = 0 \cdot c_S$ the BEM results for the minimum displacements are significantly larger than the equivalent minima from the FEM results. When v_1 approaches null the solution of the time integrals of the Green functions become unstable and therefore, the results for very small velocities are not reliable.

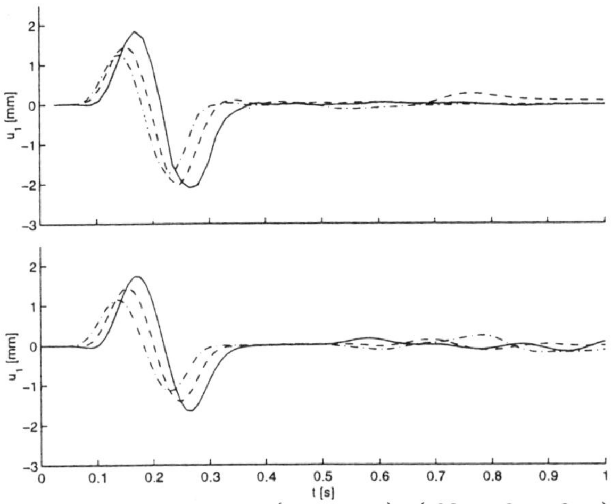

Figure 5. Horizontal displacements at (x_1, x_2, x_3)=(-20 m, 0 m, 0 m) obtained using BEM (top) and FEM (bottom) for $v_1 = 0 \cdot c_S$ –, $v_1 = 0.2 \cdot c_S$ -- and $v_1 = 0.5 \cdot c_S$ – · –.

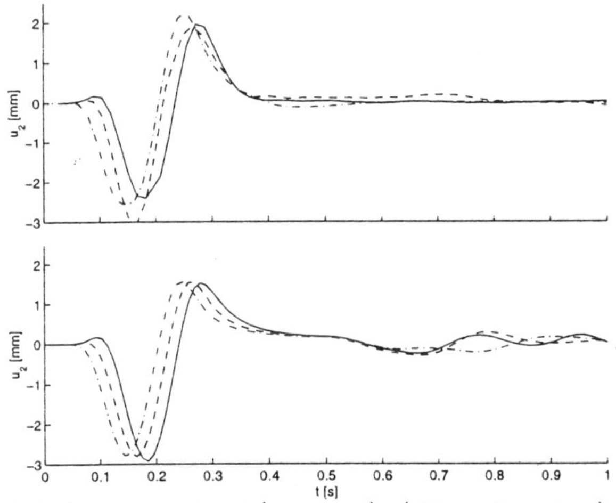

Figure 6. Vertical displacements at (x_1, x_2, x_3)=(-20 m, 0 m, 0 m) obtained using BEM (top) and FEM (bottom) for $v_1 = 0 \cdot c_S$ –, $v_1 = 0.2 \cdot c_S$ -- and $v_1 = 0.5 \cdot c_S$ – · –.

In all the figures 3-6, but especially in figure 6, the FEM results show tendencies to reflect Rayleigh waves from the boundaries. However, the BEM results do not show any signs of reflection in any of the figures. Furthermore, notice that the amplitude of the displacements do not increase for $v_1 \to c_R$. The calculation time for the FEM solution was 62 seconds on a 200 MHz PC while the calculation time for the BEM solution was several hours on a SiliconGraphics Indy R4400.

7. Conclusion

FEM and BEM formulations in a moving coordinate system have been formulated. The formulation for the FEM uses an impedance boundary condition as absorbing boundary. The BEM formulation introduces Green's functions for a moving force and shows how the normal BEM formulation for a non moving coordinate system can be modified to a formulation in the moving coordinate system by using Green's functions for a moving force.
Finally, 2D results from the formulations have been compared and good agreement is found.

References

P. K. Banerjee, S. Ahmad and G. D. Manolis. Advanced Elastodynamic Analysis, Computational Methods in Mechanics, Volume III. Boundary Element Methods in Mechanics. ed. D. E. Beskos, Elsevier Science Publishers B.V, 1987.

A. C. Eringen and E. S. Suhubi. *Elastodynamics, Volume II, Linear Theory.* Academic Press, 1975.

A. S. M. Israil and P. K. Banerjee. Advanced Time Domain Formulation of BEM for Two Dimensional Transient Elastodynamics. *International Journal for Numerical Methods in Engineering.* 29:1421-1440, 1990.

S. Krenk, L. Kellezi, S. R. K. Nielsen and P. H. Kirkegaard. Finite Element and Transmitting Boundary Conditions for Moving Loads. *Proceedings of the 4th European Conference on Structural Dynamics, Eurodyn'99*, Praha. 1:447-452, 1999.

S. Krenk and P. H. Kirkegaard. Radiation Conditions for Elastic Eaves. *Technical University of Denmark.* (to be published)

V. Krylov. Generation of Ground Vibrations by Superfast Trains. *Applied Acoustics.* 44:149-164, 1995.

J. T. Nelson. Recent Developments in Ground-Borne Noise and Vibration Control. *Journal of Sound and Vibration.* 193(1):367-376, 1996.

K. M. Rasmussen. Stress Wave Propagation in Soils Modelled by the Boundary Element Method. Ph.D. Thesis, *Aalborg University*, 1999.

STRONG MOTION IN ABSORBING NONLINEAR MEDIUM AND PROBLEMS OF THEIR REGISTRATION

V.ZAALISHVILI

Center of Applied Geophysics, Engineering Seismology and Seismic Protection of Structures, Georgian Geophysical Society (Tbilisi, Georgia);
Joint Institute of Physics of the Earth (Moscow, Russia)
Institute of Applied Mathematics and Informatics of North Ossetian Scietific Center (Vladikavkaz, Russia);
1, M.Aleksidze str., Tbilisi 380093, Georgia
email: agesas@ip.osgf.ge; vzaal@wdcb.rssi.ru

Abstract

In the paper some physical phenomena in natural soils are considered, provided by intensive seismic excitation. On the basis of regressive analysis of Racha earthquake instrumental data and so–called "dimensionality method", the new parameters, which are tightly connected with absorbing indexes and non-linearity of natural soils have been obtained. The consideration of corresponding correlation links has shown that in epicentral zone of earthquake (30-40 km) the absorption of seismic energy in soft soil is inversely proportional to earthquake magnitude and not significantly depends on acceleration. In rocks the absorption is directly proportional to the acceleration and slightly depends on magnitude. This fact explains considerably the difference degree of buildings' and constructions' damage on soft and rock soils.

It was detected the degrees of non-linearity and non-elasticity of soils. Also, it was considered the usage of offered parameters during solving of practical issues of seismic microzonation.

It was shown the uncertainty of differentiation of strong and weak motions. The properties of strong motion is being analyzed, based on corresponding behavior of geomaterial at linear-elastic, nonlinear-elastic, non-elastic deformation.

Partially, it was obtained the empirical quantity for account of intensity level of seismic impact. Following leads to a conclusion about necessity of registration strong motion not only in maximum narrow band of spectrum, but also in wide frequency band of spectrum.

M. Erdik et al. (eds.), Strong Motion Instrumentation for Civil Engineering Structures, 561–572.

1. Introduction

It is known, that damage of structures, buildings (and any physical system) during earthquake is caused by their inability to absorb corresponding seismic energy. Then, after partial damage, the system obtains an extra absorption ability, which preserves the system from future damages at the given level of impact. During the exceeding impact, the system will be damaged in bigger degree or absolutely destroyed. Another type of system's behavior is the well-known building's or structure's leaving from resonance condition by the way of partial damage followed by changes of rigidity and correspondingly of natural vibration frequency of buildings (structures). In this connection it is very important the issue of appraisal adequacy of seismic energy absorption by different types of physical systems (soils, buildings, structures, etc.). The theoretical solutions in this field is often differed from the real results. Such uncertainty appraisals are especially significant at strong motions. This leads to big scientific and practical interest to the results of analyses of strong motion instrumental records.

Not less important is the phenomena of non-linearity during strong motion. The quantitative assessment of non-linearity degree of materials or any physical system presents sufficiently complex and uncertain problem.

Non-linear distortions of wave shapes, occurred in so-called "bimodular" medium, is shown itself in "spreading" of spectrum in the area of low and high frequencies. These mediums are real soils, the compression module of which is significantly differed from strain one. The simple quantitative degree of "spreading" process is the width or area of normalized spectrum.

2. Soft and rock soils

What's the difference between soft and hard soils? On the basis of analysis it was obtained that area of normalized spectrum in rock soils (Oni station) is increasing, and in soft soils (Ambrolauri station) insignificantly decreasing, as the average weighted vibration frequency of soils is increasing (Fig.1a). Amplitude of spectra of soft and hard soils is directly proportional to acceleration of soil vibration (Fig.1b).

The value of acceleration is widely used in very different analyses of engineering seismology and theory of dynamics of structures. At the same time it is well-known its often poor correlation with seismic energy, which is inverse proportional to the frequency of oscillation (easily obtained from the recordings). This fact provided the research of dependence of oscillations maximal frequency on acceleration. It should be noted very fastidious property of this dependence at some areas (Zemo Bari, Iri stations). So, it is significant the conclusion about opportunity of existence of high frequencies as for low accelerations, as for high ones (Fig.1a), appreciating the fact that the frequencies themselves are linked tightly, as a rule, with values of magnitude, it is clear the dissimilarity of correlation between magnitudes and accelerations. During high accelerations the characteristic of dependence of areas frequencies, contained in soft soils is close to rock ones. So, at low initial values of acceleration, as they are increasing, the curve (acceleration – maximal frequency) declines and the vibrations are characterized by low frequency. Further, as the acceleration is increasing, the frequency of corresponding vibrations increases. This fact explains the existing concept [4], that during low frequencies, the building is damaged by acceleration (e.g. the amplitude level of acceleration is significantly high), and the absorption of energy is minimal, because the frequency of vibrations is low. "Average amplitude level" of acceleration can cause the significant damages in sufficiently elastic buildings, and at the same time accelerations of high frequency (considerably high), have exceeded some "elastic area" of its quantity, are fading fast because of high absorption. Perhaps, by this fact can be explained, sometimes, little damages during high-frequent vibrations.

In Fig. 2 dependence of real spectra areas on acceleration for rock soil vibration (Fig. 2a) and for soft soil (Fig. 2b) are shown. It is evident, that in rock soils area of spectrum is increased with increasing of accelerations and is decreased in soft soils.

It should be noted that during research it was obtained for the behavior of viscous-elastic body that the real part of motion equation solution corresponds to the increasing of rigidity on frequency [2]. Imaginary part of equation's solution corresponds to insignificant decreasing of value of parameters on frequency (Fig.2c).

Comparison of experimental and theoretical data shows their good coincidence. Thus, area of the real spectrum of vibration is, firstly, reliable indicator of physical state of medium and, secondly, characterizes its deformability or degree of its behavior's deviation from Hoock's linear-elastic law.

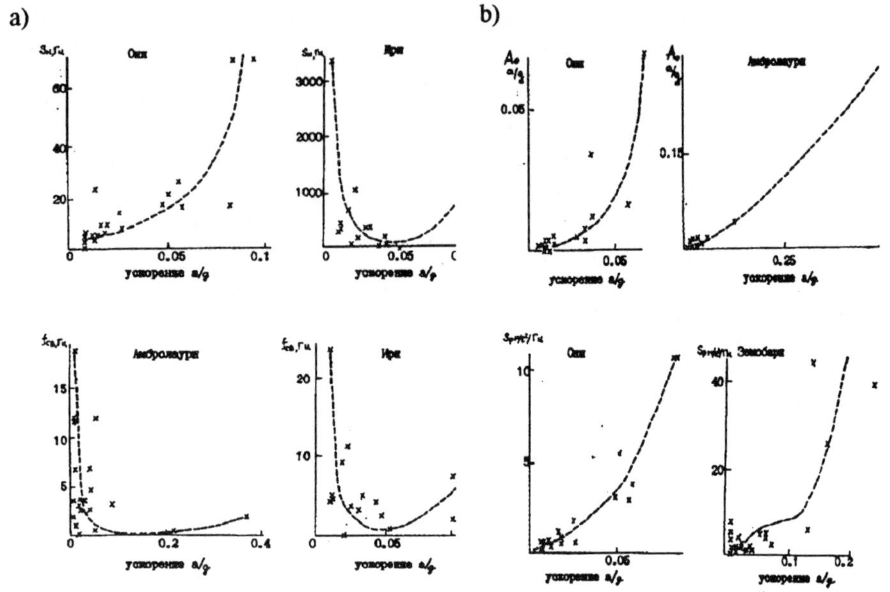

Figure 1. a) Dependence of normalized spectrum area and vibrations' average-weighted frequency on acceleration (Racha Earthquake, 1991); b) Dependence of spectrum maximal amplitude and vibration real spectrum area on acceleration.

In Fig.3 is shown the dependence of real spectrum's area of vibration on acceleration (Array SMART-1, Taiwan). It is clearly shown a sharp turns at acceleration a=0.1g. It is interesting to note, that according to [1] the change of shear to longitudinal waves velocity ratio (this quantity directly characterizes physical condition of medium) has two sharp turns at acceleration 0.1g and 0.2g (Fig.4). Comparing with Fig.1b (Zemo Bari) shows clear correspondence.

c)

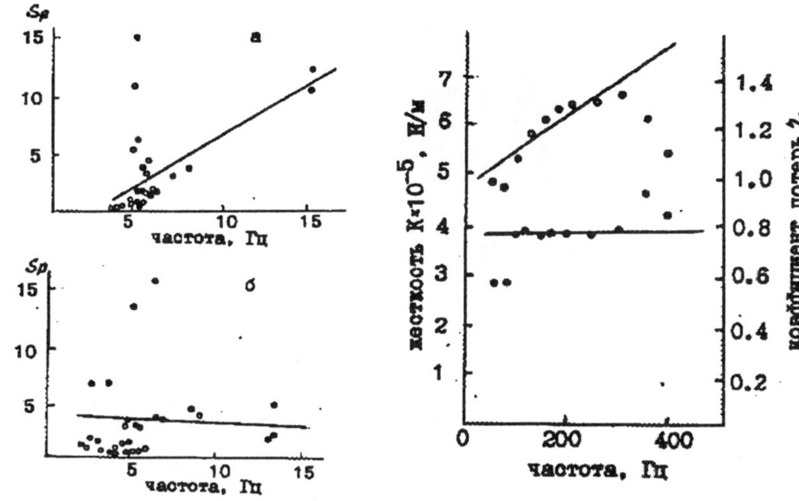

Figure 2. Dependence of real spectrum area for rock (a, Oni) and soft (b, Ambrolauri) on soils' vibration frequency. c) Dependence of rigidity and material loss coefficient on frequency of vibrations..

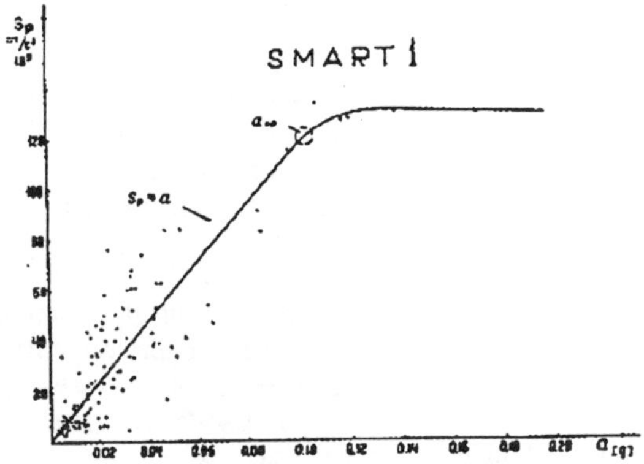

Figure 3. Dependence of area of "real" spectrum of vibrations on acceleration (Taiwan).

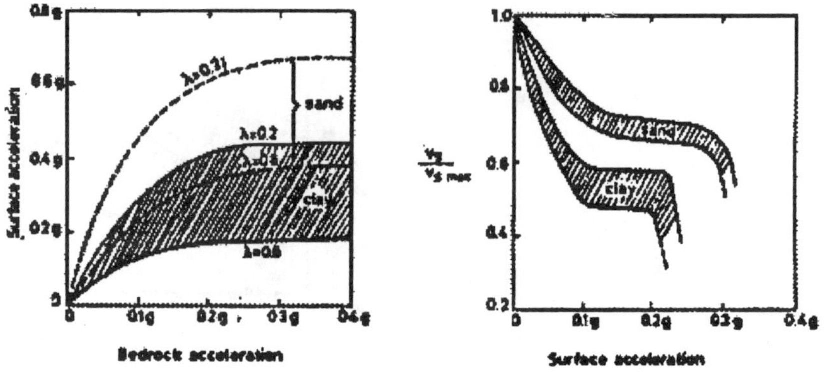

Figure 4. Dependence of shear wave to longitudinal wave velocity ratio on soil acceleration [1].

So, the area of real spectrum is important indicator of soils' behavior properties at different level of excitation.

3. Ground Strong Motion

What is strong motion and how it differs from weak? Why the rigid, weight building destroys ? Currently is well known the significant influence of non-linear phenomena on seismic effect or intensity of earthquake.

The first research in this field, in fact, was the investigation, conducted under direction of A.Nikolaev from Joint Institute of Physics of the Earth (Moscow) in the end of fifties in Russia. Much later, he has published with the great difficulties the first theoretical research dedicated to the problem of non-linearity [3]. This was explained by the fact that the non-linear effects are presented by most of seismic events as insignificant because of number of factors, determining the intensity of earthquake.

In 1988 K.Aki wrote, that except of case of liquefaction, nonlinearity practically is not significant in seismological data. Later he recognized the necessity of existence of visible non-linearity in soils' behavior and the necessity of non-linearity accounting during seismological analysis of events [9]. So, non-linearity of soils is expressed in dependence of all parameters of soil vibration on intensity of earthquakes.

In 1987 I carried out an experiment on study of soils' behavior under high dynamic loading. This experiment was carried out in the region of the new by-pass highway near Tbilisi city.

Soils presented an artificial embankment with 40 m thickness. Seismographs were set – one at a depth of 1m and second at a depth of 5m into the embankment soil. On the surface of the embankment an impulse impact of a moving part of the scraper was applied. In the seismogram one can see a predominant peak of high frequency, which attenuates quickly with depth. It is known that nonlinear distortion of signal is appeared in its enrichment in high-frequency spectrum. Thus, high frequency peak is due to nonlinear deformations in the embankment. In other words, at intensive effect in near zone of earthquake a high frequency excitation can predominate in soft soils. At that it is clear, that a rigid structure can be destroyed because of resonance phenomena in a high frequency diapason of spectrum.

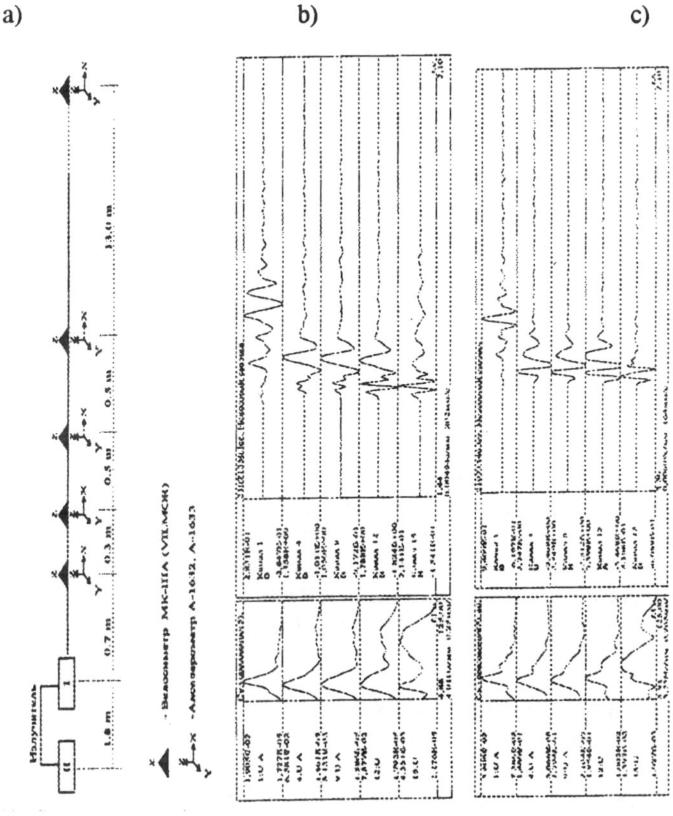

Figure 5. a) Profile of measurements. b) Record of soil vibrations (I exciter) c) Record of soil vibrations (II exciter).

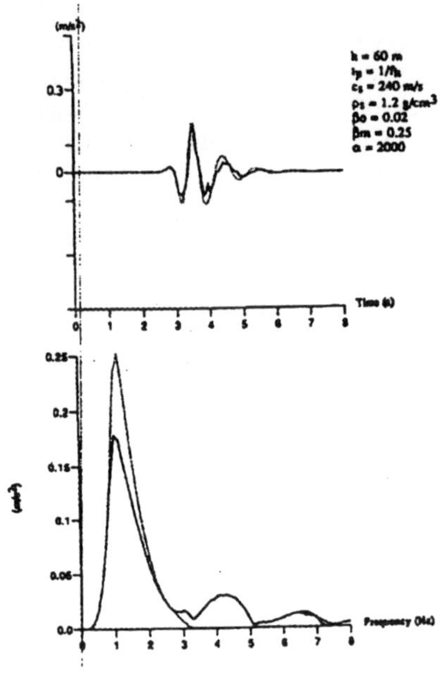

Figure 6. Comparison of accelerograms and Fourier spectra with or without taking into account the non-linear effect. The dotted line is for the linear solution.

At the sector contained from powerful sediments of sands, in 1996, Russia, we made experimental research dependence of spectrum changing on level of influence [5]. As a source of vibration we have experienced the powerful pulse, source with two exciter. In Fig. 5 is presented the oscillations of soils caused by near (b) and far (c) exciters. During close strike HF peak is domain in oscillation spectrum, and which is absorbed upon increasing of distance. As the seismo-recievers we used Russian, special developed, three-component accelerometers and English one-component Willmore accelerometer. Recording and data processing was carried out with special GEMIS-WIN software [7].

In this connection, it is interesting to compare the obtained data with calculations (Fig.6). It is clearly shown, that during elastic non-linear deformations, the energy "downloaded" to HF area of spectrum. It should be noted, that during change from linear-elastic deformations to non-linear-elastic ones, the square of real spectrum remains the same. At the same time the square of real spectrum is decreasing during non-elastic deformations, characterized buy losses of energy.

At the basis of buildings' damage analyses during earthquakes, some engineers explain damages by significantly high accelerations, much bigger than 2g. In every case, it can be said, that in some conditions in soft and especially in rock soils, the main energy can be situated in HF spectrum area.

4. Seismic coercion and reaction of soil-structures.

As a result of analysis of instrumental entries of strong earthquakes and special experiments the following equation was obtained [5,7,8]:

$$\alpha \sim S_n / f_w \tag{1}$$

where: α is the parameter of an absorption of oscillations by a soil;
S_n is the area of the normalized spectrum of oscillations.

It has been determined that the area of a peak Fourier spectrum S_R, is related to the normalized area by a following relationship:

$$S_R = S_n A \tag{2}$$

where A, being a peak amplitude of a spectrum of oscillations, characterizes a parameter of nonlinearity of grounds. From (1) and (2) we receive, that:

$$S_R \sim \alpha A f_w \tag{3}$$

A spectrum area is integrated performance of the phenomena of a absorption and nonlinearity. A parameter of absorption, excluded, we shall obtain "pure" nonlinearity (parameter of nonlinearity of a soil):

$$S_R / \alpha \sim A f_w \tag{4}$$

Thus, the product of maximum amplitude of a spectrum of vibrations of a ground by average weighted frequency of oscillations is an elementary and easily measurable quantity describing the special property of soil nonlinearity.

The formula for calculating of intensity modification on the basis of account of nonlinear properties of grounds with obligatory use of high-power impulse or vibration sources were obtained.

On the basis of account is of nonlinear - elastic behaviors of grounds, the following relationship is obtained:

$$\Delta I = K \lg \frac{A_i f_{wi}}{A_0 f_{w0}} \tag{5}$$

As a result of account of the inelastic phenomena, the expression is obtained [6]:

$$\Delta I = K \lg \frac{(A_i f_{wi})_n (A_0 f_{w0})_f}{(A_i f_{wi})_f (A_0 f_{w0})_n} \tag{6}$$

where $(A_i f_{wi})_{n,f}$, $(A_0 f_{w0})_{n,f}$ are parameters of nonlinearity of researched and standard soils in near and far zones of effect.

By special researches was obtained, that in soft soils the absorption within 10-40 km

$$\alpha \sim \frac{f_w t}{\sqrt{M}} \tag{7}$$

where t is duration of oscillations, M- magnitude of earthquake.

from a source (of earthquake aftershocks with M = 1.2-6.3) is proportional.
In rocky or rigid grounds the absorption is proportional:

$$\alpha \sim f^2 t \sqrt{ar} \tag{8}$$

where: a is a peak ground acceleration;
r - epicentral distance.

Thus, the absorption in soft soils practically does not depend on acceleration. Its level completely is determined by magnitude - the more magnitude, the less absorption. In rocky grounds the absorption is directly proportional to acceleration. The relations are valid within the given soils statistical number.

On the other hand, on the basis of comparison of parameters of nonlinearity for soils as S_p / f (as $\alpha \sim f$), we shall obtain from (3), taking into account that $A \sim M^{2.5} \sqrt{a}$:

$$\Delta I = K\,lg\frac{f_{wi}^2 M_i^2 t_i}{f_{w0}^2 M_0^2 t_{0i}}\sqrt{\frac{a_i}{a_0}} \tag{9}$$

The formula allows to evaluate influence of a modification of power to the intensity modification.

Let $M \sim E$ (energy of effect of an artificial source). Taking into account, that for a vibrator $M_i \approx M_0$, $t_i \approx t_0$, $a \approx A^2$, we obtain:

$$\Delta I = K\,lg\frac{A_i f_{wi}}{A_0 f_{w0}} \tag{10}$$

which is similar to expression (5).

If frequencies do not differ much, then $f_{wi} \approx f_{w0}$, also we shall obtain an expression

$$\Delta I = K\,lg\frac{A_i}{A_0} \tag{11}$$

Thus, seismic hazard or earthquake intensity is defined by magnitude, seismic focus depth, epicentral distance. This is the so-called "averaged" intensity of the given territory. At the same time, variations of intensity on a territory are defined by other, no less important parameters. These are a technical state of buildings and constructions, their orientation in relation to the arriving seismic impact, underground relief of bedrock, power of soil thickness, surface relief, soil types, their physical condition, etc. Damages of buildings and constructions, caused by soil conditions, may take place due to amplification of the arriving oscillations (from the bedrock) by loose substance or due to the uneven settling of buildings and constructions. In the first case, soil-construction interaction presupposes elastic deformation, and in the second case-unelastic one. Amplification of vibrations will be the highest in case of coincidence of natural frequencies of vibrations of buildings and constructions with natural frequency of soil substance under the foundations. Damages will also increase in case of the above mentioned liquefaction.

5. Conclusions

1. At the bases of strong motion instrumental recording analyses, have been obtained parameters, tightly linked with absorption and non-linearity of soils (and any physical system).

2. It was shown the difference of absorption phenomena of seismic energy in soft and hard (rock) soils. In soft soils absorption is mainly caused by magnitude of earthquake and in rock soils by acceleration. By the way, in common types of soils the absorption is directly proportional to average-weighted frequency and duration of oscillations.
3. During strong motion the main energy in soft and especially in rock soils is situated in high-frequent part of spectrum. During changing linear-elastic deformation to non-linear-elastic deformation the energy of oscillations remains constant. During non-elastic deformation the energy of oscillations decreases. It is clear, that in conditions of large strain it should be the most significant decreasing.
4. The modern level of strong motion instrumentation is considerably high (USA, Japan, Switzerland, etc.). Its opportunities are used non-fully. The obstacle is from one hand the low information density of analyses of data, that have traditionally used, and from the other hand -- properties of measuring- resolution of system itself.
5. The results of the work give opportunity to conclude that in spite of using type of equipment, the main properties of ground strong motion is similarly reflected in corresponding recordings, from earliest to the recent. From the other hand it is necessary to pay special attention to the range of resolution in high-frequency area of spectrum of oscillations. Here it should be upped the level of soils acceleration registration from widely used value of 2g. It gives opportunity to register significantly high accelerations, if they accompany strong motion. As it seems they accompany them. Besides, it is necessary to increase radically the level of resolution of record of strong motion for obtaining momentary and "short-living" ultra-high accelerations. And, at last, it is preferred creation of these strong motion instrumentation, that will give opportunity to register directly passing seismic energy, e.g. to create so-called "energometer". This will give opportunity to exclude the necessity using the big number of parameters of instrumental recordings and simplify calculations of buildings and constructions upon seismic influence. In last years, it was a number of trials to present in such calculation the seismic influence in energy type.

References

1. Finn Liam W.D., Semih S., Tercan, Mussafer Ipek. A parametric study of soil amplification. Proceedings of the 5th European Conference on Earthquake Engineering. vol. 1, № 20, Istanbul, 1975.
2. Nashif A., Jouj D., Henderson J. Dempferation of vibration. M., Mir,1988, 448 pp. (Russian).
3. Nikolaev A.V. Seismic propertious of soft medium. Physics of the Earth № 2, 1967, pp.23-31 (in Russian).
4. Sheidegger A.E. Physical Aspects of Natural Catastrophes. Amsterdam, Oxford, New-York, 1975, 232 pp.
5. Zaalishvili V., Chelidze T., Nikolaev A. Seismic Hazard and Nonlinear Properties of Soils. The Second International Conference on Earthquake Hazard and Seismic Risk Reduction., Sept., 15-21,1998, Yerevan.
6. Zaalishvili V.B., Nikolaev A.V. Insrumental tool of seismic microzonation. Patent of Russian Federation №209075.(20.12.1997).
7. Zalishvili V.B. Insrumental method of seismic microzonation. Vladikavkaz, 1998, 76 pp.
8. Zaalishvili V.B. Modern concept of seismic microzonation. Proceedings of the 11th European Conference on Earthquake Engineering. Paris, 1998, 11 pp.
9. E.H.Field, S.Kramer, A.-W.Elgamal at al. Nonlinear Site Response: Where we're at (a report from a SCEC/PEER seminar and workshop). Seismological research letters, Vol. 69, Numb.3, May/June 1998.

SEISMIC RESPONSE STUDY OF TWO-STOREY BUILDING IN EILAT USING WEAK AND STRONG MOTION DATA

Y. ZASLAVSKY, J. LEONOV , A. SHAPIRA
Seismology Division, The Geophysical Institute of Israel, P.O. Box 2286, Holon 58122, Israel. Tel. 972-3-5576060,
e-mail: jossleon@iprg.energy.gov.il

Abstract
A two-storey building was temporarily instrumented with two horizontal seismograph stations installed on the roof, basement and at the free field. Local earthquakes with magnitude 3.0-4.0 at a distance of 45-80 km were recorded. These week records are utilized to derive the building transfer function. The fundamental frequencies of the longitudinal and transverse vibrations of the buildings are found to be within 8-10 Hz and the corresponding damping ratios between 6-7%. The ambient vibration data were not used since the low-intensity ambient excitations do not put the building into resonant motion. The estimated parameters were then checked by cross-wise convolution - the input from one earthquake record is convoluted with the transfer function derived from the other in order to derive the building response caused by the former earthquake.

The linear dynamic characteristics, determined from the low-amplitude test, are used to forecast the buildings response, convoluting them with selected strong ground motion (the available accelerograms from Eilat/Aqaba Gulf earthquake with magnitude M_W=7.1). The building response is assessed from the derived accelerograms. The values obtained are compared with their counterparts, as determined from structural analysis, according to the guidelines of the Israeli Design Code for this type of structure.

Introduction

The town of Eilat is located in a seismically active segment of the Dead Sea Transform (DST) which, according to [15], could experience earthquakes of magnitudes greater than 6.0 on the Richter scale at close proximity to the town. The largest earthquake measured on the DST occurred on November 22, 1995 (Mw=7.1) and its epicenter was located about 100 km south of the town. The registered peak ground surface accelerations in EW, NS and vertical directions are 0.11, 0.122 and 0.11g,

M. Erdik et al. (eds.), Strong Motion Instrumentation for Civil Engineering Structures, 573–591.

respectively. Recent ongoing paleoseismicity investigations in the area [1] reveal the repeated occurrence of high magnitude earthquakes on a time scale of thousands of years, within a few kilometers of the center of Eilat.

Earthquake risk analysis is currently very limited owing to insufficient information regarding the vulnerability of existing buildings to ground shaking. An important step toward better evaluation of the earthquake risk to Eilat is associated with this study which focuses on determining the dynamic parameters of existing buildings [20, 21]. Reliable estimates of modal frequencies, stiffness and damping of structures are essential to the prediction of dynamic response under loading conditions associated with serviceability or structural safety. This is particularly important in areas where most of the buildings have not been subjected to seismic effects and their weaknesses are not yet clearly understood.

Studies of recorded responses of instrumented structures constitute an integral part of earthquake-risk reduction programs leading to improved design in many countries. Such investigations have not yet been carried out in Israel. The extensively instrumented structures facilitate a comprehensive study of their response [2, 4, 6, 8, 16]. The dynamic characteristics of real structures can be determined experimentally by simply monitoring the shaking of the structure caused by ambient excitation [5, 6, 7, 9, 11, 12, 13, 17, 18, 19]. However, this method has not been applied to determine a building frequency response function because of the unknown system input characteristics.

In this paper, our objectives are: (1) to assess the building dynamic characteristics from its response due to the recorded weak ground motions; (2) to forecast the building response under another recorded ground motion by means of the derived dynamic characteristics from the low-amplitude test and (3) to compare the estimated peak response values with their counterparts but calculated in accordance with the present Israeli Design Code 413 (IDC-413) requirements.

Data Acquisition

Ground motion was recorded using PC-SDA, a multi-channel, PC-based, digital data acquisition system [14]. The seismometers used are sensitive velocity transducers with a natural frequency of 1.0 Hz. The data are amplified, filtered within 0.2-25.0 Hz bandpass and sampled at a rate of 100 samples per second with a 12-bit/word digitizer. The instrument transfer function is checked and determined prior to performing measurements in order to facilitate the record signal transformation into true particle velocity data. The calibration is carried out by placing all seismometers at the same location and parallel to each other. These measurements also provide relative calibrations between the channels of the entire monitoring system.

The building was instrumented by placing the seismometers on the roof and in the basement – Figure 1. Each station was equipped with a pair of horizontal seismometers, placed perpendicular to each other and parallel to the outer walls of the structure. We defined the longitudinal axis as the NS direction and the transverse axis as the EW direction. An additional vertical seismometer was placed adjacent to the station in the basement.

During the observation period, two seismic events with good signal to noise ratio were recorded. Both events are local earthquakes which occurred in the Gulf of Eilat approximately 40km (M_L=2.8) and 80km (M_L=4.1) from the town. Microtremors (seismic noise) were also recorded during the observation process. Each recording session was 3 minutes long and automatically repeated every 30 minutes. Spectra estimation of the ambient noise was performed using the same procedure as for the earthquake recordings.

Building Description

The building is a low-rise, two storey, reinforced concrete structure (see figure 1). It is an irregular convex polygon in plan with an external staircase, contiguous to one of its edges. The roof and the second floor slabs protrude slightly over the first floor contour. In the vertical direction, two well-defined stories could be distinguished, as well as a basement located under part of the first/ground floor. The first floor is designed as service premises while the second floor is a large hall housing control devices. An architectural drawing from the preliminary design phase is the only information at our disposal; estimation of the structural scheme is, therefore, just an assumption. Most probably, the lateral resisting system of the building is a spatial two-storey irregular frame formed by columns and both floor slabs. The columns of the first/ground floor level are placed rather unevenly whilst, on the next level, they are, or could be, distributed evenly along the floor contour sides. This fact together with the multiple infill wall layout render the first floor stiffer and, hence, introduce irregularity into the vertical stiffness distribution.

Analysis of Low-Amplitude Data

Figure 2a shows typical particle velocity motions measured on the roof and in the basement with their corresponding Fourier amplitude spectra under ambient excitation (Figure 2b). The spectral amplitude graphs do not exhibit distinct peaks. Apparently the low-intensity wind excitation does not put the building into resonant motion (the energy from this type of ambient excitation is concentrated in the low frequency range). We could, however, detect the characteristics of the motion by using the spectral ratio between simultaneous measurements made on the roof and in the basement. The spectral ratio (Figure 2c) shows enhanced levels in the frequency ranges of 0.6-3.0 Hz and 8-10 Hz. The first band corresponds to the horizontally "pushing" wind, while the second band corresponds to the fundamental mode of the building.

During measurements, the building was exposed to some forced excitation (machine ignition or other sources). We have not identified this source, but it could be used to determine the building responses as shown in Figure 3a for both NS and EW directions. The spectra of these records are shown in Figure 3b. The highest spectral amplitude at the basement is observed within the frequency range of 6-12 Hz, whereas, on the roof, the dominant peak spectral amplitudes were consistently observed at frequency 8.4 Hz (NS direction) and 9.7 Hz (EW direction). Figure 3c displays the transfer function of the building inferred from the spectral ratio of the roof vibrations with respect to those at the basement. Hence, the dominant frequencies in NS and EW directions were detected 8.4 Hz and 9.9 Hz, respectively.

The Fourier spectra and the respective transfer functions of the building from the unidentified forced excitations are shown in Figure 4. Two distinct peaks at 8.4 Hz (for the NS direction) and at 9.5 Hz (for the EW direction) can also be identified. These frequencies are interpreted as the fundamental modes along both horizontal axes of the building. At this point, it should be emphasized that similar frequencies were concluded from different groups of records obtained at different times. Figure 5 illustrates NS components velocity seismograms of first earthquake which occurred in the Gulf of Eilat and was recorded on the roof and in the basement. The figure also shows the Fourier spectra and the estimated transfer functions of the building obtained from this event. The dominant feature of the roof spectra is high amplitudes in the frequency range about 6-10 Hz, whereas for the basement, a relative "bump" in the frequency range 2 to 5 Hz is evident. However, the transfer functions (spectral ratio "roof to basement") clearly exhibit peaks at 8.4 and 9.9 Hz. These peaks are associated with translational modes for NS and EW directions, respectively. For comparison, Figure 6 shows the estimated transfer functions using the earthquake excitation and unknown forced excitation. Note that the transitional modes for both excitations have the same frequencies and shape. Significant differences are observed in the frequency range 0.6-2 Hz; however, these do not affect accurate determination of the dynamic characteristics. We suggest that the observed differences are the result of the low signal-to-noise ratio of the measured unknown forced excitation and the noise contaminated the signals.

The consistency of the transfer/frequency response function that could be assessed from the base-roof records from both weak ground motions is checked by cross-wise convolution expressed in the following relations:

$$H_1(f)=\frac{S_1^{base}(f)}{S_1^{base}(f)},\ H_2(f)=\frac{S_2^{roof}(f)}{S_2^{base}(f)}$$

$$\Downarrow \tag{1}$$

$$\tilde{A}_1^{roof}(f)=H_2(f)\cdot A_1^{base}(f),\ \tilde{A}_2^{roof}(f)=H_1(f)\cdot A_2^{base}(f)$$

where $S_i^{base/roof}$ - power spectral density of the base, respectively the roof; $A_i^{base/roof}$ - Fourier amplitude spectra of the base, respectively the roof; $H_i(f)$ - frequency response function, derived from each of both earthquakes; $\tilde{A}_1^{roof}$ estimated roof amplitude spectrum, using $H_2(f)$; $\tilde{A}_2^{roof}$ - estimated roof amplitude spectrum, using $H_1(f)$, $i=1,2$.

Both estimated roof amplitude spectra are then inversely transformed into the time domain and compared with their recorded counterparts as shown on Figures 7 and 8. The results are quite encouraging with a slight advantage to the transfer function derived from the first earthquake deconvolution. This conclusion is also confirmed by the calculated cumulative energy indexes of the original and estimated signals and depicted as Husid graphs in Figure 9.

The consistency of the frequency response function derived independently from each of the recorded earthquakes is further checked through the coherency function:

$$\gamma_i^2=\frac{|S_{br_i}(f)|^2}{S_{bb_i}(f)\cdot S_{rr_i}(f)} \tag{2}$$

with S_{br_i} - base-roof cross-spectral density functions; S_{bb_i} - input/base*(b)*;power spectral density; S_{rr_i} - output/roof*(r)* power spectral density, *i*=1,2.

From Figure 10 it is evident that the measurements are reasonably reliable with a fair amount of small noise effect. The deconvoluted transfer function could, thus, be utilised in further estimation of the building response.

In addition, the damping ratio, ξ, of this building, is estimated using the so-called dynamic amplification factor D [3] evaluated at resonance

$$\xi=\frac{1}{2D} \tag{3}$$

The damping at the fundamental mode is 6% and 7% for NS and EW directions, respectively.

Building Response to Strong Motion

Having verified the consistency and accuracy of the derived frequency response function, we then proceeded with the estimation of the building response under other recorded ground motions. We selected the Eilat/Aqaba Gulf earthquake as the strongest ground motion ever recorded in Israel. The procedure is a straightforward convolution in frequency domain and successive inversion in the time domain. The time history of the estimated roof response along both building axes is given in Figure 11. The peak acceleration values show that the structure would have been slightly to moderately excited by this earthquake.

IDC-413 Seismic Design Forces

Bearing in mind the sparsity of information about the structural resistance elements, the only analysis that might be performed is the equivalent static analysis. According to the code description, the building is regular in both plane and height. The lateral loads in IDC-413 are specified in terms of the base shear, defined according to:

$$F_H = C_d \sum_i G_i , \tag{4}$$

where C_d is the design seismic coefficient and G_i is the effective dead load of the i-th floor.
The design seismic coefficient is in turn given by the expression

$$C_d = \frac{R_a \cdot I \cdot Z}{K} , \tag{5}$$

in which R_a - the site coefficient, specified by $R_a = 1.25 \cdot \frac{S}{T^{2/3}}$, S - spectral magnification factor depending on the local soil conditions, T.- fundamental vibration period of the structure determined by $T = 0.049 \cdot H^{3/4}$, H - overall building height, I - building importance/significance factor, Z - seismic zone factor, taken from the seismic zone map, K - factor accounting for the provided ductility level.
For the current case and specific conditions, the parameters take the following values: $R_a = 2.75$, $S = 1.2$, $I = 1$, $Z = 2$, $K = 3.5$, $H = 2$ h, $h = 3.00$, $G_1 = \nu G$, $G_2 = G$.
Here, because information is scarce, the floor dead loads are taken implicitly via their ratio - $\upsilon = \frac{G_1}{G_2}$.

After a sequence of straightforward algebraic operations and assuming the first mode ordinates to be proportional to the floor mass elevations, the following expression is obtained for the normalised roof acceleration

$$\frac{a_2}{g} = \frac{2(\upsilon+1)}{\upsilon+2} \cdot \frac{0.25\upsilon+1}{0.5\upsilon+1} \cdot \frac{R_a IZ}{K} ; \tag{6}$$

Figure 12 shows the normalised root acceleration as a function of the floor dead loads ratio factor - υ for different ductility levels. As may be observed the roof acceleration values derived in accordance with the design code prescription (equivalent static analysis), are much greater (even for high ductility) than the peak values derived through standard convolution operations (Figure 11). Such a relationship between the values obtained by the two methods suggests linear building response which is confirmed by the inspection carried out shortly after the considered event.

Conclusions

Our major conclusions are as follows:

1. AS the low-intensity ambient excitations do not put the building into resonance motion the ambient vibration data were not used.

2. The use of relatively simple spectral analysis techniques of low amplitude seismic motions facilitates determination of the significant dynamic characteristics of the analysed building. The fundamental transitional modal frequencies, for the building for either orthogonal axis, are 8.5-10 Hz. The damping percentages are 6-7%.

3. The fundamental frequency of a reinforced concrete structure calculated according to the Israeli Design Code predicts natural frequency of approximately 5 Hz for two storey buildings. The difference between the calculated and the observed natural frequency is more critical regarding estimation of structure vulnerability to seismic loads, particularly for Eilat.

4. According to the results of the comparative analysis the examined building should has behaved linearly, which is indeed the conclusion reached after it was visually screened after the selected event.

5. Dynamic characteristics of low-rise buildings obtained from seismic measurements, combined with knowledge for the dominant site frequency, form the basis for seismic risk zonation and earthquake scenarios for the Eilat urban area.

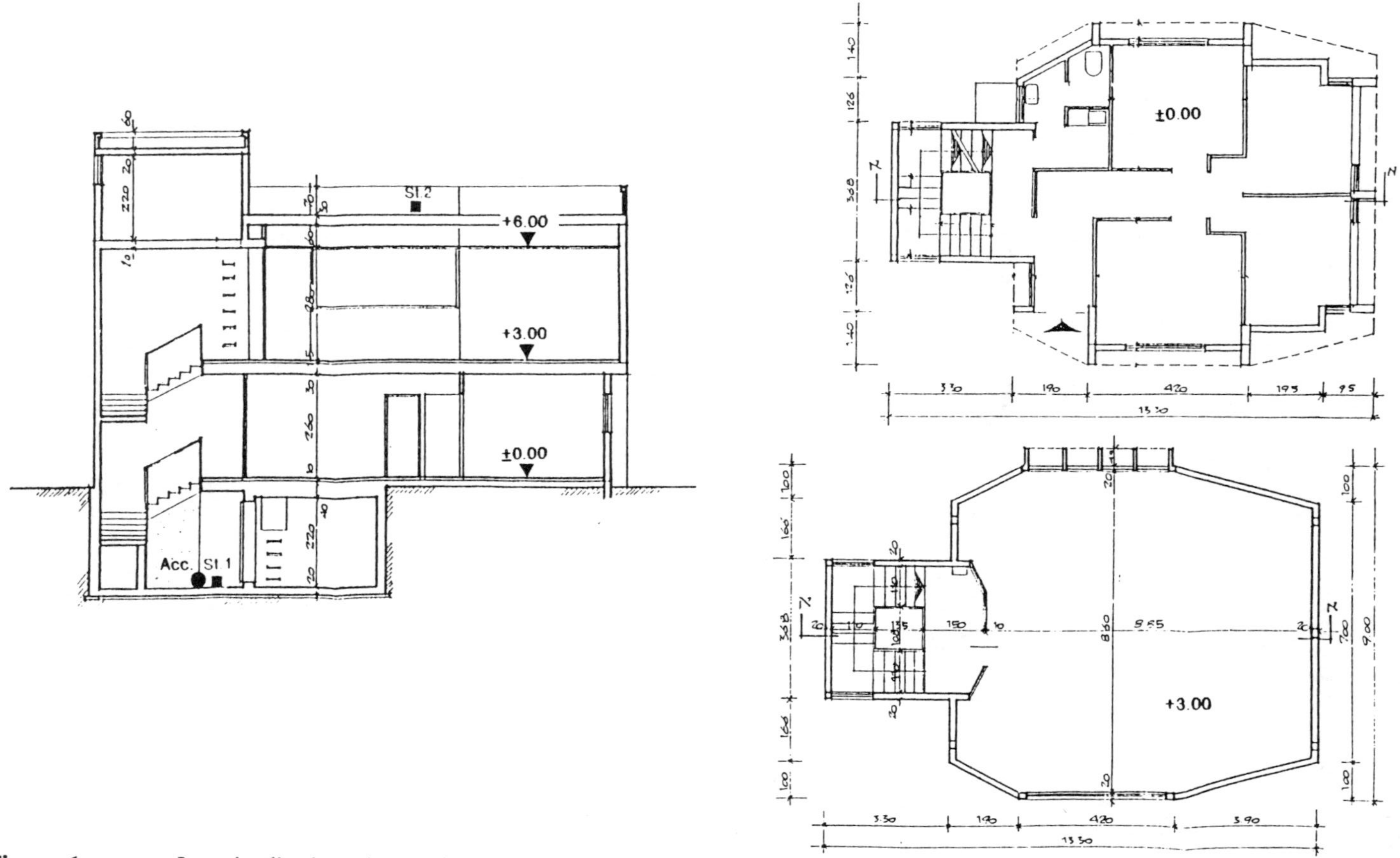

Figure 1: Longitudinal section and Vertical plans of the building.

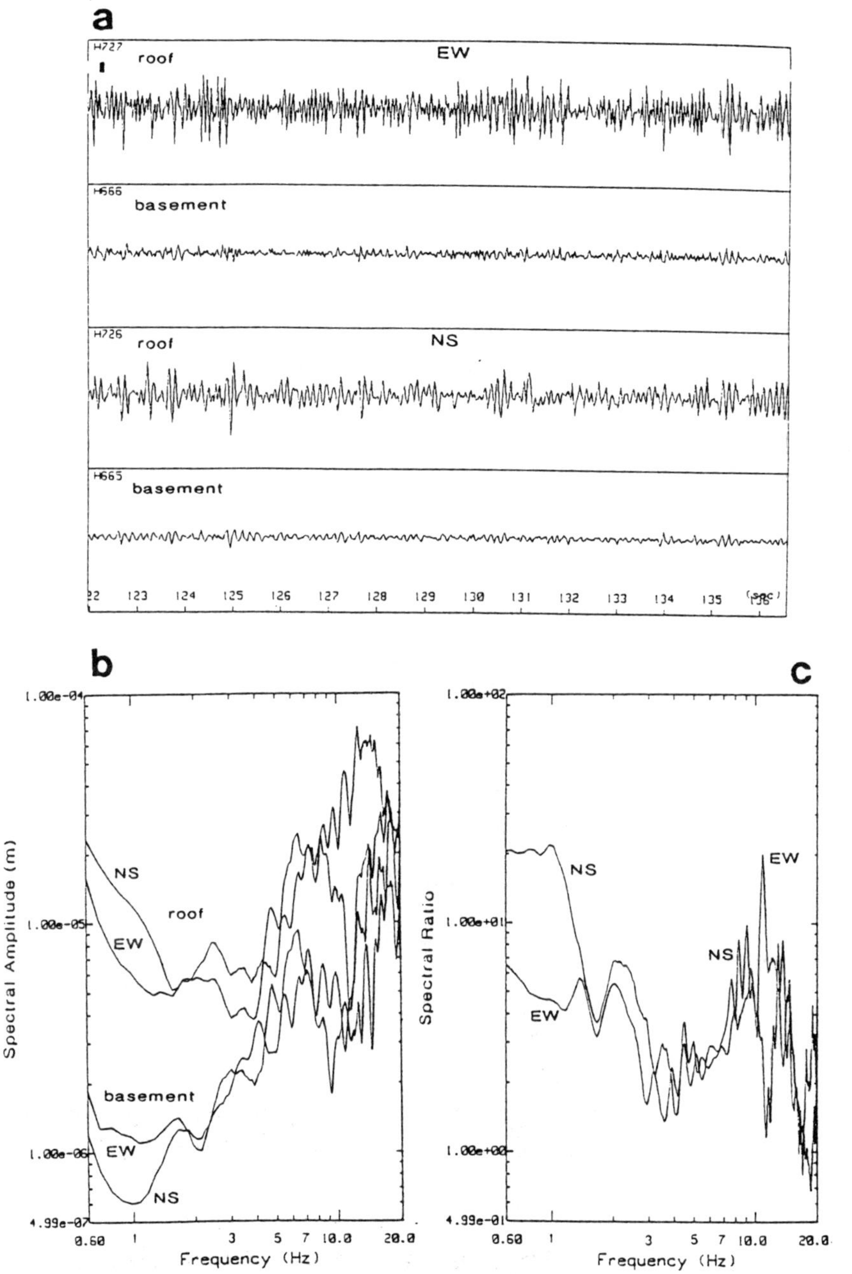

Figure 2: (**a**) Response of the building under ambient excitation; (**b**) the corresponding *Fourier* spectra obtained on the roof and in the basement; (**c**) spectral ratio (roof over basement).

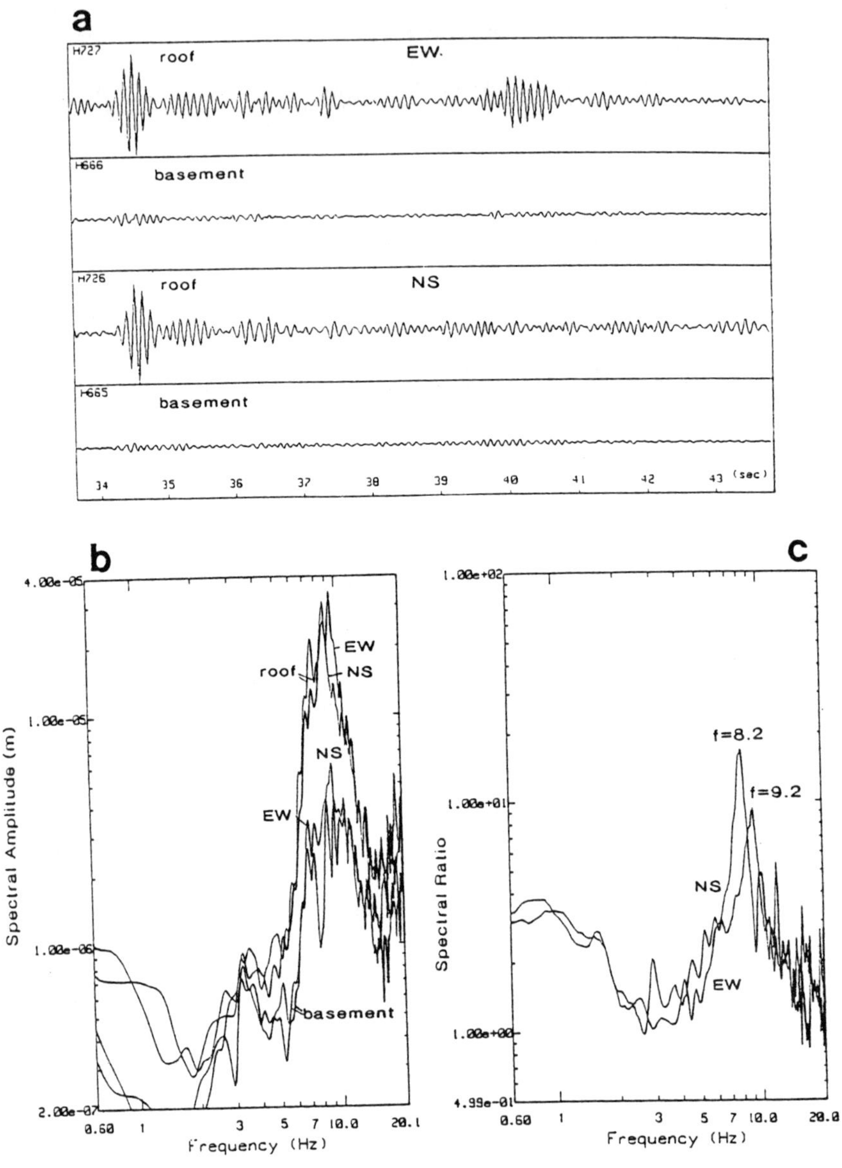

Figure 3: (**a**) Response of the building under unidentified forced excitation; (**b**) the corresponding *Fourier* spectra on the roof and in the basement; (**c**) Transfer functions of building (spectral ratio of the roof with respect to the basement).

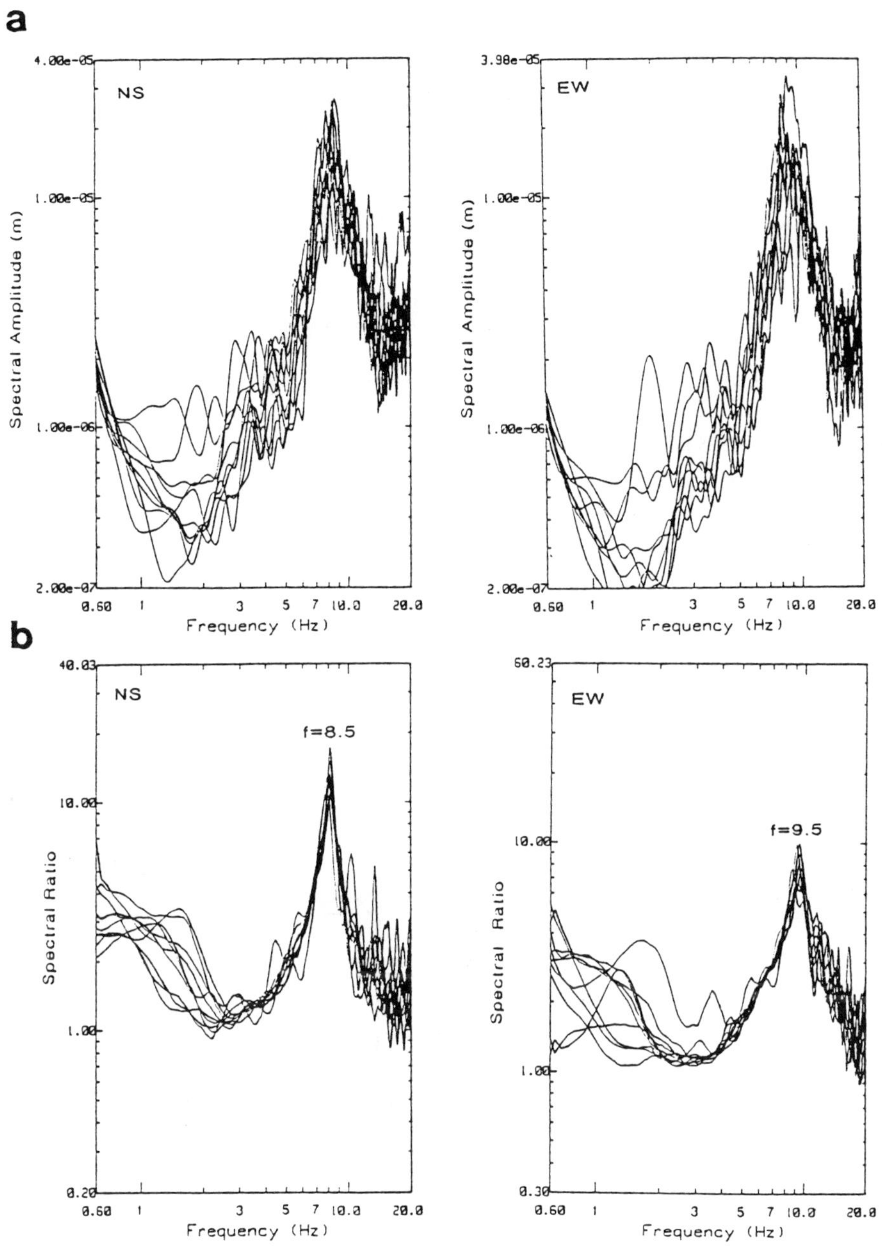

Figure 4: (**a**) Individual *Fourier* spectra on the roof and (**b**) transfer functions from an unidentified forced excitation of the building.

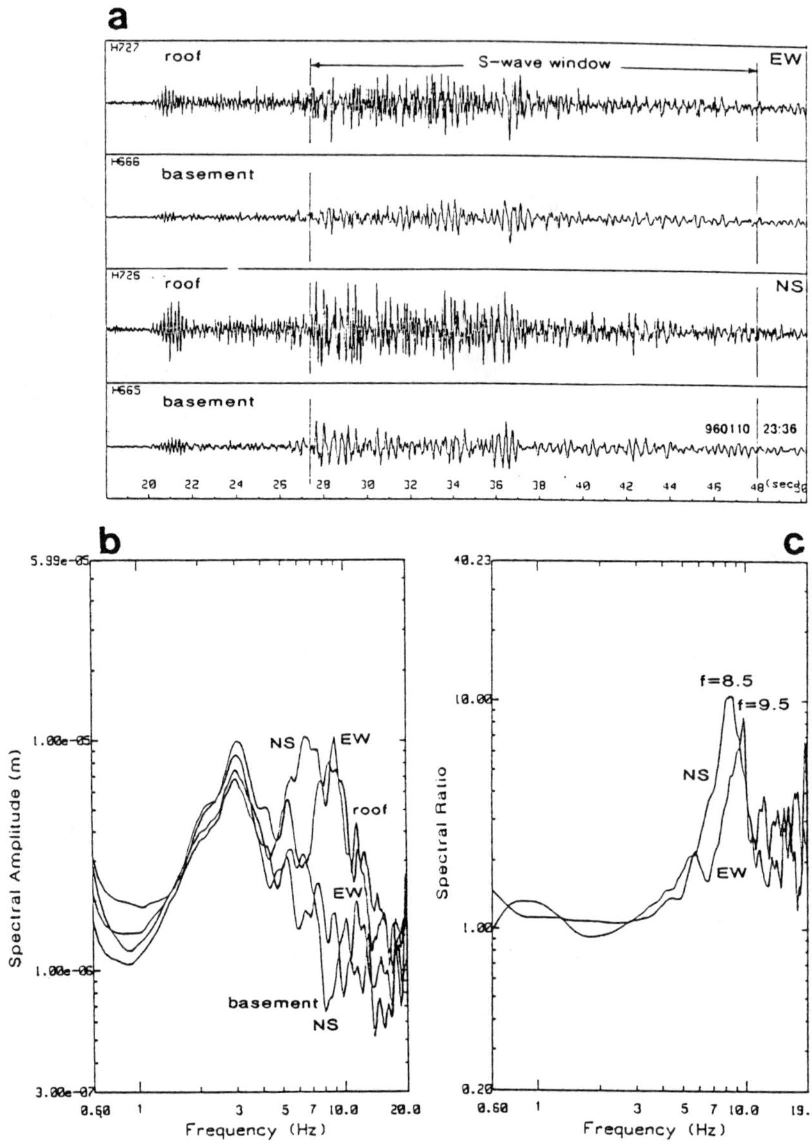

Figure 5: (**a**) Horizontal components records of an earthquake in the *Gulf of Eilat* recorded at the building (M_L=2.8 R=45 km); (**b**) the corresponding *Fourier* spectra and (**c**) transfer functions of the building.

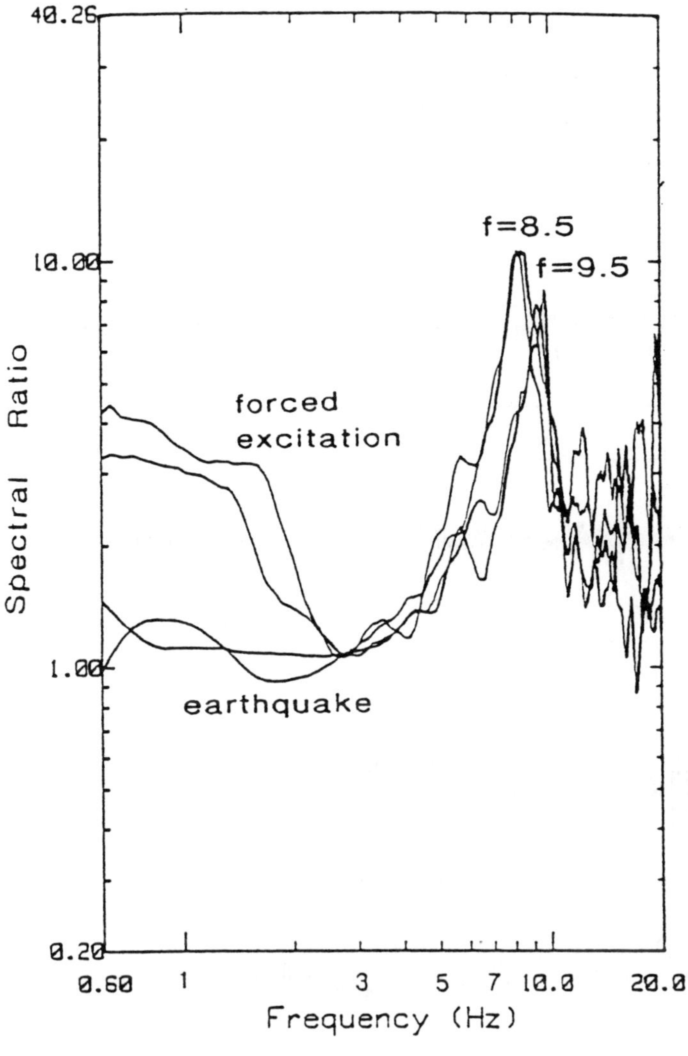

Figure 6: (**a**) Records of the horizontal components of an earthquake in the Gulf of Eilat, recorded at the building (M_L= 4.1 R=80 km); (**b**) the corresponding *Fourier* spectra and (**c**) transfer functions of the building.

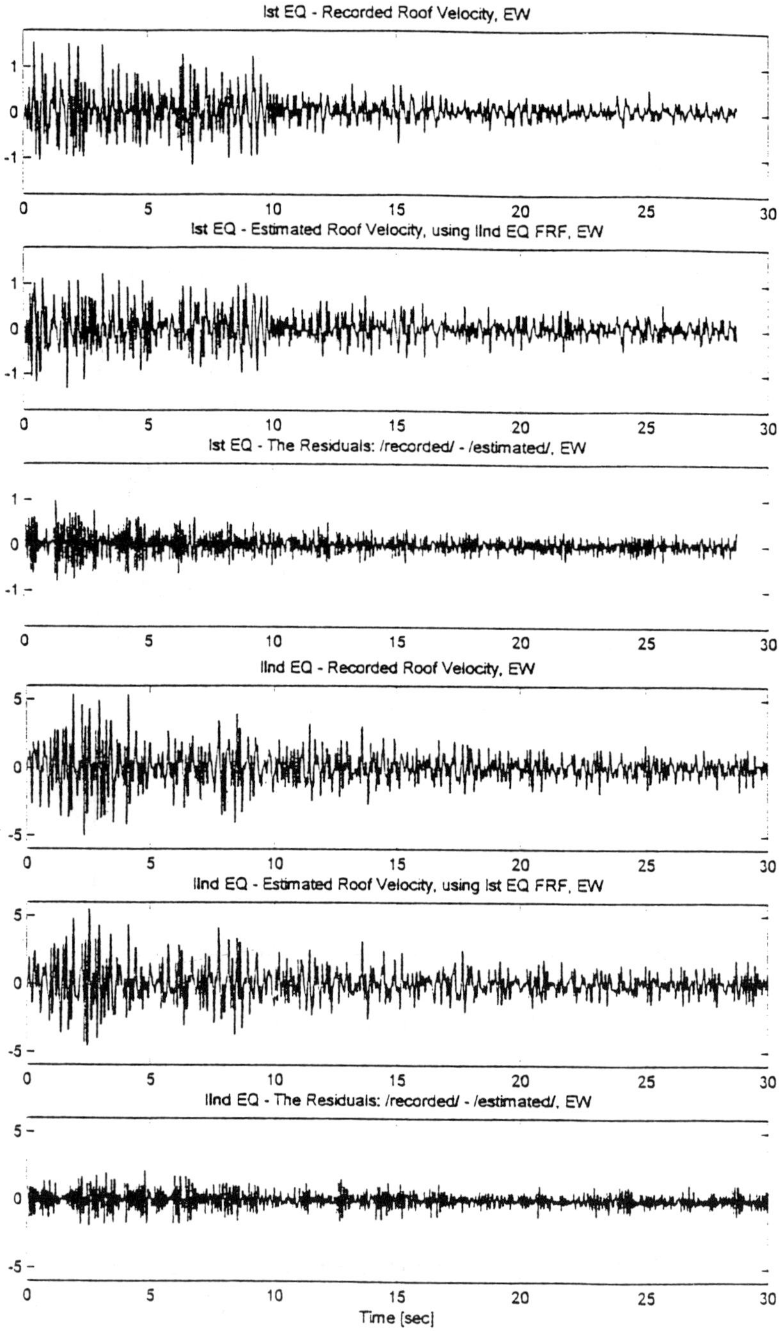

Figure 7: Recorded and Estimated Roof response in EW direction

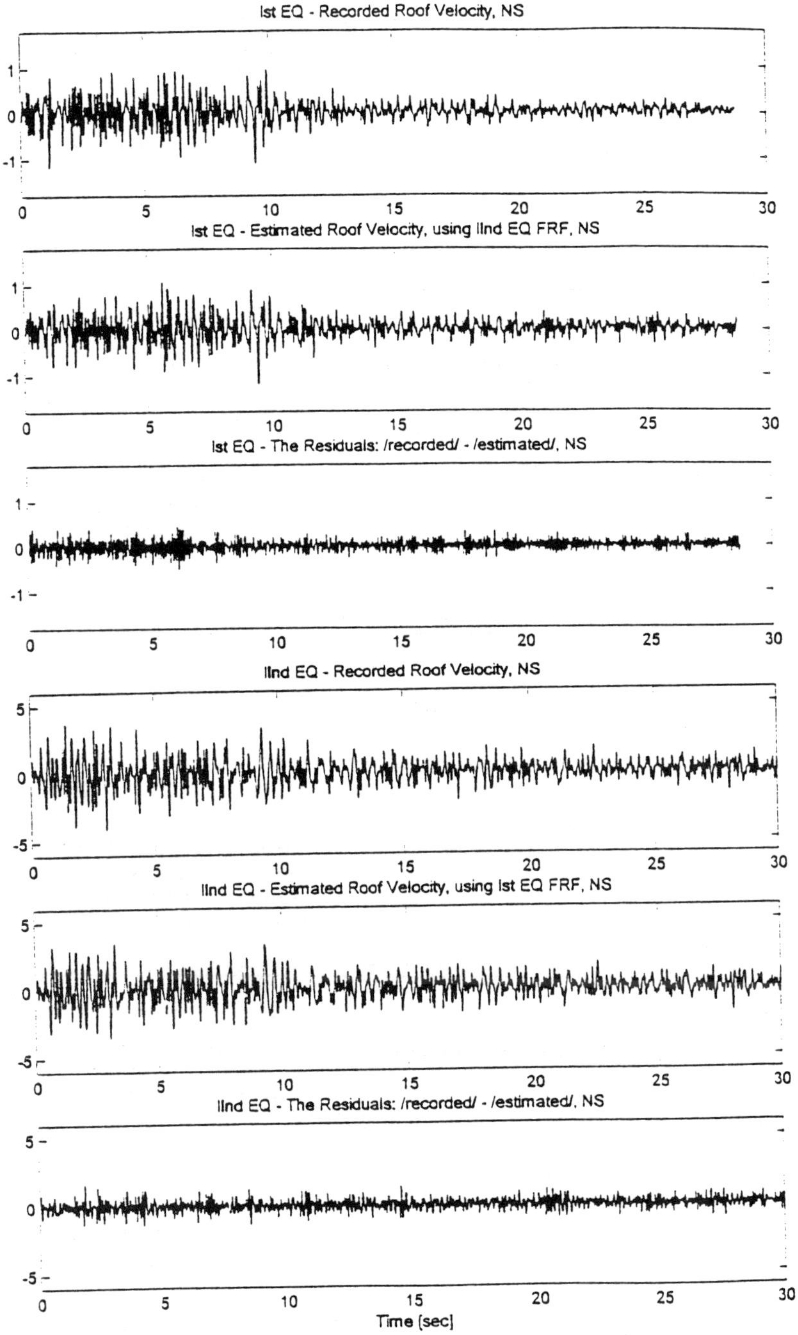

Figure 8: Recorded and Estimated Roof response in NS direction

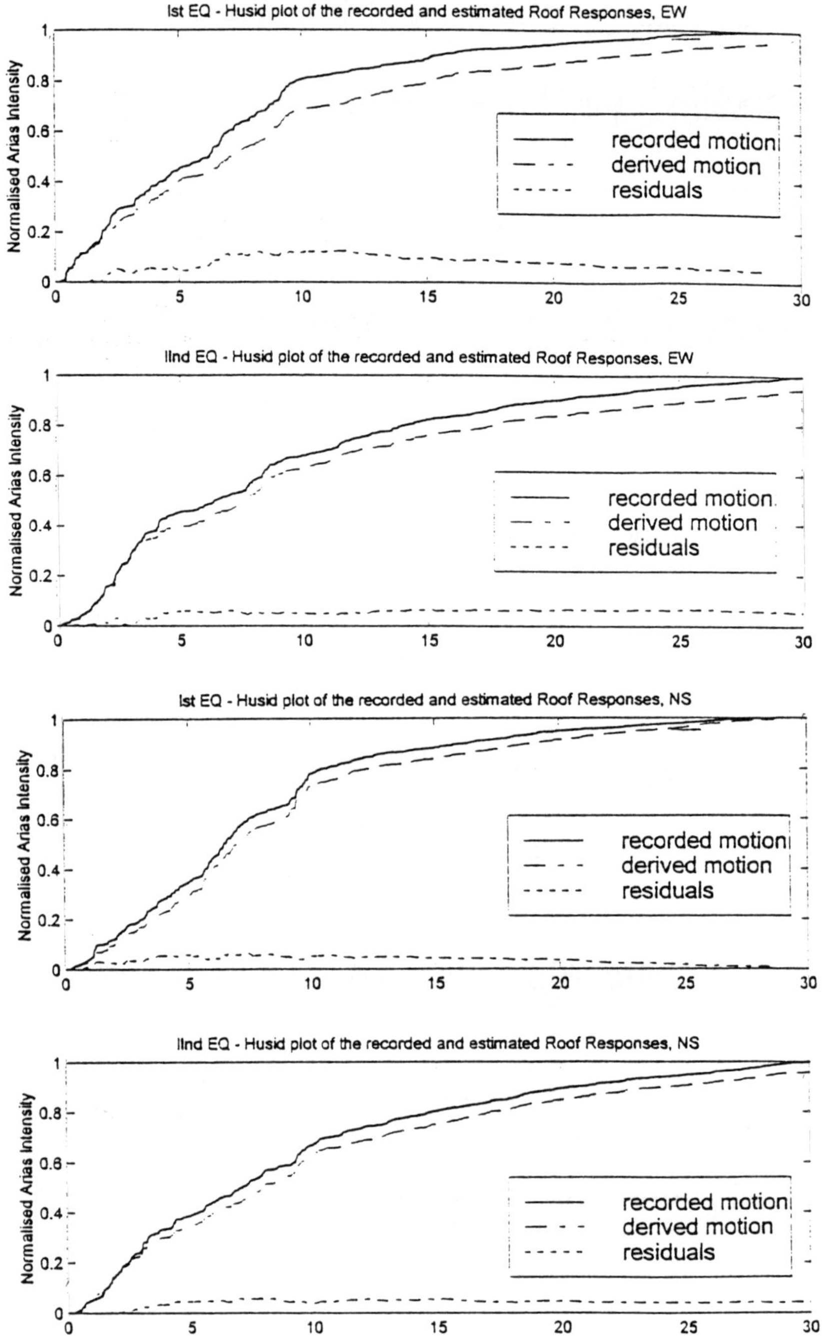

Figure 9: Arias Intensities of the recorded and estimated roof responses

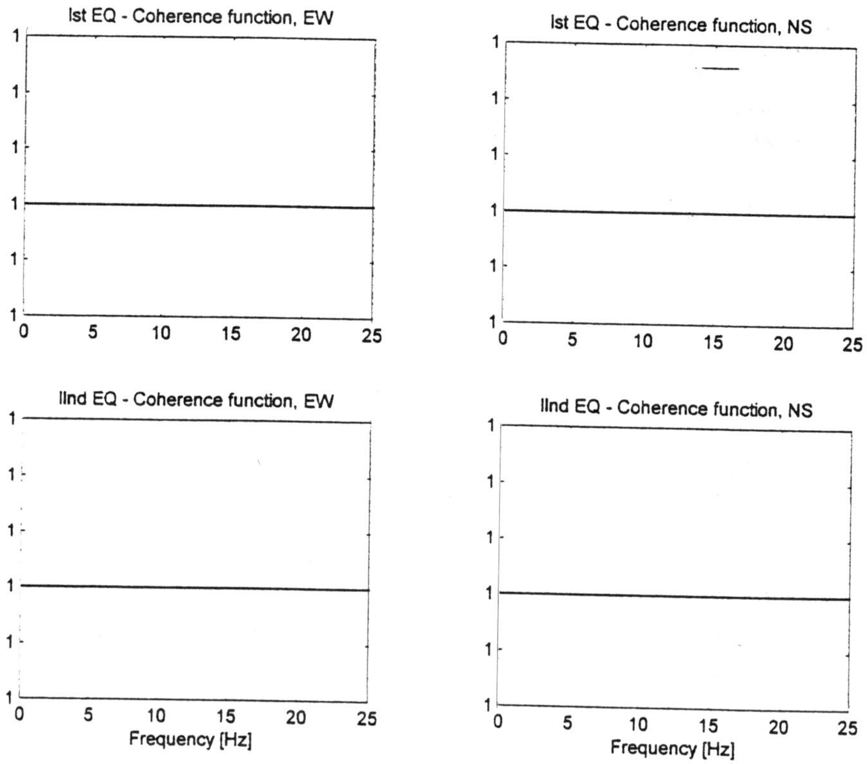

Figure 10: Coherence functions of the recorded earthquakes

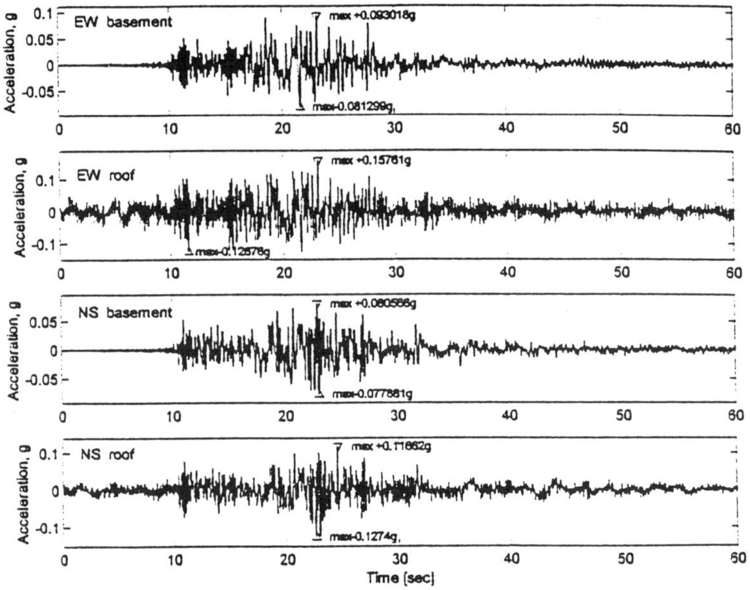

Figure 11: Basement (recorded) and Roof (estimated) acceleration under the *Gulf of Eilat* earthquake

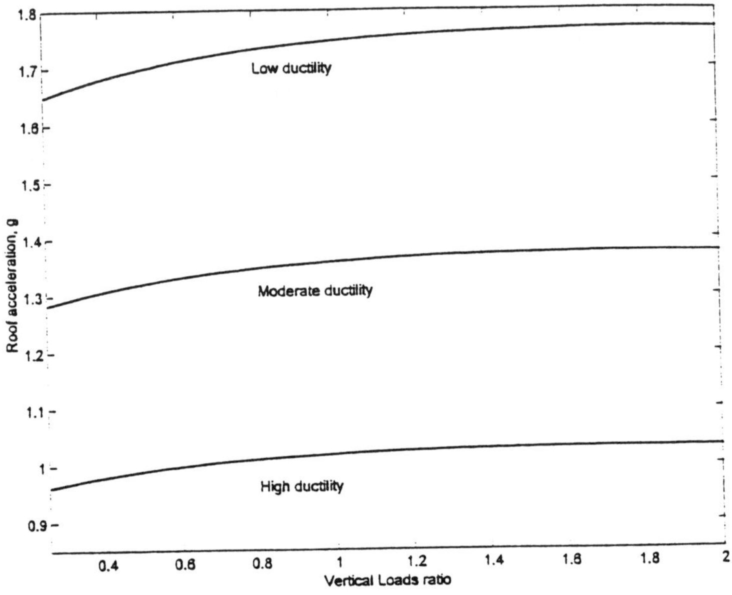

Figure 12: Roof peak acceleration as a function of the floors dead load ratio and the provided ductility level, calculated according to the equivalent static analysis procedure from the IDC - 413.

References

1. Amit, R., Harrison, J.B.J., Enzel, Y. and Parat, N. (1994) Paleoseismicity in the southern Arava rift, Israel, USGS, Open File Report 94-568.
2. Bard, P., Afra, H. and Argoul, P. (1992) Dynamic behaviour of buildings: experiment results from strong motion data, Recent Advances in Earthquake Engineering and Structural Dynamics, 441-478.
3. Bendat, J. and Piersol, A. (1986) Random data: analysis and measurement procedures, John Wiley, New York.
4. Bongiovanni, G., Celebi, M. and Safak, E. (1987) Seismic rocking response of a triangular building founded on sand, Earthquake Spectra, Vol. 3, No. 4, 793-809.
5. Carydis, P. and Mouzakis, P. (1986) Small amplitude vibration measurements of buildings undamaged, damaged and repaired after earthquakes, Earthquake Spectra, Vol. 2, No. 3, 515-535.
6. Celebi, M., Phan, L.T. and Marshall, R.D. (1993) Dynamic characteristics of five tall buildings during strong and low-amplitude motions, The Structural Design of Tall Buildings, Vol. 2, 1-15.
7. Gavin, H., Yuan, S., Grosman, E., Pekelis, E. and Jacob, K. (1992) Low-level dynamic characteristics of four tall flat-plate buildings in New York City, NCEER-92-0034, 156pp.
8. Luco, J.E., Trifunac, M.D. and Wong, H.L. (1987) On the apparent change in dynamic behavior of a nine storey reinforced concrete building, Bull. Seis. Soc. Am., 77, 1961-1983.
9. Marshall, R.D., Phan, L.T. and Celebi, M. (1994) Full-scale measurement of building response to ambient vibration and the Loma Prieta earthquake, 5th U.S. National Conference on Earthquake Engineering, Illinois, 661-670.
10. Meli, R., Faccioli, E., Muria-Vila, D., Quaas, R. and Paolucci, R. (1998) A study of site effects and seismic response of an instrumented building in Mexico City, Journal of Earthquake Engineering, Vol.2, No. 1, 89-111.
11. Mendoza, L., Reyes, A. and Luco, J.E. (1991) Ambient vibration test of the Mexical General Hospital, Earthquake Spectra, Vol. 7, 281-300.
12. Pardoen, G.C. (1983) Ambient vibration test of the Imperial County Services Building, Bull. Seism. Soc. Am., 73, 1895-1902.
13. Schuster, N.D., Ventura, C.E., Felber, A. and Pao, J. (1994) Dynamic characteristics of a 32 storey high-rise building during construction, 5th U.S. National Conference on Earthquake Engineering, Illinois, 701-710.
14. Shapira, A. and Avirav, V. (1995) PC-SDA Operation manual, IPRG Report Z1/567/79, 24pp.
15. Shapira, A. and Shamir, G. (1994) Seismicity parameters of seismogenic zones in and around Israel, IPRG Report Z1/567/79(109), 20pp.
16. U.S. Geological Survey Professional Paper 1552-C, (1998) U. S. Government Printing Office, Washington.
17. Zaslavsky, Y. and Alexandrova, N. (1982) The Method of Definition of the Resonance Frequency Value for an Earth Dam, Author Certificate No. 10-0.546 AE 02 B 7/00.
18. Zaslavsky, Y. and Shapira, A. (1994) Empirical determination of the dynamic characteristics of a reinforced concrete construction, Proceedings, 17th European Seminar on Earthquake Engineering, Haifa, Israel, 499-508.
19. Zaslavsky, Y., and Shapira, A. (1997) Empirical estimates of modal parameters of full scale structures, European Earthquake Engineering, No 1, 26-36
20. Zaslavsky, Y., and Shapira, A. (1997) Dynamic characteristics of low rise building in Eilat using seismic measurements, GII Report 550/87/96(118), 44pp.
21. Zaslavsky, Y., Shapira, A. and Pinsky, V. (1998) Dynamic characteristics of two-three-storey building obtained by seismological measurements, Proceedings of XXVI General Assembly of the European Seismological Commission (ESC), Tel Aviv, Israel, 213-217.

STRONG MOTION INSTRUMENTATION FOR STRUCTURES OF CIVIL ENGINEERING AND ECONOMICAL ASPECTS OF PLANNING OF TERRITORY OF BIG CITIES

V.ZAALISHVILI, I.TIMCHENKO, V.KACHARAVA, Z.ZAALISHVILI

Center of Applied Geophysics, Engineering Seismology and Seismic Protection of Structures of Georgian Geophysical Society; Institute of Structural Mechanics and Earthquake Engineering of Georgian Academy of Sciencies. 1, M.Aleksidze str., Tbilisi 380093, Georgia

1. Abstract

A number of strong earthquakes have taken place in last several years, and they were characterized by suddenly high peak values of accelerations (0.4-2.0 g). The existing position of structural engineers and scientists at the usage of so-called design accelerations is based on rather formal connection between acceleration and seismic intensity. So, according to Mercalli seismic scale (MMI), the design acceleration, which corresponds to earthquake intensity 9 degrees, has changed 5 times (from 0.1 to 0.5g) over the last 40 years. From the other side during high accelerations (2.0g, Nortridge, 1995 etc.) the intensity has reached even above-mentioned intensity of 9. At the same time a buildings designed at the intensity of 7 (e.g. 0.1g), very often can endure the seismic intensity of 9 degrees (0.4g, Sakhalin, 1995).

The existing factors of seismic hazard are correlated badly with instrumental characteristics of earthquakes. All above-mentioned sets the problem for selection of parameter, which is directly connected with destructive seismic effect.

Based at this and other views, the paper shows necessity of such type of equipment (strong motion instrumentation) which will give opportunity to test the reaction of buildings at seismic effect accurately, based on direct energy parameters or more closely on parameter which characterizes seismic effect (acceleration, velocity etc.).

In the regions with low seismic intensity (Georgia), which means the country where the return period of strong earthquakes is rather long, it should be used the equipment (strong motion instrumentation) which would have widen working characteristics, in order to accept the weak signals.

The paper contains economical aspects of planning of urban areas, based on hazard mitigation and vulnerability of buildings and structures in the case of their provision by

M. Erdik et al. (eds.), Strong Motion Instrumentation for Civil Engineering Structures, 593–602.

corresponding strong motion instrumentation. The following part contains considerations about the creation of monitoring instrumental systems in buildings and at major structures of urban areas of Georgia.

2. Introduction

The obtaining the high quality field records is in direct connection with quality and regularity of receiving and recording type of equipment. The equipment traditionally used in former USSR, is very well known (SSRZ etc.). The majority of the CIS countries specialists, in their researches over the last few years, have used foreign equipment. The well-known Kinemetrics Inc. (Pasadena, California) has made significant contributions to the development of world instrumental networks, and offers variety of systems, differing by their function and characteristics: Accelerograph of the strong motions "Etna" (3 channels, 18 bits, volume range 108 Db) etc. Also it should be noted Swiss accelerograph of strong motions "SMACH" which has been used successfully during Spitak (Armenia, 1988) and Racha (Georgia, 1991) earthquakes. During establishment of ARRAY SMART-1 (Taiwan, 1982-1987) and during Chernogorya (Yugoslavia, 1979), it were used "SMA-1" and "SMA-2" accelerographs. The result of different indicators analyses of the ground motion, which we obtained after elaboration of instrumental recordings of strong earthquakes, have shown the high reliability of the mentioned-above equipment.

The hardware in former USSR was not so lacked as few years before. But at the same time the known examples of modern seismographs in USSR traditionally lacked to the foreign analogues even from the external point of view (exterior, overall dimensions, weight etc.) and require a radical restructuring.

It should be noticed, that Soviet equipment, used in researches (for example in seismic microzonation), often hasn't corresponded to the purposes to which it was designed. Examples: big weight, necessity of manual or semi-automatic way of convention of received data for calculation of oscillation spectrum, irregularity of the technical parameters, causing the necessity of complex works for calibration of equipment (Fig. 1).

Nevertheless, the set problems, were generally solved at sufficiently high level. The characteristics of seismic microzonation methodology, based on confrontation of the same-type parameters of soil movement gives opportunity to bypass many difficulties and receive outcomes authentic enough.

Figure 1. Calibration of instrumentation on a vibration table

3. Instrumental observations at strong earthquakes

The resultant ground motion is divided by energy potential formally into two groups: weak motion from remote or weak earthquake and strong motion from close or strong earthquake. During the elaboration of seismic hazard they generally operate by parameters of the strong ground motion [13]. Therefore in absence of strong earthquakes it is necessary to obtain records in the near-field region of earthquake [5, 6], because acceleration peaks damping as strong as we move from epicenter [10, 11].

In early researches they traditionally consider that the application of accelerations, is the best way to characterize the earthquake hazard. Earthquake engineers and specialists preferred especially these applications, because it gives them the possibility to obtain the seismic forces. Generally they used horizontal values of acceleration, although the vertical acceleration can reach significant values. The high values of vertical accelerations (1.3-1.4 g) were registered in 1976 (f. USSR). There was only one such high horizontal acceleration registered before - 1.15-1.25g (USA, 1971). Concerning the relationship of peak accelerations variations range with destructive potential, we can consider the following example.

At Mexico earthquake (1985) with M=8.1 the maximal value of acceleration reached 0.17g, besides in that region a big number of buildings were damaged and thousands of people died. The earthquake at northeast of Ohio State (USA), four months later, has made only mild damages (ceilings, window-glasses). The ground motion recorded at NPS in town Perri (17 km from epicenter) has reached 0.18g. But there were no damages and even no sensations of the event. The recordings in Mexico-city were got by long-period motion, which was continuing for two minutes, in Perri- by short-period motion for only one second [19,20].

Deriving from this, as a destructive potential we can use the value of ground motion duration [15] the increasing of which, during equal accelerations, depends upon peak ground accelerations. So, the displacement and velocity of particles in soft soils are greater than ones in rock soils, but the accelerations in both coincide. The destructive potential of earthquake isn't very reliable at different levels [11]. So, during acceleration of 0.44g only weak damages were observed. [8, 18, 21].

During Neftegorsk earthquake (Russia, 1995) in Sakhalin, over the 2/3 of the population in little town, who lived in the buildings constructed without seismic protection measures, died. At the same time, buildings constructed according these measures and designed at the earthquake intensity of 7 degree (0.1g) endured the earthquake with the degree of 9-10 (0.4g). This fact delights the formality of usage the peak accelerations as characteristic of seismic effect, and also showed the high potential of seismic protection measures, and at last the high quality of code buildings in former USSR.

It was increased the number of oscillation velocity usage in calculations over the last year, because their application gives opportunity to estimate directly the kinetic energy absorbed by the considered object. It is completely not clear, which acceleration is the borderline between weak and strong motions. If we judge from above mentioned MM scale, we can see that value of peak accelerations increased at least five times and reached respectively degree of 7-8 (0.1-0.5g). at the same time the registrations of accelerations, which exceed the gravity force, show the infinity and illegality of the application of this parameter in seismic constructions at the first sight. It should be noted that although peak values of acceleration play the important role in seismic constructions, the application of only some effective value is more suitable for this case. For instance for Pakoma segment, the effective acceleration was determined at 0.75g.

Based at theoretical researches of Ambraseys, it was shown possibility of earthquake with 2.0g. Generally, the engineers during examinations of epicentral zone of strong earthquake, have concluded that there should be a strong acceleration. Before appearance of first recordings of strong motion, the corresponding accelerations was calculated using displacements of some objects etc [1].

The calculations of Ambraseys were confirmed 20 years later. In Iran (1990), they registered the peak accelerations of 0.5g and the epicenter accelerations of 0.7g [17]. During Hokaido earthquake (Japan, 1993) the acceleration reached 0.4-0.9g, with earthquake magnitude M=7.8, and during aftershocks of the same earthquakes (M=6.4) the acceleration reached 1.6g. It is necessary to mark, that the second horizontal part showed the four times less acceleration- 0.4g (Kudo, 1995). It gave the reason to doubt about the verity of the registrations. The registration of acceleration during Nortridge earthquake (1994) with value of 1.8g, finally approved the conclusion about absence of sharp outer border of acceleration scale [9]. At the same time, the experience showed, that it is impossible to determine the seismic coercion force at certain point by one constant value of acceleration, depended upon magnitude and epicentral interval, during estimation of seismic danger. It should be noted that damage of buildings were very minor at the segment, where was registered the high value of acceleration 1.6g [16]. Author explains this phenomena by unlinearity of soil motion.

In the case of earthquake of given magnitude and distance, the distribution of acceleration is held under normal logarithmic law. The vacillation of accelerations strongly influences the probability of exceeding the given threshold by forecasted acceleration (Bender, 1984). From the engineer point of view, the frequency characteristics of soil vibrations of building bases and buildings itself, are very interesting. As it was mentioned above, many specialists consider velocity as more reliable parameter of damage. Unfortunately non-of the parameters has become such indicator (even velocity). The different measuring based on combination of parameters characterized by variable success [2].

The analysis of 250 recordings established that the best correlation between beginnings of damage and soil movement can be processed with the help of Arias intensity [4]. But it should be noticed that coincidence of results is very seldom achievement. In such conditions, the practical and scientific interest presents the usage of early accepted into seismic microzonation experience parameter, which directly related with amplitude, frequency, and duration of oscillations. Under this parameter we can count the area of sub-spectral segment of the area of "real" spectrum [24] of strong earthquakes.

a)

b)

Figure 2. Racha earthquake (Georgia, 29.04.1991): a) general view and damages after main shock of hostel building of technical training college in Ambrolauri; b) collapse of middle part of a building after an after-shock (3.05.1991)

Among the misunderstandings of seismicity and models of seismic risk, the major role play the estimations of magnitude and intensity that aren't following the same distributions. Deriving from this, the distribution of acceleration peaks can not be equally distributed neither in near-field zone, because of independence from magnitude, nor in distant zone, because of infinite distributions of magnitudes themselves. In other words the usage of acceleration peaks, as index of damaging is not quite sufficient [7,12,14].

In other words, usage of peak ground accelerations, as a parameter of damage is not sufficient. Their usage should be grounded also on full information picture of statistical correlation. In such sense instead of peaks of accelerations, probably, it is much more expedient to use not an efficient value of acceleration (Newmark), being attempt to cobble damaged building of aged representations, but a quantitative expression of a spectrum. Then the allocation of more "thin" parameters of reaction of a ground strata or system will have other, quite practical output.

The positive experience of international scientific cooperation, received at the Spitak earthquake (Armenia, 1988), has facilitated arrival of international epicentral expedition to Georgia after the main shock of the Racha earthquake, organizing a network of digital Swiss recording accelerometers SMACH on May 2, 1991 and recorded the following day an earthquake of M = 5.3 with peak ground acceleration - 0.53g. The network included 5 digital tree-component autonomous recording accelerometers established in different sites of this region [3]. The operational frequency of instrumentation was 0.3-30 Hz.

It is known, that the damage of a building, structure and any physical system is anticipated by absorption of seismic energy. After excess of capabilities of absorption of a structure, local damages occur, which one augment absorption etc.

The analysis of instrumentation recordings of a number of strong earthquakes by the help of the multifactor regression analysis for a near-field region of earthquakes (Armenia, 1988; Georgia, 1991) has allowed to receive expression for a parameter of absorption of seismic energy by a soil massif [24]:

$$\alpha \approx 0.82 \frac{f_{CB} t}{\sqrt{M}} \tag{1}$$

where: f_{CB}- an average weighted oscillation frequency of a soil;

$$f_{CB} = \frac{\sum f_i A_i}{\sum A_i}; \tag{2}$$

t - duration of ground vibration; M - magnitude of earthquake.

Thus, the absorption of seismic energy in soft soils is directly proportional to average-weighted frequency and duration of vibrations and in inverse proportion to a magnitude of earthquake. Last circumstance is well known from the data of engineering macroseismic examination of territories, subjected to destructive earthquakes and according instrumentation data.

In rocky soils ratio [24] the relationship is:

$$\alpha \approx 0.02 f^2{}_{c\theta} t\sqrt{ar} \tag{3}$$

where: a – ground acceleration; r - epicentral distance.

Thus in rocky soils the absorption is directly proportional to acceleration and does not depend obviously on a magnitude.

This circumstance apparently is the cause of poor comprehension of connection of acceleration with the intensity of earthquake. Confrontation of the indicated outcomes to the data, obtained on instrumentation recordings of ARRAY SMART 1 (Taiwan) and other earthquakes has shown good conformity [22]. It is necessary to continue researches in this direction. Necessity of analysis of databases of digital records of ground strong motion follows from above-mentioned. In this connection serious attention should be given to more full information output of records of strong motions. Still there are complexities of data retrieval from international centers especially for the users from CIS countries. At the same time not all data cam be registered now with the help of available strong motion instrumentation. In particular, for allocation of nonlinearity a condition of underground systems [21] sometimes is used.

In conditions of regions with not high seismicity (for example, Georgia) i.e. in country, where the repeatability of strong earthquakes is rather low, usage of such equipment (strong motion instrumentation) is desirable, which have amplitude-frequency characteristic, enhanced in the party of reception of slight signals. It will allow to store the seismological data in absence of strong earthquakes, to investigate a different kind of features of a region, to control activity of stations etc.

The planning of urban territories represents an attempt of a linkage of the different parties of activity of the population in conditions of space, saturated with buildings, structures, environmental control systems and equipment of a definite functional orientation. One of the parties of human activity is the noticeable change of environment, geologic and hydro-geological conditions of a habitation (cut of declines, moving of soils, waterlogging etc.). Thus, population of urban territories considerably changes its seismic hazard. Thus, the adequate estimation of a vulnerability of buildings and structures to seismic effect is in close connection with equipment them with strong motion instrumentation. Now vulnerability of buildings and structures (equipment, environmental control system etc.) estimates enough by reference level of effect characterized also by a conditional oscillation frequency. The indicated parameters are elected on the basis of a restricted statistical sequence of data of tests on other buildings or structures (even of the same type). In conditions of Georgia (and CIS countries) with the definite technical equipment of building industry and culture of building, such approach is extremely unsafe, dispersion of dynamic parameters of structures is rather high, specially in conditions of intensive loads. In other words, there is a concreteness and accuracy of researches. Many of indicated problems are removed automatically at the analysis of predictable motion of buildings or structures on the basis of the instrumental data of observations.

The participation (1969-1981) of one of the authors [23] in activities on organization and subsequent exploitation of engineering seismometric stations, arranged in habitation and public buildings, and also installation of similar temporary stations in epicentral zone of strong and destructive earthquakes demonstrates, that the equipment of a building with strong motion instrumentation allows to forecast behavior of buildings and structures enough confidently in conditions of varied seismic effect. A data analysis of vibrations of buildings and structures at strong ground motion (Chernogory, 1976; Dmanisi, 1978; Gavazi, 1981; Spitak, 1988; Racha, 1991 and etc.) demonstrates, that buildings and structures notably change indicated technical parameters, that can serve for a reliable parameter (index) of their amortization and accordingly of augmentation of their vulnerability. It is known, that the real objects are characterized by capacity of accumulation of damages. The effect, due by weak earthquakes, is not taken into account, as a rule, by engineers. At the same time, it is known, that the integral effect of several weak earthquakes can correspond to effect of one strong earthquake.

The seller's high cost of instrumentation also hampers the development of researches. The efforts are indispensable on a broad intrusion of new engineering solutions with the purpose of cost reduction of strong motion instrumentation and its availability for the users from less developed countries. It is necessary to raise the level of advertising of the equipment and necessity of its acquisition for government officials, businessmen, scientists etc. This will allow to provide more efficiently distribution of state financing.

The instrumentation will really allow to estimate changes of parameters of buildings and structures under seismic effects and to estimate seismic risk of residing or activity in the indicated objects. Thus, installation of instrumentation in definite quantity of modern standard buildings (for example, in one per urban quarter) and in responsible structures (hydroelectric power station Inguri, chemical plant in Rustavi etc.) is recommended. The collocation of data of observations will allow to determine a vulnerability of buildings, the seismic risk and will serve as the reliable basis for their cost equivalent in the market of the real estate, and also in insurance policy. The indicated data can be utilized together with capabilities of GIS technology permitting visually demonstrate technical state of building. On the other hand, the exploratory capabilities of the indicated technologies are not used completely. In particular, they can be directly utilized for the characteristic of features of urban territory. The availability of a set of different records of strong motion will allow to make reliable extrapolation of the data and to model a combination of definite seismic effects with anticipated reacting of building. In other words, a possibility occurs of creation of the real scenarios of earthquakes. As a rule, such scenarios are not enough accurate, and the losses of the population estimated by different algorithms, for example, vary within limits of 1-30 and more percents. Last circumstance is due by not very reliable initial information and, apparently, by absence of the data of strong motion instrumentation. Installation of strong motion instrumentation in urban territories will allow to lower anticipated economical damage and will facilitate to urban authorities to accept the adequate solutions to decrease seismic risk.

4. Conclusions

1. The efforts are indispensable on a broad intrusion of new engineering solutions with the purpose of cost reduction of strong motion instrumentation and availability for the users from less developed countries. It is necessary to raise the level of advertising of the equipment and necessity of its acquisition for government officials, businessmen scientists etc.

2. Introducing of new parameters of absorption of seismic energy in soils will allow to design buildings and structures on energy based methodology, determining a degree of energy dissipation in corresponding elements of construction. Standardization of seismic effect provides an increasing of statistical series on the basis of valuable usage of databases of digital records of strong earthquakes.

3. It is represented expedient to organize in Georgia installation of strong motion instrumentation on standard buildings of modern construction and on responsible structures, and also directly on soils. The equipment of objects in Georgia with standard instrumentation will allow to include country into civilized scientific space, to raise considerably the status of routine seismological observations and to bring doubtless economic effect.

4. The data retrieval with strong motion instrumentation will allow to create the own databases and to exchange with large international Centers. The participation in the international projects will allow to use more essentially capabilities of scientific potential of Georgia.

5. References

1. Ambraseys N.N. Dynamics and response of foundation Materials in Epicentral Regions of Strong Earthquakes. Proc. World, Conference Earthguake Eng., 5 th, Rome 1973, pp . 115-119.
2. Aptikaev F.F. On the Correlation of MM Intensity with Parameters of Ground Skating.Proc. VII Europ. Gonf. Earth. Eng¦ Greece, v.2 ,1981, pp. 117-126 .
3. Arefiev S., Parini I., Romanov A., Mayer-Rosa D., Smith P. The Ratchi (Georgia, USSR) - Earthquake of 29 April 1991: Strong Motion Data of Selected Aftershocks. 3 May - 30 June 1991, Aug. 1991, vol. 1, Zurich, 211p.
4. Arias A.A. Measure of Earthquake Intensity. - Seismic Design of Nuclear Power Plants.MIT. PRESS. USA, 1970, pp. 438-483.
5. Bender B. Incorporating acceleration variability into Seismic Hazard Analysis. Bulletin of the Seismological Society of Ameri ca, 74, N 4, 1984, pp.1451-1462.
6. Castellany A., Petrini V. Research Activity on Decign Response Spectra for Italian Sites. Proc. World. Conf. Earth. Engineering, 5-th, Rome 1973, pp. 1210-1213.
7. Chinnery M.A. Earthquake Magnitude and Source Parameters . Bulle tin of the Seismological Society of America, 59,1969, pp. 1969- 1982.
8. Evernden J.F., Hibbard R.R., Schneider J.F. A model for Predic ting Seismic Intensity. Proc. World. Conf. Earthq. Engineering 5-th, Rome, 1973, pp.1684-1687.
9. Finn Liam W.D., Iai S., Matsunavga Y. Effect of Site Conditions of Ground Motions. Proc.10-th European Gonference on Earthquake Engineering, Vienna, abstract,vol.2,1994.
10. Fukushima I., Tanaka T. A new attenuation relation for Peak Horizontal Acceleration of Strong Earthquake Ground Motion in Japan. Shimizu Technical Research Bulletin, No 10, Tokyo, March 1991, pp.1-11.
11. Fukushima I.,Tanaka T. A new attenuation relation for Peak Horizontal Acceleration of Strong Earthquake Ground Motion in Japan. Bull. of the Seismological Society of America, 1990, 80, N 4, pp. 757-783 ¦
12. Grandori E.G., Tagliani A. A method for the magnitude Distribu tion Estimate. 10-th European Gonference on Earthquake Enginee ring, Vienna, abstract, vol.1,1994.

13. Idriss I.M., Seed H.B. An Analysis of Ground Motions During the 1957 San. Francisko Earthquake. Bull. Seism. Soc. Amer. 58, 1968 pp. 2013-2032.
14. Iida K.Earthquake Magnitude, Earthquake Fault and Source Dimensions. Journal Earth Science, Nagoya Univ., 13, 1965, pp. 115- 132 .
15. Kiremidjian A., Shah H.C.Probabilistic SIte-Dependent Response Spectra. Structural Division Proceedings of Society Civil Engineering, 1980, 106, No. 1, pp. 69-86.
16. Kudo K.Topics of Effects of Surface Geology on Strong-Ground Motion from the Recent Earthquakes in Japan and the activity of Japanes Working Group on Effects on Surface Geology. Proc. 10-th EuropeanConference on Earthquake Engineering, Vienna, vol.4, 1995, p.2635-2641.
17. Moinfar A.A., Nadersadeh A. Strong Motion Characteristics and Acceleration Distribution During the Manjil, IRAN Earthquake of 20 June 1990. Proc. 10-th European Conf. on Earthguake Enginee ring, Vienna, abstract, vol.1, 1994.
18. Omote S., Yoshimura K. Considerations on Earthquake Force Evalu ation. Proc. World Conf. Earthq. Eng., 5-th, Rome, 1973, pp. 1688-1691.
19. Reiter L.Earthquake Hazard Analysis. Golumbia Univ. Press, New York, 1991, 245 p.
20. Seed H.B., Romo M.P., Sun J.I., Jaime A., Lysmer J. The Mexico Earthquake of September 19, 1985 - Relationships Between Soil Conditions and Earthquake Ground Motion. Earthquake Spectra - 4, N 4, 1988, pp. 687-789.
21. Trifunac M.D. Characterization of Responce Spectra by Parameters Governing. The Cross NANURE of Earthquake source mechanisms. Proc. World Conf. Earth. Eng. 5-th, Rome, 1973, pp. 1688-1691.
22. Wen K.L., Yeh Y.T Seismic Velocity Structure Beneath the SMART 1 ARRAY. Bulletin of the Institute of Earth Sciences, Academia Sinica, vol. 4, Dec.1984, pp. 51-72.
23. Zaalishvili V.B. The Influence of Engineering and Geologic Features of Soil Layer on the Formation of Wave Field Created by Impulse and Vibration Sources. Proceedings of the 9-th European Conference on Earthquake Engineering, vol. 4 A, Moscow, 1990, pp. 169-175.
24. Zaalishvili V.B. Modern Concept of Seismic Microzoning. Proceedings of the 11-th European Conference on Earthquake. Engineering, Sept. 6-11, Paris, 1998

SUBJECT INDEX

A

Absorbing boundaries, 551
Acceleration, 179,383
Accelerograms, 33,505
Accelerograph, 157,195
Accelerograph Array, 195
Accelerograph network, 195
Accelerometers, 229
Ambient vibration, 343
Analytical evaluation, 481

B

Band-pass filter, 417
Base isolated test building, 369
Borehole arrays, 167
Boundary element method, 551
Bridges, 481,533
Broad band, 1
Broadband seismometers, 331
Building, 179,369,383,397,545,573
Building codes, 33
Building response, 573
Building response records, 33

C

Cable-stayed bridge, 109,481

D

Dam, 195,275
Damage assesment, 33
Damage detection, 91
Damage identification, 447
Data communication, 1
Data dissemination, 1
Data processing, 91
Data storage, 1
Dense instrumentation, 469
Discrete – time filters, 91
Displacement, 179,383
Dynamic behaviour, 61,369
Dynamic earth pressure, 519
Dynamic performance, 481
Dynamic response, 195

E

Early warning system, 293
Earthpressure, 519
Earthquake, 61,81,293,519
Earthquake alert, 53,81
Earthquake record, 195
Earthquake response, 195
Experimental analysis, 481

F

Finite element method, 551
Free-field instrumentation, 505
Free-field motion, 195
Free-field strong motion, 505
Frequency, 179,383
Frequency domain analysis, 313
Full-scale testing, 131

G

Global positioning system, 383,407
Ground motion, 433,545
Ground motion station, 433
Ground response station, 17

H

Hazard mitigation, 593
Health monitoring, 53
Historical monuments, 209,303

I

Impediance, 551
Implementing technologies, 229
Inclination sensor, 397
Instrumentation, 53,157,179,331,481
Instrumentation data, 243
Instrumented building, 33,313
Input uncertainty, 495
Intensive seismic excitation, 561

L

Linear dynamic characteristics, 573
Load uncertainty, 495
Low-amplitude test, 573

M

Material non-linearity, 495
Micromachined accelerometers, 229
Microtremor method, 303
Modal Identification, 91
Modelling, 131
Monitoring, 109,179,229,253,275,383
Monumental building, 61
Movement of building, 397
Moving force, 551

N

Near-real time, 17,495
Network, 195,545
Non-linear dynamic behaviour, 313
Non-linearity of soil, 561
Non-elasticity of soil, 561
Numerical models, 481

O

One-dimensional wave propagation theory, 519
Optimal control, 243

P

Performance evaluations, 33
Post earthquake damage evaluation, 17
Post earthquake response, 17,81
Processing of accelerogram, 417
Pushover analysis, 313

R

Real time monitoring, 293
Response, 195
Response spectra, 17,417,433
Retrofit, 533
Rubber bearings, 369

S

Satellite data transmission, 1
Seismic instrumentation, 533
Seismic interaction, 243
Seismic monitoring, 275
Seismic network, 505
Seismic records, 91
Seismic vulnerability, 61
Seismograph station, 573
Seismology, 1
Soil-structure interaction, 131,369,469
Spatial uncertainty, 495
Strong earthquakes, 275
Strong-motion, 1,179,383,545,561
Strong-motion accelerographs, 505
Strong-motion data, 17,81,331
Strong motion instrumentation, 17,157,195,303,331,505,533,593
Strong-motion measurements, 81
Strong-motion network, 209
Strong-motion recorders, 81
Structure health assesment, 495
Structural health monitoring, 343
Structural monitoring, 253

Structural response, 131,167,179,383
Suspension bridges, 109,343
System Identification, 91,209

T

Three-dimensional non-linear, 313
Time history, 417
Travel time, 447

U

Utilities, 81
Utilization, 179

V

Vibration monitoring system, 343
Viaduct, 407
Vulnerability of building and structures, 593

W

Wireless sensor system, 253
Wave propagation, 447